Y0-IWE-933

Toxic Substances and Human Risk

Principles of Data Interpretation

LIFE SCIENCE MONOGRAPHS

SELF-ORGANIZING SYSTEMS
The Emergence of Order
Edited by F. Eugene Yates

TOXIC SUBSTANCES AND HUMAN RISK
Principles of Data Interpretation
Edited by Robert G. Tardiff and Joseph V. Rodricks

A Continuation Order Plan is available for this series. A continuation order will bring delivery of each new volume immediately upon publication. Volumes are billed only upon actual shipment. For further information please contact the publisher.

Toxic Substances and Human Risk

Principles of Data Interpretation

Edited by

Robert G. Tardiff

and

Joseph V. Rodricks

Environ Corporation
Washington, D.C.

Plenum Press • New York and London

Library of Congress Cataloging in Publication Data

Toxic substances and human risk.

(Life science monographs)
Includes bibliographies and index.
1. Toxicity testing—Statistical methods. 2. Poisons—Dose-response relationship. 3. Health risk assessment—Statistical methods. I. Tardiff, Robert G. II. Rodricks, Joseph V., 1938- . III. Series. [DNLM: 1. Environmental Exposure. 2. Epidemiologic Methods. 3. Probability. 4. Toxicology—methods. QV 600 T7542]
RA1199.4.S73T69 1987 615.9′07 87-18660
ISBN 0-306-42529-7

© 1987 Plenum Press, New York
A Division of Plenum Publishing Corporation
233 Spring Street, New York, N.Y. 10013

All rights reserved

No part of this book may be reproduced, stored in a retrieval system, or transmitted in any form or by any means, electronic, mechanical, photocopying, microfilming, recording, or otherwise, without written permission from the Publisher

Printed in the United States of America

Contributors

Larry S. Andrews ARCO Chemical Co., Newtown Square, Pennsylvania 19073. *Present address:* Agriculture Research Division, American Cyanamid Co., Princeton, New Jersey 08540

Charles C. Brown National Cancer Institute, National Institutes of Health, Bethesda, Maryland 20892-4200

Stephen L. Brown Environ Corp., Washington, D.C. 20007

David Brusick Division of Molecular Toxicology, Hazelton Laboratories, Kensington, Maryland 20895

Edward J. Calabrese Division of Public Health, University of Massachusetts, Amherst, Massachusetts 01003

Gary P. Carlson Department of Pharmacology and Toxicology, School of Pharmacy and Pharmacal Science, Purdue University, West Lafayette, Indiana 47907

James E. Cone Division of Occupational Medicine, Medical Service, San Francisco General Hospital, San Francisco, California 94110

John Doull Department of Pharmacology, Toxicology, and Therapeutics, University of Kansas, College of Health Sciences, Kansas City, Kansas 66103

Kurt Enslein Health Designs Inc., Rochester, New York 14604

Ronald W. Estabrook Department of Biochemistry, Southwestern Medical School, University of Texas Health Science Center, Dallas, Texas 75235

Edward R. Garrett The Beehive, College of Pharmacy, J. Hillis Miller Health Center, University of Florida, Gainesville, Florida 32610-0494

David W. Gaylor Biometry, National Center for Toxicological Research, Jefferson, Arkansas 72079

James R. Gillette Laboratory of Chemical Pharmacology, National Heart, Lung, and Blood Institute, Bethesda, Maryland 20892

Ronald W. Hart National Center for Toxicological Research (HFT-1), Jefferson, Arkansas 72079

Rolf Hartung Toxicology Program, Department of Environmental and Industrial Health, University of Michigan, Ann Arbor, Michigan 48109

David G. Hoel Division of Biometry and Risk Assessment, National Institute of Environmental Health Sciences, Research Triangle Park, North Carolina 27709

Philip J. Landrigan Division of Environmental and Occupational Medicine, Mt. Sinai Medical Center, New York, New York 10029

Stephen L. Longacre Toxicology Department, Rohm & Haas Co., Spring House, Pennsylvania 19477

Robert E. Menzer Department of Entomology, University of Maryland, College Park, Maryland 20742-5575

Gordon R. Reeve Epidemiology, City of Houston Health Department, Houston, Texas 77030

Joseph V. Rodricks Environ Corp., Washington, D.C. 20007

Robert Snyder Rutgers University College of Pharmacy, Busch Campus, Piscataway, New Jersey 08854

Robert G. Tardiff Environ Corp., Washington, D.C. 20007

James R. Withey Environmental and Occupational Toxicology Division, Environmental Health Directorate, Ottawa, Ontario, Canada K1A OL2

Jeannee K. Yermakoff Amoco Corporation, Environmental Affairs and Safety Department, Chicago, Illinois 60601

Preface

As society has become increasingly aware of the potential threats to human health due to exposures to toxic chemicals in the environment and the workplace and in consumer products, it has placed increased demands upon the still-fledgling science of toxicology. As is often the case when science is called upon to supply firm answers when pertinent information and fundamental knowledge are lacking, both the scientific and the social issues become confused and new tensions develop. One of the major purposes of this book is to focus on those aspects of the science of toxicology that pertain most to social issues—namely, analysis of risk for purposes of human health protection.

Although it is apparent that the discipline of toxicology is not yet prepared to provide firm answers to many questions concerning human risk, it is important that the rigorously derived information be used in the most objective and logical way to yield the closest approximation to the truth. This book is designed to supply as much guidance for such tasks as is permitted by the current state of our knowledge. Its emphasis is thus placed on *interpretation* of toxicity data (broadly defined) for assessing risks to human health. In this way, it differs from other basic toxicology texts, most of which emphasize methods for performing studies or describe various toxicological endpoints and classes of toxic agents.

After placing the science in its historical context, we have attained our objective by considering the various sources of evidence—from clinical, epidemiological, and laboratory data—that are used to identify toxic hazards and to infer their implications for human health. This volume also features a summary of principles pertaining to an often neglected but critical aspect of risk analysis—assessment of human exposures to toxic substances.

The chapters provide a judgmental context for synthesizing all information and methodologies that are used to analyze human risk. Risk analysis is depicted in its broadest sense, with the emphasis on its role in "safety" assessment. One of the underlying themes of this volume—that the science of toxicology serves only to assess risks and not to decide questions of risk toleration—surfaces in the final section, in the analysis of the traditional and neotraditional roles of toxicology in standard settings and in the lessons of the past decade insofar as they have increased awareness of the limitations inherent in toxicology's contributions to public policy.

The book is designed especially for professional toxicologists, students of the science, scientists in regulatory agencies, and the regulated community.

Robert G. Tardiff
Joseph V. Rodricks

Washington, D.C.

Contents

I

Historical Perspectives and General Concepts

1

Introduction

John Doull

Most regulators have long recognized the conceptual distinction between risk assessment and risk management and have utilized both concepts in establishing tolerances and standards for protecting human health. In 1983, these concepts were redefined and expanded by a National Research Council committee in a report entitled, *Risk Assessment in the Federal Government: Managing the Process* (National Research Council, 1983). In this report, risk assessment was considered to include some or all of the following four steps: hazard identification, dose–response assessment, exposure assessment, and risk characterization. Risk management was defined as the process of weighing policy alternatives and selecting the most appropriate regulatory action, and integrating the results of risk assessment with engineering data and with social, economic, and political concerns to reach a decision. More recently, a third concept has been added to the process: risk communication (Thomas, 1986). These developments have had a significant impact on the role of the toxicologist in the decision-making process and focused attention on the interactions among science, law, and regulation in the arena of safety assessment. It has been suggested, in fact, that a Risk Institute is needed to investigate the scientific and policy aspects of the total process and determine the appropriate contributions of all involved parties (Press, 1984).

Since toxicology and epidemiology have traditionally provided the scientific or factual basis for the evaluation and characterization of risk, it is useful to consider the evolution of the approach of these two disciplines to the protection of public health. In the subsequent chapters of this book, both the traditional and the current approaches of these disciplines are considered along with some of the current problems in creating the data base and interpreting both toxicological and epidemiological data.

JOHN DOULL • Department of Pharmacology, Toxicology, and Therapeutics, University of Kansas, College of Health Sciences, Kansas City, Kansas 66103.

Toxicology, like medicine, is both a science and an art (Doull, 1984). The science of toxicology includes all the observational or data-gathering activity, and the art of toxicology includes the predictive or risk-estimating phase of the discipline. For example, the observation that chronic dietary exposure of mice to chloroform produces hepatomas is a factual observation, but the prediction that chloroform will produce similar effects in humans is a hypothesis. By distinguishing between the science and art of toxicology, we recognize that the validity of our predictions or hypotheses depends not only on the quality of the scientifically derived toxicology data base, but also on the relevance of these data to the situation for which we are predicting. When we fail to recognize the distinction between the science and the art of toxicology, we tend to confuse our "facts" with our "hypotheses" and to argue that they have equal validity, which they clearly do not. Another major advantage of distinguishing between the science and the art of toxicology is that it is easier to resolve problems in risk assessment if the source of the problem can be identified with the science (e.g., protocol inadequacies, audit failure, and inappropriate animal models) or with the art (e.g., lack of relevance in interspecies comparisons and extrapolation of data beyond data limits).

Ideally, it is the science rather than the art of toxicology that characterizes the adverse effects of a chemical agent. This characterization includes as key elements the identification and description of all the various adverse effects that the agent may produce, the dose–response relationships for each of these effects, and, we hope, the mechanism(s) responsible for producing these effects. In many cases, the toxicity data base for a chemical agent will also provide information pertaining to different exposure conditions (e.g., duration, frequency, rate, route, and site), different subjects (e.g., species, strain, age, sex, genetic status, nutritional status, and presence of disease), and agent-related factors (e.g., purity, stability, solubility, chemical and physical characteristics, effect of formulation and vehicle changes, and kinetics).

The toxicity data base on most chemical agents consists of the results of studies conducted in animals. Frequently these studies are conducted according to standardized protocols designed to detect specific toxic effects, such as teratogenesis, reproductive toxicity, cancer, neurotoxicity, and specific organ damage. These protocols are often referred to as safety evaluation studies, implying that a test chemical can be judged to be "safe" if it yields negative results in a series of properly conducted tests. This concept presents several problems—some scientific or toxicological and others philosophical.

First, no chemical is free of toxic effects under all circumstances. Therefore, none can be regarded as completely safe. As Paracelsus (a Swiss alchemist and physician) pointed out over 400 years ago, "All substances are poisons, there is none which is not a poison. The right dose differentiates a poison and a remedy." Thus, a negative result in a safety evaluation test protocol does not identify the inherent toxic potential of the chemical, nor does it provide dose–response information. However, the dosage levels required to produce such information may by unrealistic for human safety evaluation purposes or they may be so large that toxic effects are induced by secondary mechanisms rather than primary toxicity.

Studies designed to characterize the toxicity of a chemical agent and safety evaluation tests on the same agent may be complementary, but they are not interchangeable, since they have different goals and are often based on different methodology.

The safety evaluation test protocol, or checklist approach, demands yes-or-no answers to such questions as: Is the chemical safe? Does it cause cancer, fetotoxicity, or blindness? To some questions of this nature, the toxicologist's answer is "yes" at high doses and "no" at low doses, but for many other questions, the scientific answer is "maybe." Thus, in toxicology, as in most other scientific disciplines, answers are rarely black or white, but, rather, are found in various shades of gray. Defenders of this yes-or-no approach to the safety evaluation of chemicals argue that it provides a regulatory convenience by avoiding the need to exercise and justify the scientific judgment that would otherwise be required. This approach appeals to the legal mind, since most of the difficult toxicological problems are ultimately decided by the courts and a yes-or-no decision is more likely to be sustained on the basis of the record than is the scientific answer "maybe." Industry also prefers this approach because it simplifies the registration of new chemicals.

With this checklist approach, negative results from studies that departed in any detail from an agency's guidelines might be considered unacceptable to that agency, whereas positive studies would lead to a classification of the chemical as unsafe, regardless of the number of departures from sound toxicological practices. Toxicologists from both industry and regulatory agencies would undoubtedly feel uncomfortable with such decisions.

The checklist approach can also result in unnecessary testing, wasting scarce resources, such as toxicology manpower and facilities. Furthermore, even with the most extensive battery of test protocols now available, there is no way to ensure that all the right questions have been asked in order to avoid the recurrence of a situation as serious as that caused by thalidomide. Toxicologists and other scientists have expressed concern about growing reliance on this kind of "cookbook" toxicology and have repeatedly argued that more professional judgment should be brought to bear in selecting elements for inclusion in the toxicology data base and that there should be more flexibility in the selection of methods used to produce these data. In a publication entitled, *Principles for Evaluating Chemicals in the Environment,* a committee of the National Research Council (1975, p. 11) pointed out, "There is no substitute for the vigilance of an inquiring and skeptical mind, which has assumed the full responsibility for making . . . safety assessments. If that responsibility is lessened by an exclusive dependence on a checklist approach, the major assurance has been lost that a responsibly perceptive and efficient investigation will be conducted." Similar statements can be found in other publications (Doull, 1981; Institute of Medicine, 1979; National Research Council, 1981; Scientific Committee, Food Safety Council, 1978).

Compared to other biological sciences, toxicology is still in its infancy and is still evolving as a science. Following the introduction of teratology as a new focus in toxicology came the use of short-term tests for genetic toxicology—a phase still undergoing rapid development as more of their potential is learned. Neurotoxicology and immunotoxicology are already on the horizon, and short-term tests for reproductive toxicity are not far behind.

Our current test protocols will lead to the development of a toxicity data base that will give us quick answers to questions of safety. In this approach, a distinction is made between the information that would be "nice to know," such as that on metabolism and mechanism of action, and the information that we "need to know" for regulatory purposes. As the science of toxicology matures, we should develop procedures that not only meet the scientific criteria for characterizing toxicity, but also provide a more reliable basis for safety predictions.

During the past decade, the emphasis in toxicology has been on the development of new or revised test protocol guidelines, such as those stipulated by the Federal Insecticide, Fungicide, and Rodenticide Act (FIFRA), the Interagency Regulatory Liaison Group (IRLG), the Toxic Substances Control Act (TSCA), the Office for Economic Cooperation and Development (OECD), and the Food Safety Council; the formulation of controls for the conduct of such studies, e.g., the Good Laboratory Practice (GLP) regulations; and the development of models for the extrapolation of test results from high to low doses, e.g., the log-probit, linear, multihit, multistage, and logit models. Partly in response to these developments, toxicologists are being asked by the industry, legislators, professional associations, and others if these new tests and guidelines really increase our ability to protect our health and the environment and, if so, whether or not the amount of additional protection is justified on the basis of the added cost. The directions in which the science of toxicology are likely to develop in the next decade will depend on our answer to this and similar questions, such as the following:

1. Can procedures be developed to characterize the toxicity of mixtures of chemical agents while avoiding the necessity of investigating all possible combinations of ingredients?
2. Can sophisticated computer-based structure analogy programs be used to predict the adverse effects of classes of chemicals and/or mixtures of chemical agents?
3. Can short-term assays that have been validated and shown to be predictive tools for mutagenesis and teratogenicity be used as alternatives for oncogenicity and fetotoxicity studies in animals?
4. What is the minimum requirement that a toxicity data base should have (e.g., number and type of data elements) to establish priorities for additional testing?
5. Can we develop methods for animal testing that will enable us to detect and predict subjective adverse effects, such as headache, anxiety, or depression?
6. How can we deal with the situation in which replicated studies using the same protocol produce conflicting results? Are there circumstances in which negative findings can balance positive findings?
7. Does the induction of the microsomal enzyme system by a chemical

agent constitute a toxic effect? How should we interpret stress-induced effects and other secondary responses that occur in our test systems?
8. Recognizing that toxicokinetic and, especially, metabolism studies in both the test species and target species would improve the validity of our predictions, how can we obtain the data when humans are the target species?
9. Do current chronic animal tests and short-term assays provide adequate evaluation of promoters, activators, and cocarcinogens?
10. Can we improve our ability to detect low-incidence effects without vast increases in the size of the test group, e.g., megamouse studies?

In contrast to the science of toxicology, which is the observational or evidence-gathering process, the art of toxicology is best characterized as a judgmental or reasoning process. In fact, toxicological predictions are frequently called "best judgments" or "best estimates," and we attach to them safety factors or other expressions of uncertainty that reflect the level of confidence that we have in their validity. Even under optimal conditions, however, our predictions are still conjectures or "educated guesses," and they remain so until verified by subsequent experience or additional study. Rodricks and Tardiff (1983) suggest that the most urgent need in risk assessment is the ability to present a comprehensive array of risk conclusions with appropriate ranges of estimates, notations of severities, indications of the strength of the scientific support, and boundaries of uncertainties.

Two general principles of toxicology are fundamental parts of both the science and the art of toxicology. The first is that effects in animals, when properly qualified, are applicable to humans; the second is that toxic effects are dose-related. These principles are not unique to toxicology; they apply to all areas of experimental biology and medicine, but their application in toxicology is sometimes more controversial than in other areas of biology. It can be argued, for example, that the $B6C3F_1$ strain of mice used for the National Cancer Institute (NCI) oncogenicity assay does not correctly predict whether a hydrocarbon will produce lung or liver cancer, or both, in humans, because of the normally high incidence of lung and liver tumors in this strain. Similarly, Oser (1981) has pointed out several major differences between rats and humans (e.g., different types of placenta, lack of gallbladder and emetic reflex, ascorbic acid requirements, and tissue enzyme levels) that could invalidate rat-to-human comparisons. Traditionally, toxicologists have attempted to avoid this type of problem by using several species and strains and basing their predictions on the most sensitive animal model. If better data are available, such an approach is neither scientific nor appropriate as a basis for making relevant predictions. Whenever possible we should base our predictions on results obtained in the model system that most nearly resembles the target species (humans) in the manner in which it responds to the test chemical.

A controversy surrounding the second principle of toxicology, i.e., dose response, involves the question of whether there is a dose below which adverse effects will not occur in humans. Traditionally, this threshold or no-observed-effect level (NOEL) in the laboratory animal has been used by toxicologists as the basis of their predictions (see Chapter 2). When the number of responses in the test group is equal to or less than that in the control group, a NOEL or operational threshold is believed to have been achieved.

Historically, it has been assumed that the NOEL is a population threshold in humans. Although the NOEL concept has been, and probably still is, the most widely used basis for establishing acceptable exposure levels for chemical agents, it is generally recognized that some adverse effects, such as cancer and mutagenesis, may not have population thresholds. Carcinogens that produce their effect through a secondary mechanism (e.g., thiourea, which causes thyroid adenoma by pituitary stimulation, and nitrilotriacetic acid, which is reputed to cause urinary tract neoplasms through crystal formation) appear to have demonstrable thresholds and, consequently, no-observed-effect levels. Others clearly do not.

A classification system for carcinogens that separates genotoxic agents (direct-acting substances and procarcinogens) from epigenetic agents (e.g., hormones, solid state substances, immunosuppressors, and carcinogens) has been proposed by Weisburger and Williams (1980), who also suggest that "safe" thresholds can be established for the epigenetic group once their mechanism of action is elucidated. Most toxicologists agree that there may be no "true" threshold for genotoxic carcinogens, and all realize that human population thresholds are impossible to measure or predict. Such practical thresholds for carcinogens could result from the body's normal detoxification and repair mechanisms (e.g., glutathione binding and repair of methylated DNA) or from immune system responses.

In the "real world" decision-making process, the issue is more complex than the scientific question of whether there are thresholds for carcinogens under the experimental conditions of our animal test protocols. The "real world" issue is whether such thresholds exist for the target or human species and whether they can be scientifically measured and defined. It is clear that as our understanding of the complex process of chemical carcinogenesis develops, we will be increasingly able to control such agents using a case-by-case approach in which potency, mechanism of action, and other toxicological properties serve as critical criteria. There is a need for risk assessment processes to incorporate both potency and exposure information. Thus, for saccharin, although the potency may be low, the exposure is so extensive that the actual risk may be sizeable; whereas for vinyl chloride, the potency may be much greater, but the exposure is so much less that the risk to public health is actually lower than for saccharin.

The development of these two principles (i.e., dose response and test to target species prediction) mark the origin of toxicology. The principle of dose response was first related to toxicology by Paracelsus in the middle of the 16th century, and the concept of using the response of test animals as a basis for predicting responses in the target species was added to toxicology by Orfila some 300 later. Early human societies and extant nonliterate peoples were and are aware of the toxic effects of animal venoms and poisonous plants and use this knowledge for hunting and warfare. On this basis, it can be argued that toxicology is as old as medicine itself. Indeed, our earliest medical records, such as the Ebers papyrus (ca. 1500 B.C.), list many well-recognized poisons, for example, hemlock, aconite, and heavy metals (Casarett and Bruce, 1980). Poisoning flourished during the Greek and Roman periods of history and became a recognized profession at least in Italy during the early Renaissance. However, it was not until the middle of the 19th century that Orfila defined toxicology as the science of poisons and linked analytical chemistry with jurisprudence to create the area known today as forensic toxicology.

Thus, the earliest toxicologists were chemists and physicians. They were soon joined, however, by pharmacologists, pathologists, biochemists, and scientists in many areas of the biological sciences. Thus, toxicology is truly multidisciplinary not only in origin, but also in its present scope. For this reason, it has taken time for toxicology to be recognized as a distinct scientific discipline, and despite the burgeoning of interest and activity in the field during the last few decades, few programs offer training and degrees in toxicology. Most practicing toxicologists today received their early scientific training in a variety of fields and gathered their experience in toxicology along the way. According to a leading pioneer in toxicology, Dr. Arnold Lehman, "Anyone can become a toxicologist in three lessons, each of which takes ten years" (A. Lehman, personal communication, 1965). Nonetheless, toxicology has emerged as a discrete discipline and is represented by professional societies (national and international), journals, meetings, and, most recently, an accreditation process for board certification.

The dose–response concepts used in toxicological studies were initially proposed by Trevan (1927), who described the characteristic sigmoid response of biological systems to chemicals and introduced the LD_{50} (median lethal dose) concept. These early studies, which primarily involved responses to drugs, demonstrated the importance of a normally distributed population in using this technique, the usefulness of expressing dosage in log rather than linear units, and the utility of the probit, which was introduced by Bliss (1935). It is not surprising, therefore, that initial attempts to establish tolerances for food additives and pesticides involved the use of the quantal dose-response and that the anticipated distribution susceptibilities were accommodated by dividing the no-effect level by a safety factor to obtain an acceptable exposure level or tolerance—a practice that persisted with little major change until the late 1970s.

Although the limitations in this procedure were soon recognized, there were few practical alternatives. It was recognized, for example, that the no-effect level was dependent on the number of animals in the test groups (a zero response in ten animals is not the same as a zero response in 100 animals) and that the selection of the safety factor was an arbitrary decision. It is common practice to use a safety factor of 100 in ADI (acceptable daily intake) calculations on the assumption that it provides one order of magnitude for interspecies variation and another for intraspecies variation. However, in calculating suggested no-adverse-response levels (SNARLs) for noncarcinogenic organic solutes in drinking water, the National Research Council Safe Drinking Water Committee used an uncertainty factor (safety factor) of 10 for chemicals having an adequate data base for toxicity in humans and supporting data from laboratory animals, and a factor of

1000 for chemicals with a poor toxicity data base with a toxic effect that elicited particular concern (National Research Council, 1980, 1982). Safety factors for carcinogens have been proposed, but are not generally accepted (World Health Organization, 1978).

The publication of the first volume of *Drinking Water and Health* (National Research Council, 1977) was a major factor in stimulating renewed interest in alternatives to the safety factor approach for nonthreshold chemicals. In that report the Safe Drinking Water Committee stressed the idea that materials should be assessed in terms of human risk factor rather than as "safe" or "unsafe" and proposed a risk extrapolation procedure for use with carcinogens. Since the safety factor approach cannot be used to regulate chemicals that do not exhibit a no-effect level, the risk extrapolation approach was supported by the IRLG (Interagency Regulatory Liaison Group) as a practical method for handling nonthreshold chemicals with major health or economic benefits.

During the last few years, numerous mathematical models have been proposed for extrapolating the results of high-dose animal studies to low-dose human exposures, and some are used as regulatory devices. In the review of this area, Fishbein (1980) discussed five dichotomous models for describing dose–response curves (one-hit, gamma multihit, multistage, log-probit, and logit models) and an equal number of low-dose risk assessment methods (Mantel–Bryan, which was first suggested in 1961, linear extrapolation, multistage estimation, time to occurrence, and a statistical-kinetic, or Cornfield approach). The report of the Scientific Committee, Food Safety Council (1978) includes a comparison of the predictive validity of several of the current models using chemicals with extensive animal test data. In addition to these and other general references (Brown *et al.,* 1978; Gaylor and Shapiro, 1979; Hoel, 1979; Office of Technology Assessment, 1981), the reader is referred to Chapter 10 of this book for a thorough discussion of this topic by C. Brown.

From the preceding discussion, it is evident that there are two major problems in the art of toxiocology: (1) the quality and relevance of data provided by the science of toxicology and (2) the interpretation of the data and the assessment of how they should be used to formulate hypotheses or predictions. In the ideal situation, the test species and the target species would be the same. This may occur when making predictions in environmental toxicology for fish and wildlife. Toxiocologists are well aware of Alexander Pope's aphorism, "The proper study of mankind is man," and recognize the potential of epidemiological studies for attaining this goal. Unfortunately, it is difficult to get answers quickly or cheaply from either the case–control study or the more conventional concurrent or nonconcurrent cohort studies. We need an epidemiological "Ames test" that is both predictive and safe and that can be correlated in some fashion with the toxicology data base on animals. We also need to look more closely at the methods that we use to gather the data in human studies (i.e., the science of epidemiology) and how we utilize these data to develop the risk estimate (i.e., the art of epidemiology). It is appropriate, however, to include both the science and the art of toxicology and epidemiology in risk assessment or hazard evaluation, and it is the purpose of this book to describe the principles that can be used for this purpose.

REFERENCES

Bliss, C. L. 1935. The calculation of the dose–mortality curve. *Ann. Appl. Biol.* **22:**134–167.

Brown, C. C., T. R. Fears, M. H. Gail, M. A. Schneiderman, R. E. Tarone, and N. Mantel. 1978. Models for carcinogenic risk assessment. *Science* **202:**1105–1106.

Casarett, L. J., and M. C. Bruce. 1980. Origin and scope of toxicology. Pp. 3–10 in J. Doull, C. D. Klaassen, and M. O. Amdur, eds. *The Basic Science of Poisons.* Macmillan, New York.

Doull, J. 1981. Food safety and toxicology. Pp. 295–316 in Howard R. Roberts, ed. *Food Safety.* Wiley, New York.

Doull, J. 1984. The past, present and future of toxicology. *Pharmacol. Rev.* **36:**15S–18S.

Fishbein, L. 1980. Overview of some aspects of quantitative risk assessment. *J. Toxicol. Environ. Health* **6:**1275–1296.

Gaylor, D. W., and R. E. Shapiro. 1979. Extrapolation and risk estimation for carcinogenesis. Pp. 65–87 in M. A. Mehlman, R. E. Shapiro, and H. Blumenthal, eds. *Advances in Modern Toxicology,* Volume 1, Part 2. Hemisphere, Washington, D.C., distributed by Halsted Press, New York.

Hoel, D. 1979. Low-dose and species-to-species extrapolation for chemically induced carcinogenesis. Pp. 135–145 in V. K. McElheny and S. Abrahamson, eds. *Banbury Report,* Volume 1. *Assessing Chemical Mutagens: The Risk to Humans.* Cold Spring Harbor Laboratory, Cold Spring Harbor, New York.

Institute of Medicine. 1979. *Food Safety Policy: Scientific and Societal Considerations.* Part 2 of a two-part study of the Committee for a Study on Saccharin and Food Safety Policy. National Academy of Sciences, Washington, D.C. 419 pp.

National Research Council. 1975. *Principles for Evaluating Chemicals in the Environment.* A report of the Committee for the Working Conference on Principles of Protocols for Evaluating Chemicals in the Environment. National Academy of Sciences, Washington, D.C. 454 pp.

National Research Council. 1977. *Drinking Water and Health.* A report of the Safe Drinking Water Committee. National Academy of Sciences, Washington, D.C. 939 pp.

National Research Council. 1980. *Drinking Water and Health,* Volume 3. A report of the Safe Drinking Water Committee. National Academy Press, Washington, D.C. 415 pp.

National Research Council. 1981. *Strategies to Determine Needs and Priorities for Toxicity Testing.* Volume 1: *Design.* A report of the Steering Committee on Identification of Toxic and Potentially Toxic Chemicals for Consideration by the National Toxicology Program. National Academy Press, Washington, D.C. 143 pp.

National Research Council. 1982. *Drinking Water and Health.* Volume 4. A report of the Safe Drinking Water Committee. National Academy Press, Washington, D.C. 299 pp.

National Research Council. 1983. *Risk Assessment in the Federal Government: Managing the Process.* A report of the Committee on the Institutional Means for Assessment of Risks to Public Health. National Academy Press, Washington, D.C. 192 pp.

Office of Technology Assessment. 1981. *Technologies for Determining Cancer Risks from the Environment.* Congress of the United States, Washington, D.C.

Oser, B. L. 1981. The rat as a model for human toxicological evaluation. *J. Toxicol. Environ. Health* **8:** 521–542.

Press, F. 1984. Keynote Address, Symposium on Safety Assessment. *Fund. Appl. Toxicol.* **4:**S257–S260.

Rodricks, J. V., and R. G. Tardiff. 1983. Biological bases for risk assessment. Pp. 77–84 in *Safety Evaluation and Regulation of Chemicals.* S. Karger, Basel.

Scientific Committee, Food Safety Council. 1978. Proposed system for food safety assessment. *Food Cosmet. Toxicol.* **16**(Suppl. 2):109–120.

Thomas, L. 1986. Risk communication. *Environment* **28**:4–5.

Trevan, J. W. 1927. The error of determination of toxicity. *Proc. R. Soc. Long. (Biol.)* **101**:483–514.

Weisburger, J. H., and G. M. Williams. 1980. Chemical carcinogens. Pp. 84–138 in J. Doull, C. D. Klaassen, and M. O. Amdur, eds. *Casarett and Doull's Toxicology: The Basic Science of Poisons.* Macmillan, New York.

World Health Organization. 1978. *Environmental Health Criteria,* No. 6: *Principles and Methods for Evaluating the Toxicity of Chemicals,* Volume 1. World Health Organization, Geneva.

2

Toxicologic Units

Jeannee K. Yermakoff

Toxicologic units have become an integral part of communicating information for the toxicologist. In human and animal studies, these units are used to conveniently summarize study results as a single value (e.g., the SMR and LD_{50}, respectively). When describing exposure to a chemical, toxicologic units provide an indication of the purpose for selection of a given dose (e.g., the MTD) or an acceptable level of human exposure (e.g., the TLV). However, effective communication using this unique vocabulary of toxicologic abbreviations requires that the definitions be clearly understood. This is particularly important when these toxicologic units are used to communicate toxicologic information to individuals outside the realm of the practicing toxicologist.

This chapter is oriented toward those who have limited experience with the vocabulary of toxicologic units. It defines these units of toxicologic measurements, describes how they are generated, and briefly discusses their application. It is intended as a reference to terms used and discussed in greater detail in other chapters of this book.

UNITS USED IN ANIMAL STUDIES

Lethal Dose

LD_{50}

The LD_{50}, or median lethal dose, was first reported by Trevan (1927) over 50 years ago as a measure of acute toxicity. Since that time, different methods have evolved for determining the LD_{50} and many applications of the results have been proposed. Nonetheless, current definitions of the LD_{50} (National Research Coun-

JEANNEE K. YERMAKOFF • Amoco Corporation, Environmental Affairs and Safety Department, Chicago, Illinois 60601.

cil, 1975; Organization for Economic Cooperation and Development, 1981; National Institute for Occupational Safety and Health, 1983) remain similar to the one first reported by Trevan (1927). The definition adopted by the National Institute for Occupational Safety and Health (1983) is representative: "A calculated dose of a substance which is expected to cause the death of 50% of an entire defined experimental animal population."

The LD_{50} is determined by administering graduated doses of test material to groups of experimental animals (Organization for Economic Cooperation and Development, 1981). Generally, three or more dose levels are selected to produce mortality rates bracketing 50%. When information to estimate proper dose levels is not available, a limit test or LD_{50} screen may be conducted using a single dose level (usually 5 g/kg body weight for oral and 2 g/kg body weight for dermal administration). If mortality is not observed in the limit test, a study to determine an LD_{50} would generally be unnecessary since mortality above these dose levels would have limited toxicologic significance. If mortality is observed, this information could be used to set dose levels for a complete LD_{50} study. Alternatively, an LD_{50} can be estimated by the up-and-down method (Brownlee *et al.,* 1953; Murray and Gibson, 1972). With this procedure, test groups are dosed sequentially, and the results from each group are used to predict for succeeding groups the doses that will most efficiently bracket the LD_{50}.

Selection of a given route of administration for an LD_{50} study is generally based on anticipated human exposure. For example, drugs and food additives are commonly tested by oral administration, whereas chemicals associated with occupational exposures would most appropriately be tested dermally or by inhalation. Other parenteral routes might include subcutaneous, intraperitoneal, intravenous, and intratracheal administration.

An LD_{50} is described most often as weight of the test substance administered per animal body weight (e.g., mg/kg), but alternatives may be used as situations warrant. Savini (1968) recommended expressing LD_{50} values as mmole/kg to eliminate the contribution of molecular weight differences and show only changes in biological activity. In a homologous series of *n*-alcohols, the toxicity across the series increased 2.7 times as molecular weight increased when LD_{50} values were reported as mg/kg and 8.8 times when expressed as mmole/kg. Values of LD_{50} may also be expressed as weight of test substance administered per animal surface area (e.g., mg/m^2). Freireich *et al.* (1966) reported higher correlations of data between animals and humans when dose units were converted from mg/kg to mg/m^2.

Studies used to estimate the LD_{50} should include several endpoints in addition to lethality to maximize the toxicity data generated by test animals. In addition to mortality, the evaluation of acute toxicity should include clinical observations (e.g., behavioral abnormalities) and pathological analyses of experimental animals (Boyd, 1968; Code of Federal Regulations, 1985a; Sperling, 1976). While the observation period should be at least 14 days beyond dosing, observations in survivors beyond 14 days can provide information on delayed toxicity and reversibility of effects.

In recent years, LD_{50} studies have been criticized for their excessive use of experimental animals relative to the value of information obtained, and pressure for alternatives has increased. As a result, regulatory agencies are changing some of their requirements for acute toxicity studies. The Environmental Protection Agency's Health Effects Testing Guidelines (Code of Federal Regulations, 1985a) recommend the use of a limit test, where possible, as an alternative to an LD_{50} study, and encourage the use of data from structurally related chemicals. Yet, change comes slowly, and despite criticism, the absence of a validated alternative necessitates that regulatory agencies continue to classify the degree of acute toxicity for various chemicals according to LD_{50} values. For examples, the LD_{50} forms the basis of toxicity categories for pesticides (see Table 2-1) and for hazardous chemicals in the workplace (European Economic Community, 1983; Code of Federal Regulations, 1985b).

LD_{50} values for components of a chemical mixture can be used to generate an estimate of the acute toxicity of the mixture. For example, the European Economic Community (1984) has proposed a mathematical weighting of component LD_{50} values to classify acute toxicity for warning labels on hazardous mixtures. Smyth *et al.* (1969) used the following formula based on component LD_{50} values to predict LD_{50} for chemical mixtures:

$$\frac{1}{\text{predicted } LD_{50}} = \frac{P_A}{LD_{50} \text{ component A}} + \frac{P_B}{LD_{50} \text{ component B}}$$

where P_A and P_B are the respective proportions of components A and B. The mixtures were tested for acute toxicity with P_A and P_B at 0.5 and compared to the

TABLE 2-1. Pesticide Toxicity Categories Requiring Warning Labels[a]

Hazard indicator	Criteria for given toxicity category			
	I	II	III	IV
Oral LD_{50}	Up to and including 50 mg/kg	From 50 to 500 mg/kg	From 500 to 5000 mg/kg	Greater than 5000 mg/kg
Inhalation LC_{50}	Up to and including 0.2 mg/liter	From 0.2 to 2 mg/liter	From 2 to 20 mg/liter	Greater than 20 mg/liter
Dermal LD_{50}	Up to and including 200 mg/kg	From 200 to 2000 mg/kg	From 2000 to 20,000 mg/kg	Greater than 20,000 mg/kg
Eye effects	Corrosive; corneal opacity not reversible within 7 days	Corneal opacity reversible within 7 days; irritation persisting for 7 days	No corneal opacity; irritation reversible within 7 days	No irritation
Skin effects	Corrosive	Severe irritation at 72 hr	Moderate irritation at 72 hr	Mild or slight irritation at 72 hr

[a] From Code of Federal Regulations (1985c).

predicted values. Although this formula did not account for potentiation or antagonism, it was useful for predicting a large percentage of the LD_{50} values for the pairs of chemicals investigated. Accordingly, the authors concluded that the soundest hypothesis for joint action of untested pairs is that of additive action.

In these and other applications, the LD_{50} provides a measure of acute toxicity independent of the target organ and mechanism. However, the LD_{50} does not provide information on the slope of the dose versus percent response curve, and therefore does not indicate the size of the dose range in which lethality is likely to occur. Also, there are numerous sources of variability for the LD_{50}, which are reflected only in part in the reported confidence limits, but are evident in the often disparate LD_{50} values determined for the same material by different laboratories. Experimental factors that may contribute to variation in LD_{50} values include animal age, animal weight, species, strain, housing conditions, diet, sample size, and dose levels (National Research Council, 1977a; Zbinden and Flury-Roversi, 1981).

LD_{01} and LD_{99}

In addition to the LD_{50}, other measures of lethality can be derived from dose versus percent lethality curves generated from LD_{50} studies. Of these, the LD_{99} has been used to estimate the minimum dose that is likely to be lethal to an entire animal population. Conversely, the LD_{01} approximates the maximum sublethal dose. The LD_{99} and LD_{01}, when reported together with the LD_{50}, give an estimate of the size of the dose range in which lethality might occur.

The greatest limitation to the LD_{01} and LD_{99} is their lack of precision. At the extreme ends of dose versus percent lethality curves, from which these values are derived, the confidence limits of lethal dose percentages increase (Finney, 1971). For example, Hodge (1965) reported that the potential error for LD_{05} values is likely to be approximately eight times greater than that for LD_{50} values.

LDLo

The lowest lethal dose (LDLo) of a substance is that dose administered "over any given period of time in one or more divided portions and reported to have caused death in humans or animals" (National Institute for Occupational Safety and Health, 1983). As indicated by this definition, the LDLo differs from lethal dose percentages in three respects. First, it is not calculated from dose–response data; in fact, it is often an isolated observation. Second, it is not restricted to acute exposure. Third, the LDLo may be based on animal or human data. Generally, the use of LDLo values for the prediction of human health risk is quite limited and will depend on the source of the LDLo and associated toxicity information. Human LDLo values, which are commonly obtained from accidental exposure or drug overdose, can provide valuable information by eliminating the need to extrapolate toxicity information from animals to humans.

Effective/Toxic Dose

ED_{50} and TD_{50}

Dose versus percent response curves are not restricted to lethality as an endpoint. In addition to dichotomous responses, such as lethality, graded responses may be used by identifying a value above which the response is defined as positive. Any clearly defined measure of efficacy or toxicity may be used. In such cases, the median effective dose (ED_{50}) or median toxic dose (TD_{50}) can be obtained from a dose versus percent-response curve. The ED_{50} and TD_{50} are often used together to estimate the margin of safety, or therapeutic index, which is a measure of the selectivity of a drug or other chemical for a given response. This safety margin is generally expressed as the ratio TD_{50}:ED_{50}. It is preferred to TD_{01}:ED_{99} because of the greater variability in these latter values.

ED_{01} and TD_{01}

The dose that is effective or toxic in only 1% of treated animals is referred to as the ED_{01} or TD_{01}, respectively. Like the LD_{01}, these values are associated with poor precision because of the large confidence limits at the extreme ends of the dose versus percent response curve (Finney, 1971).

Applications of the ED_{01} and TD_{01} values have generally been restricted to studies of carcinogenesis, where the biological response at extremely low doses is of interest. The National Center for Toxicological Research (NCTR) has addressed some of the issues associated with investigations of chemical effects at low doses in the so-called ED_{01}, or megamouse, study (Cairns, 1979). This study was designed to determine the ED_{01} for 2-acetylaminofluorene-induced carcinogenesis with increased precision by using sample sizes numbering in the thousands.

Lethal Concentration

LC_{50} and LCt_{50} (Inhalation Exposure)

The median lethal concentration of an inhaled chemical has been defined as a "statistically derived concentration of a substance that can be expected to cause death during exposure or within a fixed time after exposure in 50% of the animals exposed for the specific time" (Organization for Economic Cooperation and Development, 1981). This parameter is abbreviated as LC_{50} or LCt_{50}, where LC is the lethal concentration and t is the time of exposure (MacFarland, 1976). The exposure period is usually 4 hr, but shorter or longer exposures are appropriate if the acute effects for these time periods are of interest.

The LC_{50} differs from the LD_{50} in that it is based on the concentration and duration of exposure to the test material rather than the dose administered. It does not reflect the quantity of the test substance that is inhaled and retained by the exposed animal. The actual dose administered by an inhalation exposure (excluding any test material ingested during animal grooming) can be estimated using the following formula:

$$D = \frac{VCt\alpha}{w}$$

where V is the respiratory volume per minute, C is the concentration of agent in mg/liter, t is the duration of exposure in minutes, α is the percentage of agent retained, w is the body weight in kg, and D is the inhaled dose in mg/kg (Weston and Karel, 1946).

The above formula may be used to estimate an inhalation LD_{50} from an LC_{50} when the variables V and α are known. However, calculation of the inhalation LD_{50} is usually not practical or necessary since the LC_{50} is widely used and generally accepted as a measure of acute inhalation toxicity. Also, airborne concentration is more suitable than dose for comparisons with environmental and industrial exposure concentrations as well as with measures of safe exposure, e.g., threshold limit values.

LC_{50} (Exposures Other Than Inhalation)

The LC_{50} is also an integral part of the hazard evaluation scheme for aquatic toxicity (Maki and Duthie, 1978). In this context, the LC_{50} is the concentration of a chemical in water "killing 50% of a test batch of fish within a particular period of exposure" (Organization for Economic Cooperation and Development, 1981). Fish or other aquatic organisms are exposed to five or more concentrations of the test substance in water usually for a 96-hr exposure period. Exposure systems may be static (no change in the test solution during exposure) or dynamic (test water and test material continually renewed). Methods of calculating aquatic LC_{50} values are comparable to those used for calculating inhalation LC_{50} values (Stephan, 1977; Stora, 1974).

The LC_{50} has also been used as a means of standardizing *in vitro* toxicity. For example, Styles (1977) applied the LC_{50} to studies of cultured hamster kidney cells. The objective of the study was to investigate transformation of these cells in response to carcinogens without toxicity-induced selection of subpopulations. The LC_{50} served as an independent test to establish the presence of equitoxic concentrations of test substances during transformation assays.

Maximum Tolerated Dose

MTD

The maximum tolerated dose (MTD) has been defined as the "highest dose of the test agent given during the chronic study that can be predicted not to alter the animal's longevity from effects other than carcinogenicity" (Sontag *et al.*, 1976). The MTD should not produce greater than 10% inhibition of weight gain, produce clinical evidence of toxicity or pathological lesions, nor alter longevity except as a result of carcinogenesis (Sontag *et al.,* 1976). The dietary MTD is also limited by the test animal's capability to absorb the test substance and by effects on nutrition (Interagency Regulatory Liaison Group, 1979). To prevent nutritional impairments, a maximum dietary concentration of 5% is recommended for the test compound unless it is a nutrient (Organization for Economic Cooperation and Development, 1981).

Information from acute and subchronic toxicity studies is generally used to estimate the MTD for chronic studies. Page (1977) recommended that acute, 14-day repeated-dose and 90-day subchronic studies be conducted to obtain a reliable estimate of the MTD. The metabolism and pharmacokinetics of the test compound, if known, should also be used to predict the MTD.

The use of the MTD for carcinogenicity bioassays is justified by the difficulty in detecting cancer incidence at lower exposure levels when using typical sample sizes of 50–100 animals per group. For example, with a sample size of 40 and a 95% confidence interval, excess tumor incidence would be significant only if at least 10% of treated animals developed tumors, and no tumors were observed in controls (Cairns, 1979). Therefore, carcinogenicity bioassays are conducted at the MTD to decrease the possibility of false negative results.

Although the use of the MTD in carcinogenicity bioassays is a generally accepted and, in most cases, necessary practice, it has limitations. It complicates the assessment of cancer risk in humans by adding the uncertainty associated with high- to low-dose extrapolation to the uncertainty associated with interspecies extrapolation. If the MTD exceeds metabolic capability, the profile of active metabolites of the test compound will be different from that at low doses, and the animal's susceptibility to carcinogenesis will be altered (Munro, 1977). For example, in a study with tetrachlorvinphos in the mouse, doses exceeding the MTD produced oncogenicity in association with hormonal imbalance and metabolic overload, while none of these effects occurred at dose levels not exceeding the MTD (Parker *et al.,* 1985). Uncertainty caused by this or other mechanisms is minimized, but not eliminated, by using multiple dose levels in carcinogenicity bioassays. Accordingly, the Environmental Protection Agency's Health Effects Testing Guidelines (Code of Federal Regulations, 1985a) recommend that a minimum of three doses be used and that the low dose be no less than 10% of the highest dose.

No-Effect Level

NEL

The no-effect level (NEL) has been defined by the Food and Agriculture Organization/World Health Organization (FAO/WHO) Expert Committee on Food Additives (1974) as "the level of a substance that can be included in the diet of a group of animals without toxic effects." Although that definition is appropriate for the purpose of the FAO/WHO committee, a definition with wider applicability is provided by the Organization for Economic Cooperation and

Development (1981): "The maximum dose used in a test which produces no adverse effects." It may be expressed as the weight of test substance administered per weight of test animal or, when administered in food or water, as mg/kg of food, mg/liter of water, or as ppm in either food or water. For the purposes of the toxicologist, the NEL is generally used interchangeably with the no-adverse-effect level (NAEL), the no-observed-effect level (NOEL), and the no-observed-adverse-effect level (NOAEL).

Accurate determination of the NEL is highly dependent on the adequacy of the toxicity studies used to identify this parameter. Since poorly designed studies will generate false negative results and overestimate the NEL, studies should include state-of-the-art methodology with a multiplicity of endpoints. Gross pathology, histopathology, clinical biochemistry, hematology, and behavioral observations have been recommended for inclusion in subchronic toxicity studies used to generate NELs (Organization for Economic Cooperation and Development, 1981). Furthermore the statistical power of a test should be taken into consideration when statistically significant effects are not observed.

In addition to the adequacy of the toxicity study, the NEL depends on the judgment of an effect's toxicologic significance. Toxic effects are not often presented as dichotomous and distinctly adverse responses. For example, chemically induced fatty infiltration of the liver can be both benign and reversible, and may not be considered a toxic effect. However, it can also progress to cirrhosis. The stage at which liver disease becomes irreversible is difficult to clearly identify histologically (Popper, 1975). The FAO/WHO committee (1974) considers toxicologically significant effects as those not "fully attributable to normal physiologic adjustment and . . . reversible." Such definitions serve as useful guides, but the assessment of "normal" and "reversible" ultimately depends on sound toxicologic judgments.

Uncertainty in the determination of NELs in animal studies, which contributes to the uncertainty in estimating permissible human exposure levels, has been dealt with by applying safety factors to animal data. The FAO/WHO committee (1958) discussed the use of a safety factor almost 30 years ago to extrapolate food-additive toxicity data from animals to humans. The arbitrary factor of 100 was designed to "allow for any species difference in susceptibility, the numerical differences between the test animals and the human population . . . , the greater variety of complicating disease processes in the human population, the difficulty of estimating the human intake and the possibility of synergistic action among food additives." A committee of the National Research Council (1977b) suggested the use of a flexible uncertainty factor based on the quality and quantity of the available data. This uncertainty factor may range from 10 when long-term data on humans are available to 1000 when there are no data on humans and few data on animals, and should also be modified in accordance with the nature and severity of the observed toxicity. In addition to the routine application of safety factors to animal toxicity data for predicting human health risk (Code of Federal Regulations, 1985d), mathematical modeling of dose–response data (Food Safety Council, 1980) or a combination of both methods (National Research Council, 1980b) may be informative when the necessary data are available.

Minimum Effective Dose/Lowest Effect Level

MED/LEL

The minimum effective dose (MED), also referred to as the lowest effect level (LEL), is sometimes used as an alternative to the NEL. It is the minimum dose that produces an observed effect.

UNITS USED TO ESTIMATE ACCEPTABLE HUMAN EXPOSURE LEVELS

Threshold Limit Value

TLV

Threshold limit values (TLVs) are recommended atmospheric concentrations of workplace substances to which workers may be exposed without adverse health effects. They are expressed as either ppm or mg/m^3, and may indicate time-weighted average (TLV-TWA, or simply TLV), short-term (TLV-STEL), or ceiling (TLV-C) concentration limits (American Conference of Governmental Industrial Hygienists, 1985). Many of the TLVs have been adopted by the Occupational Safety and Health Administration as industrial air standards (Code of Federal Regulations, 1985e,f) and, as such, are referred to as permissible exposure limits (PELs). Unlike the TLVs, Occupational Safety and Health Administration PELs are legal standards.

The TLV-TWA (TLV) represents a time-weighted average concentration for an 8-hr workday and a 40-hr workweek. It can be exceeded for short periods during the workday without producing adverse health effects as long as the average concentration is at or below the TLV. The degree of acceptable deviation is governed by the nature of a substance's effects at concentrations greater than the TLV (Stokinger, 1962). For example, chemicals such as chlorine, with exposure levels based on high acute toxicity, may produce adverse effects when the TLV is exceeded for only a brief period. Such chemicals usually have short-term exposure limits (TLV-STEL) for use in conjunction with the TLV to prevent brief, but potentially harmful, deviations from the TLV. The TLV-STEL is a time-weighted concentration limit for 15-min. No more than four such exposure periods are permitted per day, and a maximum of 60 min must elapse between exposures.

The TLV-C is a stringent maximum permissible exposure that may not be exceeded even for short periods. It is applied to fast-acting highly toxic, or extremely irritating substances, e.g., acetic anhydride, for which even brief exposure periods may cause serious toxicity. This value is similar to the maximum acceptable (formerly allowable) concentrations (MACs) recommended by the American National Standards Institute (1974). The TLV-C is used in place of rather than in conjunction with a time-weighted average.

TLVs are based on the assumption that there is a level below which adverse effects will not occur, even when subjects are exposed to a chemical for 40 hr per week (Stokinger, 1972). Although TLVs may be determined from human data obtained retrospectively, they are usually derived by applying safety factors to apparent thresholds for effects observed in animal studies. The size of the safety factor selected is weighted heavily by the type of toxicity expected. For example, if lacrimation is the first adverse reaction and serious effects occur only at much higher doses, the safety factor would be based on lacrimation and would be small.

Although the existence of a threshold has been questioned with respect to chemical carcinogens, it has not precluded the establishment of TLVs for some known and suspected carcinogens in the workplace. TLVs have been listed by the American Conference of Governmental Industrial Hygienists (1985) for asbestos, bis(chloromethyl) ether, certain chromium compounds, coal tar pitch volatiles, nickel sulfide roasting fumes and dusts, and vinyl chloride. For others, such as 4-aminodiphenyl, benzidine, β-naphthylamine, and 4-nitrodiphenyl, exposures by any route and at any level are considered unacceptable.

Acceptable Daily Intake

ADI

The acceptable daily intake (ADI) has been defined by the FAO/WHO Expert Committee on Food Additives (1974) as "the amount of food additive that can be taken daily in the diet, even over a lifetime, without risk." As this definition indicates, the ADI was designed to indicate safe levels of food additives, but the concept is also applicable to contaminants in food, air, and water. It is determined by applying a safety factor to the NEL derived from animal toxicity data. The safety factor is determined by professional judgment and is intended to account for differences in variability of response among species, differences in sensitivity among humans, and the quality of the toxicity data underlying the NEL.

Suggested No-Adverse-Response Level

SNARL

Suggested-no-adverse-response levels (SNARLs) are levels of drinking water contaminants below which no adverse effects are expected to occur following a specified exposure period. SNARLs for 24-hr, 7-day, or chronic (lifetime) exposure to contaminated drinking water were first determined by the Safe Drinking Water Committee of the National Research Council (1980a) by applying safety factors ranging from 10 to 1000 to animal and human toxicity data. SNARLs are not legal standards for drinking water, but are intended to assist efforts to protect the public health from contaminated drinking water supplies until pollutants can be reduced to acceptable levels.

RISK INDICES USED IN EPIDEMIOLOGY

Relative Risk

RR

Relative risk (RR) is the ratio of the risk in a group of subjects exposed to a chemical of interest to that in an unexposed group (Rogan and Brown, 1979). In epidemiology, it is used as a measure of association (not causation) between a chemical and a disease. As discussed below, relative risk can be determined directly or indirectly, depending on the method of sampling used in the study.

When a random sample is taken from groups of subjects with and without exposure to the chemical of interest, relative risk is determined directly (White and Bailar, 1956). In such cases, the risk is estimated by the proportions of exposed and unexposed persons that develop the disease.

Relative risk can also be determined indirectly in retrospective epidemiological studies by sampling from groups of subjects with and without the disease and subsequently determining the exposure of subjects to the chemical of interest. This procedure has the advantage of being less expensive and less time-consuming than that used in prospective studies. It is briefly summarized below as it has been described by Cornfield (1951).

If the proportion of exposed subjects among those with the disease is represented as P_1, the proportion of exposed subjects among those without the disease as P_2, and the proportion of the general population with the disease as X, then the proportions of the general population in each of four possible exposure and disease categories are as shown in Table 2-2. The proportion of subjects with and without exposure who have the disease are then $P_1X/[P_2 + X(P_1 - P_2)]$ and $(1 - P_1)X/[(1 - P_2) - X(P_1 - P_2)]$, respectively. If X is small, or is assumed to be small, relative to the proportions of the nondiseased group with and without the exposure, this term can be eliminated and these formulas become P_1X/P_2 and $(1 - P_1)X/1 - P_2$, respectively. The ratio of these can then be used to obtain the following simplified formula for estimating relative risk:

$$\frac{P_1(1 - P_2)}{P_2(1 - P_1)}$$

This formula is also referred to as the cross-product ratio, because when X is neglected, it is equivalent to the ratios of products diagonally across in Table 2-2. It has been widely used to estimate the relative risk of disease in retrospective

TABLE 2-2. The Proportions of a Population in Different Exposure and Disease Categories

Exposure category	Proportion with disease	Proportion without disease
Exposed	P_1X	$P_2(1 - X)$
Not exposed	$(1 - P_1)X$	$(1 - P_2)(1 - X)$

studies, but it may be applied only when the proportion of the population with the disease is small.

Population Attributable Risk

PAR

Population attributable risk (PAR) has been defined as "the proportion of a disease attributable to a given etiologic agent" (Levin, 1953) and is represented as follows:

$$\frac{b(r-1)}{b(r-1)+1}$$

where r is the relative risk and b is the proportion of the general population exposed to the etiologic agent. PAR is used as measure of the contribution of chemical exposure to a population's disease burden. It is particularly suitable for studying long-term disease and mortality but has some limitations when applied to short-term, recurrent disease (Park, 1981).

Standardized Mortality Ratio

SMR

The standardized mortality ratio (SMR) is widely used as a measure of chemical-related mortality. It has been defined by Symons and Taulbee (1981) as "the ratio of the observed to the expected total number of deaths, where the expected number is computed by applying the schedule of age-specific mortality rates in the comparison population to the distribution of person-years in the study population." These authors reported that the SMR can be used as an approximation of relative risk if excess mortality is consistent across age bands and if the age bands are not too large (10 years or less).

The procedure used to determine the SMR is not limited to mortality as an endpoint or to age as a confounding factor. Disease incidence can be substituted for mortality to determine standardized disease ratios. The ratio may be adjusted for variables other than age that correlate with the endpoint selected.

ACKNOWLEDGMENTS. The author is grateful to Herbert E. Stokinger for his invaluable advice and to John C. Bailar for sharing his expertise in epidemiology.

APPENDIX: ABBREVIATIONS OF TOXICOLOGICAL UNITS

ADI	acceptable daily intake
C	ceiling
ED_{01}	effective dose/1% response
ED_{50}	effective dose/50% response
ED_{99}	effective dose/99% response

LC_{50}	lethal concentration/50% response
LCt_{50}	lethal concentration × time/50% response
LCLo	lowest lethal concentration
LD_{01}	lethal dose/1% response
LD_{50}	lethal dose/50% response
LD_{99}	lethal dose/99% response
LDLo	lowest lethal dose
LEL	lowest effect level
LOAEL	lowest-observed-adverse-effect level
LOEL	lowest-observed-effect level
MAC	maximum acceptable concentration
MED	minimum effective dose
MTD	maximum tolerated dose
NAEL	no-adverse-effect level
NEL	no-effect level
NOAEL	no-observed-adverse-effect level
NOEL	no-observed-effect level
NTEL	no-toxic-effect level
PAR	population attributable risk
PEL	permissible exposure limit
RR	relative risk
SMR	standardized mortality ratio
SNARL	suggested-no-adverse-response level
STEL	short-term exposure limit
TD_{01}	toxic dose/1% response
TD_{50}	toxic dose/50% response
TD_{99}	toxic dose/99% response
TLV	threshold limit value
TWA	time-weighted average

REFERENCES

American Conference of Governmental Industrial Hygienists. 1985. *Documentation of the Threshold Limit Values,* 4th ed. American Conference of Governmental Industrial Hygienists, Cincinnati, Ohio. 486 pp.

American National Standards Institute. 1974. American National Standard Acceptable Concentration of Acetic Acid, No. 237.39. American National Standards Institute, New York. 8 pp.

Boyd, E. M. 1968. Prediction of drug toxicity: Assessment of drug safety before human use. *Can. Med. Assoc. J.* **98**:278–293.

Brownlee, K. A., J. L. Hodges, Jr., and M. Rosenblatt. 1953. The up-and-down method with small samples. *J. Am. Stat. Assoc.* **48**:262–277.

Cairns, T. 1979. The ED01 study: Introduction, objectives, and experimental design. *J. Environ. Pathol. Toxicol.* **3**:1–7.

Code of Federal Regulations. 1985a. Title 40, part 798. Office of the Federal Register, Washington, D.C.

Code of Federal Regulations. 1985b. Title 29, Part 1910.1200. Office of the Federal Register, Washington, D.C.

Code of Federal Regulations. 1985c. Title 40, Part 162.10. Office of the Federal Register, Washington, D.C.

Code of Federal Regulations. 1985d. Title 21, Part 170.22. Office of the Federal Register, Washington, D.C.

Code of Federal Regulations. 1985e. Title 41, Part 50-204.50. Office of the Federal Register, Washington, D.C.

Code of Federal Regulations. 1985f. Title 29, Part 1910.1000. Office of the Federal Register, Washington, D.C.

Cornfield, J. 1951. A method of estimating comparative rates from clinical data: Applications to cancer of the lung, breast, and cervix. *J. Natl. Cancer Inst.* **11**:1269–1275.

European Economic Community. 1983. Commission Directive of July 1983 adapting to technical progress for the fifth time Council Directive 67/548/EEC on the approximation of the laws, regulations and administrative provisions relating to the classification, packaging, and labelling of dangerous substances, 83/467/EEC. *Off. J. Eur. Commun.* **26**(L257):1–33.

European Economic Community. 1985. Proposal for a Council Directive on the Approximation of the Laws, Regulations and Administrative Provisions of the Member States Relating to the Classification, Packaging, and Labelling of Dangerous Preparations. 85/C211/03. *Off. J. Eur. Commun.* **C211**:3–15.

FAO/WHO (Food and Agriculture Organization/World Health Organization Expert Committee on Food Additives). 1958. Procedures for the Testing of Intentional Food Additives to Establish Their Safety for Use. Technical Report Series No. 144. World Health Organization, Geneva. 19 pp.

FAO/WHO (Food and Agriculture Organization/World Health Organization Expert Committee on Food Additives). 1974. Toxicological Evaluation of Certain Food Additives in a Review of General Principles and Specifications. Technical Report Series No. 539. World Health Organization, Geneva. 40 pp.

Finney, D. J. 1971. Estimation of the median effective dose. Pp. 19–49 in *Probit Analysis,* 3rd ed. University Press, Cambridge.

Food Safety Council. 1980. *Proposed System for Food Safety Assessment.* Food Safety Council, Washington, D.C. 160 pp.

Freireich, E. J., E. A. Gehan, D. P. Rall, L. H. Schmidt, and H. E. Skipper. 1966. Quantitative comparison of toxicity of anticancer agents in mouse, rat, hamster, dog, monkey and man. *Cancer Chemother. Rep.* **50**:219–243.

Hodge, H. G. 1965. The LD_{50} and its value. *Am. Perfum. Cosmet.* **80**:57–60.

Interagency Regulatory Liaison Group. 1979. Scientific bases for identification of potential carcinogens and estimation of risks. Report of the Work Group on Risk Assessment. *Fed. Reg.* **44**:39858–39879.

Levin, M. L. 1953. The occurrence of lung cancer in man. *Acta Unio Intern. Contra Cancrum* **9**:531–541.

MacFarland, H. N. 1976. Respiratory toxicology. Pp. 121–154 in W. J. Hayes, Jr., ed. *Essays in Toxicology,* Volume 7. Academic Press, New York.

Maki, A. W., and J. R. Duthie, Jr. 1978. Summary of proposed procedures for the evaluation of aquatic hazard. Pp. 153–163 in J. Cairns, Jr., K. L. Dickson, and A. W. Maki, eds. *Estimating the Hazards of Chemical Substances to Aquatic Life,* ASTM ATP 657. American Society for Testing and Materials, Philadelphia.

Munro, I. C. 1977. Considerations in chronic toxicity testing: The chemical, the dose, the design. *J. Environ. Pathol. Toxicol.* **1**:183–197.

Murray, R. E., and J. E. Gibson, 1972. A comparative study of paraquat intoxication in rats, guinea pigs, and monkeys. *Exp. Mol. Pathol.* **17**:317–325.

National Institute for Occupational Safety and Health. 1983. *Registry of Toxic Effects of Chemical Substances,* Volumes 1–3. U.S. Government Printing Office, Washington, D.C.

National Research Council. 1975. *Principles for Evaluating Chemicals in the Environment.* A report of the Committee for the Working Conference on Principles of Protocols for Evaluating Chemicals in the Environment. National Academy of Sciences, Washington, D.C. 454 pp.

National Research Council. 1977a. *Principles and Procedures for Evaluating the Toxicity of Household Substances.* A report of the Committee for the Revision of National Academy of Sciences Publication 1138. National Academy of Sciences, Washington, D.C. 130 pp.

National Research Council. 1977b. *Drinking Water and Health.* A report of the Safe Drinking Water Committee. National Academy of Sciences, Washington, D.C. 939 pp.

National Research Council. 1980a. *Drinking Water and Health,* Volume 3. A report of the Safe Drinking Water Committee. National Academy Press, Washington, D.C. 415 pp.

National Research Council. 1980b. Risk Assessment/Safety Evaluation of Food Chemicals. A report of the Committee on Food Protection, Subcommittee on Food Technology. National Academy Press, Washington, D.C. 36 pp.

Organization for Economic Cooperation and Development. 1981. *Guidelines for Testing of Chemicals.* Organization for Economic Cooperation and Development, Paris.

Page, N. P. 1977. Current concepts of a bioassay program in environmental carcinogenesis. Pp. 87–171 in H. F. Kraybill and M. A. Mehlman, eds. *Advances in Modern Toxicology,* Volume 3. Hemisphere, Washington, D.C.

Park, C.B. 1981. Attributable risk for recurrent events. *Am. J. Epidemiol.* **113**:491–493.

Parker, C. M., G. A. Van Gelder, E. Y. Chai, J. B. M. Gellatly, D. G. Serota, R. W. Voelker, and S. D. Vesselinovitch. 1985. Oncogenic evaluation of tetrachlorvinphos in the $B6C3F_1$ mouse. *Fund. Appl. Toxicol.* **5**:840–854.

Popper, H. 1975. Introduction to the problem, hepatic fibrosis and collagen metabolism in the liver. Pp. 1–14 in H. Popper and K. Becker, eds. *Collagen Metabolism in the Liver.* Stratton Intercontinental Medicine Book Co., New York.

Rogan, W. J., and S. M. Brown. 1979. Some fundamental aspects of epidemiology: A guide for laboratory scientists. *Fed. Proc.* **38**:1875–1879.

Savini, E. C. 1968. Estimation of the LD_{50} in mol/kg. *Proc. Eur. Soc. Study Drug Toxicity* **9**:276–278.

Smyth, H. F., C. S. Weil, J. S. West, and C. P. Carpenter. 1969. An exploration of joint toxic action: Twenty-seven industrial chemicals intubated in rats in all possible pairs. *Toxicol. Appl. Pharmacol.* **14**:340–347.

Sontag, J. M., N. P. Page, and U. Saffiotti. 1976. *Guidelines for Carcinogen Bioassay in Small Rodents.* NCI Technical Report Series No. 1, DHEW Publication No. (NIH) 76-801, U.S. Government Printing Office, Washington, D.C. 465 pp.

Sperling, F. 1976. Nonlethal parameters as indices of acute toxicity: Inadequacy of the acute LD50. Pp. 177–191 in M. A. Mehlman, R. E. Shapiro, and H. Blumenthal, eds. *Advances in Modern Toxicology,* Volume 1, Part 1. Hemisphere, Washington, D.C.

Stephan, C. E. 1977. Methods for calculating an LC_{50}. Pp. 65–84 in F. L. Mayer and J. L. Hamelink, eds. *Aquatic Toxicology and Hazard Evaluation.* American Society for Testing and Materials, Philadelphia.

Stokinger, H. E. 1962. Threshold limits and maximal acceptable concentrations. *Arch. Environ. Health* **4**:115–117.

Stokinger, H. E. 1972. Concepts of thresholds in standard setting: An analysis of the concept and its applications to industrial air limits (TLVs). *Arch. Environ. Health* **25**:153–157.

Stora, G. 1974. Computation of lethal concentrations. *Mar. Pollut. Bull.* **5**:69–71.

Styles, J. A. 1977. A method for detecting carcinogenic organic chemicals using mammalian cells in culture. *Br. J. Cancer* **36**:558–563.

Symons, M. J., and J. D. Taulbee. 1981. Practical considerations for approximating relative risk by the standardized mortality ratio. *J. Occup. Med.* **23**:413–416.

Trevan, J. W. 1927. The error of determination of toxicity. *Proc. R. Soc. Lond. B.* **101**:483–514.

Weston, R. E., and L. Karel. 1946. An application of the dosimetric method for biologically assaying inhaled substances. *J. Pharmacol. Exp. Ther.* **88**:195–207.

White, C., and J. C. Bailar. 1956. Retrospective and prospective methods of studying association in medicine. *Am. J. Public Health* **46**:35–44.

Zbinden, G., and M. Flury-Roversi. 1981. Significance of the LD_{50} test for the toxicological evaluation of chemical substances. *Arch. Toxicol.* **45**:77–99.

3

Dose–Response Relationships

Rolf Hartung

INTRODUCTION

One of the most powerful tools in the analysis of toxicological events is the description of the intensity of the exposure in terms of dose, dose rate, or concentration for a specified duration of exposure in relation to the frequency or intensity of the observed responses. Dose–response relationships have received a generic treatment in all general textbooks of toxicology, such as those by Loomis (1974), Doull *et al.* (1980), and Hapke (1975). The statistical risk of experiencing an effect from an exposure to a chemical can be attributed to the interaction of several important factors. As noted in Chapter 1, this complex relationship of dose to response was already recognized as early as the 16th century by Paracelsus. To restate his observations as they apply in contemporary toxicology—a sufficiently high dose of any compound will produce severe adverse effects in all exposed organisms. Conversely, a sufficiently low dose of any compound will produce no significant effects that can be experimentally determined, no matter how sophisticated or extensive the experiment.

This second point requires some amplification. As stated, it avoids the question of the presence or absence of theoretical thresholds for toxicological responses. Rather, it addresses itself to the existence of practical thresholds. This issue assumes importance when one considers extrapolation models used for safety evaluations. Thus, if one assumes that certain toxicological events, especially carcinogenicity and mutagenicity, exhibit no threshold on a theoretical

ROLF HARTUNG • Toxicology Program, Department of Environmental and Industrial Health, University of Michigan, Ann Arbor, Michigan 48109.

basis, then zero response can only occur at zero dose. For some types of toxicological responses, that statement may merely be an expression of the experimental limitations that failed to uncover a response that should theoretically have occurred. For other types of responses, perhaps target organ toxicity, there may be real thresholds, based on theoretical considerations, and the absence of a response at a specific dose regimen may reflect both the existence of a real threshold or experimental limitations.

In addition to dose, an equally important parameter is the inherent toxicity of the compound. This parameter depends on the specific interactions of the chemical administered at a specific dose with the normal biochemical and physiological processes in an individual organism. These characteristics have quantitative as well as qualitative aspects. Thus, it is clearly understood by everyone that arsenic and cyanide are "more toxic" than sugar or table salt. It is implicit that the more toxic compounds produce qualitatively more severe effects at relatively small doses. However, the classification of chemicals into extreme categories, such as toxic and nontoxic, seems to be a very subjective undertaking.

Another major factor that influences the expression of toxic effects is the duration of exposure. For short-term exposures, very high concentrations may be required to produce an effect that would result only after much longer exposures to much lower concentrations. Although these relationships are much more complex than this introductory section would seem to indicate (Figs. 3-1 and 3-2), it is clear that the traditional dose–response relationship should be more appropriately considered as a dose rate (or concentration) versus duration of exposure versus response surface (Hartung, 1981; Society of Toxicology, ED_{01} Task Force, 1981).

This concept improves our understanding of the many interactive factors that influence to occurrence of an effect due to exposure. Regrettably, reports of

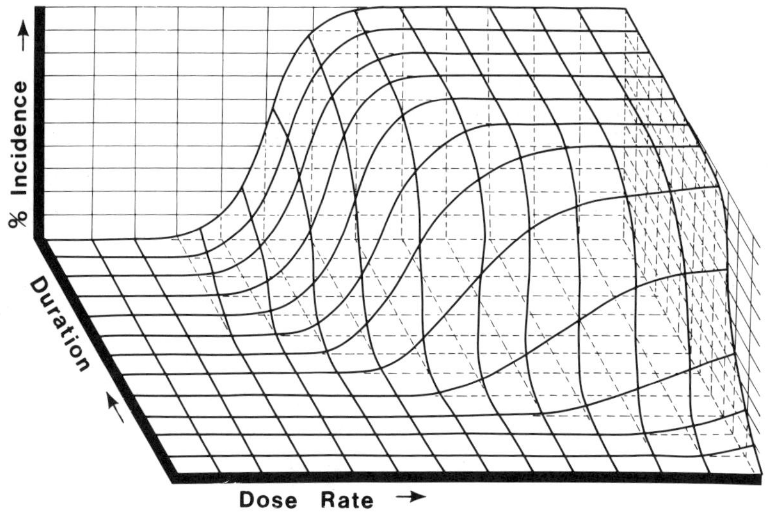

FIGURE 3-1. The generalized dose–response surface.

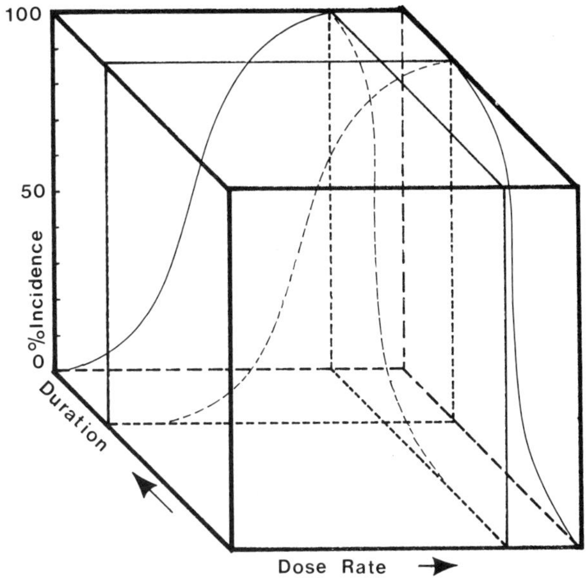

FIGURE 3-2. Sections of the dose–response surface to generate dose–response curves and time-to-effect curves.

most toxicological experiments conducted to date do not contain sufficient data to allow construction of the dose–duration–response surfaces.

When the dose–duration–response surface is sectioned by a plane parallel to the dose and response axes and perpendicular to the duration of exposure axis at time *t*, then the resulting curve represents the conventional dose–response curve for a duration *t* of exposure. Conversely, when the dose–duration–response surface is sectioned parallel to the duration of exposure and response axes and perpendicular to the dose axis at dose *d*, then the resultant curve represents the time-to-effect responses at dose *d*.

DOSE–RESPONSE CURVES

The classical dose–response curve is restricted to presenting the observed responses in relation either to dose, dose rate, or concentration. The duration of the exposure is implied by the conditions prevailing in the experiment that is being described.

The response axis has received different treatments by different disciplines. In toxicology the response axis is most frequently used to denote the percentage of exposed organisms that respond at various dose levels, so that the response axis charts the incidence of responses at specific dose regimens. In that case, the slope of the dose–response curve is related to the degree of individual variability of the exposed organisms in their response to the toxicant. Dose–response curves with a shallow slope would indicate a high degree of variability of the population with respect to the response, while a steep slope would imply low variability and therefore a relatively uniform response (Fig. 3-3). The confidence intervals or

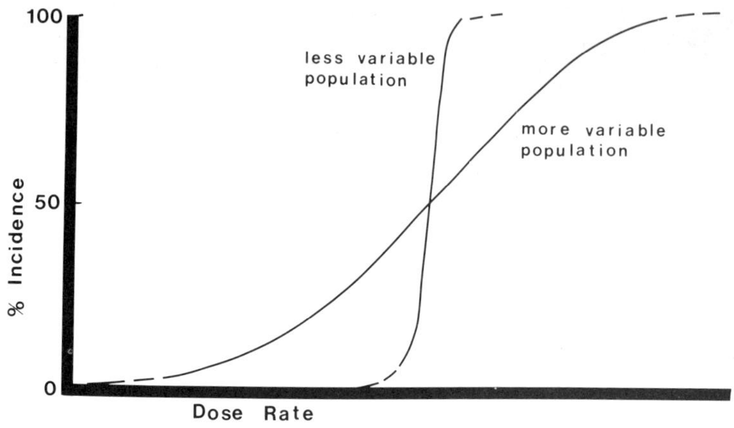

FIGURE 3-3. Slopes of dose–response curves indicate the extent of individual variability of response to a toxicant.

other statistical measures of scatter around the individual points of the dose–response curve are indices of experimental precision. In practice there are some tradeoffs between slope (as an indicator of population variability) and confidence or fiducial limits around the curve (as indicators of experimental variability). Thus, the slopes of dose–response curves also tend to flatten somewhat when experimental procedures lack precision.

The slopes of experimentally determined dose–response curves have no inherent relationship to the concepts of "potency," "intrinsic toxicity," or "relative toxicity." These latter concepts, when used in conjunction with the slope of the dose–response relationship, are useful only when they are applied to specific mathematical dose–response models. Beyond those specific applications, these terms have some utility in a more generalized context. Thus, when these terms are considered independently of any mathematical model, then potency relates the incidence and severity of effects to the dose–duration regimen. When the incidence is high and severity great after a relatively low-dose or short-term exposure, then the potency is high. Numerically, potency is a variable whose value usually changes as the dose regimen changes. It is a constant only in special linear dose–response models, such as the "single-hit" model. The term "relative toxicity" is especially useful when one compares the toxicity of one compound to that of another under very specific exposure conditions.

When the primary goal of a dose–response curve is to display the incidence of responses, then these must be classified as all-or-none, or quantal responses. The statistical analysis of quantal responses in toxicity tests reflects initial concerns with mortality or survival as the most easily measured responses (Bliss, 1935). Many other events, such as an elevation of blood pressure or elevated serum enzyme levels, represent types of data sets that are part of a continuum of toxicity and therefore need to be converted to all-or-none responses. This is done by selecting a criterion against which a continuous response is tested. Thus, one might select a criterion for elevated systolic blood pressure at 180 mm Hg, so that all exposed organisms that meet or exceed that criterion are considered to be responders for that particular dose regimen. The criterion could readily be chosen at a lower level to detect smaller deviations from normal blood pressure. However, as the criterion approaches normal background levels, a greater number of responses will occur in the absence of any exposure. This situation, in turn, would require a larger number of observed organisms per dose level in order to differentiate dose-related responses from background responses for the specific effect for which the criterion has been established.

In pharmacology, the response axis has often been used to denote the intensity of a response, so that responses by individuals are represented in terms of the maximal response observable (Levine, 1973). Such responses have been classified as graded responses in contrast to the quantal or all-or-none responses. When graded responses are represented in relation to maximal responses, the issue of individual variability becomes submerged in the data. However, when plotting the percentage of animals responding at each dose, it is difficult to represent graded responses, since each response has to be coded initially as an all-or-none or quantal response. For either convention, each type of response has its own dose–response curve and therefore, by implication, its own dose–duration–response surface. Thus, the interaction of a chemical with a group of organisms would be expected to generate a large family of dose–response surfaces. In the subsequent discussions, the emphasis is placed on those dose–response relationships that emphasize the importance of incidence and therefore individual variability. The quality and the intensity of the response are considered in the context of quantal responses.

A close examination of a generalized set of dose–response data reveals that the relationship of dose to response is nonlinear in a complex fashion, especially when the data are presented with linear response and linear dose axes (Fig. 3-4). Both 0 and 100% responses are commonly observed for experimental data at the extremes of the dose–response curve. The derivation of such dose–response curves requires a relatively large number of animals, since the points of the curve represent the percentages of the organisms that respond at each dose level. The

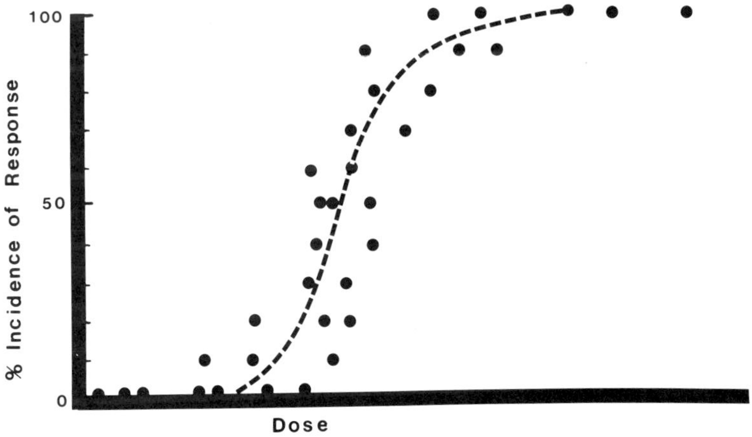

FIGURE 3-4. Generalized dose–response curve, model-independent.

mathematical resolution of any one point on the curve to the closest 10% would require ten animals per dose level, and to resolve 1% would require 100 animals per dose level. To achieve a statistically sound resolution to these levels of sensitivity, a multiple of that number would be required.

Curves of the type presented in Fig. 3-4 tend to be asymmetrical in addition to being nonlinear. The symmetry of the curve is often improved by transforming the dose axis to the logarithm of the dose (Fig. 3-5).

Up to this point, the extremes of the curves have been omitted from the dose–response curves, even though extreme responses (0 and 100%) are commonly observed and can often be demonstrated repeatedly in replicate experiments. Dose–response curves based on single doses, repeated doses, dose rates, or concentrations of chemicals in food, water, or air appear to have similar shapes (Gaddum, 1953). Likewise, dose–response curves based on continuous data, such as enzyme concentrations, that have been transformed to quantal responses using some criterion appear to have similar shapes (Hager and Punnett, 1973).

The statistical and mathematical analysis of dose–response curves becomes especially important when it is necessary to compare dose–response curves, or when one attempts to utilize dose–response curves in risk assessments and safety evaluations. The selection of a specific mathematical model applied to the resolution of a dose–response curve is not very critical when one is mainly interested in the central region of the dose–response curve; however, dose–response models diverge greatly at the extremes of the dose–response relationship, particularly when extrapolating outside the range of observed data. None of the presently used mathematical dose–response models (e.g., log-probit, log-logit, multistage, multihit, or Weibull) is fully satisfactory. The models are either too simplistic from a

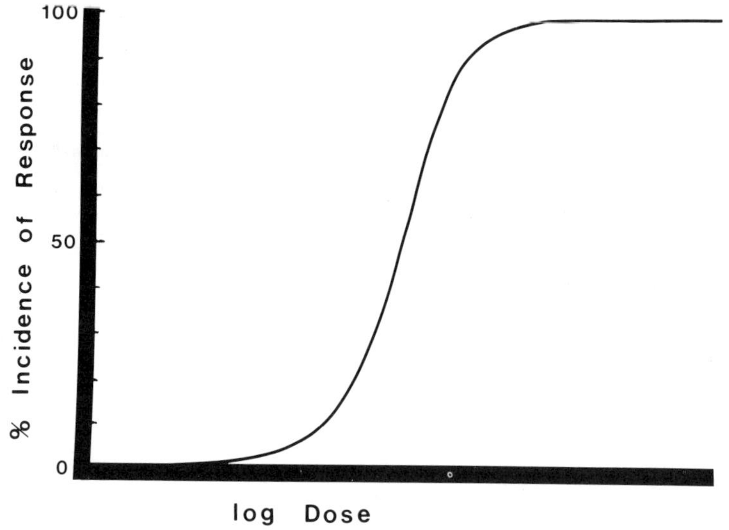

FIGURE 3-5. Dose–response curve with logarithmically transformed dose axis.

biological point of view or, if they are multivariate models, the individual statistical variables probably lack biological meaning. In addition, when the number of variables in the model begin to approach the number of data points for a dose–response experiment, then the solution becomes trivial, i.e., the resultant dose–response curve will pass through all individual data points, regardless of where they may lie. The probit model has enjoyed the widest application to a wide range of toxicological problems; however, its application to carcinogenicity data has been questioned.

THE PROBIT MODEL

The basic assumption of the probit model is that the probabilities that individuals exposed to a certain dose will actually respond are normally or log-normally distributed. This implies that the dose–response relationship can be satisfactorily explained on the basis of statistical variability. Biological factors, such as toxicokinetics, repair mechanisms, and enzyme kinetics, are not included in this model.

As was discussed previously, the asymmetrical, generalized dose–response curve can usually be made more symmetrical by transforming the dose axis to the logarithm of the dose. In addition to performing that transformation, the probit model (Finney, 1971) transforms the response axis from percentages to probits. This changes the dose–response curve to a straight line, which is more amenable to mathematical and statistical analysis (Fig. 3-6).

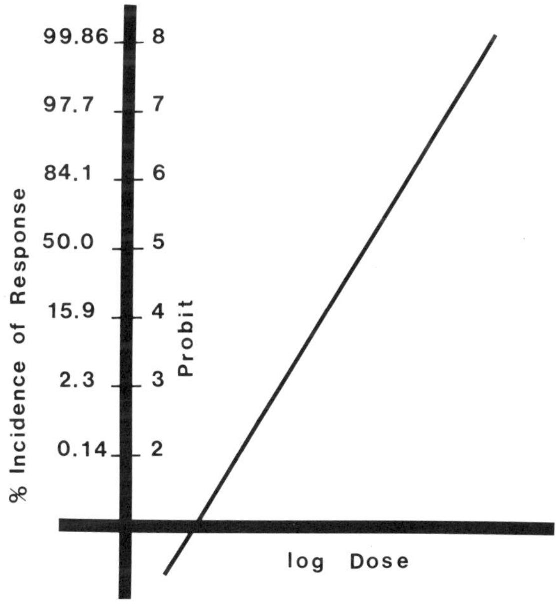

FIGURE 3-6. Log-probit transformation of a dose–response curve.

Probits are related to the proportion of individuals that respond by the indefinite integral

$$P = \frac{1}{(2\pi)^{1/2}} \int_{-\infty}^{Y-5} \exp\left(-\frac{1}{2}u^2\right) du \tag{1}$$

where P is the proportion of organisms responding, Y is the probit corresponding to P, and u is the mean of the distribution. The relationship is clearly related (1) to the Gaussian curve, which can be further illustrated by investigating the relationships of the percentiles of the Gaussian distribution (or percent of the test organisms responding), (2) to the "normal equivalent deviant" (NED), and (3) to the corresponding probits (P) in Table 3-1.

This transformation results in an excellent fit for most experimental data, except for responses at the 0 and the 100% levels. Since the Gaussian bell-shaped curve is asymptotic at both its extremes, the probit of 100% equates with positive infinity and the probit of 0 with negative infinity. Therefore, these two response levels effectively cannot be readily included in any subsequent calculations, unless the methodology for probit analysis is modified to include the artifice of changing 0 and 100% responses to greater than 0 and less than 100% responses.

There are a variety of probit-based methods for the mathematical solution of dose–response relationships. A number of computer methods, based on publications by Finney (1971) and Bliss (1935), have been reported by Spratt (1966) and Daum (1969).

Several simpler probit methods have various advantages and disadvantages. The "up and down method" by Brownlee *et al.* (1953) can calculate a median lethal dose or a median effective dose from tests on very few animals. However, the method cannot be used to calculate the slope of the dose–response curve, and under some conditions the fiducial limits for the LD_{50} or EC_{50} cannot be estimated.

The tabular method developed by Weil (1952) is among the simplest to use,

TABLE 3-1. Relationship Between Percentile, Probit, and the Normal Equivalent Deviant

Percentile	Normal equivalent deviant	Probit
0.14	−3.0	2.0
2.3	−2.0	3.0
6.7	−1.5	3.5
15.9	−1.0	4.0
30.9	−0.5	4.5
50.0	0.0	5.0
69.1	0.5	5.5
84.1	1.0	6.0
93.3	1.5	6.5
97.7	2.0	7.0
99.38	2.5	7.5
99.86	3.0	8.0

but it has certain restrictions. The dose levels must be geometrically spaced; the same number of test organisms must be used at each dose level; tables are provided for only 2, 3, 4, 5, 6, and 10 animals per dose level; and only four dose levels can be used for the calculation. The method calculates the LD_{50} or its equivalent, plus its fiducial limits, but cannot be used to calculate the slope of the dose–response curve.

The graphic method reported by Lichtfield and Wilcoxon (1949) is probably the most detailed of the noncomputer methods and is one of the most widely used. In this method, a dose–response curve is fit to the data by eye. The "goodness of fit" is subsequently evaluated and adjusted, if necessary. The spacing of the doses is not critical, and most of the other restrictions of the Brownlee and Weil methods do not apply. The method derives a variant of the formula of the probit curve and all applicable statistics. Since part of the solution of a dose–response data set by this method depends on a fit of data "by eye," the solutions are somewhat variable and depend in part on the skill of the analyst. However, the analytical variations tend to be much lower than the biological variations in the test.

When the probit model is used for comparative purposes, utilizing the central region of the dose–response curve, then the difficulties encountered by the model at the extremes of the relationship can normally be ignored, because the calculated curve tends to fit the experimental data in the central response region well. However, when the probit model is applied to measurements at the extremes of the curve, then a number of problems arise. The model predicts zero response only at zero dose, even if the data strongly suggest the existence of a threshold. The model also suggests that a 100% response only occurs at infinite doses. The latter is illogical, and the former may well be equally untenable, although it is more resistant to experimental analysis or argument on biological principles. Despite these shortcomings, the probit dose–response model is the most widely accepted model for the evaluation of toxicological effects other than carcinogenicity.

In addition to solutions of dose–response curves, the probit transformation has also been applied successfully to the solution of time-to-effect curves (Lichtfield, 1949). The cited method provides for graphic solutions that are similar to the graphic solution of dose–response curves by Lichtfield and Wilcoxon (1949).

APPLICATIONS: THERAPEUTIC INDEX AND MARGIN OF SAFETY

Since it is possible to plot the dose–response curves for adverse effects and therapeutic effects on the same graph (Fig. 3-6), it is possible to visualize and calculate the distance between the dose–response curve for adverse effects and the dose–response curve for therapeutic effects. The therapeutic index has been traditionally defined as follows:

$$\text{therapeutic index} = \frac{LD_{50}}{ED_{50}} \qquad (2)$$

The magnitude of the therapeutic index can give a false feeling of security when the two dose–response curves have significantly different slopes. This is partly overcome by the use of the "margin of safety" in relation to drug toxicity, which is defined as follows:

$$\text{margin of safety} = \frac{ED_{01}(\text{adverse effect dose})}{ED_{99}(\text{therapeutic dose})} \tag{3}$$

Neither of these two concepts has received much consideration in public policy decisions about safety. Possible reasons for this may be that the use of lethality and the median comparisons for the therapeutic index allow for too many chances of error (Klaassen and Doull, 1980). Moreover, the "margin of safety" has many varied definitions, involving different percentiles and different endpoints, including mortality (Fig. 3-7).

EXPLORATION OF THE "THRESHOLD" REGION

Some compounds exert their biological action through a direct interaction with DNA, resulting in the faulty coding of genetic material, which can be propagated through subsequent cell divisions. These compounds may lack a threshold for the production of toxicological effects. Among such substances may be some that have carcinogenic, mutagenic, or teratogenic activity.

When xenobiotics act through mechanisms such as enzyme inhibition, depletion of required substances, or inhibition of transport mechanisms, then the pro-

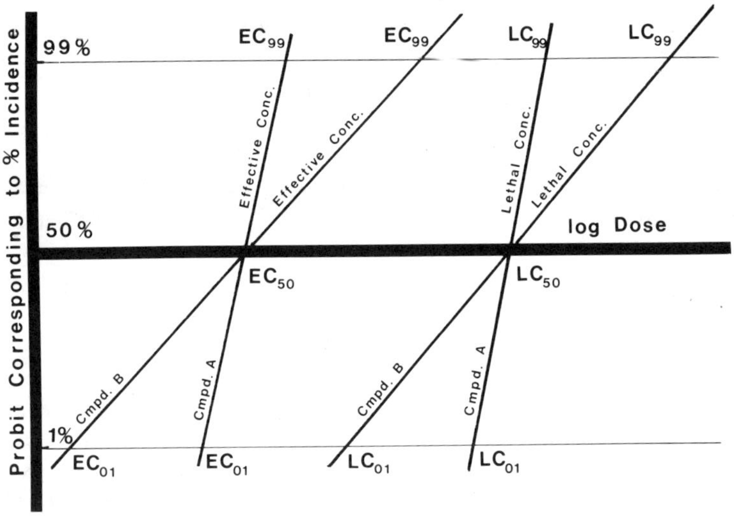

FIGURE 3-7. Influence of slope on the evaluation of the therapeutic index and the margin of safety. The therapeutic indices are equal, but the margins of safety differ considerably. The LC_{99}, LC_{50}, and LC_{01} represent the 99th, 50th, and first percentile of the lethal concentration. The percentiles of the effective concentrations are shown using the same convention.

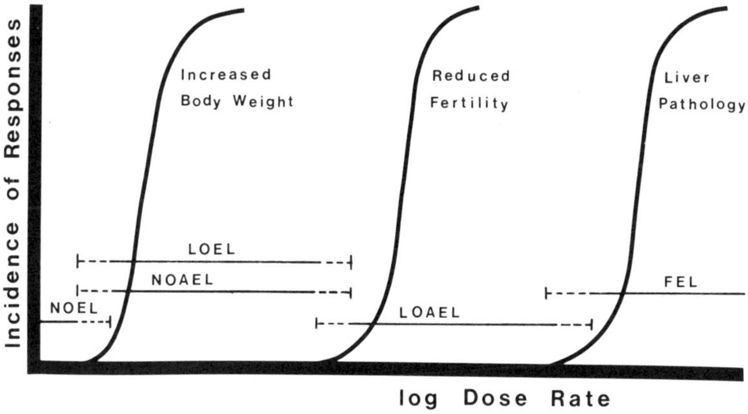

FIGURE 3-8. The "threshold" region for chronic dose–response curves. The illustrated dose–response curves represent idealized "true" dose–response curves; the brackets illustrate the ranges for individual experimental results that might be categorized according to this scheme. NOEL, no-observed-effect level; NOAEL, no-observed-adverse-effect level; LOEL, lowest-observed-effect level; LOAEL, lowest-observed-adverse-effect level; FEL, frank-effect level. Depending upon the quality of the data, some of these categories (e.g., NOAEL and LOEL) may overlap.

duction of an effect is thought to depend on the interaction among the concentration of the xenobiotic, reactivity with the receptor, reserve capacity in the affected system, the turnover time of affected enzymes, or the regenerative capacity of the repair systems.

The interpretation of these interactions is that much more than one molecule of the xenobiotic is required to elicit an effect, and therefore that absolute thresholds exist for many kinds of target organ toxicity. However, even though one may stipulate the existence of a threshold on the basis of biological arguments, the quantitation of such a threshold is usually not possible because of the analytical uncertainties at the low-response extreme of the dose–response curve (National Research Council, 1977).

In chronic toxicity studies, the incidence of severe adverse effects become scarcer as the dose is lowered. Subtle effects tend to predominate as the dose rate is lowered further, and those subtle effects tend to disappear at even lower dose rates until a dose regimen is reached where the experimental effects can no longer be distinguished from the controls. This latter region constitutes the closest approach to the threshold that could be made under the specific experimental conditions used. Obviously, changes in experimental design, such as the selection of other endpoints and changes in the number of test organisms, could significantly affect the outcome.

In the absence of generally accepted methods to calculate thresholds, "no-observed-response levels" (NOELs) and "no-observed-adverse-effect levels" (NOAELs) interpreted from experimental data sets have been utilized as approximators of the threshold region. The circumstances surrounding their use and the justification for their use are best discussed with the aid of an illustration (Fig. 3-8).

In long-term studies designed to elicit subtle effects and to help define the threshold or "no-response" region, one usually finds a failure to test at a sufficient number of dose levels with enough animals per dose level to define the dose–response curves accurately. Characteristically, dose levels are so widely spaced that the slopes of such curves cannot be readily determined. Although Fig. 3-8 implies the existence of "true" dose–response curves, the actual position of those curves usually remains obscure to the experimenter, who therefore has to use the sparse data available to judge how close the experimental data lie in relation to the elusive "threshold" dose regimen. In this process, data derived from particular dose levels are commonly assigned to various categories (e.g., NOEL, NOAEL, LOEL, etc.), which reflect the relationship of the data to the location of the suspected threshold region. In any experimental protocol, the location of the apparent threshold can be influenced by experimental variables, such as the number of animals used in the experiment, the frequency of occurrence of the effect among control animals, and the sensitivity of the test. Thus, the apparent experimental threshold can only approximate the threshold that might occur when a population of the same species used in the experiment is exposed under similar conditions in the environment. If one intends to extrapolate these results to populations of other species, such as humans, then one needs to account for interspecific differences in responses at equivalent doses. Since this combined process of objective and judgmental evaluation of data is subject to many uncertainties, its subsequent application to safety evaluations has usually been combined with the use of "uncertainty factors" or "safety factors."

UNUSUAL DOSE–RESPONSE CURVES

Dose–response curves derived under special conditions may take on shapes that are considerably different from the normally expected sigmoid relationship. In all the figures presented in this section, the ordinate will be the logarithm of the dose or dose rate and the abcissa will be the percent of the tested population that responds.

Unusually Sensitive Groups

Although the traditional dose–response curve represents the range of sensitivities in a population, encompassing both the most sensitive and the most resistant individuals, the individual sensitivities are assumed to be unimodally distributed. If a significant proportion of the population, say one-half, should be ten times more sensitive to a chemical, then the resultant bimodal distribution of the dose–response curve would appear as in Figs. 3-9 and 3-10. The curve never takes on a negative slope, since all the sensitive individuals that have reacted at a lower dose will also respond at the higher dose. The shape of this dose–response curve may be attributed to two different populations—the sensitive and the resistant. Since the two populations (like different strains or species) were not recognized

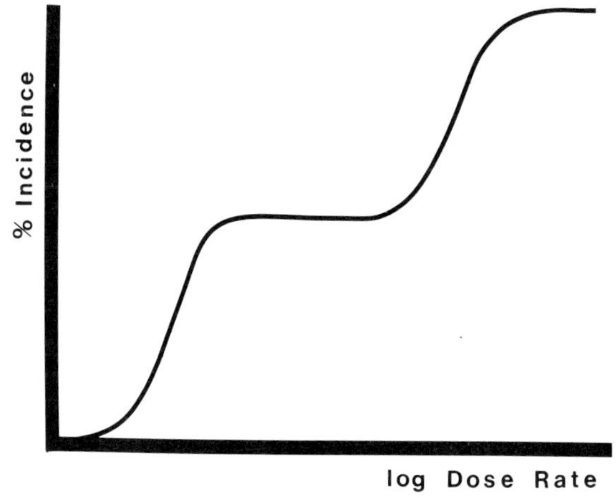

FIGURE 3-9. Dose–response curve with a bimodal distribution of sensitivities.

or separated before the exposure, the curve represents the summation of the sensitivities of the two groups. As the proportion of an unusually sensitive group decreases, the differences in sensitivities must be correspondingly greater before they can be differentiated from the normal spectrum of individual variations within the main group.

Following are two examples of genetic deficiencies that lead to differences in sensitivities. Succinylcholine chloride, a commonly used muscle relaxant, is normally metabolized by pseudocholinesterase. A small proportion of the human population has a genetic deficiency of pseudocholinesterase, making them prone to prolonged relaxation of skeletal muscles, which may lead to respiratory arrest. This minor segment of the population responds at very much lower doses than the bulk of the population, in a fashion similar to that illustrated in Fig. 3-10 (Kalow, 1965). Another genetically determined sensitivity is due to glucose-6-phosphate dehydrogenase deficiency, which predisposes individuals to hemolytic anemia when exposed to primaquine, phenylhydrazine, naphthalene, and many

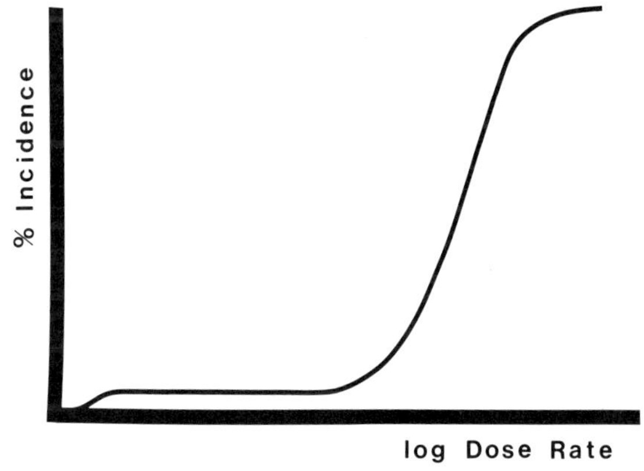

FIGURE 3-10. Unusually sensitive groups (equal spacing of dose levels to Fig. 3–9).

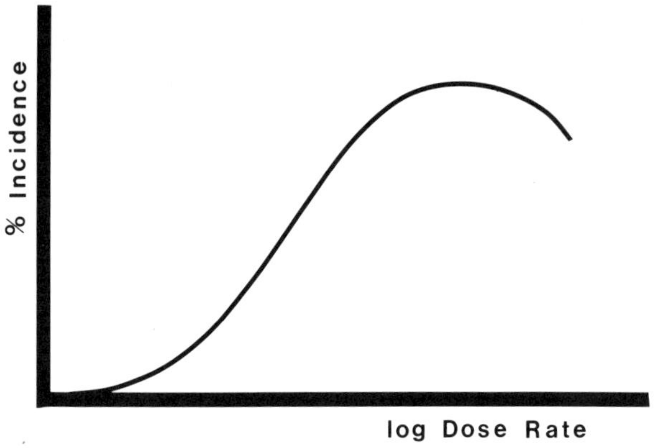

FIGURE 3-11. Competing effects can alter the shape of the dose–response curve.

other substances (Kalow, 1965). In some human populations the occurrence of this deficiency approaches the range of sensitivities illustrated in Fig. 3-9.

Changes in Dose–Response Curves Due to Competing Effects

The types of effects that one investigates during a toxicological study obviously do not occur in isolation from one another. Thus, when one is studying severe liver damage due to carbon tetrachloride in rats, for instance, some of the animals exposed to high dose levels may die from the more immediate effects on the central nervous system without ever exhibiting measurable severe liver damage. Similarly, in a carcinogenicity assay the selected dose levels may be so high that animals exposed to the highest doses succumb before sufficient time elapses for tumors to develop. In both instances, the resulting dose–response curve, which summarizes the events through time *t*, will show a segment at high dose rates with negative slope (Fig. 3-11).

Essentiality versus Toxicity

Substances that perform essential functions have dose–response curves with a negative slope at low dose rates (reversed dose–response curves), which are below the region of minimum requirements. At higher dose rates, these essential substances can reach levels that produce toxic responses, resulting in the traditional dose–response curve (Fig. 3-12). For some substances, such as vitamin C and some amino acids, the range of doses between essentiality and toxicity may be great; but for others, such as selenium and vitamin A, the range may be small, so that only a small window is available for optimal exposure.

The complex dose–response relationships that occur in cases of essentiality versus toxicity, or when there are competing effects or unusually sensitive subgroups in the population, can present additional difficulties during safety evaluations. The mathematical extrapolation models in common use cannot accommodate data involving these phenomena (see also Figs. 3-8 through 3-12) without significant modification.

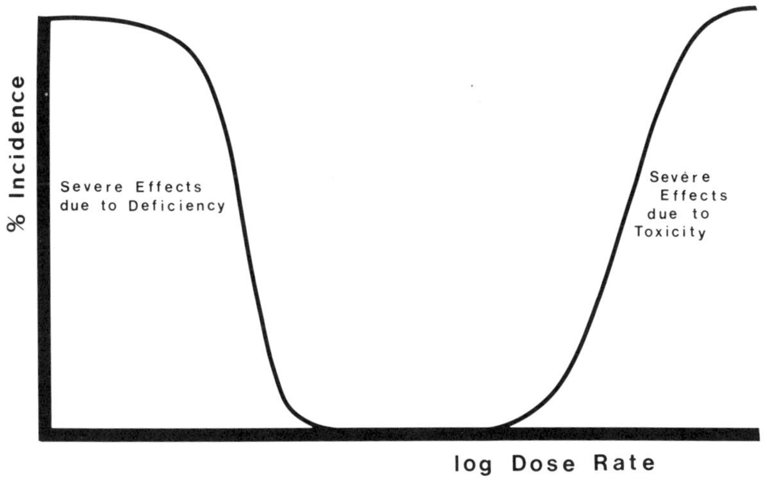

FIGURE 3-12. Essentiality versus toxicity.

OTHER DOSE–RESPONSE MODELS

The following discussions of specific dose–response models are brief, because the subject is covered in much greater detail in Chapter 10.

The Logistic (Logit) Model

In the logistic (or log–logit, or logit) model, the dose axis is transformed into the logarithm of the dose, and the response axis is transformed into logits (L), which are defined as

$$L = \ln [(1 - p)/p] \quad (4)$$

where p is the proportion of the tested population that is responding at a specific dose (Berkson, 1944) and ln is the natural logarithm. The basic concept of using the logistic function rests on concepts of laws that govern chemical reactivity, including the law of mass action and autocatalytic processes. The computation of the logistic dose–response relationship is significantly easier than the probit analysis. Within the range of nearly all experimental data, both models fit the data equally well.

The Single-Hit Model

There are many other dose–response models. Many of them have been largely inspired by the need to generate dose–response models for extrapolation of the incidence of carcinogenic events. One such model is the "single-hit" model (Hoel *et al.*, 1975). Although this model seems elegant at first sight, it characteristically provides very poor fits for most experimental data.

The single-hit model is based on the hypothesis that a single molecule of a xenobiotic may interact with another molecule (e.g., DNA) in an organism and produce an irreversible effect. The mathematical formulation of this concept is

$$P = 1 - e^{-aD} \tag{5}$$

where P is the proportion of organisms demonstrating the effect, e is the base of the natural logarithms, D is the dose, and a is a constant that has sometimes been equated with potency. By design, this mathematical formulation can only fit through dose zero when the proportion affected is zero, and one other point, usually the lowest concentration at which a significant number of tumors has been encountered.

The Mantel–Bryan Extrapolation Model

The Mantel–Bryan model (Mantel and Bryan, 1961) has been designed exclusively for extrapolations. Using this approach, one also selects the lowest concentration at which significant effects were found, calculates the upper 99% confidence bound on that value, and then extrapolates downward at a slope of 1 probit per order-of-magnitude change in dose rate.

The Multistage Model

The multistage model by Armitage and Doll (1961) can fit experimental data well. The basic concept for the model is that disease processes, especially carcinogenesis, require several stages in order to express themselves. The model is in fact an elaboration of the single-hit model. Its mathematical formulation is

$$P = 1 - \exp\left[-(a_0 + a_1 D + a_2 D^2 + \cdots + a_n D^n)\right] \tag{6}$$

Since this is an exponential polynomial, it is able to fit very complex dose-response relationships. However, the various terms of the exponential polynomial may not correspond to any biological processes.

Three-Dimensional Dose–Response–Duration Models

The availability of large data sets, such as the ED_{01} study conducted by the National Center for Toxicological Research (Staffa and Mehlman, 1980), has led to interest in modeling dose–response surfaces that depict dose rate, time to effect, and response rate. The two models that have received the greatest use in this area are the Hartley–Sielken model (Hartley and Sielken, 1977) and a modified Weibull model (Society of Toxicology, ED_{01} Task Force, 1981). The Hartley–Sielken model is essentially a modified multistage model, and the modified Weibull model is essentially a two-parameter double exponential. The Hartley–Sielken model was found to be able to fit the ED_{01} data more precisely than the modified Weibull model (Society of Toxicology, ED_{01} Task Force, 1981).

SUMMARY AND CONCLUSIONS

The analysis and interpretation of dose–response relationships are major tools in toxicology. The term "dose–response relationship" has been used to connote a general concept concerned with the analysis of two major classes of responses following a specific dose regimen applied for a specific length of time. The first class of responses consists of all-or-none (quantal) responses, plotted in terms of the percentage of the animals that respond. This may be thought of as presenting the spectrum of sensitivities in the population. The other class consists of graded responses that may also be regarded as a measure of the intensity of response. Since all studies involve a dosing regimen for a specific duration, a proper presentation of dose–response relationships should be three-dimensional, involving dose–duration–response as the individual axes, thereby defining a dose–response surface. Although most dose–response relationships are still being analyzed in traditional two-dimensional representations, the dose–response surface concept is receiving increasingly greater recognition.

The shape of the dose–response curves and surfaces can become very complex when several biological phenomena interact, e.g., essentiality versus toxicity, hypersusceptible versus normal population subgroups, obscuration of chronic effects due to early mortality.

After many years of relative inactivity in this field, rapid advances are now being made. We may expect numerous changes in the uses of dose–response relationships in the near future, especially in consideration of dose–duration–response surfaces and in safety evaluations.

REFERENCES

Armitage, P., and R. Doll. 1961. Stochastic models for carcinogenesis. Pp. 19–38 in *Proceedings of the 4th Berkeley Symposium on Mathematical Statistics and Probability*. University of California Press, Berkeley.

Berkson, J. 1944. Application of the logistic function to bioassay. *Am. Stat. Assoc. J.* **39**:357–365.

Bliss, C. L. 1935. The calculation of the dose–mortality curve. *Ann. Appl. Biol.* **22**:134–167.

Brownlee, K. A., J. L. Hodges, Jr., and M. Rosenblatt. 1953. The up-and-down method for small samples. *Am. Stat. Assoc. J.* **48**:262–277.

Daum, R. J. 1969. A Revision of Two Computer Programs for Probit Analysis. U. S. Department of Agriculture, Pest Control Division, Hyattsville, Maryland.

Doull, J., C. D. Klaassen, and M. O. Amdur, eds. 1980. *Casarett and Doull's Toxicology,* 2nd ed. Macmillan, New York. 778 pp.

Finney, D. J. 1971. *Probit Analysis,* 3rd ed. Cambridge University Press, Cambridge. 333 pp.

Gaddum, J. H. 1953. Bioassays and mathematics. *Pharmacol. Rev.* **5**:87–134.

Hager, W. G., and T. R. Punnett. 1973. Probit transformation: Improved method for defining synchrony of cell cultures. *Science* **182**:1028–1030.

Hapke, H. J. 1975. *Toxicologie für Veterinärmediziner.* F. Enke Verlag, Stuttgart. 408 pp.

Hartley, H. O., and R. L. Sielken, Jr. 1977. Estimation of "safe doses" in carcinogenic experiments. *Biometrics* **33**:1–30.

Hartung, R. 1981. The use of animal toxicity data. Pp. 4-77 to 4-94 in *Conference Proceedings: Environmental Risk Assessment.* Electric Power Research Institute, Palo Alto, California.

Hoel, D. G., D. W. Gaylor, R. L. Kirchstein, U. Saffiotti, and M. A. Schneiderman. 1975. Estimation of risk of irreversible, delayed toxicity. *J. Toxicol. Environ. Health* **1**:133–151.

Kalow, W. 1965. Dose–response relationships and genetic variation. *Ann. N. Y. Acad. Sci.* **123**:212–218.

Klaassen, C. D., and J. Doull. 1980. Evaluation of safety: Toxicologic evaluation. Pp. 11–27 in J. Doull, C. D. Klaassen, and M. O. Amdur, eds. *Casarett and Doull's Toxicology,* 2nd ed. Macmillan, New York.

Levine, R. R. 1973. *Pharmacology.* Little, Brown & Co., Boston. 412 pp.

Lichtfield, J. T. 1949. A method for rapid graphic solution of time-per cent effect curves. *J. Pharmacol. Exp. Ther.* **97**:399–408.

Lichtfield, J. T., and F. Wilcoxon. 1949. A simplified method of evaluating dose–effect experiments. *J. Pharmacol. Exp. Ther.* **96**:99–113.

Loomis, T. A. 1974. *Essentials of Toxicology,* 2nd ed. Lea & Febiger, Philadelphia. 223 pp.

Mantel, N., and W. R. Bryan. 1961. "Safety" testing of carcinogenic agents. *J. Natl. Cancer Inst.* **27**:455–470.

National Research Council. 1977. *Drinking Water and Health.* A report of the Safe Drinking Water Committee. National Academy of Sciences, Washington, D.C. 939 pp.

Society of Toxicology, ED_{01} Task Force. 1981. Re-examination of the ED_{01} study. *Fund. Appl. Toxicol.* **1**:26–128.

Spratt, J. L. 1966. Computer program for probit analyses. *Toxicol. Appl. Pharmacol.* **8**:110–112.

Staffa, J. A., and M. A. Mehlman. 1980. Innovations in cancer risk assessment (ED_{01} Study). *J. Environ. Pathol. Toxicol.* **3**:1–246.

Weil, C. S. 1952. Tables for convenient calculation of median effective dose (LD_{50} or ED_{50}) and instruction in their use. *Biometrics* **8**:249–263.

4

Factors Modifying Toxicity

Gary P. Carlson

In the evaluation of hazards associated with exposure of humans to chemicals, whether they be encountered as drugs, through occupational exposure, or as the result of environmental contamination, the use of animal data is of obvious importance. In the usual safety evaluation studies of a compound in laboratory animals, environmental conditions are controlled as carefully as possible, animal bedding and food are selected with great diligence, and the utmost care is taken to maintain the health of the animals in order to minimize confounding variables. In order to institute the proper controls, one must first recognize the factors that affect the responses of animals to exposures to toxic chemicals. Some of these are discussed in this chapter.

Only by recognizing the influence of various factors on toxicity can one decide on the validity of data obtained in any particular study before using them to extrapolate to humans. This is especially true where conflicting data are observed in two or more studies on the same compound. No attempt will be made here to examine such factors exhaustively.

In the past decade, much research has focused on these factors, and many excellent reviews have been written about differences in xenobiotic metabolism related to sex (Kato, 1974), species (Williams, 1974), age (Short *et al.*, 1976), and pathological or abnormal physiological state (Hoyumpa *et al.*, 1978; Kato, 1977; Vesell, 1978). The literature also contains papers concerning the influence on drug metabolism and responses in animals exerted of specific environmental factors, such as housing (Brain and Benton, 1975) and temperature (Weihe, 1973), and by multiple factors (Lindsey *et al.*, 1978; Newton, 1978; Ringler and Dabich, 1979; Savordeker and Lambert, 1974; Vesell *et al.*, 1976). Some of these considerations have also been examined in reviews on the selection of the proper species

GARY P. CARLSON • Department of Pharmacology and Toxicology, School of Pharmacy and Pharmacal Science, Purdue University, West Lafayette, Indiana 47907.

for toxicity testing (Fancher, 1978). This chapter does not cover many of the classical factors discussed in other texts, such as the resistance of rabbits to atropine in belladonna plants due to high levels of atropinase and the use of red squill as a rodenticide because of rodents' inability to vomit.

This chapter does provide examples demonstrating how factors shown to influence the laboratory animal's response to chemical toxicity may also play a role in the responses of humans. It becomes quite clear that the assessment of the hazard associated with the exposure of normal persons to a chemical under one set of conditions may need to be modified for another set of conditions. In addition, consideration must be given to modification of responses by genetic variations.

The factors of interest have been divided somewhat arbitrarily into sections of species, strain, age, sex, nutrition, timing, lifestyle, and health status. Although these subjects are discussed separately, the reader should recognize that the *summation* of differences in response must be considered when judging the risk to humans associated with exposure to a particular agent.

SPECIES

The bases for species differences are multiple. In some cases they may be anatomical; in others, biochemical. If one considers the normal exposure to a compound, the first consideration may simply be one of absorption. For example, absorption of an oral dose of caffeine varies considerably among species (Burg, 1975). The absorption half-time in the rat is approximately 0.1 hr; in the rabbit 0.7 hr; in the human 0.07 hr. In the pig, the peak plasma concentration is not reached for 5 hr. In both the pig and squirrel monkey, caffeine has a very long half-life. Thus, when species differences appear, consideration might first be given to this possibility.

Differences in absorption have also been observed with other portals of entry, including skin. This has been evaluated in comparative *in vivo* studies by Bartek *et al.* (1972), who applied a number of radiolabeled compounds to the skin of rats, rabbits, pigs, and humans and then followed the urinary excretion of these materials, adjusting for excretion via other routes and storage within the body. The compounds included haloprogin, *N*-acetylcysteine, testosterone, cortisone, caffeine, and butter yellow. Skin permeability decreased from rabbit to rat to pig to human. In some cases, the differences were very large. For example, absorption of haloprogin was essentially complete in the rat and rabbit, but only 11% was absorbed in humans. Cortisone was 30% absorbed in the rabbit, but only 3.4% in humans. These differences have led to the use of the rhesus monkey to assess the possible absorption of chemicals by humans, since the absorption of hydrocortisone, testosterone, and benzoic acid in that species has been shown to resemble that in humans (Wester and Maibach, 1975). The absorption of materials from the lungs also varies among species. Lipid-insoluble compounds are absorbed 2–3 times faster in the mouse than in the rat, but 1.3–3 times faster in the rat than in the rabbit (Schanker, 1978). On the other hand, lipid-soluble drugs appear to be absorbed at a similar rate in all three species.

Once absorbed, the tissue uptake of toxic materials becomes a concern. Studies by Yang *et al.* (1975) demonstrated that this distribution may vary from species to species and from compound to compound. In studying the comparative metabolism of hexachlorobenzene, these investigators noted that the whole-blood to plasma ratio of this compound was 6 in the rat but only 1 in the rhesus monkey. Their study included the *in vitro* binding of hexachlorobenzene to red blood cells in 17 species. The whole-blood to plasma ratio for hexachlorobenzene varied from 0.72 in female Japanese quail to 5.97 in the female mouse. In the human, the ratio was approximately 0.9. The difference in the binding of hexachlorobenzene in the rat (4.1) and human (0.9) was not observed with DDT or Mirex. This study supports the need for examining such factors for individual compounds when considering the use of animals for extrapolation to humans.

Another good example of the importance of distribution in the assessment of chemical hazard is given by the study of Willes (1977), who compared the toxicity of methylmercury in the cat and humans, which tend to show similar behavioral and histopathological responses to this compound. The whole-body half-life clearance for methylmercury was the same in both species, and the brain concentrations that caused comparable signs of toxicity were similar, but the blood to brain ratio was ten times as high in the cat (1:1) as in humans (1:10). This suggested that the cat should be ten times more resistant than humans. Indeed, cats demonstrated ataxia at blood methylmercury levels of 6–8 ppm, whereas this effect in humans was calculated to be manifested at levels of 0.6–0.8 ppm.

Extensive studies have been conducted in species ranging from bacteria and invertebrates to vertebrates, including mammals and humans, to determine their ability to metabolize chemicals into more or less toxic forms. The results have shown that interspecies differences in metabolism may be quantitative, qualitative, or both. In humans, rats, and guinea pigs, less than 10% of a cyclohexylamine dose is metabolized and the routes vary from deamination to hydroxylation to a combination of both routes, respectively (Renwick and Williams, 1972). In the rabbit, approximately one-third of the compound is metabolized, through both ring hydroxylation and deamination.

A classic study on species variations in animals, including humans, involves amphetamine. Dring *et al.*, (1970) found that humans excreted approximately 30% of an administered dose as unchanged drug, 21% as total benzoic acid, and 3% as the 4-hydroxy compound. A similar profile was observed in dogs and rhesus monkeys. The rat was the only species in which extensive hydroxylation of the compound in the aromatic ring was observed. In the rabbit, there was extensive deamination of the compound. Thus, when considering metabolism, humans could be compared to the dog or monkey, but certainly not to the rabbit or rat.

The importance of epoxide formation in the metabolism of chemicals is well recognized. This event leads to the formation of reactive species whose fates are determined by their progression through a number of pathways, including glutathione conjugation and enzymatic hydration by epoxide hydrase. In attempting to determine the risk to humans posed by such compounds, it is important to determine which species might most resemble humans in the formation and disposition of such epoxides. Pacifici *et al.* (1981) compared a number of species for their ability to conjugate styrene oxide with glutathione and also measured styrene oxide hydrase activity in the liver, kidney, and lung. The baboon had the highest hepatic epoxide hydrase activity, but the lowest glutathione transferase activity. The mouse had the lowest hepatic hydrase activity, but very high glutathione transferase activity, although this was still only about one-half that found in the guinea pig. In general, hepatic epoxide hydrase activity in rodents (i.e., the rat, mouse, hamster, and guinea pig) was somewhat lower than in nonrodents (i.e., the dog, baboon, pig, and human). However, the glutathione transferase activity in the livers of the rodents was on the average nine times greater than in the other species. Thus, the differences in the competition of the two pathways studied plus differences among tissues make it very difficult to select one species that would be expected to mimic humans and serve as a model in predicting the eventual disposition of a compound that is metabolized through an active epoxide intermediate.

The evaluation of differences in metabolism may be very important in assessing whether they may account for observed differences in response. However, this relationship between species differences in metabolism and toxicity can become very complex. In humans, for example, a very serious hazard is methanol poisoning, which often leads to metabolic acidosis and death or recovery with attendant blindness. These effects also occur in the monkey, but not in the rat. The difference appears to be related to the pathways for metabolism of the formate in these two species, since the human condition can be mimicked in the monkey, but not in the rat, unless special measures are taken to cause folate deficiency in the rat, in which case the acidosis does occur (McMartin *et al.,* 1977, 1979).

Although the basis for interspecies differences in susceptibility to insult may not be completely clear, it is prudent to attempt to understand them as well as possible since they may be quite large in some cases. For example, monensin was originally introduced as a control for coccidiosis in poultry and has subsequently been shown to increase feed efficiency in beef cattle (Whitlock *et al.,* 1978). In the chicken, the LD_{50} is 200 mg/kg; in mice, 125 mg/kg; in cattle, 50–80 mg/kg; and in rats, 35 mg/kg. In horses, however, the LD_{50} is only 2–3 mg/kg. Evidence of poisoning is associated with both skeletal and cardiac myopathy. The reason for the differences in sensitivity to this agent, which inhibits adenosine triphosphate (ATP) hydrolysis and oxidation of certain substrates, is not known.

In some cases, various species may show different responses to the same chemical. This may again be very important in determining which species most closely resembles humans. McConnell and Moore (1975) compared case reports on humans with results of laboratory studies on the toxicity associated with halogenated aromatics, primarily dibenzo-*p*-dioxins, dibenzofurans, biphenyls, and naphthalenes, in the usual laboratory animals and nonhuman primates. Certain signs appeared to be common to most species. These included lesions of the thymus and other lymphoid tissues and effects on the bone marrow with accompanying hematological changes. However, the monkey and perhaps to a lesser extent the rabbit, but not rodents, readily incurred the cutaneous and gastrointestinal lesions that have been observed in humans. It is curious, however, that observed increases on porphyrin levels associated with 2,3,7,8-tetrachlorodibenzo-*p*-dioxin in humans are not found in monkeys or guinea pigs. The observed differences may be partly related to the fate of the compounds in the various species. For example, the disposition of 2,3,7,8-tetrachlorodibenzofuran is similar in the monkey and rat, but it is cleared much more slowly in the monkey (Birnbaum *et al.,* 1981).

When assessing health risk, great care must be taken in accepting data from one species over another in its applicability to humans. For example, in a mouse infectivity model, air pollutants can be tested for their capacity to modify experimentally induced infection. Female pathogen-free mice are exposed to pollutants and then to aerosolized microorganisms such as *Streptococcus pyogenes.* The attendant mortality is then assessed. Nitrogen dioxide can increase this response in this species (Ehrlich, 1980; Gardner, 1980). Such data would strongly support a role for this compound in increased infections in humans exposed to environmental air pollution, as has been evidenced by epidemiological studies. However, studies on the squirrel monkey indicate that nitrogen dioxide concentrations of approximately 50 ppm are necessary to cause 90–100% mortality, whereas in the mouse, this is observed at 2.5–3.5 ppm. Although one might conclude that the monkeys, and perhaps humans, are not susceptible to such interactions, the most likely basis for the discrepancy is that the death rate for control mice is 41%, whereas the squirrel monkey has a greater resistance to this bacterium and the death rate is zero. Thus, the data on mice are probably much more applicable than would at first be assumed by a cursory glance at the mortality data on monkeys exposed to nitrogen dioxide.

Interspecies differences are also of considerable interest in studies of teratogenicity, mutagenicity, and carcinogenicity. Their importance in teratology is pointed out by Mulvihill (1973) in relation to caffeine. He noted that despite the positive mutagenic effects identified for bacteria, fungi, and higher plants, conflicting data on rodents make it difficult to assess the risk to humans. Nonetheless, the Food and Drug Administration took action, primarily because skeletal abnormalities were found in Osborne-Mendel rats intubated with 6–125 mg/kg doses of caffeine during gestation. There was a very high incidence of skeletal abnormalities (90% had four or more) in the animals at the highest level; fewer were seen at the lower levels (Collins *et al.,* 1983). Despite the differences in species metabolism (Burg, 1975) and the conclusion of an advisory group that evidence was insufficient at that time for any conclusion concerning human teratogenicity, a warning was issued concerning the consumption of caffeine-containing beverages by pregnant and potentially pregnant women (Anonymous, 1980). In considering which species might give the most reliable information on the carcinogenicity of chemicals in humans, Purchase (1980) reviewed the literature dealing with carcinogenicity studies in mammalian species. Of the 250 chemicals studied, 38% were noncarcinogenic in either the rat or mouse, whereas 44% were found to be carcinogenic in both species. In these cases, 64% produced cancer at the same site in both species. Eight percent of the compounds were carcinogenic in the mouse but not the rat, and 7% were carcinogenic in the rat but not the mouse. This variability is important to consider when attempting to extrapolate from the animal data to humans.

STRAIN

Strain differences have been extensively investigated, especially in mice and rats. Particular interest has focused on the appropriate strain to be used in studies of carcinogenicity (see Chapter 8). Considerations have been given to the advantages and disadvantages of inbred and hybrid mice with particular attention to uniformity of response. Consideration must also be given to qualitative differences in response.

In a study of uterine and thymic weight responses to various dietary levels of diethylstilbesterol, Greenman *et al.* (1979) examined BALB/cStCrl$_f$C3H$_f$/Nctr, C57BL/6J$_f$C3H$_f$/Nctr, and B6CF$_1$ (C57BL/6 × BALB/c)F$_1$ hybrid and monohybrid crosses (F$_2$) from the breeding of the F$_1$ mice. The slope of the dietary response line for the uterine weights of the BALB/c mice was significantly less than for the other strains. Thymic weight variances were smallest for the BALB/c and F$_1$ mice, but were higher for the C57BL/6. BALB/c mice developed testicular lesions, whereas C57BL/6 mice developed pituitary lesions, and the F$_1$ and F$_2$ hybrids developed a high incidence of both. The study points out that generalizations cannot normally be made concerning the variability of inbred versus hybrid mice. The variations observed among the hybrids may very well depend not only upon the parent strains, but also upon the parameter being examined.

In many studies, strain differences have been correlated with genetic responsiveness at the *Ah* locus; however, this is not the only factor to bear in mind when choosing a strain for a study. The induction of tumors in mouse skin by topical application of 7,12-dimethylbenz(*a*)anthracene and promotion by phorbol ester resulted in no tumors in 13 C57BL/6N mice, but produced 19 tumors in 21 DBA/2N mice. This genetic difference in susceptibility to tumor formation is not related to levels of aryl hydrocarbon hydroxylase (AHH) in liver or skin nor to levels of epoxide hydrase (Nebert *et al.*, 1972). The AHH was inducible by 3-methylcholanthrene in the resistant, but not in the susceptible, strain. Similar experiments using benzo(*a*)pyrene also indicated that the AHH levels were not important in discerning the difference in susceptibility between the resistant and susceptible strains of mice (Benedict *et al.*, 1973).

The ordering of mouse strains by their susceptibility to tumor induction must be made with care. In a study of susceptibility among ten strains of inbred and hybrid mice to sarcoma development in response to the subcutaneous implanting of disks containing 3-methylcholanthrene, Prehn and Lawler (1979) found that the rank order at a low concentration of 0.05% 3-methylcholanthrene was the reverse of that found when the concentration was 5%. The basis of this paradoxical effect is not clear.

The susceptibility to injury other than tumor induction in relation to the "aromatic hydrocarbon responsiveness" in mice has also been studied in detail by Robinson *et al.* (1975). This characteristic is associated with the degree of cytochrome P448 induction and a variety of monooxygenase activities. In the responsive strains (C57BL/6,C3H/HeN, and BALB/cAnN), there is a shortened survival time following large intraperitoneal doses of benzo(*a*)pyrene, 7,12-dimethylbenzanthracene, 3-methylcholanthrene, β-naphthoflavone, or polychlorinated biphenyls. However, in nonresponsive strains (DBA/2 and AKR/N), shortened survival times are seen with low oral doses of benzo(*a*)pyrene, 7,12-dimethylbenzanthracene, 3-methylcholanthrene, or lindane, but not with the polychlorinated biphenyls. The apparent sensitivity of nonresponsive animals to the benzo(*a*)pyrene results from increased depression of bone marrow. Genetic differences in susceptibility were not found in association with a number of other agents, including bromobenzene, zoxazolamine, phenytoin, dichlorodiphenyltrichloroethane (DDT), butylated hydroxytoluene (BHT), or hexachlorobenzene. These studies demonstrate the complexity of predicting susceptibility to exposure to one environmental chemical based on studies of another. More importantly, they show that genetic traits not only influence life span, but also exert either beneficial or detrimental effects, depending on whether the detoxification processes are normal, inhibited, or induced.

The importance of strain differences extends beyond the concept of the optimum model for carcinogenicity testing. For example, the toxicity of chloroform has been show to differ in various strains of mice (Clemens *et al.,* 1979; Hill *et al.,* 1975). The C57BL/6J strain was found to be four times more resistant to the lethal effects than was the DBA/2J strain. B6D2F_1 (C57BL/6J × DBA/2J) animals were found to be in between. The results suggested that either single-gene intermediate inheritance or multifactorial genetic control over chloroform toxicity was operating. Twice as much chloroform accumulated in the kidney of the sensitive strain as in the resistant strain. Hepatic changes were not correlated with accumulation of chloroform in the liver. Covalent binding to the renal microsomes of the sensitive strain was greater than in the resistant one. Such strain differences in sensitivity to chloroform may be important in evaluating the data linking chloroform to cancer in both animals and humans.

Exposure to pentobarbital also demonstrates strain differences. Susceptibility to its effects, including both sleep induction and lethality, in various strains has been found to decrease in the following order: DBA/2J > C57BL/6J > Crl:(ICR) (Nabeshima and Ho, 1981). This appears to be a reflection of the longer half-life in the brain and serum of the DBA/2J animals, which also have lower levels of cytochrome P450 and b_5 as well as reduced NADPH dehydrogenase and NADPH cytochrome *c* reductase activities when compared to the more resistant strains. Such genetic variations may be important not only in acute toxicity, but also, more importantly, in the development of dependence and tolerance.

It is sometimes difficult to detect whether an effect such as hepatotoxicity is due to a particular drug or to some other factor. For example, liver damage has been reported to be associated with heroin addiction, but it is not clear whether this is due to the heroin itself. Mice can be used as a test species, but there are strain differences (Needham *et al.,* 1981). A single intraperitoneal injection of a 100 mg/kg dose of morphine sulfate caused only a doubling of serum glutamic-oxaloacetic transaminase (SGOT) in female CXBK mice, but a 13-fold increase in B6AF_1 (C57BL/6J × A/J) mice and a 25-fold increase in C57BL/By female mice. A significant sex difference was seen in C57BL/6By and BALB/cBy mice, but not in B6AF_1 (C57BL/6J × A/J) or CXBK mice. In the C57BL/6By animals, it was more toxic in the females, whereas in the BALB/cBy mice, it was more toxic to the males. Thus, the degree of hepatotoxicity associated with the use of morphine depends on both the sex and the strain of mouse.

Finally, strain differences may sometimes prove useful in investigating the toxicological risk that may be associated with particular groups of humans. For example, it is well known that individuals with erythrocyte glucose-6-phosphate dehydrogenase deficiencies are at greater risk of the hemolytic action of certain oxidant drugs and chemicals than are normal individuals. The possibility of such an effect can be demonstrated using mouse strains with high or low activity. Calabrese *et al.* (1979) demonstrated that exposure to sodium chlorite caused significantly greater decreases in the red blood cell count, hematocrit, and hemoglobin level in C57L/J mice with low glucose-6-phosphate dehydrogenase activity than in A/J mice, in which such activity about three times as high.

AGE

The influence of an animal's age on responses to chemicals is very important in toxicological studies. In recent years considerable interest has been focused on the relationship between age and development of the capacity of the experimental animal to metabolize the chemical compound of interest. Therefore, it is not surprising that the relationship of age to maturation of drug-metabolizing enzymes has been studied in various ways by a number of investigators. In one study, MacLeod and co-workers (1972) made a systematic evaluation of both drug-metabolizing activity and level of components of the microsomal electron transport system in both male and female Crl:(LE) rats. Both aminopyrine and aniline metabolism increased during the first 5 weeks in both sexes. By the end of 3 weeks, males demonstrated a greater activity for metabolizing aminopyrine than did females, but aniline hydroxylase activity was the same in both sexes. NADPH cytochrome P-450 reductase and cytochrome P-450 content also showed sex-related differences, which developed between 2 and 3 weeks of age. The development of full potential for drug metabolism thus appears to correlate with maturation of the components of the microsomal electron transport system.

Many of the classical studies on the importance of the development of these enzyme systems have centered on the anticholinesterase and lethal effects of organophosphorous pesticides. For example, Alary and Brodeur (1970) showed a good correlation between the activity of liver enzymes in older rats and a decrease in the LD_{50} of parathion. This decrease in sensitivity with age is apparently due to the increasing ability of the animal to detoxify the parathion and its oxygen analog paraoxon (Benke and Murphy, 1975; Gagne and Brodeur, 1972). Such correlations have also been made for a number of other chemicals.

Immature animals are not always more susceptible to the toxicity of chemicals than are mature animals. The wide range of susceptibilities among immature animals in comparison to adults was demonstrated clearly by Yeary (1967). Although there was often consistency within a class of drugs, there was often a wide disparity among classes, with newborns showing both increased and decreased susceptibility (Table 4-1). This illustrates the importance of examining immature animals to determine the toxicity of chemicals with which newborns may have had either direct or indirect contact.

Age differences in response have been shown not only for pesticides and drugs, but also for solvents (Kimura *et al.*, 1971). As indicated in Table 4-2, one-half of the solvents tested for acute toxicity in 14-day-old and adult male rats were more toxic in the younger animals. This increased susceptibility of younger animals must be considered in determining the hazard associated with the use of solvents.

Aflatoxins also show an age dependence. When Newberne and Butler (1969) examined the effect of age on the hepatotoxicity and lethality of aflatoxin B_1, they found that the younger animals were much more susceptible. The LD_{50} for neo-

TABLE 4-1. Comparisons of Acute Toxicities in Newborn and Adult Rats[a]

		$LD_{50} \pm SE$ (mg/kg)	
Compound	Route	Newborn	Adult
Compounds more toxic in the newborn			
Acetaminophen	Oral	420 ± 23	2404 ± 95
Acetylsalicylic acid	Oral	558 ± 24	1528 ± 156
Arsanilic acid	Oral	216 ± 13	1000
Chloral hydrate	Oral	284 ± 22	475 ± 42
Desipramine	Oral	128 ± 10	320 ± 18
Dicoumarin	Oral	68 ± 9	711 ± 20
Iron dextran (Fe)	Oral	2983 ± 134	Nonlethal
Mepenzolate	Oral	186 ± 10	742 ± 47
Mephenoxalone	Oral	645 ± 32	3820 ± 197
Meprobamate	Oral	353 ± 33	1522 ± 16
Neomycin	Subcutaneous	360 ± 7	668 ± 41
Phenobarbital	Oral	121 ± 5	317 ± 24
Compounds essentially similar or less toxic in the newborn			
d-Amphetamine	Oral	80 ± 6	38 ± 5
Fencamfamin	Oral	52 ± 1	83 ± 4
Chlormerodrin (Hg)	Oral	93 ± 34	~82
Ferrous sulfate (Fe)	Oral	237 ± 11	319 ± 21
Iron dextran	Intraperitoneal	~2500	~3000
Isoproterenol	Oral	1310 ± 23	2221 ± 93
Lead arsenate	Oral	1545 ± 98	2400 ± 140
Menadiol sodium diphosphate	Intraperitoneal	313 ± 17	231 ± 26
Menadione	Intraperitoneal	66 ± 3	75
Meralluride (Hg)	Subcutaneous	42 – 50	28 ± 7
Neomycin	Oral	2557 ± 112	~2750
Pentylenetetrazol	Subcutaneous	192 ± 28	85 ± 2
Protokylol	Per os	1163 ± 38	935 ± 36
Strychnine sulfate	Subcutaneous	~6	1.7 ± 0.05
EX 4883 (MAOI)	Per os	~72	95.9 ± 7

[a] From Yeary (1967).

TABLE 4-2. Acute Oral LD_{50} Values for 16 Solvents in Four Age Groups of Rats[a]

	Acute Oral LD_{50} (ml/kg) (95% CL)		
Solvent	14-Day-old	Young adult	Older adult
Acetone	5.6 (3.9–8.0)	9.1 (6.8–12.1)	8.5 (7.8–9.3)
Acetonitrile	0.2 (0.1–0.3)[b]	3.9 (3.3–4.6	4.4 (2.8–6.8)
Benzene	3.4 (2.0–5.7)	3.8 (2.9–4.8)	5.6 (4.0–7.8)
Chloroform	0.3 (0.2–0.5)[b]	0.9 (0.8–1.1)	0.8 (0.7–0.9)
Cyclohexane	8.0 (6.6–9.8)[b]	39.0 (32.5–46.8)	16.5 (11.0–23.8)[b]
Diethyl ether	2.2 (1.2–3.3)	2.4 (2.0–2.7)	1.7 (1.5–1.9)
Dimethyl formamide	1.5 (0.9–2.7)	4.2 (2.7–6.6)	7.2 (6.0–8.6)
Ethyl alcohol	7.8 (6.3–9.7)[b]	22.5 (18.8–27.0)	14.6 (12.8–16.7)
Hexane	24.0 (22.5–25.8)[b]	49.0 (35.5–68.0)	43.5 (31.4–58.8)
Isopropyl alcohol	5.6 (4.2–7.4)	6.0 (5.2–7.0)	6.8 (5.3–8.8)
Isopropyl ether	6.4 (4.7–8.6)[b]	16.5 (15.3–17.7)	16.0 (14.1–18.4)
Methyl alcohol	7.4 (6.0–9.1)[b]	13.0 (11.9–14.2)	8.8 (7.2–10.8)[b]
Methylene chloride	1.8 (1.3–2.3)	1.6 (1.3–1.9)	2.3 (1.7–3.2)
Methyl ethyl ketone	3.1 (2.5–3.9)	3.6 (2.9–4.4)	3.4 (2.6–4.4)
Tetrahydrofuran	2.3 (1.4–3.8)	3.6 (3.0–4.5)	3.2 (2.6–4.0)
Toluene	3.0 (2.5–3.7)[b]	6.4 (5.7–7.3)	7.4 (6.2–8.8)

[a] From Kimura *et al.* (1971). For the newborn, the LD_{50} for acetone was 2.2 ml/kg (1.7–3.8). For the remaining solvents, the LD_{50} could not be determined due to volume limitations; thus, they have been given a value of < 1.0 ml/kg.

[b] Significantly different from LD_{50} of adult rats, $p < 0.05$.

nates was only 0.56 mg/kg, whereas for weanlings it was 5.5–7.4 mg/kg and for the 150-g rat it was 17.9 mg/kg.

In some cases, differences between infants and adults other than metabolism can greatly alter susceptibility to chemical injury. Nitzan *et al.* (1979) reported a small outbreak of methemoglobinemia in a hospital nursery. This was traced to the soybean formula fed to the infants, which contained three phenolic antioxidants (butylated hydroxyanisole, butylated hydroxytoluene, and propyl gallate). The NADH-diaphorase system in the erythrocytes of the infants was within the normal range. The toxicity was probably the result of the fact that fetal hemoglobin is considered to have a greater tendency for being oxidized to the ferric state than does the adult form. This example again illustrates the difficulty of assessing the safe use of a compound across broad age groups.

The effect of age on the response to carbon tetrachloride-induced heptatotoxicity was studied by Cagen and Klaassen (1979). They found that following a 1 ml/kg dose, 4-day-old and adult rats responded with the same degree of rise in plasma aspartate aminotransferase and triglyceride accumulation. However, the degree of binding to hepatic microsomal protein was less in 4- and 14-day-old animals than in adults. Thus, although the young rats cannot metabolize the compound as fast as adults, they are still sensitive. The reason for the lack of correlation is not clear, although it might be due to such factors as enhanced toxicity in the younger animals in the presence of acetoacetate, the concentration of which is five times higher in the blood of 14-day-old rats than in the adults.

It is difficult to relate age difference to toxic response of humans to data on animals. As noted above, despite the preponderance of data indicating that young animals are more sensitive to the toxic effects of chemicals, particularly those requiring metabolic activation to exert their effects, children are not necessarily at greater risk in all cases. In this area of comparison, most of the more complete and reliable studies have been conducted with drugs. Indeed, neonates are more at risk for the toxicity of morphine, coumarin derivatives, curare, and thiobarbiturates than are adults, but they appear to tolerate better the toxicity of digitalis, atropine, succinylcholine, halothane, and meperidine (Mirkin, 1978).

Although the decreased capacity of the newborn to metabolize drugs is well recognized, relatively few studies have been conducted on young infants to investigate the relationship of metabolism to age when compared to animal studies. In one study, fresh liver samples from premature and full-term newborns were compared with liver tissues from adults (Aranda *et al.*, 1974). Although lower than in adults, hepatic microsomal drug-oxidizing activities, namely aminopyrine *N*-demethylase and aniline hydroxylase, as well as the components of the microsomal electron transport system were present in the premature and newborn infants. A progressive increase was noted during the last trimester of pregnancy and during the neonatal period. Thus, as in animals, there is a steady progression of development that must be considered in estimating the ability of humans of different ages to detoxify compounds. In a review, Short *et al.* (1976) concluded that there appears to be little difference in the development of the drug-metabolizing ability among various species from birth to maximum activity at 30–40 days. However, this development may be affected by the substrate of interest, and very large differences for activities reported by various laboratories make interlaboratory comparisons difficult.

In humans, age differences in drug metabolism can be complicated. In a study by Alam *et al.* (1977), salicylamide and acetaminophen were administered to adults and children (age 7–10 years). Although no age difference was found in the half-lives of salicylamide conjugates appearing in the urine, there was a difference in the type of conjugate. The percentage of the compound excreted as the sulfate was 78% in children and only 36% in adults, where the major conjugation product was the glucuronide. Similar observations were made for acetaminophen conjugation. Care must be taken, therefore, in assuming that age differences in the ability to carry out one reaction will mean that definite differences in toxicity will exist in response to the compound of interest.

In addition, one must be aware that the drug-metabolizing system in the human neonate is perturbable. Sereni *et al.* (1973) observed that phenobarbital administration was capable of inducing the metabolism of diazepam in human newborns. This was true whether the phenobarbital was administered to the newborn or to the mother during the last few days of pregnancy.

SEX

Differences in response between males and females have been observed for a variety of compounds, especially in rats. Generally, such between-sex differences have been shown to result from differences in metabolism. For example, parathion is more toxic to the female rat than to the male, but its oxygen analog, paraoxon, does not show this sex difference (DuBois *et al.,* 1949). This difference is due to the male's ability to detoxify the parathion at a faster rate (Neal and DuBois, 1965). In another example, sex differences in hepatic carcinogenesis have been observed in rats exposed to aflatoxin B_1. Newberne and Wogan (1968) reported that both males and females developed tumors, but that the females required a longer time before the tumors appeared. When high doses were administered, mortality was high (Righter *et al.,* 1972). However, prepubertal males were much more resistant, and castration performed on rats less than 10 weeks of age offered protection, which could be overcome by administration of testosterone. Such a sex difference may be related to the metabolism and transport in the two sexes, since under similar conditions of administration much less of the material is bound to liver cell fractions in the females than in the males (Mainigi and Campbell, 1981).

Sex differences have also been shown in toxic responses to heavy metals. For example, cadmium causes a greater inhibition of drug-metabolizing enzymes in the male rat than in the female (Hadley *et al.,* 1974). This may be due to cadmium's ability to inflict testicular damage, which might in turn alter the stimulating effect of testosterone on drug metabolism.

In some cases, the reasons for the sex differences are not as readily apparent. Hexachlorobenzene induces porphyria in both male and female rats when administered in the diet at low doses for 2–3 months, but the effects are greater in females (San Martin de Viale *et al.,* 1970). The porphyrin excretory pattern was similar to that observed in humans who suffered from an outbreak of hexachlorobenzene-induced porphyria (DeMatteis *et al.,* 1961; Ockner and Schmid, 1961). In humans, acute intermittent porphyria occurs predominantly in women. Although extrapolation cannot be made directly from rats to humans, it is interesting to note that δ-aminolevulinic acid synthetase varies with the estrous cycle of the rat and is twice as great during proestrus as during diestrus (Held and Przerwa, 1976).

The major difficulty in evaluating the importance of sex with regard to toxic responses comes not from the lack of examples in animals, which are numerous (Kato, 1974), but from their application to potential importance in assessing risk in humans. In humans, when adjustments are made for drug dosages, the primary consideration between men and women (other than pregnancy, which is a special case) is generally one of body size and, perhaps, the ratio of lean body mass to fat (Levine, 1983).

There are a few indications of sex differences in humans. In studies concerned with the metabolism of benzo(*a*)pyrene in cultured lymphocytes and antipyrine plasma half-life, Kellermann *et al.* (1976) made the interesting observation that the menstrual cycle may have some effect on antipyrine half-life, although no such dependence was founded in the AHH activity of the lymphocytes.

In a study of the metabolism of trichloroethylene in humans, Nomiyama and Nomiyama (1971) found that females excreted 2–3 times more trichloroacetic acid than did males in the first 24 hr after exposure, but in the first 12 hr, they excreted only about one-half the amount of trichloroethanol. Although no known sex difference in susceptibility to poisoning by this solvent has been reported, it illustrates the fact that monitoring of exposure to assess risk should take into account this difference in profile of metabolites.

NUTRITION

The influence of nutrition on toxicity is multifaceted. It is made particularly difficult in humans by the heterogeneity of the diet consumed by populations with vast differences in nutritional status or by individuals with peculiar eating habits, who consume large amounts of a particular type of food, such as fish, or who follow some fad diet. In contrast, in laboratory studies on animals, particular attention is usually paid to maintaining a well-balanced diet; only occasionally is the diet manipulated to facilitate the study of diet and chemical interactions. Even under rigid conditions, problems are sometimes encountered, especially in assessing the potential hazard of food additives by replacing a large portion of the diet by nonnutritive material. For example, in a study of guar gum in rats, a dietary level of 15% for 90 days caused significant decreases in body weight and some decreases in glucose levels (Graham *et al.,* 1981). Such effects may be partially overcome by paired feeding.

In some studies, percentages of constituents in "normal" diets have been changed to examine the influence of such changes on chemical toxicity. Boyd (1972) conducted a detailed investigation of the effect of changes in protein, especially to determine whether protein deficiency, such as in kwashiorkor, places humans, particularly infants, at a greater risk of toxicity from pesticides than humans on a normal diet. In a series of experiments, rats were fed diets differing in protein concentration for 28 days following weaning. A diet containing 26% casein was considered a normal diet. Protein was decreased to one-third (9% casein) or one-seventh (3.5% casein) to simulate restricted protein intake. In some cases, a protein source was completely lacking. Comparisons were also made with animals maintained on standard laboratory chow. As indicated in Table 4-3, the acute toxicity of a variety of pesticides was increased following the administration of a number of pesticides, including representative organochlorine insecticides, organophosphorous insecticides, herbicides, and fungicides. In some cases, e.g., with endosulfan, diazinon, malathion, and monuran, restriction to one-third the normal protein resulted in an approximate doubling of the toxicity. In all cases presented in the table, there was a significant increase in toxicity when protein was severely restricted to one-seventh of the acceptable protein level. Thus, for a variety of pesticides, protein content of the diet can change the toxicity, and this factor must be considered in assessing the hazard of a particular pesticide or other chemical to be used on a worldwide basis.

The influence of dietary factors on the production of tumors was recognized very early (Tannenbaum and Silverstone, 1949). When casein content in the diet fed to C3H mice was varied from 9 to 45%, there was no change in spontaneous carcinomas or skin tumors induced by 3-methylcholanthrene or benzo(*a*)pyrene.

TABLE 4-3. Oral LD_{50} Values of Pesticides in Albino Rats Fed Diets Containing Varying Amounts of Protein (Casein) for 28 Days Following Weaning[a]

Insecticide	Oral LD_{50} (mg/kg) for diets containing different percentages of casein: 0	3.5%	9%	26%	Laboratory chow
Chlordane	—	137 ± 30	—	267 ± 44	311 ± 46
Endosulfan	5 ± 1	24 ± 10	57 ± 4	104 ± 16	121 ± 16
Endrin	—	6.7 ± 0.8	—	16.6 ± 3.0	27.2 ± 6.5
Lindane	15.4 ± 3.9	95 ± 33	187 ± 37	184 ± 16	157 ± 37
Toxaphene	—	80 ± 19	—	293 ± 31	220 ± 33
Demeton	—	2.1 ± 0.4	—	7.6 ± 0.2	10.4 ± 0.8
Diazinon	56 ± 12	215 ± 26	223 ± 26	415 ± 39	466 ± 87
Malathion	539 ± 42	599 ± 138	759 ± 91	1,401 ± 99	—
Parathion	—	4.9 ± 0.3	—	37.1 ± 4.9	23.4 ± 5.4
Chlorpropham	1.2 ± 0.1	2.6 ± 0.5	6.0 ± 1.0	10.4 ± 1.6	4.4 ± 0.5
Diuron	—	437 ± 139	—	2,390 ± 1440	1,017 ± 222
Monuron	0.25 ± 0.05	0.95 ± 0.24	1.58 ± 0.47	2.88 ± 0.31	1.48 ± 0.21
Captan	6.15 ± 3.1	480 ± 110	10,670 ± 2800	12,600 ± 2100	—

[a] From Boyd (1972).

However, spontaneous hepatoma development was significantly less in mice fed the diet containing 9% casein than in those fed either 18% or 45% casein.

The effects of dietary manipulation have also been shown in humans. When volunteers were taken off their normal diets and placed on low-carbohydrate, high-protein diets, the plasma half-life of antipyrine decreased from 16.2 to 9.5 hr and the half-life of theophylline decreased from 8.1 to 5.2 hr (Kappas *et al.*, 1976). When a high-carbohydrate, low-protein diet was given, the half-lives of the two compounds increased from 9.5 to 15.6 hr and from 5.2 to 7.6 hr, respectively. Such dietary manipulations may occur either involuntarily in individuals suffering from inadequate nutrition or voluntarily in humans consuming fad diets.

A confounding factor in considering the influence of diet on the toxicity of chemicals is the presence in certain foods of substances that may very well have endogenous effects on drug metabolism and chemical toxicity. In a test on healthy volunteers, the inclusion of brussels sprouts and cabbage in a test diet resulted in a 13% reduction in the plasma half-life of antipyrine. The plasma concentration of phenacetin was decreased by 34–67% between 30 min and 7 hr after its administration, although its plasma half-life was not altered (Pantuck *et al.*, 1979). In a similar experiment, the administration of charcoal-broiled beef lowered the plasma concentration of orally administered phenacetin, but did not decrease its plasma half-life (Pantuck *et al.*, 1976). Such studies indicate that dietary factors in humans can alter responses. Although the actual magnitude appears small, which makes their application in the assessment of increased risk difficult, such factors as this must be considered in combination with other conditions that may alter the metabolism and toxicity of chemicals.

Yet another consideration concerning diet and toxicity relates to the influence of vitamins and essential minerals. The effect of vitamin C on drug metabolism is of special interest in guinea pigs because of their need, as in humans, for an exogenous source. Extensive studies were conducted by Zannoni and co-workers (1972) in animals made deficient in vitamin C by reducing the dietary content of the vitamin to one-half its normal level. They observed decreases in cytochrome P450, cytochrome b_5, and cytochrome P450 reductase, which were reflected in decreases in aniline hydroxylation, aminopyrine *N*-demethylation, and *p*-nitroanisole *O*-demethylation (Table 4-4). These decreases were dependent upon the degree of decrease in ascorbic acid levels. The observed effects could be reversed by administering ascorbic acid in the diet, but it took at least 6 days for drug-metabolizing activity to return to control levels despite repletion of ascorbic acid levels in 3 days. Such changes in drug metabolism have also been found *in*

TABLE 4-4. Effect of Vitamin C Deficiency (10 and 21 days) on Drug Enzymes and Electron Transport Components in Guinea Pig Liver Microsomes[a]

Drug enzymes and electron transport components	Activity: Normal	Vitamin C-deficient (10 days)	Vitamin C-deficient (21 days)	Decrease (%)
Aniline hydroxylase (μmole *p*-aminophenol/hr per 100 mg microsomal protein)	1.6 ± 0.2 ($p < 0.001$)	1.3 ± 0.1	0.8 ± 0.2 ($p < 0.001$)	50
Aminopyrine *N*-demethylase (μmole HCHO/hr per 100 mg microsomal protein)	3.9 ± 0.1 ($p < 0.001$)	3.3 ± 0.4	1.7 ± 0.3 ($p < 0.001$)	56
p-Nitroanisole *O*-demethylase (μmole *p*-nitrophenol/hr per 100mg microsomal protein)	3.2 ± 0.4 ($p < 0.001$)	3.0 ± 0.2	1.1 ± 0.2 ($p < 0.001$)	66
Cytochrome P450 (μmole/100 mg microsomal protein)	0.05 ± 0.01 ($p < 0.01$)	0.05 ± 0.001	0.03 ± 0.003 ($p < 0.01$)	40
NADPH cytochrome P450 reductase (μmole cytochrome P450 reduced/hr per 100 mg microsomal protein)	0.80 ± 0.2	0.87 ± 0.33	< 0.10[b]	85
NADPH cytochrome *c* reductase (μmole cytochrome *c* reduced/hr per 100 mg microsomal protein)	124 ± 21 ($p < 0.05$)	167 ± 20	83 ± 11 ($p < 0.05$)	33
Cytochrome b_5 (μmole/100 mg protein)	0.03 ± 0.004 ($p > 0.05$)	0.03 ± 0.003	0.02 ± 0.006 ($p > 0.05$)	33
Liver ascorbic acid supernatant fraction 15,000 × *g* (μg/g wet weight)	194 ± 29	62 ± 15	25 ± 15	NR[c]
Microsomal fraction (μg/g wet weight)	11 ± 3.8	6 ± 0.9	3.5 ± 2.0	NR[c]

[a] From Zannoni *et al.* (1972).
[b] This value is a lower limit of detection. Mean ±SE of ten animals per group.
[c] Not reported.

vivo: in guinea pigs placed on an ascorbic acid-free diet, there was a decrease in the biological half-lives of acetanilide, aniline, and antipyrine as compared to animals whose diets were supplemented with 20 mg of ascorbic acid daily (Axelrod *et al.,* 1954). Theophylline was more toxic in the scorbutic guinea pigs.

The effect of nutrients on toxicity may be quite complex. For example, the administration of toxic levels of cadmium to quail resulted in iron deficiency due to inhibition of iron absorption (Fox *et al.,* 1980). This deficiency can be overcome by the administration of iron. Ascorbic acid appears to protect against cadmium toxicity by improving iron absorption. Thus, a high level of vitamin C intake with what would be considered a normal level of iron intake for humans can alter cadmium toxicity. On the other hand, it is suspected that low iron stores or uptake would increase the cadmium toxicity. In assessing the hazard associated with cadmium, therefore, both iron stores and diet would have to be considered.

The toxicity of cadmium is also greatly influenced by the presence of essential elements such as calcium, zinc, and copper. It has been shown that cadmium can disturb the metabolism of these essential metals, especially if the cadmium levels are high. These metals may also alter the disposition and, thus, the toxicity of cadmium. The effect of zinc, copper, and cadmium may be quite complex. The picture is complicated in that species differences may exist. For example, Bremner and Campbell (1980) have reported that zinc tended to decrease the toxicity of cadmium with regard to copper deficiency in lambs, but zinc supplementation enhanced the severity of the copper deficiency induced in rats by cadmium administration. Thus, dietary influences are important with regard to essential metal intake and body stores.

Although not really a matter of diet, some consideration should be given to the influence of gut flora on the metabolism and toxicity of exogenously administered compounds. The importance of the gastrointestinal flora in altering liver tumors in mice has been demonstrated by Mitzutani and Mitsuoka (1979), who found that liver tumorigenesis in C3H/He mice was enhanced by the presence of most of the intestinal bacteria studied. The combination of *Escherichia coli* M66, *Streptococcus faecalis* M266TA, and four strains of *Clostridium paraputrificum* resulted in a tumor incidence of 95% (19/20 mice), whereas in the germ-free mice, the incidence was only 39% (19/49 mice). This effect may be related to the production of carcinogens or cocarcinogens produced by these intestinal bacteria. Moreover, the greater activity seen with combinations of strains, as opposed to single strains, suggests the possibility of interactions. The importance of the bacteria in altering the background level of tumors in controls becomes very crucial, therefore, and might account for differences seen among laboratories or at different times and could be important in assessing the true tumorigenicity of a compound.

In some cases the effect of gut flora on metabolism can have a profound influence on acute toxicity. In normal rats, nitrobenzene administration readily results in methemoglobin formation. In rats treated with the antibiotics neomycin, tetracycline, and bacitracin to sterilize the gut, the effect does not occur (Facchini and Griffiths, 1981; Reddy *et al.,* 1976). In other cases, however, similar compounds, notably the dinitrobenzenes, are capable of inducing methemoglobinemia, regardless of the microflora of the gut (Facchini and Griffiths, 1981). This suggests that care must be exercised when extrapolating from one compound to another, despite similarities in their structures. Such differences in effects have been reported for drugs as well as for other chemicals. Koch *et al.* (1980) have shown that the radiation sensitizer misonidazole is metabolized to its amino derivative by cultures of intestinal microflora. This derivative appears in the excreta of normal rats, but not in the excreta of germ-free rats.

Adamson *et al.* (1970) studied the role of gut bacteria in metabolism in different species. When quinic acid was given orally to humans and the rhesus monkey, it was extensively converted to benzoic acid and excreted as hippuric acid. When injected intraperitoneally or given orally to the monkey following neomycin, it was excreted largely unchanged. Such aromatization was also seen in the baboon and green monkey, but not in New World monkeys (e.g., the spider, squirrel, or capucin monkeys) or in nonprimates such as the dog, cat, ferret, rabbit, mouse, guinea pig, or hamster.

TIMING

The term "timing" includes a number of factors. First among these are changes in toxic responses corresponding to the seasons of the year (circannual) or to the time of day (circadian). The latter is of particular interest. The importance of daily rhythmic variation in the metabolism of drugs has been amply demonstrated. Radzialowski and Bousquet (1968) showed that hepatic fractions from male and female mice showed circadian rhythms in their ability to metabolize a number of compounds, including aminopyrine, *p*-nitroanisole, hexobarbital, and dimethylaminoazobenzene. Except for the latter compound, this rhythm could be abolished by adrenalectomy, indicating the importance of the adrenal gland in this short-term regulation of activity. Similarly, Nair and Casper (1969) examined the influence of the light cycle on hexobarbital oxidase and *p*-nitroanisole demethylase in the rat and demonstrated that the rhythm could be abolished by continuous illumination or darkness.

When the involvement of metabolism is not evident, sensitivity may depend on other factors. Romero (1976) examined the relationship between sensitivity to isoproterenol and its reflection of the basic cycles associated with various hormones and receptors, with specific regard to their regulation by the pineal gland. Isoproterenol caused a twofold increase in cyclic adenosine monophosphate (cAMP) in the pineal gland at the end of 12 hr of darkness, but there was nearly a tenfold increase in response at the end of the 12-hr light period. The acute toxicity of a benzodiazepine (diazepam) has also been shown to follow a circadian pattern (Ross *et al.*, 1981). The maximum toxicity occurred at 9:00 a.m. and the minimum at 9:00 p.m. The fluctuations of this central nervous system depressant may be related to levels of biogenic amines or other neurotransmitters whose concentrations are also time-dependent.

In other studies, the normal lighting cycle has been altered. The effect of such shifts in photoperiod on the responses of animals to the toxicity of chemicals is complex. Different strains of mice have been shown to respond in different ways. Random complete reversal of the photoperiod appears to be very stressful in mice, as judged by the loss of their ability to adapt to cold stress and a decrease in secobarbital-induced sleeping time (Sakellaris *et al.,* 1975). It is tempting to speculate that such changes in humans, as a result of shiftwork or other events leading to altered photoperiods, might also prove stressful to the extent that the response to toxicity is altered.

Temporal variations in drug metabolism and, possibly, response are known to occur in humans. In a study by Vessel *et al.* (1977), antipyrine disposition was studied in normal volunteers following a single oral dose given at 7:00 a.m. Measurements of antipyrine half-life did not differ between noon and midnight time periods. Between these two times, however, 12 of the 19 subjects tested had changes in half-life: seven had increases and five had decreases. The changes in some of the subjects were dramatic. In one individual, for example, the increase in antipyrine half-life was 173%; in another there was a 42% decrease. These changes were reproducible in both direction and degree and were not reduced by administration of the drug at 7:00 p.m. instead of 7:00 a.m. Thus, there appears to be a temporal rhythm in metabolism in humans that may differ considerably among individuals. This variation within a variation may depend on many other factors discussed in this chapter, including diet, lifestyle, and exogenous exposure to chemicals.

Timing may also be related to other factors, such as feeding. Jaeger *et al.* (1974) found that rats fasted overnight (18 hr) were more susceptible to the hepatotoxicity of 1,1-dichloroethylene than were rats allowed free access to food. The estimated 24-hr LC_{50} for fed rats was 15,000 ppm, compared to 600 ppm for the fasted animals. As indicated in Table 4-5, measurements made 4 hr after exposure to 1000 ppm dichloroethylene indicated that there was only a small increase in serum alanine α-ketoglutarate transaminase levels in fed rats, but much greater elevations in fasted animals. Since a major pathway for the detoxification of 1,1-dichloroethylene is via conjugation with liver glutathione (McKenna *et al.,* 1978), it is not surprising that there was a good correlation between the degree of hepatotoxicity and liver glutathione levels (Table 4-5). This study demonstrates well the dramatic differences that can be seen between fed and fasted animals and the fact that these can be correlated with biochemical changes.

Another aspect of this close link between glutathione levels and 1,1-dichloroethylene was shown when Jaeger *et al.* (1973) demonstrated the presence of a diurnal rhythm for hepatic glutathione levels. The toxicity of 1,1-dichloroethylene, measured either in terms of serum enzymes or lethality, varied inversely with the reduced glutathione (GSH) levels. This illustrates how even the time of day can alter the toxicity associated with a compound, based on factors affecting its disposition.

LIFESTYLE

The term lifestyle may mean different things in animals and in humans. Yet, in both cases it is important to understand the influence that such factors may

TABLE 4-5. Effect of 18-hr Fast on Hepatic Reduced Glutathione (GSH) and Serum Alanine α-Ketoglutarate Transaminase (SAKT) from a 4-hr Exposure to 1000 ppm 1,1-Dichloroethylene[a]

Treatment	Number of rats	Liver GSH concentration[b] (mg/g) ± SE	SAKT activity (mg pyruvate/ml serum per hr) ± SE
I Fed; air control	5	1.75 ± 0.08	0.27 ± 0.02
II Fasted; air control	5	1.50 ± 0.06[c]	0.19 ± 0.02
III Fed-1; 1-DCE exposed	5	0.89 ± 0.06[d]	0.62 ± 0.12[f]
IV Fasted-1; 1-DCE exposed	5	0.59 ± 0.07[e]	17.55 ± 4.13[g]

[a] From Jaeger *et al.* (1974).
[b] Total liver GSH in controls (I and II): 18.82 ± 1.07 mg for fed rats versus 11.78 ± 0.62 mg for fasted rats.
[c] $p < 0.05$, II versus I.
[d] $p < 0.05$, III versus I.
[e] $p < 0.05$, IV versus II.
[f] $p < 0.05$, III versus I or II.
[g] $p < 0.05$, IV versus I, II, or III.

exert in altering the toxicity of a compound. For animals, the suspected toxicity may be quite different under various "lifestyles." For humans, individual habits, such as smoking and the consumption of alcoholic beverages, may change responses.

Considerable interest has been focused on the concept of stress in animal experimentation. The effects of isolation on the ability of the liver to metabolize drugs has been studied by Capel *et al.* (1980a). They observed that isolation of rats for up to 8 days resulted in an expected rise in plasma cortisol. Although antipyrine metabolism was not altered *in vivo,* benzo(*a*)pyrene metabolism was increased, as was its binding to DNA. The effects of overcrowding have also been studied (Capel *et al.,* 1980b). Increasing the number of rats per cage from 6 to 40 for 6-hr periods for as long as 8 days also increased corticosteroid levels. Antipyrine half-life was not altered. Benzo(*a*)pyrene hydroxylation and epoxide hydrase were increased, but binding of benzo(*a*)pyrene to DNA was decreased. It is difficult to apply these results to a situation that might be encountered by humans; however, the findings do point out that stress does have an effect on the ability of the liver to activate and detoxify procarcinogens. Stress may also play a role in tumor growth or regression. Mice subjected to experimental stress (rotation on a turntable) had a greatly reduced capacity for inhibiting the growth of implanted lymphosarcomas as compared to normal control animals (Riley, 1981).

In animal housing, the dramatic influence of stress from a variety of conditions such as noise is demonstrated by the finding that mice housed under low-stress conditions have plasma corticosterone levels ranging from 0 to 35 ng/ml, whereas these levels may be 10–20 times higher in conventional facilities (Riley, 1981).

The effect of temperature upon toxicity in experimental animals has been known for many years. The LD_{50} for procaine in mice was shown to be relatively constant at 800 mg/kg between 7 and 24°C, but it rapidly decreased with increasing temperature to approximately 200 mg/kg at 43°C (Sievers and McIntyre,

1937). On the other hand, rats exposed to the cold for 1 week showed an increased sensitivity to isoproterenol that was 1000–10,000 times higher than that of controls (Balazs *et al.*, 1962). Greater sensitivity was seen in the females than in the males.

In some cases, the effect of temperature may interact with other factors. For example, Chance (1946) noted that aggregation of mice greatly increased the toxicity of amphetamine. Moreover, this effect was markedly reduced by lowering the temperature from 26.7 to 15.6°C. At the lower temperature, there was little difference between the solitary and aggregated mice. Numerous other drugs also have been shown to have a varied toxicity depending upon the temperature (Weihe, 1973). Thus, it is essential that temperature be controlled during experimental animal studies because its influence may be substantial.

The degree to which temperature affects chemical toxicity in humans is not as well understood (Weihe, 1973). Although the toxicity of ethanol and probably certain neuroleptics is altered by temperature, the evidence that this is an important factor is not as great as would be suspected from animal data and theoretical considerations.

Even the capturing and handling of mice may have an effect on response. Riley (1981) demonstrated that increased plasma corticosterone levels were evident within 5 min after mice were handled.

The effects of other conditions under which animals are housed should also be considered. The importance of bedding was recognized when Vesell (1967) demonstrated that softwood beddings could enhance microsomal mixed-function oxidase activities. Later studies indicated that animals held under dirty conditions (accumulation of urine and feces under wire mesh pans for 1 week) also had an influence resulting in a decrease in metabolism (Vesell *et al.*, 1973). Filter tops on the cages also decreased drug-metabolizing activity, suggesting that trapped ammonia might have had an effect.

The importance of such environmental factors was amply illustrated by Sabine and co-workers (1973), who found that C3H-A^{vy} and C3H-A^{vy}fB mice, which had been reported to develop virtually a 100% incidence of liver and mammary tumors in the United States, had almost zero incidence when they attempted to use that strain in Australia. However, when these animals were provided with feed and bedding (cedar shavings) from the United States, the incidence of mammary tumors was again near 100%. Most of this appeared related to the cedar.

Perhaps not surprisingly, the importance of such findings in animals may indeed have relevance to humans. Acheson and co-workers (1968) investigated the incidence of nasal cancer in workers employed in the furniture industry in southern England. Their study suggested that the exposure of workers to dust from hardwoods such as oak and beech was related to the induction of these cancers, and that polishes, lacquers, and varnishes were not a cause of the problem. The incidence of 0.7 ± 0.2 per 1000 annually is approximately 500 times higher than the incidence of adenocarcinoma in adult males. Similarly, the risk of nasal carcinoma in Danish woodworkers has been calculated to be 0.5 per 1000 annually (Andersen *et al.*, 1977). In this study, the variety of woods used made it impossible to single out any particular type.

Consideration might also be given to the influence of such seemingly mundane factors as contamination upon toxicity experiments. This includes not only such sources as food, water, and perhaps even air, but the difficulties in avoiding contact of the controls with the toxicant of interest. For example, Sansone *et al.* (1977) used a fluorescein dye to demonstrate how difficult it is to prevent the spreading of contamination via the air, equipment, and personnel. The problems associated with caging have also been noted in other studies. In one study, Cagen and Klaassen (1979) found that control animals housed with littermates treated intraperitoneally with carbon tetrachloride had elevated serum aspartate aminotransferase levels when compared to littermates housed with untreated animals. This finding indicated that the first group of animals were probably exposed by inhalation to the carbon tetrachloride expired by the treated animals.

Certain factors that might be regarded as strictly applicable to the lifestyle of humans have been studied in both humans and animals. One of them—smoking—has been shown to have a significant effect on drug metabolism in humans (Pantuck *et al.,* 1974). Plasma levels of phenacetin in smokers were considerably lower than those in nonsmokers given an oral dose. The ratio of the plasma concentration of *N*-acetyl-*p*-aminophenol to phenacetin was severalfold higher in the smokers than in the nonsmokers. Stimulation of drug metabolism by cigarette smoke has also been demonstrated in the rat (Welch *et al.,* 1972). Lung aryl hydrocarbon hydroxylase (AHH) activity was found to be elevated up to 186% after 4 hr of exposure to cigarette smoke. In pregnant rats exposed to cigarette smoke for 5 hr daily for 3 days, AHH activity was elevated not only in the lung (12-fold) but also in the placenta (four-fold). Thus, the effect of smoking could be mimicked in laboratory animals, but cigarette smoking is a human habit that may influence the biotransformation and response of toxicants.

Factors may also interact. For example, when the effects of cigarette smoking were examined in men of different ages, the inducing effect of smoking as measured by antipyrine clearance was observed in younger (less than 40 years of age) but not in older individuals, suggesting that the inducting effect diminishes with age (Wood *et al.,* 1979). The clearance of indocyanine green also decreased with age, regardless of smoking habits, indicating that hepatic blood flow decreased with age. Thus, in any individual, both age and environmental factors may be important in assessing the hepatic drug clearance.

The influence of smoking cannot be universally extended, however. For example, the metabolism and effects of the corticosteroids used in the treatment of a variety of diseases are often associated with interactions with well-known inducers of drug metabolism such as phenobarbital and phenytoin, and one would suspect that smoking might also influence the reaction of an individual. However, it has been shown that smoking does not alter the pharmacokinetics of prednisone or dexamethasone (Rose *et al.,* 1981).

A lifestyle factor of considerable interest in humans is the interaction of ethanol with other chemicals. Classically, there has been concern about such agents as chloral hydrate, which along with ethanol inhibits metabolism and thus leads to profound depression of the central nervous system; inhibitors of acetaldehyde metabolism, e.g., disulfiram; and a variety of sedative-hypnotic, neuroleptic, and other central nervous system depressant drugs with which ethanol may have additive effects. Such interactions are usually obvious and are often predictable. Over the last decade, more emphasis has been placed on the interactions of ethanol with other chemicals with regard to their influences on drug metabolism. Rats given ethanol for 2–4 weeks develop increased drug-metabolizing activity toward a number of substrates (Rubin and Lieber, 1968; Singlevich and Barboriak, 1971). Induction also occurs in humans following chronic ingestion (Rubin and Lieber, 1968), but it is difficult to predict interactions because significant decreases in drug metabolism occur in humans after acute exposures to ethanol, possibly as the result of substrate competition for microsomal enzymes (Rubin *et al.,* 1970). In addition, alcoholics develop liver cirrhosis and parenchymal damage, which decreases their ability to metabolize drugs and other chemicals (Sotaniemi *et al.,* 1977).

HEALTH STATUS

The maintenance of healthy animals is obviously of primary importance in any safety evaluation of new compounds and need not be discussed here. When considering the diversity of responses among the human population, it is critically important to know the health status of the animals in order to assess their usefulness in studies to determine whether diseases might alter toxicity in humans.

In some cases the effect of disease may be predicted based on knowledge of the metabolism of the compound of interest, e.g., a drug whose termination of activity is dependent upon metabolism by the liver. In early studies, Levy *et al.* (1968) found that patients with liver disease had a decreased ability to metabolize both phenylbutazone and isoniazid. Their results also suggest that treatment with other drugs may alter the half-life of phenylbutazone, indicating that multiple effects play a role in determining the metabolic rate and the subsequent response. In a series of patients with chronic liver disease, the metabolism of antipyrine, lidocaine, and paracetamol was significantly reduced in a large percentage of patients (Forrest *et al.,* 1975). Hepner *et al.* (1977) studied the elimination of a number of drugs in patients with a variety of liver diseases. They found that the elimination of aminopyrine, diazepam, antipyrine, and indocyanine green was inhibited in patients with hepatocellular disease or neoplasms of the liver, but that the elimination of antipyrine and indocyanine green, but not aminopyrine or diazepam, was depressed in cholestatic disease. These findings illustrate well the problem of extrapolating from one drug or chemical to another (Vessell, 1978).

The role of biotransformation alone may not be the determining factor. Mawer *et al.* (1972) examined the influence of chronic liver disease in humans on the metabolism of amobarbital and found that the patients could be divided into two groups. In those with normal protein binding, there was no decrease in metabolism, but in those with decreased serum albumin levels, there was a decrease. However, the clinical responses of these two groups of patients were not different.

In addition to studies on the half-lives or elimination rates of drugs, there have been investigations using biopsy specimens from human liver (Schoene *et al.,* 1972). In subjects with severe hepatitis and cirrhosis, cytochrome P450 levels were approximately one-half the control value; in those with only moderate hepatitis, there was no change. There were similar changes in the demethylation of aminopyrine and *p*-nitroanisole. Reduced nicotinamide-adenine dinucleotide phosphate (NADPH) cytochrome *c* reductase was unaltered, and pseudocholinesterase was decreased only in subjects with the severe disease.

The difficulty of demonstrating consistent results for liver disease in humans was also encountered in studies in animals conducted under less vigorous conditions than are often imposed in studies of hepatotoxic agents administered in high doses that cause extensive cellular damage. Willson and Hart (1977) reported generally dose-dependent changes in rats given an acetaminophen–dimethylsulfoxide mixture. At low doses there were slight increases in aminopyrine demethylation, bilirubin glucuronyl transferase, and cytochrome P450; higher doses produced decreases in these as well as in aminopyrine demethylation. However, the decreases were not large and did not correlate well with changes in histopathology and serum enzymes. In animals as well as in humans, therefore, there is much variation in response, depending on the enzyme or other parameter being assessed.

A number of animal models have been used in studies to determine whether diabetics are especially susceptible to chemical-induced toxicity. Male rats made diabetic with alloxan demonstrated a decrease in drug-metabolizing enzymes, as evidenced by decreases in hexobarbital metabolism and a concomitant increase in sleeping time (Dixon *et al.,* 1961). This effect was reversed by insulin treatment. Streptozotocin treatment leading to a diabetic state has also been shown to cause decreases in the metabolism of hexobarbital and aniline in rats (Ackerman and Leibman, 1977). However, Reinke *et al.* (1978) found that male rats treated with streptozotocin had a decrease in aminopyrine *N*-demethylation, whereas the female had an increase. In both cases, insulin antagonized the effect of the streptozotocin. It is difficult to predict, therefore, whether the male or female rat would provide more accurate information on the possible response of human diabetics to drugs that are metabolized.

The limited number of studies in humans are not all that clear-cut. Ueda *et al.* (1963) compared normal subjects with diabetics and those with liver or kidney impairment. The half-life of tolbutamide in the diabetic group was not different from that in controls, but it was increased in one-half of the patients with liver cirrhosis. It was also markedly increased in most of the patients with impaired renal function. On the other hand, Nelson (1964) found that the values for patients with hepatic dysfunction were within the control range, suggesting that the disease state might not alter the rate of metabolism of this drug.

The factors affecting the metabolism and disposition of a compound may not always be clearly understood, yet they may be important and need consideration in evaluating the hazard of a compound within a specific group of individuals. For example, tumors in animals may influence the metabolism of foreign compounds. It has been shown that tumor-bearing animals generally demonstrate a decrease in drug-metabolizing activity (Kato *et al.,* 1970; Rosso *et al.,* 1971). In humans, Ambre *et al.* (1974) reported that in patients with lung cancer, there was a shortened plasma elimination half-life of antipyrine when compared with controls. Tschanz *et al.* (1977) reported that there were no differences either quantitatively or qualitatively between patients with lung cancer and matched controls, and Higuchi *et al.* (1980) reported that cancer patients demonstrated a decrease in drug-metabolizing ability. These divergent results may be the result of improper matching of the groups. However, as Higuchi *et al.* (1980) noted, the presence of the tumor itself may not be the only important factor; one must also consider liver function, nutritional status, and the possible involvement of anticancer drugs.

SUMMARY AND CONCLUSIONS

In considering the importance of the various factors discussed in this chapter as they may affect the outcome of toxicological evaluations of chemicals, it is clear that they all need to be better understood than they are today. These factors need not be studied in one category, but may be somewhat arbitrarily divided into three groups.

The first group of factors are those that are likely to influence the safety evaluation of materials, but are normally controlled. These would include environmental factors, such as caging, bedding, temperature, and humidity, and the assurance of an adequate diet free from contaminating substances.

The second group of factors would be those that do indeed account for conflicting results among laboratories, but when used properly can actually be advantageous in studies on mechanisms of toxic action and on the choice of models to simulate human responses. Studies on this group would include determination of the bases for differences among species or among strains of the same species. Even sex differences may provide clues concerning the importance of sex hormones with regard either to biotransformation or to response. Differences in toxicity varying with the time of day or from season to season may reflect hormonal interactions.

The third group would include factors that may be manipulated to gain a better understanding of how they might be important in different human populations. Thus, studies with diets containing low concentrations of protein may mimic better the conditions in impoverished countries than do those employing only the usual laboratory chow. Similar arguments could also be made for vitamin deficiencies. Much is known with regard to abnormal physiological states and the disposition and both pharmacological and toxicological actions of therapeutic agents. However, relatively little has been done in this area for environmental or occupationally important chemicals. Similarly, the area of interactions among chemical entities as the result of the concomitant administration of drugs is well known, but with few exceptions there is a dearth of such knowledge with regard to chemicals in general. Age has been studied as a factor in some areas, e.g., biotransformation, but more needs to be done to ascertain the increased risks to children and the bases for the differences.

REFERENCES

Acheson, E. D., R. H. Cowdell, E. Hadfield, and R. G. Macbeth. 1968. Nasal cancer in woodworkers in the furniture industry. *Br. Med. J.* **2**:587–596.

Ackerman, D. M., and K. C. Leibman. 1977. Effect of experimental diabetes on drug metabolism in the rat. *Drug Metab. Dispos.* **5**:405–410.

Adamson, R. H., J. W. Bridges, M. E. Evans, and R. T. Williams. 1970. Species differences in the aromatization of quinic acid *in vivo* and the role of gut bacteria. *Biochem. J.* **116**:437–443.

Alam, S. N., R. J. Roberts, and L. J. Fisher. 1977. Age-related differences in salicylamide and acetaminophen conjugation in man. *J. Pediatr.* **90**:130–135.

Alary, J. G., and J. Brodeur. 1970. Correlation between the activity of liver enzymes and the LD50 of parathion in the rat. *Can. J. Physiol. Pharmacol.* **48**:829–831.

Ambre, J., F. Butes, D. Haupt, D. Graeff, and K. Deason. 1974. Antipyrine metabolism in patients with lung cancer. *Clin. Res.* **22**:598A.

Andersen, H. C., I. Andersen, and J. Solgaard. 1977. Nasal cancers, symptoms and upper airway function in woodworkers. *Br. J. Ind. Med.* **34**:201–207.

Anonymous. 1980. Caffeine and pregnancy. *FDA Drug Bull.* **10**:19–20.

Aranda, J. V., S. M. MacLeod, K. W. Renton, and N. R. Eade. 1974. Hepatic microsomal drug oxidation and electron transport in newborn infants. *Pediatr. Pharmacol. Ther.* **85**:534–542.

Axelrod, J., S. Udenfriend, and B. B. Brodie. 1954. Ascorbic acid in aromatic hydroxylation. III. Effect of ascorbic acid on hydroxylation of acetanilide, aniline and antipyrine *in vivo*. *J. Pharmacol. Exp. Ther.* **111**:176–181.

Balazs, T., J. B. Murphy, and H. C. Grice. 1962. The influence of environmental changes on the cardiotoxicity of isoprenoline in rats. *J. Pharm. Pharmacol.* **14**:750–755.

Bartek, M. J., J. A. LaBudde, and H. I. Maibach. 1972. Skin permeability *in vivo:* Comparison in rat, rabbit, pig and man. *J. Invest. Dermatol.* **58**:114–128.

Benedict, W. F., N. Considine, and D. W. Nebert. 1973. Genetic differences in arylhydrocarbon hydroxylase induction and benzo(*a*)pyrene-produced tumorigenesis in the mouse. *Mol. Pharmacol.* **9**:266–277.

Benke, G. M., and S. D. Murphy. 1975. The influence of age on the toxicity of methyl parathion and parathion in male and female rats. *Toxicol. Appl. Pharmacol.* **31**:254–269.

Birnbaum, L. S., G. M. Decad, H. B. Matthews, and E. E. McConnell. 1981. Fate of 2,3,7,8-tetrachlorodibenzofuran in the monkey. *Toxicol. Appl. Pharmacol.* **57**:189–196.

Boyd, E. M. 1972. *Protein Deficiency and Pesticide Toxicity.* Thomas, Springfield, Illinois. 468 pp.

Brain, P., and D. Benton. 1975. The interpretation of physiological correlates of differential housing in laboratory rats. *Life Sci.* **24**:99–116.

Bremner, I., and J. K. Campbell. 1980. The influence of copper intake on the toxicity of cadmium. *Ann. N.Y. Acad. Sci.* **355**:319–332.

Burg, A. W. 1975. Physiological disposition of caffeine. *Drug Metab. Rev.* **4**:199–228.

Cagen, S. Z., and C. D. Klaassen. 1979. Hepatotoxicity of carbon tetrachloride in developing rats. *Toxicol. Appl. Pharmacol.* **50**:347–354.

Calabrese, E. J., G. Moore, and R. Brown. 1979. Effects of environmental oxidant stressors on individuals with a G-6-PD deficiency with particular reference to an animal model. *Environ. Health Perspect.* **29**:49–55.

Capel, I. D., M. Jenner, M. H. Pinnock, H. M. Dorrell, and D. C. Williams. 1980a. The effect of isolation stress on some hepatic drug and carcinogen metabolizing enzymes in rats. *J. Environ. Pathol. Toxicol.* **4**:337–344.

Capel, I. D., M. Jenner, M. H. Pinnock, H. M. Dorrell, and D. C. Williams. 1980b. The effect of overcrowding stress on carcinogen-metabolizing enzymes of the rat. *Environ. Res.* **23**:162–169.

Chance, M. R. A. 1946. Aggregation as a factor influencing the toxicity of sympathomimetic amines in mice. *J. Pharmacol. Exp. Ther.* **87**:214–219.

Clemens, T. L., R. N. Hill, L. P. Bullock, W. D. Johnson, L. G. Sultatos, and E. S. Vesell. 1979. Chloroform toxicity in the mouse: Role of genetic factors and steroids. *Toxicol. Appl. Pharmacol.* **48**:117–130.

Collins, T. F. X., J. J. Welsh, T. N. Black, and D. I. Ruggles. 1983. A study of the teratogenic potential of caffeine ingested in drinking water. *Food Chem. Toxicol.* **24**:763–777.

DeMatteis, F., B. E. Prior, and C. Rimington. 1961. Nervous and biochemical disturbances following hexachlorobenzene intoxication. *Nature* **191**:363–366.

Dixon, R. L., L. G. Hart, and J. R. Fouts. 1961. The metabolism of drugs by liver microsomes from alloxan-diabetic rats. *J. Pharamacol. Exp. Ther.* **133**:7–11.

Dring, L. G., R. L. Smith, and R. T. Williams. 1970. The metabolic fate of amphetamine in man and other species. *Biochem. J.* **116**:425–435.

DuBois, K. P., J. Doull, P. R. Salerno, and J. M. Coon. 1949. Studies on the toxicity and mechanism of action of *p*-nitrophenyl diethyl thionophosphate (parathion). *J. Pharmacol. Exp. Ther.* **95**:79–91.

Ehrlich, R. 1980. Interaction between environmental pollutants and respiratory infections. *Environ. Health Perspect.* **35**:89–100.

Facchini, V., and L. A. Griffiths. 1981. The involvement of the gastro-intestinal microflora in nitro-compound-induced methaemoglobinaemia in rats and its relationship to nitro-group reduction. *Biochem. Pharmacol.* **30**:931–935.

Fancher, O. E. 1978. Species selection and animal models for toxicological study. *Clin. Toxicol.* **12**:239–247.

Forrest, J. A. H., N. D. C. Finlayson, K. K. Dejepon Yamoah, and L. F. Prescott. 1975. Antipyrine, lignocaine and paracetamol metabolism in chronic liver disease. *Gut* **16**:828–829.

Fox, M. R. S., R. M. Jacobs, A. O. L. Jones, G. E. Fry, and C. L. Stone. 1980. Effect of vitamin C and iron on cadmium metabolism. *Ann. N.Y. Acad. Sci.* **355**:249–261.

Gagne, J., and J. Brodeur. 1972. Metabolic studies on the mechanisms of increased susceptibility of weanling rats to parathion. *Can. J. Physiol. Pharmacol.* **50**:902–915.

Gardner, D. E. 1980. Influence of exposure patterns of nitrogen dioxide on susceptibility to infectious respiratory disease. Pp. 267–288 in S. D. Lee, ed. *Nitrogen Oxides and Their Effects on Health.* Ann Arbor Science Publishers, Ann Arbor.

Graham, S. L., A. Arnold, L. Kasza, G. E. Ruffin, R. C. Jackson, T. L. Watkins, and C. H. Graham. 1981. Subchronic effects of guar gum in rats. *Food Cosmet. Toxicol.* **19**:287–290.

Greenman, D. L., R. R. Delongchamp, and B. Highman. 1979. Variability of response to diethylstilbesterol: A comparison of inbred with hybrid mice. *J. Toxicol. Environ. Health.* **5**:131–143.

Hadley, W. M., T. S. Miya, and W. F. Bousquet. 1974. Cadmium inhibition of hepatic drug metabolism in the rat. *Toxicol. Appl. Pharmacol.* **28**:284–291.

Held, H., and M. Przerwa. 1976. Effect of the female sexual cycle on the activity of δ-aminolevulinate synthetase in rat liver. *Eur. J. Clin. Invest.* **6**:411–413.

Hepner, G. W., E. S. Vesell, A. Lipton, H. A. Harvey, G. R. Wilkinson, and S. Schenker. 1977. Disposition of aminopyrine, antipyrine, diazepam, and indocyanine green in patients with liver disease or on anticonvulsant therapy: Diazepam breath test and correlations in drug elimination. *J. Lab. Clin. Med.* **90**:440–456.

Higuchi, T., T. Nakamura, and H. Uchino. 1980. Antipyrine metabolism in cancer patients. *Experientia* **36**:236–238.

Hill, R. N., T. L. Clemens, D. K. Liu, E. S. Vesell, and W. D. Johnson. 1975. Genetic control of chloroform toxicity in mice. *Science* **190**:159–161.

Hoyumpa, A. M., R. A. Branch, and S. Schenker. 1978. The disposition and effects of sedatives and analgesics in liver disease. *Annu. Rev. Med.* **29**:205–218.

Jaeger, R. J., R. B. Conolly, and S. D. Murphy. 1973. Diurnal variation of hepatic glutathione concentration and its correlation with 1,1-dichloroethylene inhalation toxicity in rats. *Res. Commun. Chem. Pathol. Pharmacol.* **6**:465–471.

Jaeger, R. J., R. B. Conolly, and S. D. Murphy. 1974. Effect of 18 hr fast on 1,1-dichloroethylene-induced hepatotoxicity and lethality in rats. *Exp. Mol. Pathol.* **20**:187–198.

Kappas, A., K. E. Anderson, A. H. Conney, and A. P. Alvares. 1976. Influence of dietary protein and carbohydrate on antipyrine and theophylline metabolism in man. *Clin. Pharamacol. Ther.* **20**:643–653.

Kato, R. 1974. Sex-related differences in drug metabolism. *Drug Metab. Rev.* **3**:1–32.

Kato, R. 1977. Drug metabolism under pathological and abnormal physiological states in animals and man. *Xenobiotica* **1/2**:25–92.

Kato, R., A. Takanaka, and A. Takahashi. 1970. Decrease in the substrate interaction with cytochrome P-450 in drug hydroxylation by liver microsomes from rats bearing Walker 256 carcinosarcoma. *Gann* **61**:359–365.

Kellermann, G., M. Luyten-Kellermann, M. G. Horning, and M. Stafford. 1976. Elimination of antipyrine and benzo(*a*)pyrene metabolism in cultured human lymphocytes. *Clin. Pharmacol. Ther.* **20**:72–80.

Kimura, E. T., D. M. Ebert, and P. W. Dodge. 1971. Acute toxicity and limits of solvent residues for 16 organic solvents. *Toxicol. Appl. Pharmacol.* **19**:699–704.

Koch, R. L., B. B. Beaulieu, and P. Goodman. 1980. Role of intestinal flora in the metabolism of misonidazole. *Biochem. Pharmacol.* **29**:3281–3284.

Levine, R. R. 1983. *Pharmacology: Drug Actions and Reactions,* 3d ed. Little, Brown and Co., Boston.

Levy, A. J., S. Sherlock, and D. Walker. 1968. Phenylbutazone and isoniazid metabolism in patients with liver disease in relation to previous drug therapy. *Lancet* **1**:1275–1279.

Lindsey, J. R., M. W. Conner, and J. J. Baker. 1978. Physical, chemical and microbial factors affecting biologic response. Pp. 31–43 in *Laboratory Animal Housing.* National Academy of Sciences, Washington, D.C.

MacLeod, S. M., K. W. Renton, and N. R. Eade. 1972. Development of hepatic microsomal drug-oxidizing enzymes in immature male and female rats. *J. Pharamacol. Exp. Ther.* **183**:489–498.

Mainigi, K. D., and T. C. Campbell. 1981. Effect of sex differences on subcellular distribution of aflatoxin in F-344 rats treated with various risk-modifying factors. *Toxicol. Appl. Pharmacol.* **58**:236–243.

Mawer, G. E., N. E. Miller, and L. A. Turnberg. 1972. Metabolism of amylobarbitone in patients with chronic liver disease. *Br. J. Pharmacol.* **44**:549–560.

McConnell, E. E., and J. A. Moore. 1975. Toxicopathology characteristics of the halogenated aromatics. *Ann. N.Y. Acad. Sci.* **320**:138–150.

McKenna, M. J., J. A. Zempel, E. O. Madrid, and P. J. Gehring. 1978. The pharmacokinetics of (^{14}C) vinylidene chloride exposure in rats following inhalation exposure. *Toxicol. Appl. Pharmacol.* **45**:599–610.

McMartin, K. E., G. Martin-Amat, A. B. Mekar, and T. R. Tephly. 1977. Methanol poisoning. V. Role of formate metabolism in the monkey. *J. Pharmacol. Exp. Ther.* **201**:546–572.

McMartin, K. E., G. Martin-Amat, P. E. Noker, and T. R. Tephly. 1979. Lack of a role for formaldehyde in methanol poisoning in the monkey. *Biochem. Pharmacol.* **28**:645–649.

Mirkin, B. L. 1978. Pharmacodynamics and drug disposition in pregnant women, in neonates and in children. Pp. 127–152 in K. L. Melmon and H. F. Morreli, eds. *Clinical Pharmacology.* Macmillan, New York.

Mitzutani, T., and T. Mitsuoka. 1979. Effect of intestinal bacteria on incidence of liver tumors in gnotobiotic C3H/He male mice. *J. Natl. Cancer Inst.* **63**:1365–1369.

Mulvihill, J. J. 1973. Caffeine as teratogen and mutagen. *Teratology* **8**:69–72.

Nabeshima, T., and I. K. Ho. 1981. Pharmacological responses to pentobarbital in different strains of mice. *J. Pharmacol. Exp. Ther.* **216**:198–204.

Nair, V., and R. Casper. 1969. The influence of light on daily rhythm in hepatic drug metabolizing enzymes in rat. *Life Sci.* **8**:1291–1298.

Neal, R. A., and K. P. DuBois. 1965. Studies on the mechanism of detoxification of cholinergic phosphorothioates. *J. Pharmacol. Exp. Ther.* **148**:185–192.

Nebert, D. W., W. F. Benedict, J. E. Gielen, F. Oesch, and J. W. Daly. 1972. Aryl hydrocarbon hydroxylase, epoxide hydrase and 7,12-dimethylbenz(*a*)anthracene-produced skin tumorigenesis in the mouse. *Mol. Pharmacol.* **8**:374–379.

Needham, W. P., L. Shuster, G. C. Kanel, and M. L. Thompson. 1981. Liver damage from narcotics in mice. *Toxicol. Appl. Pharmacol.* **58**:157–170.

Nelson, E. 1964. Rate of metabolism of tolbutamide in test subjects with liver disease or with impaired renal function. *Am. J. Med. Sci.* **248:**657–659.

Newberne, P. M., and W. H. Butler. 1969. Acute and chronic effects of aflatoxin on the liver of domestic and laboratory animals: A review. *Cancer Res.* **29:**236–250.

Newberne, P. M., and G. N. Wogan. 1968. Sequential morphological changes in aflatoxin B_1 carcinogenesis in the rat. *Cancer Res.* **28:**770–781.

Newton, W. M. 1978. Environmental impact on laboratory animals. *Adv. Vet. Sci. Comp. Med.* **22:**1–28.

Nitzan, M., B. Volovitz, and E. Topper. 1979. Infantile methemoglobinemia caused by food additives. *Clin. Toxicol.* **15:**273–280.

Nomiyama, K., and H. Nomiyama. 1971. Metabolism of trichloroethylene in human. *Int. Arch. Arbeitsmed.* **28:**37–48.

Ockner, R. K., and R. Schmid. 1961. Acquired porphyria in man and rat due to hexachlorobenzene intoxication. *Nature* **189:**499.

Pacifici, G. M., A. R. Doobis, M. J. Brodie, M. E. McManus, and D. S. Davies. 1981. Tissue and species differences in enzymes of epoxide metabolism. *Xenobiotica* **11:**73–79.

Pantuck, E. J., K.-C. Hsiao, A. Maggio, K. Nakamura, R. Kuntzman, and A. H. Conney. 1974. Effect of cigarette smoking on phenacetin metabolism. *Clin. Pharmacol. Ther.* **15:**9–17.

Pantuck, E. J., K.-C. Hsiao, A. H. Conney, W. A. Garland, A. Kappas, K. E. Anderson, and A. P. Alvares. 1976. Effect of charcoal-broiled beef on phenacetin metabolism in man. *Science* **194:**1055–1057.

Pantuck, E. J., C. B. Pantuck, W. A. Garland, B. H. Min, L. W. Wattenberg, K. E. Anderson, A. Kappas, and A. H. Conney. 1979. Stimulatory effect of brussels sprouts and cabbage on human drug metabolism. *Clin. Pharmacol. Ther.* **25:**88–95.

Prehn, L. M., and E. M. Lawler. 1979. Rank order of sarcoma susceptibility among mouse strains reverses with low concentrations of carcinogens. *Science* **204:**309–310.

Purchase, I. F. H. 1980. Interspecies comparisons of carcinogenicity. *Br. J. Cancer* **41:**454–468.

Radzialowski, F. M., and W. F. Bousquet. 1968. Daily rhythmic variation in hepatic drug metabolism in the rat and mouse. *J. Pharmacol. Exp. Ther.* **163:**229–238.

Reddy, B. G., L. R. Pohl, and G. Krishna. 1976. The requirement of gut flora in nitrobenzene-induced methemoglobinemia in rats. *Biochem Pharmacol.* **25:**1119–1122.

Reinke, L. A., S. J. Stohs, and H. Rosenberg. 1978. Altered activity of hepatic mixed function monooxygenase enzymes in streptozotocin-induced diabetic rats. *Xenobiotica* **8:**611–619.

Renwick, A. G., and R. T. Williams. 1972. The metabolites of cyclohexylamine in man and certain animals. *Biochem. J.* **129:**847–867.

Righter, H. F., W. T. Shalkop, H. D. Mercer, and E. C. Leffel. 1972. Influence of age and sexual status on the development of toxic effects in the male rat fed aflatoxins. *Toxicol. Appl. Pharmacol.* **21:**435–439.

Riley, V. 1981. Psychoneuroendocrine influences on immunocompetence and neoplasia. *Science* **212:**1100–1109.

Ringler, D. H., and L. Dabich. 1979. Hematology and clinical biochemistry. Pp. 105–121 in H. J. Baker, R. Lindsey, and S. H. Weisbroth, eds. *The Laboratory Rat,* Volume 1. *Biology and Diseases.* Academic Press, New York.

Robinson, J. R., J. S. Felton, R. C. Levitt, S. S. Thorgeirsson, and D. W. Nebert. 1975. Relationship between "aromatic hydrocarbon responsiveness" and the survival times in mice treated with various drugs and environmental compounds. *Mol. Pharmacol.* **11:**850–865.

Romero, J. A. 1976. Influence of diurnal cycles on biochemical parameters of drug sensitivity: The pineal gland as a model. *Fed. Proc.* **35:**1157–1161.

Rose, J. Q., A. M. Yurchak, A. W. Meikle, and W. J. Jusko. 1981. Effect of smoking on prednisone, prednisolone, and dexamethasone pharmacokinetics. *J. Pharmacokin. Biopharm.* **9:**1–14.

Ross, R. H., A. L. Sermons, J. O. Owasoyo, and C. A. Walker. 1981. Circadian variation of diazepam acute toxicity in mice. *Experientia* **37:**72.

Rosso, R., M. G. Donelli, G. Franchi, and S. Garattini. 1971. Impairment of drug metabolism in tumor-bearing animals. *Eur. J. Cancer* **7:**565–577.

Rubin, E., and C. S. Lieber. 1968. Hepatic microsomal enzymes in man and rat: Induction and inhibition by ethanol. *Science* **162**:690–691.

Rubin, E., H. Gang, P. S. Misra, and C. S. Lieber. 1970. Inhibition of drug metabolism by acute ethanol intoxication. *Am. J. Med.* **49**:801–806.

Sabine, J. R., B. J. Horton, and M. B. Wicks. 1973. Spontaneous tumors in C3H-A^{vy} and C3H-A^{vy}fB mice: High incidence in the United States and low incidence in Australia. *J. Natl. Cancer Inst.* **50**:1237–1242.

Sakellaris, P. C., A. Peterson, A. Goodwin, C. M. Winget, and J. Vernikos-Danellis. 1975. Response of mice to repeated photoperiod shifts: Susceptibility to stress and barbiturates. *Proc. Soc. Exp. Biol. Med.* **149**:677–680.

San Martin de Viale, C., A. A. Viale, S. Nacht, and M. Grinstein. 1970. Experimental porphyria induced in rats by hexachlorobenzene. Study of the porphyrins excreted by urine. *Clin. Chim. Acta* **28**:13–23.

Sansone, E. B., A. M. Losikoff, and R. A. Pendleton. 1977. Sources and dissemination of contamination in material handling operations. *Am. Ind. Hyg. Assoc. J.* **38**:433–442.

Sanvordeker, D. R., and H. J. Lambert, 1974. Environmental modification of mammalian drug metabolism and biological responses. *Drug Metab. Rev.* **3**:201–229.

Schanker, L. S. 1978. Drug absorption from the lung. *Biochem. Pharmacol.* **27**:381–385.

Schoene, B., R. A. Fleischmann, H. Remmer, and H. F. V. Oldershausen. 1972. Determination of drug metabolizing enzymes in needle biopsies of human liver. *Eur. J. Clin. Pharmacol.* **4**:65–73.

Sereni, F., M. Mandelli, N. Principi, G. Tognoni, G. Pardi, and P. L. Morselli. 1973. Induction of drug metabolizing enzyme activities in the human fetus and newborn infant. *Enzyme* **15**:318–329.

Short, C. R., D. A. Kinden, and R. Stith. 1976. Fetal and neonatal development of the microsomal monoxygenase system. *Drug Metab. Rev.* **5**:1–42.

Sievers, R. F., and A. R. McIntyre. 1937. The effects of temperature upon the toxicity of procaine for white mice. *J. Pharmacol. Exp. Ther.* **59**:90–92.

Singlevich, T. E., and J. J. Barboriak. 1971. Ethanol induction of microsomal drug-metabolizing enzymes in the rat. *Toxicol. Appl. Pharmacol.* **20**:284–290.

Sotaniemi, E. A., J. Ahlqvist, R. O. Pelkonen, H. Pirttiaho, and P. V. Luoma. 1977. Histological changes in the liver and indices of drug metabolism in alcoholics. *Eur. J. Clin. Pharmacol.* **11**:295–303.

Tannenbaum, A., and H. Silverstone. 1949. The genesis and growth of tumors. IV. Effects of varying the proportion of protein (casein) in the diet. *Cancer Res.* **9**:162–173.

Tschanz, C., C. E. Hignite, D. H. Huffman, and D. L. Azarnoff. 1977. Metabolic disposition of antipyrine in patients with lung cancer. *Cancer Res.* **37**:3881–3886.

Ueda, H., T. Sakurai, M. Ota, A. Nakajima, K. Kamii, and H. Maezawa. 1963. Disappearance rate of tolbutamide in normal subjects and in diabetes mellitus, liver cirrhosis and renal disease. *Diabetes* **12**:414–419.

Vesell, E. S. 1967. Induction of drug-metabolizing enzymes in liver microsomes of mice and rats by softwood bedding. *Science* **157**:1057–1058.

Vesell, E. S. 1978. Disease as one of many variables affecting drug disposition and response: Alteration of drug disposition in liver disease. *Drug Metab. Rev.* **8**:265–291.

Vesell E. S., C. M. Lang., W. J. White, G. T. Passananti, and S. L. Tripp. 1973. Hepatic drug metabolism in rats: Impairment in a dirty environment. *Science* **179**:896–897.

Vesell, E. S., C. M. Lang, W. J. White, G. T. Passananti, R. N. Hill, T. L. Clemens, D. K. Liu, and W. D. Johnson. 1976. Environmental and genetic factors affecting the response of laboratory animals to drugs. *Fed. Proc.* **35**:1125–1132.

Vesell, E. S., C. A. Shively, and G. T. Passananti. 1977. Temporal variations of antipyrine half-life in man. *Clin. Pharmacol. Ther.* **22**:843–852.

Weihe, W. H. 1973. The effect of temperature on the action of drugs. *Annu. Rev. Pharmacol.* **13**:409–425.

Welch, R. M., J. Cavallito, and A. Loh. 1972. Effect of exposure to cigarette smoke on the metabolism of benzo(*a*)pyrene and acetophenetidin by lung and intestine of rats. *Toxicol. Appl. Pharmacol.* **23**:749–758.

Wester, R. C., and H. I. Maibach. 1975. Percutaneous absorption in the rhesus monkey compared to man. *Toxicol. Appl. Pharmacol.* **32**:394–398.

Whitlock, R. H., N. A. White, G. N. Rowland, and R. Plue. 1978. Monensin toxicosis in horses: Clinical manifestations. *Annu. Conv. Am. Assoc. Equine Pract.* **24**:473–486.

Willes, R. F. 1977. Tissue distribution as a factor in species susceptibility to toxicity and hazard assessment. Example: methylmercury. *J. Environ. Pathol. Toxicol.* **1**:135–146.

Williams, R. T. 1974. Inter-species variations in the metabolism of xenobiotics. *Biochem. Soc. Trans.* **2**:359–377.

Willson, R. A., and F. E. Hart. 1977. Effect of experimental hepatic injury on *in vitro* drug metabolizing enzyme activities in the rat. *Gastroenterology* **73**:691–696.

Wood, A. J. J., R. E. Vestal, G. R. Wilkinson, R. A. Branch, and D. G. Shand. 1979. Effect of aging and smoking on antipyrine and indocyanine green elimination. *Clin. Pharmacol. Ther.* **26**:16–20.

Yang, R. S. H., F. Coulston, and L. Goldberg. 1975. Binding of hexachlorobenzene to erythrocytes: Species variation. *Life Sci.* **17**:545–550.

Yeary, R. A. 1967. Drug toxicity in newborn animals. *Appl. Ther.* **9**:918–921.

Zannoni, V. G., E. J. Flynn, and M. Lynch. 1972. Ascorbic acid and drug metabolism. *Biochem. Pharmacol.* **21**:1377–1392.

5

Statistical Interpretation of Toxicity Data

David W. Gaylor

Many of the topics discussed in this chapter pertain to experimental data in general, but the context of their use and examples given are in the field of toxicology. The discussion focuses on the statistical interpretation of data rather than on the statistical procedures used in the data analysis. For an extensive discussion of the statistical analysis of biological data, the reader may refer to a multitude of books and articles.

Before embarking on a discussion of the statistical interpretation of toxicological data, it is important to consider where statistics has an impact on scientific investigation. Descriptive statistics, such as means, standard deviations, and dose–response regression curves, help to summarize results. Statistical test procedures and confidence intervals assist in permitting decisions to be made in the presence of uncertainty.

The first stage in an experimental situation generally is the assimilation, analysis, summarization, and evaluation of biological knowledge. This knowledge could be entirely theoretical, in which case the statistician might be called upon to provide mathematical models to describe biological processes. Generally, the biological knowledge assembled is quantitative data, which must be summarized and weighed in determining the next questions to be investigated. Whether or not a formal statistical analysis is performed, a measure of the accuracy (agreement with true value) and precision (variability) of the existing data must be considered.

The second stage in an experimental situation generally is the formulation of the objective of a new experiment or a statement of the precise hypothesis to be

DAVID W. GAYLOR • Biometry, National Center for Toxicological Research, Jefferson, Arkansas 72079.

tested. Although it may seem unnecessary to consult a statistician at this point, this is in many respects the most important part of the scientific process. Not all hypotheses are testable. For example, it is not possible to perform an animal bioassay experiment to establish a threshold dose below which no toxic effect occurs. Lack of a statistically significant difference between treated and control animals does not necessarily mean that the dose level used was below a threshold dose. It may mean one of two things: (1) there is no difference in the toxic response between the controls and treated animals, i.e., a dose below the threshold was used, or (2) the resolving power of the experiment was not sufficient statistically to detect the presence of the toxic effect. Since it is impossible to distinguish between these two situations, it is impossible to demonstrate the existence of a threshold dose with this type of experiment.

The third stage in a scientific investigation generally is the formulation of an experimental plan to meet the objectives or to test the hypothesis. At this stage, the statistician is generally consulted to develop an experimental design to control and balance the effects of factors not under investigation, to supply randomization schemes to reduce the effects of uncontrolled extraneous factors, and to determine the sample sizes needed to achieve the desired precision or resolving power of the experiment.

It is important that true replication be built into an experiment, since reproducibility is one of the primary requirements of a scientific result. For example, it would be inappropriate to compare a group of animals exposed to nuclear radiation one week with a group of control (sham) animals housed in the same facility the following week. The results may be biased by subtle or not so subtle differences in the animals, in the environmental (laboratory) conditions, or in the conduct of the experiment from week to week. This bias would not be measured by the differences among animals within the groups. Hence the animals within groups are not considered true replicates of the total experimental process. In this case, it would be necessary to repeat the entire experimental process. The week-to-week replications provide the estimate of variation for determining the precision of the results for comparing the treated and control animals.

The fourth stage of an experiment is generally the actual conduct of the experiment, e.g., treating animals, making observations and measurements, and recording data. During the conduct of the experiment, it is critical that control and treated animals be handled similarly, except for the controlled factors under study.

The final stage of an experiment generally is the analysis and interpretation of the data. Here, the role of the statistician is to provide summary statistics with their associated statistical significance levels.

Although the entire experimental process may take place without the involvement of a statistician, the toxicologist must be aware that the experimental process is quantitative and is composed of uncertainty and probabilities. When the toxicologist and statistician do interact, it is necessary that they have a clear understanding of each other's discipline, particularly the underlying assumptions and principles that guide the analysis and interpretation of experimental data.

STATISTICAL TESTS FOR TOXICITY

Statistical and Biological Significance

Statistical Significance Level

Statistical significance, often associated with a statistical test of a hypothesis, is generally expressed as a p-value. For example, one may calculate that the comparison of a treated and control group is statistically significant at the $p \leq 0.05$ level. This statement means: if the control and treated animals were identically distributed, e.g., no difference in their average values, and all the mathematical assumptions and conditions of the data required for the statistical test are met, the probability of observing a difference as large or larger than the one obtained is less than 5%. That is, the probability that the difference could have arisen by chance is less than 5%, if in fact the chemical treatment has no biological effect. Sometimes the significance level is stated as the complement, i.e., $p \leq 0.05$ is stated as being significant at the 95% level. The p-value is often called the α-level, type I error probability, or probability of a false positive. The use of the descriptive term "false positive" is popular among toxicologists and those in the medical profession. Indeed, the p-value is the probability of falsely stating a positive outcome when in fact there is no true effect.

Rather than state that a difference is statistically significant if p is less than a specific value, say $p \leq 0.05$, and not significant if it is not, many statisticians prefer to report the actual level achieved, e.g., $p < 0.02$ or $p < 0.09$. Then, the toxicologist has an indication of the level of statistical significance achieved and the probability of making a false-positive statement. From a purely rigorous probabilistic approach, the significance level should be chosen prior to the calculation of the statistical test and not after the outcome is known. Many statisticians believe that the toxicologist should be allowed the flexibility to base evaluations on the level of significance achieved. This author recommends caution in the definition of statistical significance so as to avoid arbitrary distinctions such as calling $p < 0.045$ "statistically significant" but a $p < 0.055$ "not statistically significant."

The terms "highly statistically significant" for $p < 0.01$, "statistically significant" for $p \leq 0.05$, and "borderline significance" for $p < 0.10$ arose primarily before the advent of electronic computers when statistical tables of significance levels for common statistical tests only quoted these levels. With modern calculating and computing equipment, it is relatively easy to calculate the p-value for most statistical tests.

One-Sided and Two-Sided Hypotheses

An important distinction must be made between one-sided and two-sided hypotheses. A one-sided test can be used when there is interest only in a change in one direction due to a chemical treatment. For example, the experimentalist may be concerned if a drug expresses a toxicity that results in a decrease in average litter size when administered to pregnant rats, but may not be concerned if the drug causes an increase in average litter size. If an increase occurs, no statistical test is performed and such an event may be thought to be of no concern. If the only concern is a decrease in litter size, which may precipitate some further action, then the hypothesis is truly one-sided (one-directional). On the other hand, if the investigator is interested in discovering if the drug influences the reproductive processes as evidenced by a change in the average litter size, either an increase or decrease (two-directional), then a two-sided test is used.

The statistical difference between a one-sided test and a two-sided test can be substantial. For example, suppose that the average litter size is ten for the controls and eight for the animals receiving the drug, a one-sided test gives a statistical significance level of $p < 0.04$, and a two-sided test would give a doubled statistical significance level of $p < 0.08$. Obviously, the observed difference can only be in one direction, in this case, a decrease. However, for a two-sided test, a decrease is no more important than an increase. In calculating the probability of making a false-positive statement, the experimentalist must also allow for the probability that a difference of similar size in the opposite direction also may occur with a probability ≤ 0.04. Since this size difference in either direction contributes 0.04 to the probability of making a false-positive statement, the total probability of making such a statement is 0.08.

The choice of whether to use a one-sided or two-sided statistical test must be made before the data are observed, in order to maintain objectivity. That is, the decision cannot be delayed until the data have been accumulated in an effort to raise or lower the significance level. Since the choice may not be obvious, consultation among the interested parties is essential for the toxicologist to make the appropriate selection and inform the statistician ahead of time which to use.

Historical Controls

The use of historical controls is controversial. Although a formal comparison with such controls may not be performed, toxicologists and pathologists generally will consider experimental results in view of past or related studies in order to ascertain if any unusual events had occurred.

Treated groups of animals are usually compared to a concurrent control group. If a statistically significant difference is observed but the historical controls were closer in their response to the treated animals, the results of the comparison with the concurrent controls may be given less weight.

On the other hand, a rare event, such as a few rare birth defects, may occur in the treated group. A comparison with the concurrent controls may fail to show a statistically significant difference because of the typically small numbers of animals used in any one experiment. Historical data may indicate that the particular birth defect observed has never or has very rarely occurred in the past. Then, the observance of the rare event in the experiment attains more significance.

If historical data are used in a formal statistical test, they should be obtained from the same laboratory and from tests in which the same procedures and same strains of animals were used. If the toxic endpoints are tumors or birth defects, then the same pathologist or teratologist should review the historical cases.

Biological Significance

Statistical significance must not be equated with biological significance. However, its existence is needed before biological significance can be considered. In its absence, in fact, it is virtually impossible to determine biological significance.

Thus, the two types of significance must be regarded sequentially. For example, in a treated group of animals, a statistically significant average body weight of 5% less than the controls may be indicative of some unspecified or systemic toxicity. However, the toxicologist may decide that the 5% weight difference is not of biological significance. In fact, the lower body weight might even be viewed as a healthy event. This illustrates that a statistically significant difference may not necessarily be viewed as biologically significant.

A lack of statistical significance leaves the toxicologist in an uncertain position. Suppose a treated group of animals shows an average weight loss of 15%. If there is large animal-to-animal variability or the groups contain relatively few animals, this difference may not be statistically significant. That is, there is not sufficient evidence to state with much certainty that a difference really occurred. Although the observed average weight loss is large, the testing of additional animals may demonstrate no weight loss. Although a 15% weight loss may be biologically important, we cannot call it biologically significant, because it may not be reproducible. However, such a result may well initiate further experimentation.

This example also illustrates the undue emphasis that may be placed on a statistical test of hypotheses. Quite often, confidence intervals impart more information in that they show the range of uncertainty of the results. For example, suppose the 95% confidence limits on the average weight loss are $15 \pm 20\%$. This says that that we are 95% confident that the true weight change is covered by the interval from a 5% weight gain to a 35% weight loss. Thus, no biological significance can be attached to the average 15% weight loss, because the data are compatible with both a slight weight gain and a large weight loss. Of course, that is what the statistical test indicated, but the confidence interval indicates the potential for a large weight loss, which may warrant further investigation. Other aspects of confidence intervals are discussed later in this chapter in the section entitled Estimation of Toxic Effects.

False-Positive and False-Negative Statements

Definitions

Seldom are scientists concerned only about a specific experimental result under a given set of experimental conditions. They are generally interested in making inferences about a wider domain than a particular experimental circumstance. An isolated result is of little value unless it can be reproduced at a later time in another laboratory. Because bioassays are limited, it is impossible to ascertain absolutely the accuracy of conclusions that a certain positive effect (significant difference) or negative effect (no significant difference) has occurred. The "correctness" or rather "incorrectness" of such statements can be established only in a probability sense. In conclusions from comparative studies, two kinds of errors can occur: a false positive or a false negative. False positive refers to a statement that a difference exists when in fact it does not. False negative applies when a difference does exist but an investigation has led to the conclusion that it does not.

A false-negative conclusion results when a biologically significant difference is not detected by a statistical test. Of course, this requires that the toxicologist define a toxicologically significant difference in order to estimate the probability of a false-negative result.

Error Rates of Comparative Tests

The probability that a false-positive error may occur can be controlled by specifying the type I error or α level to be used in statistical tests of significance. However, as the type I error is lowered, the type II error (β error, or false-negative rate) increases. That is, as one requires larger differences for statistical significance, the probability of a false-positive is reduced but the probability of a false negative increases. Much deliberation between the toxicologist and statistician may be required to determine the proper trade-off between false-positive and false-negative error rates, since the probability of false negatives increases as the real differences between treated and control animals decrease. The probability of false-negative statements can also be reduced by increasing sample sizes (see, e.g., Cochran and Cox, 1957).

Multiplicity of Tests

It is difficult to control false-positive error rates when a large number of hypotheses are tested in one experiment. This is especially true in teratological and reproductive studies, where several endpoints for various indices of fertility, reproduction, and birth defects in various organs may be examined. The problem also arises in oncogenicity studies, in which tumor rates for a number of tissue and organ sites may be examined. For example, if comparisons are made between treated and control animals for 20 independent endpoints, each of which may lead to a false-positive error, one of these comparisons would be expected to be significant at the 5% level by chance alone. Fortunately, the situation for oncogenicity studies is not as uncertain as it might first appear. Although nearly 50 tissue and organ sites may be examined for tumors, the incidence of tumors at most of these sites is so rare that there is only a negligible probability of getting enough rare tumors by chance at such a site to achieve statistical significance.

Peto *et al.* (1980) recommend separate statistical analyses of each tissue or organ in animals bearing tumors. However, satisfactory pathological and statistical procedures have not yet been developed for analyzing the number of tumors detected in an organ or a tissue. In addition to tumors by site, analyses may be performed on histological subcategories, such as malignancies, within a tissue or organ. Tumors from distinct sites may be combined if there are valid biological reasons. Finally, analysis of all tumors, irrespective of site, is possible, but probably should not supersede the site-specific analyses.

Bioassays for carcinogenicity frequently are conducted in both sexes of two rodent species. The increase in tumors at any particular tumor site may not achieve statistical significance, but may be close in both sexes of one or both species or in both species for one or both sexes. To increase the sensitivity of this bioassay to detect carcinogenicity, the test statistics may be combined to test for carcinogenicity in one or both species for both sexes combined or for one or both sexes in both species combined. That is, for each tumor site in bioassays conducted in both sexes of mice and rats, statistical tests may be performed for the following:

- male mice
- female mice
- male rats
- female rats
- male mice and male rats
- female mice and female rats
- male and female mice
- male and female rats
- all animals

Obviously, these tests are not independent, but they provide a method of increasing the sensitivity of bioassay screens, which are generally exploratory in nature.

As the number of statistical tests on a set of data increases, the possibility of detecting a potential carcinogenic effect may increase, but the probability of false positives may also increase. Accompanying any statistical test is the possibility that relatively rare chance events result in a significantly higher tumor rate in treated animals when there is no real carcinogenic effect, i.e., a false-positive result. This can only happen if the spontaneous background tumor rate in control animals is high enough to produce enough excess tumors to be statistically significant. Fears *et al.* (1977) showed that at least five more spontaneous tumors would have to develop in a group of 50 treated animals than in 50 control animals in order to produce a false-positive result at the $p \leq 0.05$ significance level. That is, there must be at least 5 out of 50 additional spontaneous tumors in the treated group. This generally can occur only at those few tumor sites with the higher spontaneous background tumor rates, because those sites with near-zero background rates will have negligible probabilities of producing false-positive results. Fears *et al.* (1977) reported false-positive rates of 0.05, 0.03, 0.10, and 0.13 for B6C3F_1 male and female mice and Fischer 344 male and female rats, respectively, when performing statistical tests for 21 tissue sites in groups of 50 animals. The overall false-positive rate was calculated as

$$1 - (1 - 0.05)(1 - 0.03)(1 - 0.10)(1 - 0.13) = 0.28$$

Utilizing the Bonferroni correction (Miller, 1966), one can reduce the overall false-positive rate for this two-species, both-sexes bioassay to approximately the

$p \leq 0.05$ level by requiring the significance level to reach $p \leq 0.01$ for any particular test.

If the bioassay is being conducted to test specifically for an increase of tumors at a particular site in a given sex and species, then no correction in the significance level for this specific hypothesis is required. However, any other test of an exploratory nature conducted at other tumor sites will require more stringent significance levels.

When more than one statistical test is performed, the Bonferroni inequality (Miller, 1966) frequently is used to make an adjustment in the significance levels. That is, if k statistical tests are performed on a set of data, the significance level for each test should be multiplied by k. For example, if the p-value for a particular test is 0.02, but four different tests are performed on the data, then the significance level should be raised to $4 \times 0.02 = 0.08$. Conversely, if it is decided to control the overall false-positive rate from four statistical tests to 0.05, then each test should be conducted at the $0.05 \div 4 = 0.0125$ level. In this way, the false-positive error rate for the chemical will be controlled at the 0.05 level rather than this level of significance being applied to each endpoint tested.

The k tests need not be independent in order to apply the Bonferroni inequality. However, it is not necessary to adjust the significance levels by counting every test conducted. For example, in oncogenicity studies, tests of tumor rates at tissue or organ sites where tumors rarely occur spontaneously in control animals need not be counted in the number of tests k, since they have a negligible probability of producing false-positive results (Fears *et al.*, 1977). Likewise, in a teratological study, not all of the many tests conducted need to be counted to obtain the value of k for adjusting significance levels. In a given litter, for example, the number of implants equals the number of viable fetuses plus the number of dead and resorbed fetuses. A chemical administered in a teratological experiment after implantation will not affect the number of implants. Thus, any statistical test based on the litter size (number of live fetuses) should supply essentially the same information as a statistical test based on the fetal loss (number of dead and resorbed fetuses). Although these two statistical tests may be performed, they should be counted only as one test result in any adjustment for multiplicity of tests.

Multiple Comparisons and Trend Tests

Frequently, many tests are conducted because a control and k different dosages of a chemical are administered. In such a case the control group may be compared with each of the k treated groups individually. Williams (1971, 1972) discusses statistical tests in this situation. A more appropriate analysis in this case may be to test for a dose–response trend. It is assumed that the difference between most toxicological responses and controls will become greater as the dose is increased. A statistical test for a dose–response trend is not only suggested by biological theory, but also makes use of all the different dose group data simultaneously, resulting in a single statistical test and thereby circumventing the problem associated with multiple tests.

Oncogenicity Experiments

Age-Adjusted Tumor Rates

In oncogenicity experiments, it is especially difficult to compare treated groups with the controls, either individually or as a trend. The tumor incidence in a particular organ does not necessarily increase with increasing dose. For example, increasing the dose may result in general toxicity that kills the animals at higher doses before tumors can be observed. Or an increase in the incidence of one fatal tumor type at higher doses may result in a decrease in the incidence of another tumor type. For example, in studies of the incidence of bladder tumors as a function of dose of a chemical, the chemical may induce early developing, rapidly fatal lymphoreticular tumors at the higher doses. Consequently, the animals may die before the bladder tumors develop at the higher doses, thereby resulting in a disruption of the bladder tumor dose response. In such cases, time- or age-adjusted tumor rates must be obtained before statistical tests can be performed.

Context of Tumor Observation

The appropriate method of adjusting for longevity depends upon how tumors are discovered. Death rate, or actuarial, methods are appropriate for *fatal* tumors, and a slight modification is appropriate in a *mortality-independent* context for onset rates of directly observable skin or palpable tumors. For *incidental* nonfatal tumors discovered upon necropsy at the time of scheduled sacrifices or at the time of death of an animal due to a cause other than the tumor type under investigation, a different prevalence rate method is required for correcting for longevity.

Since different statistical methods are required for fatal and incidental tumors, pathologists must attempt to distinguish between these types of tumors. For some animals, there may be errors of judgment. The more accurate the distinction between incidental and fatal tumors, the more reliable will be the adjustments for longevity. Obviously, all tumors discovered at the time of a scheduled sacrifice are incidental, regardless of their stage of development. Tumors that result in the death of their hosts are classified as *fatal* tumors. Tumors observed upon necropsy for an animal that has died of some unrelated cause are classified as incidental.

Peto *et al.* (1980) recommend classifying tumors as fatal, probably fatal, probably nonfatal, and nonfatal. The effects of misclassification can be established by including and excluding the probably nonfatal tumors with or from the nonfatal tumors, etc.

A particular tumor type may be primarily nonfatal or primarily fatal, or it may affect animals in both ways. The proportion of fatal tumors at a particular site may change with age. If there are several tumors in an organ (e.g., liver cell hepatomas), they must be taken collectively as fatal or nonfatal. Although it may be difficult to determine the exact cause of death in an animal, it may be easier to determine whether or not a particular tumor contributed to death, which is all that is needed for an appropriate statistical analysis. A classification of a tumor at the time of necropsy is needed—not a prediction of whether or not a tumor would eventually cause death.

If animals are removed from the study when skin tumors or palpable tumors develop, the construction of data tables and the calculation of the expected numbers are the same as for fatal tumors. If subsequent histopathologic examination of an animal reveals only a palpable mass not fitting the definition of a tumor, then that animal is not counted as having been removed because of a tumor, but, rather, as having died from causes other than the tumor under investigation.

Animals with skin tumors or palpable tumors may be left in the experiment; however, they are treated as though they were removed from study the day that the tumor was diagnosed, because they are no longer at risk of developing that tumor. Thus, the number of animals at risk for calculating an expected number is the number of survivors *without* a previously detected tumor of interest.

The animals at risk for incidental tumors are those examined during any specific period; more accurately, it is the number of animals in which tissues or organs are examined for the tumor of interest. Animals that die due to that type of tumor are not included, since they are counted in the death rate analysis for fatal tumors. The actuarial death rate method of analysis for fatal tumors is based on the number of survivors at any specific time.

Autolysis

If a tissue is not examined because of autolysis, that animal is not counted. However, when autolysis is only partial, gross tumors can still be detected. There is some question whether or not to count tumors detected in partially autolyzed tissues. If there is a lower probability of detecting tumors in autolyzed tissues, then including animals with autolyzed tissues will bias the tumor rate downward. On the other hand, if the inclusion of animals with autolyzed tissues is restricted to those in which tumors have been found, the tumor rate will be biased upward. Thus, if the intent is to obtain an unbiased estimate of the tumor rate, animals with an autolyzed tissue should not be included in estimating the tumor rate at that site. If the proportion of such tissues is low, the biases generally will be extremely low. If the degree of autolysis is similar across dose groups, then the amount of bias will be similar in each group. Hence, if the intent is to test for a positive trend, tumors observed in autolyzed tissues can be included if the extent of autolysis is similar across doses. Also, the type of tumor or organ might influence the decision to include or exclude autolyzed samples.

ESTIMATION OF TOXIC EFFECTS

Thresholds and No-Observed-Effect Levels

Perhaps undue attention has been given by statisticians and research workers to statistical tests of hypotheses. This may be due to the apparent ease of utilizing statistical tests of hypotheses in decision-making processes. However, preselected significance levels, such as ≤ 0.05, although traditional, may be inappropriate in any given situation. Such procedures tend to focus much attention on false-positive error rates while neglecting false-negative error rates, which may be large. Furthermore, there may be little difference between a result that is significant at the $p < 0.04$ level and one that is significant at the $p < 0.06$ level. On the other hand, a confidence interval provides information on the degree of uncertainty as well as demonstrating whether or not a statistically significant difference or change has occurred.

The existence of dose–response relationships might lead one to assume incorrectly the existence of threshold doses below which no toxic effects could occur. As dosage is decreased, the prevalence of an observable toxic effect (e.g., the excess proportion of animals with tumors) diminishes to zero. Eventually, a dosage is reached below which the experiment has essentially no resolving power to distinguish between the spontaneous background rate and small induced toxic effects. The observance of no tumors in a group of animals exposed to a given dosage of a chemical does not necessarily mean that a subthreshold dosage has been found. It means only that the true prevalence rate of tumors is low. For example, even if no tumors are observed in 100 animals, the upper 99% confidence limit on the true proportion of animals with a tumor is 0.045. Thus, one can be relatively certain only that the true tumor rate is less than 4.5%.

If no toxic effects are detected at a specified dosage, this dosage often is called the no-effect, or more correctly the no-observed-effect, dosage. Because of the limitations of any given experiment, the no-observed-effect dosage is not a precise estimate of a true no-effect level. Lack of statistical significance is not equivalent to no toxic effect. It may or may not be, and further experimentation would be required to resolve this equivocal issue. Comparing the results of each treated group with the controls in a dose–response study to find the largest dose at which a statistically significant difference does not occur (i.e., the no-observed-effect level) does not establish a biological threshold dose level. Any small difference can always be made statistically significant with a larger sample. For example, a tumor rate of 2% (1 of 50 animals) in a treated group compared to a tumor rate of 0 (0 of 50 animals) in a control group would not be considered statistically significant, whereas the same tumor rates of 2% (6 of 300 animals) in a treated group compared to 0 (0 of 300 animals) in the controls is statistically significant at the $p < 0.015$ level. The no-observed-effect level is not a biological property, but, rather, a statistical property or operational threshold that is highly dependent on sample size.

Biological thresholds cannot be established statistically. Increasing the number of animals will provide a more precise estimate of the tumor rate, but can never guarantee absolute safety. For example, if no animals in 10,000 exhibit a particular type of tumor, one cannot conclude that a biological subthreshold dose has been found. The upper 99% confidence limit on the true tumor rate is 0.00045. Observing no tumors with increasing sample sizes reduces the upper limit on the potential tumor rate, but it can never reach zero to establish a biological threshold.

Parametric and Nonparametric Estimates

If a certain statistical distribution is assumed to apply to a set of data, then the sample estimates of the parameters of the distribution can be used to provide a summary of the data. For example, suppose it can be assumed that a set of measurements can adequately be described by a Guassian (normal) distribution. Then two parameters, the mean and the variance, completely describe this distribution. Similarly, a negative exponential model for a first-order reaction is completely described by the initial amount and the reaction rate. In biology it is often difficult to find a tractable mathematical distribution or model that adequately describes a complex set of data. This is often the case with time-to-tumor curves. The log-normal distribution has been used to describe time-to-tumor occurrence (e.g., Albert and Altschuler, 1973). The Weibull distribution has been used extensively for time-to-tumor occurrence (e.g., Peto and Lee, 1973). On the other hand, Kaplan and Meier (1958) proposed a nonparametric estimator for incidence rates of fatal tumors that adjusts for causes of death competing with the cause of interest. Hoel and Walburg (1972) proposed a nonparametric estimator for the prevalence rate of incidental tumors when there is independent censoring by competing risks. The advantage of nonparametric techniques is that they do not require the assumption of an underlying distribution or model. Their disadvantage is that they do not summarize data to the extent that parametric methods do.

In the application of statistical methods to the analysis of chronic animal studies, the trend has been toward analyzing disease rates and mortality rates separately, and dwelling upon only one or the other, depending upon the lethality assumption. Turnbull and Mitchell (1978; Mitchell and Turnbull, 1979) described an approach to the analysis of survival/sacrifice data that requires neither the assumption of independence nor a lethality assumption and that simultaneously estimates functions of disease and mortality rates. Kodell *et al.* (1982) propose nonparametric estimators that simultaneously estimate distributions of time to occurrence of disease (and thereby disease incidence) and time to death caused by the disease. It is assumed that the disease of interest is irreversible and that the cause of death of each animal can be determined. However, diseases need not be classified as strictly lethal or strictly nonlethal. Indeed, most long-term diseases cannot be so classified, since they will cause death in some animals but will be only incidental findings in others.

Low-Dose Extrapolation and Safety Factors

Low-dose extrapolation of toxic effects is discussed in detail in Chapter 10. The topic is introduced here since safety factors often are applied to the no-observed-effect level (NOEL) in an effort to establish allowable dose limits for potentially toxic substances. For example, suppose a NOEL is divided by 100 to establish an allowable dosage of a chemical. Since the NOEL is taken to be a relatively safe dose for the experimental animal, the additional safety factor is applied to allow for potentially increased sensitivity of humans compared to the experimental animal and for differences in sensitivity among exposed humans. The size of a safety factor may depend upon the seriousness of the biological effect. For example, a safety factor of 100 may be used for reversible disease conditions, whereas a safety factor of 1000 may be applied for an irreversible disease such as cancer.

In a typical toxicological bioassay, several dosage levels are used and biological responses are observed or measured for each animal. Some responses may be quantal, i.e., they indicate the presence or absence of a disease state; others may be continuous measurements, e.g., body weight. Most biological responses curve upward over the low-dose range. Hence, as the dose is decreased the risk will generally decrease more rapidly than the dose, where risk is the proportion of animals possessing a disease state. Thus, decreasing the dosage by a safety factor of F will generally decrease risk by more than a factor of F. For example, if the upper confidence limit on a biological effect is 10%, then the risk at 1/100 of that dose is expected to be less than 10% ÷ 100 = 0.1%, i.e., less than 1 in 1000. If this level of risk is unacceptable, then a larger safety factor can be used or a better experiment can be conducted to lower the uncertainty in the experimental results. In general, if the upper confidence limit on the risk of disease at the experimental dose is U, then at the experimental dose divided by F, the upper confidence limit on the risk R is estimated to be less than U/F.

Conversely, it is possible to calculate the size of the safety factor needed to reduce the potential risk to a given level. For example, if it is desired to restrict the risk to less than 1 in 10,000, then an experimental dose with an upper confidence limit of 10% would have to be divided by a safety factor of 1000. That is, the upper confidence limit on the risk at 1/1000 of the experimental dose is estimated to be less than 10% ÷ 1,000 = 0.01%, or 1 in 10,000. In general, the safety factor required to provide a risk equal to or less than R is $F = U/R$.

Because of limited resources, it is necessary to conduct animal experiments at dosage levels greatly exceeding human exposure levels in order to detect potential toxic effects with relatively small numbers of animals. Thus, one is faced immediately with extrapolation of results at relatively high experimental dosage levels to low levels of human exposure (see Chapter 10).

In the absence of compelling biological arguments, there is no basis for selecting the appropriate mathematical dose–response model in the low-dose region. Since it is impossible to measure the dose–response relationship precisely for very low levels of exposure, *actual levels of risk at low doses cannot be obtained.* That is, best estimates of risk at low levels cannot be obtained without knowing the true form of the dose response. Animal bioassay data cannot be expected to resolve the true mathematical nature of the dose–response relationship below the experimental range. However, by making one simple plausible assumption that the response in the low-dose region is curving upward, it is possible to obtain an estimate of an upper limit on the risk at low doses using linear extrapolation below the experimental dose range without adopting any particular mathematical model. Several authors have presented various arguments for linearity of responses at low doses (Crump *et al.,* 1976; Guess *et al.,* 1977; Peto, 1978).

As used here, linear extrapolation does not mean fitting a straight line through the experimental data. In fact, most dose–response data for toxic effects are highly curved. In contrast to procedures that use mathematical models to fit dose–response data to estimate low-dose risks, linear extrapolation does not provide a best estimate of risk in the low-dose region. Rather, linear extrapolation only attempts to place an upper limit on the potential risk at low dosages below the experimental dose range (Gaylor and Kodell, 1980).

Since linear extrapolation does not predict risks at doses below the experimental range by using a parametric mathematical model fitted to the data, the unresolved issue of selecting a valid parametric model for extrapolation is circumvented. Therefore, selection of such a model to describe the responses only in the experimental dose range is no longer a critical issue, provided a valid estimate of the upper confidence limit is obtained at the lowest dosage from which extrapolation proceeds. At best, quantal bioassays resolve mathematical models in the experimental dose range, but cannot be expected to determine precisely the nature of the true dose–response relationship below the experimental region. For this reason, the extrapolation procedure limits the use of a parametric dose–response model to obtaining an upper confidence limit in the low end of the experimental dose range from which a linear extrapolation can be made over the unobserved lower dose region (Gaylor and Kodell, 1980). Thus, the application of safety factors and linear extrapolation are similar, since both procedures assume that upper limits on risk are proportional to dose at low levels.

REFERENCES

Albert, R. E., and B. Altshuler. 1973. Considerations relating to the formulation of limits for unavoidable population exposures to environmental carcinogens. Pp. 233–253 in J. E. Ballou, R. H. Busch, D. D. Mahlum, and C. L. Sanders, eds. *Radionuclide Carcinogenesis.* AEC Symposium Series, CONF-72050. National Technical Information Service, Springfield, Virginia.

Cochran, W. G., and G. M. Cox. 1957. *Experimental Designs,* 2nd ed. Wiley, New York.

Crump, K. S., D. G. Hoel, C. H. Langley, and R. Peto. 1976. Fundamental carcinogenic processes and their implications for low dose risk assessment. *Cancer Res.* **36:**2973–2979.

Fears, T. R., R. E. Tarone, and K. C. Chu. 1977. False-positive and false-negative rates for carcinogenicity screens. *Cancer Res.* **37:**1941–1945.

Gaylor, D. W., and R. L. Kodell. 1980. Linear interpolation algorithm for low dose risk assessment of toxic substances. *J. Environ. Pathol. Toxicol.* **4:**305–312.

Guess, H. A., K. S. Crump, and R. Peto. 1977. Uncertainty estimates for low-dose-rate extrapolations of animal carcinogenicity data. *Cancer Res.* **37:**3475–3483.

Hoel, D. G., and H. E. Walburg. 1972. Statistical analysis of survival experiments. *J. Natl. Cancer Inst.* **49:**361–372.

Kaplan, E. L., and P. Meier. 1958. Nonparametric estimation from incomplete observations. *J. Am. Stat. Assoc.* **53:**457–481.

Kodell, R. L., G. W. Shaw, and A. M. Johnson. 1982. Nonparametric joint estimators for disease resistance and survival functions in survival/sacrifice experiments. *Biometrics* **38:**43–58.

Miller, R. G., Jr. 1966. *Simultaneous Statistical Inference.* McGraw-Hill, New York.

Mitchell, T. J., and B. W. Turnbull. 1979. Log-linear models in the analysis of disease prevalence data from survival/sacrifice experiments. *Biometrics* **35:**221–234.

Peto, R. 1978. Carcinogenic effects of chronic exposure to very low levels of toxic substances. *Environ. Health Perspect.* **22:**155–159.

Peto, R., and P. N. Lee. 1973. Weibull distributions for continuous carcinogenesis experiments. *Biometrics* **29**:457–470.

Peto R., M. C. Pike, N. E. Day, R. G. Gray, P. N. Lee, S. Parish, J. Peto, S. Richards, and J. Wahrendorf. 1980. Guidelines for simple, sensitive significance tests for carcinogenic effects in long-term animal experiments. Pp. 311–425 in *Long-Term and Short-Term Screening Assays for Carcinogens: A Critical Appraisal.* IARC Monographs on the Evaluation of the Carcinogenic Risk of Chemicals to Humans, Annex to Supplement 2. International Agency for Research on Cancer, Lyon.

Turnbull, B. W., and T. J. Mitchell. 1978. Exploratory analysis of disease prevalence data from survival/sacrifice experiments. *Biometrics* **34**:555–570.

Williams, D. A. 1971. A test for differences between treatment means when several dose levels are compared with a zero dose control. *Biometrics* **27**:103–117.

Williams, D. A. 1972. The comparison of several dose levels with a zero dose control. *Biometrics* **28**:519–531.

II

Interpretation of Information from Human Studies

6

Clinical and Epidemiological Studies

James E. Cone, Gordon R. Reeve, and Philip J. Landrigan

Assessment of the effects on humans of exposures to hazardous chemicals relies heavily upon the results of clinical and epidemiological studies. The great advantage of such studies over animal investigations is that they provide direct evidence of the toxic actions of chemical agents in humans. Human studies are, however, difficult to conduct properly, and frequently they are difficult for policy makers to interpret (Horwitz and Feinstein, 1979; Susser, 1977). A major source of this difficulty is that almost inevitably clinical and epidemiological studies must be conducted after the occurrence of exposure. Except in rare instances, ethical constraints preclude the prospective exposure of human subjects to toxic agents (Duncan *et al.*, 1977; U.S. Department of Health, Education, and Welfare, 1978; Rothman, 1976). Thus, the typical study of the toxicity of chemical agents in humans is observational in nature and not experimental. An additional difficulty is posed here by the complexity of the causal relationships that typically exist between toxic exposures and human disease. Not only do individual humans vary greatly in their exposure to toxicants, but they vary also in their capacity for response as well as in their exposure to factors such as alcohol and tobacco, which may modify greatly the nature or severity of their responses to toxic exposures.

Despite these difficulties, techniques for the evaluation of data from human studies have been developed extensively (Kleinbaum *et al.*, 1982; Susser, 1973).

JAMES E. CONE • Division of Occupational Medicine, Medical Service, San Francisco General Hospital, San Francisco, California 94110. **GORDON R. REEVE** • Epidemiology, City of Houston Health Department, Houston, Texas 77030. **PHILIP J. LANDRIGAN** • Division of Environmental and Occupational Medicine, Mt. Sinai Medical Center, New York, New York 10029.

The epidemiological method has matured and has withstood the criticism that it is incapable of establishing the etiology of illness (see Chapter 7). Epidemiological inferences have been sustained and corroborated by the results of toxicological and biochemical studies, and epidemiology has proven to be a powerful tool for the exploration of both qualitative and quantitative cause-and-effect relationships between chemical exposure and human disease (Kleinbaum *et al.*, 1982; Susser, 1973).

In this chapter, we review the principal types of clinical and epidemiological studies that are likely to figure in evaluations of chemical toxicity. In the next chapter (Chapter 7), we examine in further detail the logical processes that underlie epidemiological reasoning, and on the basis of those considerations, we examine techniques that have been developed to evaluate and reconcile data from multiple, often apparently conflicting human studies.

THE PRINCIPAL TYPES OF HUMAN STUDY

Case reports identify one or more cases of a disease that have been detected by clinicians, by company or union officials, or by governmental officials as a result of active surveillance or passive reporting. *Cross-sectional studies* are census-like surveys of a population intended to determine the frequency of illness or of exposure. In *case–comparison studies,* cases of illness in a population are identified and then the frequency of exposure to a suspect causal agent in those cases is compared with the frequency of exposure in a comparison group without disease. In *cohort studies* the disease rate in a population exposed to a suspected toxin is compared with that found in a comparison population without such exposure; most frequently, cohort studies have evaluated the mortality experience of exposed populations.

Case reports and case–control studies attempt to work from an observed health effect (disease or death) to a cause by identifying persons with disease and measuring their exposure. This approach contrasts with that used in cross-sectional and cohort studies, which attempt to proceed from risk factor to outcome, from cause to effect, by defining groups of exposed and nonexposed persons and then measuring the frequency of disease or death (Center for Disease Control, 1979).

Each of these methods has its advantages, and each is subject to bias, varying only in the types of bias possible. Although cohort studies are frequently thought to provide superior data, case–control studies may in some instances constitute a more appropriate study design. The following sections of this chapter outline and illustrate the principles upon which a particular methodological choice may be made. In addition, approaches to the informed evaluation of epidemiological studies are suggested. It is important that investigators report detailed information on study design, data collection, and analytical methods. Too often, such methodological information is not given and, in consequence, adequate evaluation of a study is not possible (Baumgarten and Oseasohn, 1980).

Case Reports

The first recorded case studies of occupational or environmental illness were Bernardino Ramazzini's reports of lung disease among potters and weavers and Percival Pott's observations of scrotal cancer among chimney sweeps (Pott, 1775). Publication of such case reports as these often constitutes the first recognition that a problem of environmentally induced disease exists, and subsequent risk assessment proceeds from this clinical recognition.

In the past 10 years, four case series of clinical neurotoxic disease have been reported in the United States (Landrigan *et al.,* 1980). Such neurological hazards as Kepone(Taylor *et al.,* 1976), Lucel-7 (Horan, 1981), dimethylaminopropionitrile (DMAPN) (Kreiss *et al.,* 1980), and methyl-*n*-butyl ketone (MBK) exposure (Billmaier *et al.,* 1974) all were first described as human case studies by alert physicians who had observed an unusual cluster of diseases in persons found subsequently to have been exposed to these agents.

Many of the known human carcinogens were identified in a similar fashion (Miller, 1978). For example, cases of angiosarcoma of the liver associated with vinyl chloride were first reported by a company physician (Creech and Johnson, 1974). Bis-chloromethyl ether was reported as a suspect cause of lung cancer by a community physician (Figueroa *et al.,* 1973). Brain cancers in petrochemical refinery workers were first noted by an industrial hygienist (Lewin, 1980). Case reports were followed in those instances either by cross-sectional studies, case–control studies, cohort studies, or animal studies, which attempted to confirm or disprove the validity of the initial observation.

In case series, an inference of causal association between the cases and an environmental agent must be based on the plausibility of the following considerations: clustering of the cases in a limited time; relative rarity of the types of diseases observed; history of a common occupational or environmental exposure; or the apparent strength of the association.

The most common uses of case series are hypothesis generation, surveillance, and case registries.

Hypothesis Generation

Common risk factors may be identified accurately in the histories of a series of cases if the relative risk is great enough and the mode of action is biologically plausible. In fact, the evidence for causality may be strong enough at the time of reporting to justify immediate remedial action (e.g., Kepone, Lucel-7). Other instances with less dramatic effects or lower relative risks may require additional confirmation before action is taken to eliminate the suspect causative agent.

Surveillance

The case report has historically been an important surveillance tool, especially for recognition of infectious diseases. A potentially comparable device in occupational epidemiology is disease surveillance based on the workers' compensation system. Unfortunately, however, that system has been weighted toward recognition of occupational injuries to the virtual exclusion of occupational diseases (largely due to the long latency between cause and effect and the lack of insight into the etiology of diseases with multifactorial causes). Workers' compensation data are also less useful than they might be as a source of disease surveillance because of the wide variability in reporting requirements that exists among the states (Berman, 1978).

A more recent use of case reports as a surveillance tool is for the identification of "sentinal health events" (Rutstein *et al.,* 1976)—cases of disease associated with well-characterized causes whose appearance signals a breakdown in mechanisms for disease prevention. This method has been applied with success in the reduction of maternal and infant mortality and recently has been extended to include such occupational and environmental diseases as lead poisoning, silicosis, asbestosis, or coal workers' pneumoconiosis.

Case Registries

Other surveillance systems relying on case reports include case registries, such as the Beryllium Disease Registry maintained for many years by the Harvard School of Public Health (Freiman and Hardy, 1970) and currently by the National Institute for Occupational Safety and Health (NIOSH) (Cone, 1981). Other registries, such as the NIOSH Dioxin Registry (Honchar and Halperin, 1981), contain lists of persons suspected of having been exposed to a toxicant. These exposure registries in effect perform the task of grouping high-risk populations for future epidemiological studies.

A distinct advantage of case reports over most other types of studies is their low cost. Since most of the work of assembling a case series is performed by providers during the normal course of their duties, the cost is generally borne by the medical care system itself.

In addition, a short time lag between identification of cases and dissemination of information is more typical of case reports. Description of case series can be completed and information may be disseminated within weeks to months; in contrast, a delay of 6–18 months is not unusual for more detailed epidemiological studies.

Relying on case reports as an early warning system is less useful when:

- The cases are sporadic and seen by various physicians at various times
- The relative risk is low
- The outcome is a common disease or a symptom with multiple common etiologies (e.g., heart disease or lung cancer)
- The etiology is multifactorial, resulting from the cumulative effect of multiple exposures
- There is a long latency period between exposure and effect, making recognition of such an association less likely (e.g., skin cancer following exposure to coal tar pitch volatiles)
- A continuum of disease and health exists, and no clear distinction between

cases and noncases is possible (e.g., premalignant dysplasia or *in situ* carcinoma)

In addition, case reports can provide only a rough estimate of disease frequency, in that they give no information on the size of the population at risk and thus make it impossible to determine a disease rate. Finally, case reports are difficult to generalize to a population, since the population from which the cases are drawn is usually not well defined. The most common pitfall of case reports is the attempt to combine series from different sources despite varying diagnostic criteria for defining cases and noncases.

The analysis of case report data is inherently limited to crude tabulations of such characteristics as frequencies of symptoms, signs, and laboratory abnormalities and a catalogue of exposure histories. The range of the characteristics of the cases may be determined, and the case definition, usually overly broad at the outset, may be sharpened and narrowed by examination of the characteristics of the more typical cases. This is a crucial step in focusing attention on the real cases and prevents irrelevant and possibly confounding variables from being included in later case definitions.

Analysis of the surveillance types of case studies is somewhat different, but follows the same basic outline. Registries generally have the advantage of larger numbers of cases, but are highly subject to ascertainment bias. The analysis of sentinel health events depends on an exhaustive investigation into the breakdowns of the health system, leading to identification of isolated cases rather than the tabulation of a series of cases.

The important role of case studies is demonstrated by the numerous examples in the occupational and environmental health literature. In fact, most of the older literature consists of case studies, largely of accidental poisonings or acute overexposures to toxic environments. Many of the current occupational and environmental standards in effect are still based on case series rather than on other forms of studies. However, the studies of vinyl chloride demonstrate

> It required two decades for the development of malignant tumors in human beings, when 12 months of experimentation in lower vertebrates could have established the carcinogenicity of vinyl chloride at exposures comparable to those in the working environment. Although a negative experimental result in animals cannot exclude potential carcinogenicity in man, a positive result under these circumstances is an unambiguous warning. The events described [surrounding the investigation of vinyl chloride] suggest that useful toxicologic screening can be carried out in locations other than the workplace, and in species other than man. (Jaeger, 1976)

Case reports have been the major source of index cases of new disease entities, and this role will continue in the future despite stricter premarketing evaluations and increasingly more animal studies. Case reports remain the primary method of surveillance of known occupational diseases and the principal source of hypotheses concerning the etiological factors leading to newly discovered diseases. Except when there are high-risk ratios or rare and severe life-threatening diseases, however, there is an inherently low level of confidence in any assessment of association or causation derived from case reports and there must be confirmation by other means, usually by studying a defined population by a cross-sectional study, comparing exposure histories of cases and controls by a case–control study to estimate relative risk, and establishing a cohort of exposed and unexposed persons to compare disease incidence or to calculate mortality rates.

Cross-Sectional Studies

Cross-sectional studies are the most common type of occupational health study. When case reports, laboratory animal experiments, and methodological articles were excluded, cross-sectional studies made up 66.7% of the total articles in two major occupational health journals in 1977 and 1978 (Baumgarten, 1980).

This type of study involves examination of a sample or of an entire population to evaluate the frequency of both disease and exposure. A defining characteristic of cross-sectional studies is their inability to determine whether the disease or the exposure came first: "In a cross-sectional study, persons are selected irrespective of their exposure or disease status. Exposure and disease are measured essentially at the same point in time. Further, the time sequence between the onset of exposure and the onset of disease cannot be inferred" (Monson, 1980).

Cross-sectional studies (i.e., surveys and prevalence studies) often constitute the first type of follow-up investigation undertaken following reports of newly suspected cases of occupational or environmental disease. Alternatively, a cross-sectional study may be performed to screen for cases of a disease even in the absence of known cases in a population.

These studies often rely on personal interviews or questionnaires to obtain demographic, symptomatic, and exposure data and on clinical evaluations based on physical examinations and laboratory and environmental sampling data to identify the characteristics of the sample population and to quantitate exposure to potential risk factors. Since such evaluations are typically completed in a brief, intensive field study, the time from initiation to publication is usually only weeks to months. Similarly, costs are lower for cross-sectional than for cohort studies because of the short time involved and because the sample size can be limited to that required to provide the necessary power and statistical sensitivity.

Advantages of the cross-sectional study design include the rapid estimation of numerator and denominator values for determining frequency or prevalence rates both of effects and of exposure. A crude dose–response relation may be generated. The cross-sectional survey is often the crucial step in hypothesis generation and is much more powerful than case reports in this function, since it usually involves many more people, including both persons exposed to risk factors in question and those with reduced or nonexistent exposures.

The cross-sectional approach is the basis for most of the screening programs used in company and union health and safety programs to target high-risk groups and to determine the presence of a particular disease or exposure. Preemployment physical examinations or periodic medical surveillance programs such as lead monitoring may also be accomplished by means of cross-sectional studies. Studies of this type do not entail many of the decisions increasingly faced in ongoing cohort investigations, such as determining the point at which to analyze the data or how to notify exposed workers of excess risks. When some risks of exposure are known, it becomes unethical simply to follow groups of exposed workers or other populations without doing everything possible to reduce their exposures and thus prevent any demonstrable effect.

Limitations of the cross-sectional study as a risk-assessment method follow from its nature an as estimator of the point prevalence of disease. By definition, the time sequence of exposure and effect cannot be ascertained with certainty. Prior knowledge and adequate history-taking may make this less of a problem. In occupational health surveys, however, persons with certain illnesses or handicaps may be self-selected or placed out of particular jobs or exposures. These biases can make it difficult to separate cause from effect. Also, when a long latency period exists between exposure and disease (especially for cancer or pneumoconiosis), both dilution of the study population and loss of cases are potential problems. In an industrial plant, new workers with insufficient latency may have been added, and a number of previously exposed workers may no longer work at the site.

Cross-sectional approaches have limited usefulness in cancer studies because of the usual low prevalence of cases. As Alderson (1980) has noted, however, this method becomes appropriate when a screening test indicates that the cases are associated with increased doses of known cancer-causing agents or with increased risk of cancer (e.g., a Pap smear in cervical cancer or sputum cytology after exposure to bis-chloromethyl ether).

It is often extremely difficult to quantify exposure in cross-sectional studies. Past exposure data are usually not available except by examining histories (e.g., length of employment and estimated dose), and it is often difficult to extrapolate current environmental measurements back in time due to numerous factors, such as changes in industrial processes, rate of product generation, engineering control measures, substitution of chemicals, and possible synergistic effects not heretofore identified.

Six common pitfalls may be found in the use of the cross-sectional method:

1. *Selection bias.* There may be a nonrepresentative cross section of the population being surveyed.
2. *Confounding bias.* This may result from a failure to search for historical exposures that may lead to the same outcome (for example, moonshine whiskey ingestion in lead smelter workers).
3. *Inadequate sensitivity* of the tools of the evaluation. The questionnaires, laboratory methods, and environmental methods must be of maximum sensitivity (or lowest false-negative rate) to detect most of the cases. High specificity (defined in terms of a low false-positive rate) is of less concern in cross-sectional studies, and if a tradeoff is required between sensitivity and specificity, it is wise to err in the direction of greater sensitivity.
4. *Lack of standardization.* The standardization of tools used for data collection is an essential and often neglected aspect of cross-sectional studies as well as other study designs. In a recent critique of all types of occupa-

tional medicine studies, Baumgarten and Oseasohn (1980) reported that only 32.1% of diagnostic procedures, 28.6% of environmental measures, and *none* of the interviews or questionnaires were standardized (defined as the "application of procedures in a uniform manner to all subjects in all instances where lack of uniformity of method might be a source of bias").

5. *Inadequate validation.* Only a small percentage of published studies have been validated (defined by Baumgarten and Oseasohn as "any attempt at verifying the collected data by cross-checking, consulting other individuals or records or calibration of equipment with a known standard"). This suggests a serious lack of commitment to ensuring the reliability of these studies and maximizing their effectiveness as tools for risk assessment.
6. *Inadequate description of methodology.* The information needed to judge the adequacy and appropriateness of the tools used in collecting data is often not provided in a published work. In fact, recent critiques of observational studies have called for the publication of questionnaires used in epidemiological studies along with the results or at a minimum making copies available upon request or in depositories in central libraries (Horwitz and Feinstein, 1979).

The data yielded in a cross-sectional study include both subjective and objective information. Subjective data are usually gathered by interview or self-administered questionnaire and include both medical and social histories, history of symptoms, and a history of exposures (according to residence, occupation, or other confounding exposures). Objective data may include demographic data (age, sex, race), biological measures of exposure (blood lead, urinary phenol in benzene workers), environmental measures (air, water, or skin contamination measures), or measures of physiological, pathological, or psychological effects.

Cross-sectional data are usually analyzed with the tools of descriptive epidemiology and characterized by the dimensions of time, place, and person as well as by the various parameters measured. Measures of central tendency (mean and standard deviation) may be calculated for continuous or semicontinuous data (age, length of employment, or quantity/duration of exposure). Prevalence rates for symptoms, items of historical significance, and social and exposure factors may be calculated. The data so analyzed provide the basis for hypothesis generation and for further steps in the risk assessment process. In fact, as Monson (1980) has pointed out, "The data resulting from a cross-sectional study can be treated as data from a cohort study or as data from a case–control study. That is, disease rates can be compared between exposed and nonexposed groups or exposure percentages can be compared between diseased and nondiseased groups."

The following paragraphs describe recent cross-sectional studies conducted at NIOSH, which illustrate the usefulness as well as the pitfalls of the method.

Cincinnati Sewer Workers

Following a cave-in caused by highly acidic industrial sewage, 46 sewer maintenance workers were exposed for several days to a mixture of solvent vapors in the vapor space above the wastewater effluent. Five of the acutely exposed workers were interviewed and symptoms were noted in case reports. McGlothlin *et al.* (1981) reported the results of a follow-up cross-sectional study of symptoms and laboratory abnormalities among the 46 exposed and 19 unexposed workers, indicating that 25 workers had symptoms suggestive of acute solvent exposure. A case–control analysis of the same data implicated occupational exposure on one particular workday as a significant risk factor for those symptoms. In a prospective analysis of the cohort of exposed workers, eight workers showed evidence of acute or delayed damage to the liver and six had evidence of persistent hematological abnormalities (low hematocrits and elevated reticulocyte count). Simultaneous study of a control group of sewer maintenance workers who had not been exposed at the cave-in site did not show similar symptoms or laboratory abnormalities. That difference suggested that the changes observed were most likely due to the relatively acute exposure to solvent vapor and not to the chronic repeated exposures to various industrial chemicals common in the occupation. This combination of cross-sectional study design and multiple analytical methods is common, since despite the largely cross-sectional nature of the data gathering, the knowledge of the time course of exposure and disease is reasonably defined by history of exposure and time of symptom onset.

Hyde Park Landfill

Cross-sectional studies are difficult to perform when exposures are low-level and multiple and when effects are multiple and poorly defined, as the following example illustrates. NIOSH conducted a cross-sectional medical study of 428 persons who lived or worked near the Hyde Park Landfill—a chemical disposal site north of Niagara Falls, New York (Rothenberg, 1981). This site had been used by a chemical manufacturer to dispose of an estimated 80,200 tons of chemical waste, including many chlorinated hydrocarbons. The medical study included an interviewer-administered questionnaire composed of questions excerpted from a nationally standardized health survey, a limited physical examination, and blood/urine testing. Participation rate among the eligible population was 59%. Results showed an 8% rate of miscarriage among 91 pregnancies, compared with the national rate of 14%. No higher-than-expected percentages of laboratory abnormalities were found. Three people working nearby did have detectable serum levels of lindane, one of the pesticides found in the landfill. The investigators concluded that "while the cross-sectional prevalence approach is helpful in identifying existing disease, it is most useful when targeted at a specific health effect or at the effects of a specific exposure rather than at the more diffuse issue of intermittent, relatively low exposures to multiple toxic chemicals" (Rothenberg, 1981).

Michigan Residents Exposed to Polybrominated Biphenyls (PBBs)

Cross-sectional studies may serve as the first step in identifying a cohort of exposed workers and in documenting their baseline health status as a prelude to further follow-up, as shown in Michigan in 1973, when several hundred pounds of PBBs were accidentally introduced into cattle feed in Michigan. Chemical workers, dairy farmers, and milk drinkers were among the groups subsequently exposed. A study was undertaken by Landrigan *et al.* (1979) to determine the health effects of this exposure of farm residents, farm product recipients, and chemical workers exposed. They used personal interviews to determine symptoms, serum PBB levels to quantitate exposure, and immunological studies to quantitate effect. A cross-sectional analysis revealed elevated PBB levels in families living in contaminated farms and in chemical workers. There were, however, no observed relationships between outcomes or symptoms and exposure levels. A close relationship was noted between symptom frequency and mode of entry into the study, volunteers showing highest reported symptom prevalence. The second highest rate of symptoms was found among residents on farms contaminated with PBB at a level insufficient to result in quarantine. No dose–response relationship was noted between exposure level and the frequency of alterations in immunological function. This study demonstrates the difficulties that may be encountered when using the cross-sectional method to study persons with known exposures for unknown effects that most likely are long-term or chronic with undetermined latencies. The study also demonstrates the need for prospective cohort analysis in such a situation and the use of the cross-sectional study as the initiating point of a prospective analysis.

If tools with adequate sensitivity, standardization, and validity are used, and if a reasonably representative sample of a population has been selected, the cross-sectional study is useful for defining the prevalence of a defined problem and for generating hypotheses regarding potential causation. Such hypotheses can then be tested further using either a case–control, proportionate mortality, or cohort approach.

The Proportionate Mortality Study

Cause-specific proportionate mortality is defined as the proportion of deaths due to a specific disease among the total number of deaths in a particular group (MacMahon and Pugh, 1970). A proportionate mortality study can be used to compare the cause-specific proportionate mortality of a study group with that of a comparison group. In the occupational setting, the study group usually consists of deceased employees of a particular plant or members of a trade. The comparison group is usually composed of decedents in the general U.S. population. The cause-specific proportionate mortality of the employee group is compared to that of the comparison group, adjusted for age, sex, race, and calendar period. The comparison takes the form of a ratio, the proportionate mortality ratio, or PMR (Decoufle *et al.*, 1980; Monson, 1980). The statistical significance of the PMR is determined by calculating an approximate p value using the χ^2 statistic (Mantel, 1963) or by calculating an exact p value assuming the observed deaths followed a Poisson distribution (Rothman and Boice, 1979).

The computation of a PMR is illustrated by the following hypothetical example of lung cancer among the former employees of a chemical plant. A search of the plant personnel records led to the ascertainment of 154 deaths among white male former employees. Death certificates were obtained, and the underlying cause of death coded by a trained nosologist. The race, sex, age at death, and date of death were collected for each decedent. Of these deaths, 20 were attributed to lung cancer. The distributions by age and calendar period for lung cancer deaths and for all deaths appear in Table 6-1. United States white male decedents were chosen as the comparison group. On the basis of U.S. mortality data, the proportionate lung cancer mortality for U.S. white males was calculated for each age and calendar period in which at least one death had occurred among the employees of the chemical plant. These proportions appear in Table 6-1, along with the number of lung cancer deaths that would have been expected among the deceased employees if their age- and calendar-specific proportionate mortality for lung cancer had been the same as a comparable group of U.S. white male decedents. The expected number of deaths was computed by multiplying the total number of deaths in each age and calendar period by the corresponding proportionate mortalities for lung cancer of the U.S. white male decedents. The expected deaths of each age and calendar period were summed to produce a total of 9.7 expected lung cancer deaths. The ratio of the observed (20) to the expected deaths (9.7) multiplied by 100 resulted in a PMR for lung cancer of 206 ($p < 0.05$). This would indicate that lung cancer was proportionately in excess among this group of decedents and may represent an occupational health problem.

Although the example emphasizes the computation of a single cause-specific ratio, PMRs for several causes of death are often computed simultaneously during the course of an investigation. This is done by using the same set of deaths due to all causes, but calculating separate cause-specific proportionate mortalities for each of the observed causes of death A selected set of the other cause-specific PMRs for the same hypothetical group of decedents is presented in Table 6-2.

TABLE 6-1. Computation of a Proportionate Mortality Ratio

Age at death (years)	Year of death	Lung cancer deaths: Observed	Lung cancer deaths: Expected	All deaths	Proportionate mortality from lung cancer among U.S. white males
40–44	1960–1964	0	0.8	20	0.04
	1965–1969	0	0.5	10	0.05
	1970–1974	0	0.0	0	0.06
45–49	1960–1964	2	1.0	20	0.05
	1965–1969	2	1.9	32	0.06
	1970–1974	2	1.4	20	0.07
50–54	1960–1964	0	0.6	10	0.06
	1965–1969	6	0.8	12	0.07
	1970–1974	8	2.7	30	0.09
Totals		20[a]	9.7[a]		

[a] The PMR for lung cancer would therefore be calculated as follows: observed/expected × 100, or 20/9.7 × 100 = 206.

TABLE 6-2. An Example of the Results of a Proportionate Mortality Study Where Numerous PMRs Were Computed

Cause of death	Observed		Expected		PMR
All cancer	40		26.7		150
Digestive		12		7.9	151
Liver		1		0.7	142
Lung		20		9.7	206
Brain		1		1.1	91
Other		6		7.3	82
Cerebrovascular	8		9.1		88
Circulatory	62		65.6		95
External	21		23.4		88
All other causes	23		29.2		79
All causes	154		154.0		100

The Case–Comparison Study

In a case–comparison study, the distribution of a risk factor or of a characteristic among a group of individuals who died from the disease being studied is compared to a group without the disease. This type of analysis is sometimes referred to as a case–control study or retrospective study. In the occupational setting, the characteristic of interest could be employment at a particular plant, practice of a particular trade, or presumed exposure to one or more occupational hazards. The cases—individuals who died of the disease being studied—are assembled from a population composed of people with and without the characteristic. The compeers—persons without the disease—are selected from the same population. The relative frequency of distribution of the characteristic in the case and comparison groups is usually evaluated by computing an odds ratio (Fleiss, 1973). The statistical significance of that ratio is determined by calculating an exact p value, assuming the hypergeometric distribution (Rothman and Boice, 1979).

The computation of an odds ratio is illustrated by the following hypothetical example of leukemia deaths among former employees of a tire manufacturing company. A search of company personnel records resulted in the ascertainment of 57 leukemia deaths among former employees. A comparison group of current and former employees who had not died of leukemia was randomly selected from the same set of company personnel records. Two compeers were selected for each case. The work histories of each case and compeer were examined to determine whether or not the individuals had worked in a department where exposure to benzene occurred. On the basis of this characterization, each case and compeer were assigned to one cell of a 2×2 table, as shown in Fig. 6-1. The odds ratio to determine the strength of the association between death due to leukemia and benzene exposure was computed by dividing the product of ad by the product of bc. In this example, the odds ratio was 3.33 ($p \leq 0.05$), indicating an association between leukemia and benzene exposure.

Exposure	Cases Death due to leukemia	Compeers Not known to have had leukemia
Benzene exposure	26 (*a*)	29 (*b*)
No benzene exposure	31 (*c*)	115 (*d*)
	57	144

FIGURE 6-1. An example of the computation of an odds ratio. Odds ratio = $(a \times d)/(b \times c)$ = $(26 \times 115)/(31 \times 29)$ = 3.33.

The Cohort Study

In a cohort mortality study, cause-specific death rates of a population are compared with the corresponding death rates of a comparison population. In the occupational setting, the study population usually consists of employees of a particular plant during a specified time who are presumed to have been free of the disease under study at time of hire. The comparison population is usually the U.S. general population, although other comparison populations may be used. The study population is assembled from company records, and employees are followed to determine their vital status on a specific date. The death rate of the employee population is compared to that of the comparison population, adjusted by age, sex, race, and calendar period. Generally, in the occupational setting, such adjustments are indirect and the comparison takes the form of a standardized mortality ratio, or SMR (Monson, 1980). The statistical significance is determined by calculating an exact p value assuming that the observed deaths follow a Poisson distribution (Rothman and Boice, 1979). This type of study is also known as a retrospective cohort study (Monson, 1980), a prospective study, or a follow-up study.

The computation of an SMR is illustrated by the following hypothetical example of renal cancer among white male employees of a smelter. The study population of all current and former employees hired from 1940 through 1975 was assembled from company personnel records. Selected identifying, demographic, and work history information was collected for each member of the employee population. All the employees were followed up to determine their vital status as of the end of study date. For deceased employees, death certificates were obtained, the cause of death coded by a nosologist, and the date of death determined. A period of risk of dying of renal cancer associated with employment at the smelter was determined for each employee. The periods of risk began on the date of first employment at the smelter and extended through the end of the study period or date of death, whichever occurred first. This concept of period of risk and the associated person-years at risk (PYAR) is illustrated in Fig. 6-2 for a person who began work in 1955 at age 21 and died in 1967. The distribution of

Age	20–24	(3.5)		
	25–29	(1.0)	(4.0)	
	20–34		(1.0)	(2.5)
		1955–1959	1960–1964	1965–1969
			Year	

FIGURE 6-2. Person-years at risk (PYAR) accumulated within 5-year periods by an employee who was hired on July 1, 1955, at the age of 21.5 years and died on July 1, 1967. Total person-years accumulated = 12.0.

the PYAR for the total employee population by age and calendar period is presented in Table 6-3. United States white males were chosen as the comparison population. The renal cancer death rates for U.S. white males were determined for each age and calendar period where any PYAR had been accumulated by the employee population (see Table 6-4). The number of renal cancer deaths that would have been expected among the employee population if their age- and calendar-specific death rates for renal cancer were the same as U.S. white males appears in Table 6-5. The expected deaths were computed by multiplying the PYAR of each age and calendar period by the corresponding U.S. death rate for renal cancer. The expected deaths of each stratum were summed to produce a total of 10.8 expected renal cancer deaths. In follow-up studies of the employee population, 31 deaths due to renal cancer were ascertained. The distribution of the observed deaths by age and calendar period is also presented in Table 6-5. The ratio of observed (31) to expected (10.8) deaths, multiplied by 100, resulted

TABLE 6-3. Distribution of Person-Years at Risk (PYAR) by Age and Calendar Period for a Hypothetical Employee Population

	PYAR, by calendar period					
Age (years)	1950–1954	1955–1959	1960–1964	1965–1969	1970–1975	Total
20–24	6,820.8	3,194.4	924.5	2,031.6	2,807.8	15,779.1
25–29	9,796.0	10,474.2	4,279.4	2,918.5	4,510.6	31,978.7
30–34	8,265.5	12,518.3	10,926.0	5,281.8	3,809.7	40,801.3
35–39	7,241.4	9,859.0	12,704.4	11,113.0	5,599.7	46,517.5
40–44	4,906.0	7,791.9	9,855.0	12,735.4	11,205.3	46,493.6
45–49	3,323.8	5,036.4	7,577.2	9,714.3	12,559.2	38,210.9
50–54	2,444.0	3,263.9	4,865.7	7,177.0	9,233.0	26,983.6
55–59	1,558.4	2,253.1	3,126.0	4,514.9	6,905.6	18,358.0
60–64	959.2	1,420.9	2,133.7	2,734.1	4,048.3	11,296.2
65–69	272.2	924.2	1,247.9	1,721.8	2,235.5	6,401.6
70+	195.6	429.4	1,066.9	1,543.2	2,234.0	5,469.1
Total	45,782.9	57,165.7	58,165.7	61,485.6	65,148.7	228,289.6

TABLE 6-4. Death Rates per 100,000 for Renal Cancer Among U.S. White Males by Age and Calendar Period

	Death rates per 100,000, by calendar period				
Age (years)	1950–1954	1955–1959	1960–1964	1965–1969	1970–1975
20–24	0.117	0.080	0.081	0.088	0.093
25–29	0.192	0.163	0.164	0.119	0.115
30–34	0.332	0.325	0.345	0.275	0.283
35–39	0.805	0.724	0.807	0.869	0.745
40–44	1.738	1.674	1.748	2.010	1.865
45–49	3.447	3.532	3.525	3.852	3.969
50–54	5.782	6.389	6.645	6.881	6.584
55–59	9.439	9.583	9.783	10.658	11.070
60–64	13.343	13.953	13.525	14.474	15.496
65–69	15.733	17.702	18.812	19.200	20.632
70+[a]	16.600	20.100	22.700	25.900	28.200

[a] Approximate value for ages 70–85.

in an SMR of 287 ($p \leq 0.05$) for renal cancer, indicating an association between renal cancer and employment at the plant.

Although the example emphasizes the computation of a single cause-specific SMR, others can be computed on the basis of the PYAR accumulated by the study population. All that is needed are the U.S. death rates for the other causes and the corresponding observed deaths, which would have been identified by fol-

TABLE 6-5. Observed and Expected Deaths Due to Renal Cancer for a Hypothetical Employee Population by Age and Calendar Period[a]

	Observed and expected deaths, by calendar period											
	1950–1954		1955–1959		1960–1964		1965–1969		1970–1974		Total	
Age (years)	Obs	Exp	Obs	Exp	Obs	Exp	Obs	Exp	Obs	Exp	Obs	Exp
20–24	0	0.01	0	0.00	0	0.00	0	0.00	0	0.00	0	0.01
25–29	0	0.02	0	0.02	0	0.01	0	0.00	0	0.01	0	0.06
30–34	0	0.03	0	0.04	0	0.04	0	0.02	0	0.01	0	0.14
35–39	0	0.06	0	0.07	0	0.10	1	0.10	1	0.04	2	0.37
40–44	0	0.09	1	0.13	1	0.17	0	0.26	0	0.21	2	0.86
45–49	0	0.12	2	0.18	0	0.27	0	0.37	1	0.50	3	1.44
50–54	0	0.14	1	0.21	4	0.32	2	0.49	1	0.61	8	1.77
55–59	0	0.15	0	0.22	3	0.31	1	1.48	3	0.76	7	1.92
60–64	0	0.13	1	0.20	2	0.29	3	0.40	2	0.63	8	1.65
65–69	0	0.04	0	0.16	0	0.24	0	0.33	0	0.46	0	0.23
70+	0	0.03	0	0.09	0	0.24	0	0.40	1	0.63	1	1.39
Total	0	0.82	5	1.32	10	1.99	7	2.85	9	3.86	31	10.84

[a] Note: SMR = 31/10.84 × 100 = 286.

TABLE 6-6. Selected SMRs for a Hypothetical Employee Population

Cause of death	Number of deaths: Observed		Number of deaths: Expected		SMR
All cancer	599		517.8		108
Lung		188		167.7	112
Pancreas		26		28.3	92
Kidney		31		10.8	286
Liver		3		4.5	67
Other		311		306.5	101
Cerebrovascular	183		178.1		102
Circulatory	1131		1215.6		93
External	325		275.9		117
All other	542		666.2		81
All causes	2740		2853.6		96

low-up activities. A selected set of cause-specific SMRs for the same hypothetical employee population is presented in Table 6-6.

LIMITATIONS OF EPIDEMIOLOGICAL STUDY DESIGNS

Limitations Common to All Occupational Mortality Studies

Determination of the Cause of Death

In determining the cause of death, examination of hospital records and autopsy reports should result in a more accurate determination than one based entirely on the examination of death certificates. However, such detailed records are often not available for a sizable percentage of many employee groups and for the corresponding comparison groups. Therefore, in most epidemiological studies, the cause of death is defined in terms of information contained on the death certificate and coded to the appropriate revision of the International Classification of Disease (ICD).

Since most mortality studies are based on death certificate information, they are subject to the problems associated with the use of such information, i.e., misclassification of the cause of death. In a study of 48,826 cancer deaths for which hospital records were available, only 65% of the deaths demonstrated agreement between the cause of death on the certificate and that indicated in the medical record (Percy *et al.,* 1981). However, this misclassification was not consistent for all cancer sites. Of the ten leading sites of cancer deaths, more than 80% of the deaths for seven sites (lung, breast, prostate, pancreas, urinary bladder, hemopoietic tissue, and ovary) were confirmed by the hospital records. For others, the confirmation rate was much lower. For deaths due to bone cancer, only 50% were confirmed.

A certain amount of this misclassification is due to difficulties presented by the rules for coding in the ICD and to situations in which the person filling out the death certificate is not familiar with the proper procedures. Despite accurate information regarding the cause of death, if the case is not entered on the certificate in the order specified by the ICD rules, the underlying cause of death might not be correctly coded.

In addition, the coding rules of the ICD are revised approximately every 10 years. Because of changes in disease classification and medical terminology, the codes of one ICD revision are not necessarily compatible with those of a different revision. Therefore, if a study spans more than one revision of the ICD, the causes of deaths must be reclassified so that the codes are compatible. There are specific rules for coding deaths from different eras—another potential for introducing errors into a study.

Despite misclassification, there is ample evidence that death certificate data are valuable for epidemiological purposes. With the exception of diagnostic sensitivity bias, which is discussed below in the context of comparison groups, the misclassification of death certificates should equally affect both the study and comparison groups. In most situations, this potential bias ought then to have a negligible effect on the study conclusions.

Definition of the Exposure

In occupational epidemiological studies, a critical definition is that of exposure to a specific chemical, biological, radioactive, or physical agent. Due to the complexity of most occupational environments, however, job assignment or job location often must serve as a surrogate definition of actual exposure. In other instances, the definition of exposure may be even broader, such as having ever been employed at a particular plant.

In some situations, exposure profiles can be developed on the basis of detailed company personnel records. Since the personnel record was designed for payroll purposes and is ill-suited for any epidemiological uses, this process is tedious. It requires a great deal of assistance from long-term employees, involves events that occurred 20 or 30 years ago, and is often conducted with little or no quantitative industrial hygiene information. Therefore, the extent to which exposure can be accurately determined or quantified is extremely limited. In occupational environments where the subjects have been exposed to multiple compounds, various synergistic effects may exist but are difficult to document through epidemiological study.

Despite those problems, occupational exposure profiles have proved to be useful as measures of exposure dose in occupational health studies. Still, the epidemiologist and the policy-maker must realize the inherent weaknesses of analyses based on retrospective occupational exposure data.

Selection of the Comparison Group

In either the proportionate mortality study or the cohort study, two types of comparison groups—internal or external—can be selected. An internal comparison group consists of a group of plant workers who have not been exposed to the suspect occupational hazard being studied. That group should be similar to the study group with regard to important confounding variables known to be associated with the disease. An external comparison group would consist of some general population group (Monson, 1980). For example, if the study group of a cohort study consisted of all current and former employees of the vinyl chloride department of a large chemical plant, an internal comparison group might be all other current and former employees of the same chemical plant who had never worked in the vinyl chloride area. For the same group of vinyl chloride workers, an external comparison group would be a population such as the U.S. general population or an employee population of a different industry. Similarly, if the study group of a proportionate mortality study consisted of deaths among employees of the vinyl chloride department, an internal comparison group would consist of deaths among other plant employees not exposed to vinyl chloride. The corresponding external comparison group would be decedents of a general population. Either type of comparison group, however, has its limitations.

When an external comparison group is used, there is the possibility of two general biases: a hiring selection bias (Doll, 1958) and a diagnostic sensitivity bias (Greenwald *et al.,* 1981). Hiring selection bias involves the process where only apparently healthy individuals are hired and included in the study group, and those with some health defect that may result in greater mortality are not hired, but are included in the external comparison group. In cohort studies of occupational groups, this bias is called the "healthy worker effect" (HWE), where the overall mortality of the study population is less than that expected on the basis of general population death rates. The HWE tends to affect the SMRs of nonmalignant diseases to a greater extent than malignancies (McMichael, 1976). Although the HWE is defined in terms of cohort studies, the selective hiring of healthy workers probably has an effect on cause-specific PMRs (Monson, 1980).

Diagnostic sensitivity bias refers to a situation in which certain diseases are more likely to be diagnosed among particular employee populations than in the general population. There are at least two underlying mechanisms for such a bias. One mechanism would be the preferential diagnosis of certain diseases given a particular occupation. An example is more complete diagnostic work-up of undifferentiated tumors of the pleura and peritoneum on the suspicion of mesothelioma among former asbestos workers than among persons not so employed. Another mechanism for this type of bias involves a presumed difference in the quality of health care provision between employees of certain companies and the general population (Greenwald, 1981). Presumably, employees covered by comprehensive medical insurance programs would be more likely to have better diagnostic work-ups than the general population. Consequently, rare or difficult-to-diagnose diseases would be reported on the death certificate to a greater extent. Preferential diagnosis, either by more complete medical work-ups or by undue influence of the employment history on the diagnostician, would result in an artifactual increase in the cause-specific mortality of the employee group compared to the general population. In the majority of proportionate and cohort studies, the influence of this bias should be negligible.

When internal comparison groups are used, the major limitation involves the possible absence of employee groups that did not experience the exposure of interest. If the exposure was widespread among the employee population, an appropriate unexposed internal comparison group may not exist. If one does exist, it may not be of sufficient size to produce stable age- and calendar-specific proportionate mortality values (see Table 6-1) or death rates (see Table 6-4).

In case–comparison studies, two types of comparison groups can also be distinguished. These are the comparison groups of plant-based and population-based studies. A plant-based study is one in which the cases and compeers are drawn from a specific plant or employee population. A population-based study is one in which the cases and compeers are drawn from a general population.

The comparison groups of population-based studies are probably also subject to hiring selection bias and diagnostic sensitivity bias. In case–comparison studies where the cause of death is a type of cancer and the exposure is characterized as employment at a particular plant, a hiring selection bias could strengthen any association. This could happen if the comparison group was limited to death due to nonmalignant disease. Persons who die of chronic disabling diseases who never had the opportunity of being hired at the plant would be somewhat overrepresented in the comparison group. The existence of a diagnostic sensitivity bias where a greater percentage of deaths due to a specific cause are diagnosed among cases who were employees of the plant being studied could also strengthen an association.

In reference to Fig. 6-1, the hiring selection bias would tend to increase cell *a* (cases with the exposure) and to reduce cell *c* (cases without the exposure). The diagnostic sensitivity bias would tend to increase cell *d* (compeers without the exposure) and reduce cell *b* (compeers with the exposure). In either instance, the odds ratio would be increased.

Limitations Unique to Each Type of Study Design

The Proportionate Mortality Study

The greatest limitation of a proportionate mortality study is that its measure of effect, the PMR, is based only on information on the deceased members of the group under study. The cause-specific proportionate mortalities of a study group of decedents are compared to those of a comparison group, without any consideration of the overall mortality rate of the populations from which the study and comparison decedents were derived. Therefore, proportionate differences in mortality between the study and comparison groups for a particular disease do not necessarily correspond to actual differences in death rates. This is especially true for situations where the overall mortality of the study group differs markedly from that of the comparison group (Kupper *et al.*, 1978). In practice, however, this property of the study usually does not substantially affect results, since the overall mortality of most occupational groups is slightly less than that of the most often used comparison group—the U.S. general population (Waxweiler and Haring, 1981). There is, however, always the possibility that the overall mortality experiences of the two groups are markedly different and the PMR would not be predictive of the actual mortality due to the cause of death being studied.

Another limitation of proportionate mortality studies is that the percentage for all causes of death must equal 100. Thus, any cause-specific PMR may fluctuate because of the frequency of either the numerator or the denominator, or as a function of their fluctuation with respect to each other. If the proportions of deaths due to one or more causes are greater than the expected values, then the deaths due to different causes must be less than expected to an equivalent extent. Therefore, when the PMR of a particular cause of death is elevated (greater than 100), there is always some doubt as to whether the elevation is due to another cause of death having a PMR of less than 100 (Beaumont *et al.,* 1981).

Contributing also to the potential problems of interpretation of proportionate mortality studies is the fact that very few such studies include all the deaths that occurred among the employee population. This is because proportionate mortality studies are usually conducted when follow-up of all employees is not possible or feasible (Monson, 1980). Ordinarily, the deaths constituting the study group are identified from rosters of deceased employees or deceased union members. If, for some reason, the deaths due to a particular disease had been preferentially included on those rosters, then that disease would have a spuriously elevated PMR. This circumstance would result also in the PMRs for other causes of death being reduced to less than 100 due to the previously described necessity for the total observed and expected deaths to be equal in number. Evidence that certain types of death are preferentially included on both company and union rosters of decedents has been demonstrated by comparisons of proportionate mortality and cohort studies of the same populations (Beaumont *et al.,* 1981; Waxweiler and Haring, 1981). One of the few generalizations that can be made is that this selection bias is less pronounced if the study group of a proportionate mortality study is limited to those decedents who had worked long enough to become vested in the company's employee retirement program (Waxweiler and Haring, 1981).

The Case–Comparison Study

Assembling a group of cases for a case–comparison study may be difficult under some circumstances. Sometimes a case–comparison study is conducted in concert with a standard cohort study to investigate hypotheses that require a detailed work history or exposure information. In these situations, assembling the cases should present little problem. In the cohort study, all deaths (including all those due to the disease being studied) will have been ascertained. If not, the case series is usually assembled by searches of company or union records. If assembled in this fashion, the case series may not be representative of all such deaths of the employee population.

Those deaths missed by not conducting a complete follow-up tend to be those among short-term employees. There are two schools of thought concerning the direction of the potential bias associated with not including these deaths. In at least one situation, short-term employees were determined to have higher mortality than the long-term employees (Fox and Collier, 1976). However, one might question whether the increased mortality among the short-term group was due to intense but short exposures to hazardous materials or to frequent moves from one unskilled job to another. On the basis of the principle of dose–response relation, any association between disease and a particular occupational factor should be strongest among a case group comprising predominantly long-term employees, who should have had greater cumulative exposure.

Although selection of the proper companion group has already been discussed as a potential source of difficulty in interpretation common to all types of studies, there are limitations related to this process that are unique to the case–comparison study. If the cases and compeers are selected from a general population, the potential for introducing serious bias through the comparison group is more likely than in the selection of external control groups for a proportionate mortality or a cohort study. In choosing compeers from a general population, the selection of a limited number with radical differences from the case group in smoking habits, dietary habits, lifestyle, or socioeconomic class could seriously affect the study. Granted, all such individuals in the specific general population are included in the external comparison groups of a proportionate mortality study or a cohort study. The difference, however, is that these unique individuals may constitute a greater percentage of the comparison group in a case–comparison study than in those other types of study.

Another potential limitation of case–comparison studies concerns the selection of compeers matched with certain attributes similar to those of the respective cases in order to control confounding bias. Confounding bias is an attribute of data that is associated with the risk factor, but independently can be a cause of the disease being studied (Monson, 1980). Matching is also useful in increasing the power of a case–comparison study.

Selecting matched compeers is rarely a problem if only one attribute is matched with that of the case. As the number of attributes to be controlled for by matching increases, it often becomes very difficult to find compeers. Most case–comparison studies in the occupational setting incorporate matching for age, sex, and race. In studies where the cases and compeers are drawn from the same plant population, matching criteria may include additional factors, such as calendar period of hire, length of employment, or salaried or blue-collar status.

The Cohort Mortality Study

In a cohort study, the ability to assemble the study population at risk for disease due to a given occupational exposure is essential. Proportionate mortality studies do not require any knowledge of the living members of the population at risk. Case–comparison studies, where the compeers are chosen from the same employee population as the cases, do require the ability to assemble the employee population. If this is not possible, then the study must be modified so the cases and compeers are selected from a specific general population.

This requirement for assembling and studying the mortality of an entire employee population also makes cohort studies very expensive relative to the other types of studies. The substantial costs associated with the cohort study are due to basic data collection and follow-up for each member of the employee population as well as the collection and coding of death information for all deceased employees. The cost of such a study averages between $20 and $25 for each member of the population studied (J. Whalen, National Institute for Occupational Safety and Health, personal communication, 1982).

Applications of Epidemiological Methodologies

Factors to Be Considered in Study Selection

In deciding which type of study to conduct in a given situation, several factors should be considered. The single most important issue is the purpose of the study, i.e., the question to answered. Many considerations can affect the decision, including the extent to which the population at risk can be assembled, the nature of the information regarding the exposure to be investigated, and the time and resources available to conduct the study; and the analytical limitations in the selection of a type of study can vary greatly, depending on the purpose of the study and the circumstances under which it is to be conducted. For this reason, there are situations in which each of the three types of studies is appropriate.

Appropriate Applications for Each Type of Study

The Proportionate Mortality Study. The proportionate mortality study is appropriate in situations when the records necessary to assemble the employee population at risk do not exist. This is especially true if more than one cause of death is to be evaluated among the study group. In such a case, it is simply not possible to conduct a cohort study. Although it may be possible to conduct a case–comparison study, only one cause of death could be examined per study. Even if the investigation were limited to one cause of death, the proportionate mortality study would still be more appropriate. Both types of studies have potential for introducing serious biases; however, critical examination of the numerous cause-specific PMRs resulting from the proportionate mortality study should give some indication of whether or not biases were introduced.

The proportionate mortality study can also be appropriate in preliminary stages of investigation even though it is possible to assemble the population at risk. By conducting a preliminary proportionate mortality study, a large number of specific causes of death can be surveyed relatively quickly and at reasonably low cost. The results of such a study would permit a more informed decision regarding the need, if any, for additional, more extensive studies.

The Case–Comparison Study. The case–comparison study is appropriate when it is necessary to develop extremely detailed information on work history or exposure to investigate associations between suspected occupational hazards and deaths due to a particular cause. In such situations, the case–comparison study would be plant-based and all the cases and compeers employees of the plant being studied. It is assumed that there is an adequate pool of employees from which to select compeers and that it is possible to develop detailed exposure information. Although these associations could be investigated in a cohort study, the number of persons for whom detailed exposure information would have to be collected would be many times higher than that of the corresponding case–comparison study.

In addition to lower cost, there are other advantages to the case–comparison study in this setting. One is that the study group and comparison group would all have been employees of the same plant. Therefore, the potential for hiring selection bias and diagnostic sensitivity bias should be reduced. Another advantage is that it may be possible to collect information about factors known to be associated with the disease of interest, such as smoking histories in lung cancer studies.

The Cohort Study. If there is enough time and money, the cohort study is preferred in two situations: when certain exposures can be easily identified directly from personnel records and when a sufficiently large number of employees have experienced the exposure of interest to ensure detection of the disease under study if it occurred at the expected incidence.

In the first situation, it would be possible to conduct a plant-based case–comparison study if the entire employee population had not been exposed. The cohort study, although more expensive, would be preferable because it is an actual comparison of death rates. In the second situation, a plant-based case–comparison would not be possible due to the inability to select a comparison group of employees without exposure. In either situation, if time or money is not adequate to conduct a cohort study, a proportionate mortality study generally should produce equivalent results, at least for the less common causes of death (Monson, 1980).

CONCLUSION

Risk assessment based on clinical and epidemiological studies is a relatively new art. As such, the guidelines and principles for judging the relative validity of the conclusions drawn from such studies have only recently begun to be enumerated and tested. Nevertheless, several key questions that have been identified need to be asked whenever an observational study is to be evaluated for its potential contribution to the risk assessment process:

1. What is the stated (or unstated) hypothesis tested by the study?
2. What is the overall study design: case report, cross-sectional study, case–control study, cohort study, or experimental study?
3. What are the specific procedures of the study design selected? A flow diagram of the study is often helpful to determine the answer to this question. Such a diagram might include the numbers of the population from which the cases and controls are drawn, the sequence of the study, with numbers participating at each stage, and the numbers of each type of case or other outcome identified.
4. What measures were taken to control for biases, including selection pro-

cedures, confounding variables, unequal ascertainment, latency, and source of funding?

5. Have the study methods been validated and standardized? Were data verified by cross-checking by using alternative sources, and were all subjects evaluated by comparable methods?
6. Have the tables and numbers been checked for internal consistency? Are there missing data or persons not accounted for?
7. What types of analyses are appropriate, given the type of study and nature of the data? Were these applied?
8. What conclusions are justified by the data and analysis? Are the results, especially negative ones, justified by the power of the study? Were the widths of the confidence limits taken into account in the conclusions?

Although discussing case–control studies in particular, Ibrahim and Spitzer (1979) have recently concluded as follows, in words that can be applied to the evaluation of all study types:

> Obviously, policy should not be decided upon the basis of one case–control study even if it was designed to test a hypothesis, much less if it was designed to generate one. Policy formulation requires a number of valid case–control studies which have produced consistent results and which are consistent with biological models. Such studies should preferably be based on different approaches carried out independently, or collaboratively with strict adherence to explicit definitions, assessments, and procedures of data gathering. The evidence thus produced should be helpful for policy-making especially if the issue at stake is of considerable concern to the health of the public *and* if the conduct of cohort or experimental studies is infeasible or impossible. (Ibrahim and Spitzer, 1979)

Chapter 7 examines techniques for assessing data on human health effects from multiple studies.

REFERENCES

Alderson, M. R. 1980. Epidemiological studies of occupational carcinogens. *Arch. Toxicol. Suppl.* **3**:3–12.

Baumgarten, M., and R. Oseasohn. 1980. Studies on occupational health: A critique. *J. Occup. Med.* **22**:171–176.

Beaumont, J. J., T. L. Leet, and A. H. Okun. 1981. Occupational data sets appropriate for proportionate mortality ratio analysis. Pp. 391–412 in R. Peto and M. Schneiderman, eds. *Banbury Report 9. Quantification of Occupational Cancer.* Cold Spring Harbor Laboratory, Cold Spring Harbor, New York.

Berman, D. 1978. *Death on the Job.* Monthly Review Press, New York.

Billmaier, D., N. Allen, B. Craft, N. Williams, S. Epstein, and R. Fontaine. 1974. Peripheral neuropathy in a coated fabrics plant. *J. Occup. Med.* **16**:665–671.

Center for Disease Control. 1979. *Principles of Analytical Epidemiology.* Center for Disease Control, Atlanta, Georgia.

Cone, J. 1981. Sarcoidosis and beryllium disease [letter]. *Ann. Int. Med.* **94**:822.

Creech, J. L., and M. N. Johnson. 1974. Angiosarcoma of liver in the manufacture of polyvinyl chloride. *J. Occup. Med.* **16**: 150–151.

Decoufle, P., T. L. Thomas, and L. W. Pickle. 1980. Comparison of the proportionate mortality ratio and the standardized mortality ratio risk measures. *Am. J. Epidemiol.* **111**:263–269.

Doll, R. 1958. Cancer of the lung and nose in nickel workers. *Br. J. Ind. Med.* **15**:217–223.

Duncan, A. S., G. R. Dunstan, and R. B. Welbourn, eds. 1977. *The Dictionary of Medical Ethics.* Dartoin, Longman, and Dodd, London.

Figueroa, W. G., R. Raszowski, and W. Weiss. 1973. Lung cancer in chloromethyl methyl ether workers. *N. Engl. J. Med.* **228**:1096–1097.

Fleiss, J. L. 1973. *Statistical Methods for Rates and Proportions.* Wiley, New York.

Fox, A. J., and P. F. Collier. 1976. Low mortality rates in industrial cohort studies due to selection for work and survival in industry. *Br. J. Prev. Soc. Med.* **30**:225–230.

Freiman, D. G., and H. L. Hardy. 1970. Beryllium disease: The relation of pulmonary pathology to clinical course and prognosis based on a study of 130 cases from the U.S. Beryllium Case Registry. *Hum. Pathol.* **1**:25–44.

Greenwald, P., B. R. Friedlander, C. E. Lawrence, T. Hearne, and K. Earle. 1981. Diagnostic sensitivity bias—An epidemiologic explanation for an apparent brain tumor excess. *J. Occup. Med.* **23**:690–692.

Honchar, P. A., and W. A. Halperin. 1981. 2,4,5-T, trichlorophenol, and soft tissue sarcomas [letter]. *Lancet* **1**:268.

Horan, J. 1981. Nervous system dysfunction from exposure to 2-t-butylazo-2-hydroxy-5-methylhexane. (Abstract) Epidemic Intelligence Service 30th Annual Conference, April 20–24, 1981. Center for Disease Control, Atlanta, Georgia.

Horwitz, R. I., and A. R. Feinstein. 1979. Methodologic standards and contradictory results in case–control research. *Am. J. Med.* **66**:556–564.

Ibrahim, M. A., and W. O. Spitzer. 1979. The case–control study: The problem and the prospect. *J. Chronic Dis.* **32**:139–144.

Jaeger, R. J. 1976. Toxicology. Public-health rounds at the Harvard School of Public Health: Vinyl chloride: Can the worker be protected? *N. Engl. J. Med.* **294**:653–654.

Kleinbaum, D. G., L. L. Kupper, and H. Morgenstern. 1982. Epidemiologic research: Principles and quantitative methods. Chapter 2 in *Fundamentals of Epidemiologic Research.* Lifetime Learning, Belmont, California.

Kreiss, K., D. H. Wegman, C. A. Niles, M. B. Siroky, R. J. Krans, and R. G. Feldman. 1980. Neurologic dysfunction of the bladder in workers exposed to dimethylaminoproprionitrile. *J. Am. Med. Assoc.* **243**:741–745.

Kupper, L. L., A. J. McMichael, M. J. Symons, and B. M. Most. 1978. On the utility of proportional mortality analysis. *J. Chronic Dis.* **31**:15–22.

Landrigan, P. J., K. R. Wilcox, Jr., J. Silva, Jr., H. E. Humphrey, C. Kauffman, and C. W. Heath, Jr. 1979. Cohort study of Michigan residents exposed to polybrominated biphenyls: Epidemiologic and immunologic findings. *Ann. N.Y. Acad. Sci.* **320**:284–294.

Landrigan, P. J., K. Kreiss, C. Xintaras, R. G. Feldman, and C. W. Heath, Jr. 1980. Clinical epidemiology of occupational neurotoxic disease. *Neurobehav. Toxicol.* **2**:43–48.

Lewin, R. 1980. Government/industry dispute brain tumor risk. *Science* **210**:996–997.

MacMahon, B., and T. F. Pugh. 1970. *Epidemiology: Principles and Methods.* Little, Brown, and Co., Boston.

Mantel, N. 1963. Chi-squares test with one degree of freedom: Extensions of the Mantel–Haenszel procedure. *J. Am. Stat. Assoc.* **58**:690–700.

McGlothlin, J., J. Cone, and A. Lucas. 1981. Health Hazard Evaluation Report HE 81-207-945, Metropolitan Sewer District, Cincinnati, Ohio. National Institute for Occupational Safety and Health, Cincinnati, Ohio.

McMichael, A. J. 1976. Standardized mortality ratios and the "healthy worker effect": Scratching beneath the surface. *J. Occup. Med.* **18**:165–168.

Miller, R. W. 1978. The discovery of human teratogens, carcinogens and mutagens: Lessons for the future. Pp. 101–126 in A. Hollaender and F. J. DeSerres, eds. *Chemical Mutagens—Priorities and Mechanisms for Their Detection,* Volume 5. Plenum Press, New York.

Monson, R. R. 1980. *Occupational Epidemiology.* CRC Press, Boca Raton, Florida.

Percy, C., E. Stanek, and L. Gloeckler. 1981. Accuracy of cancer death certificates and its effect on cancer mortality studies. *Am. J. Public Health* **71**:242–250.

Pott, P. 1775. *Chirugical Observations Relative to the Cataract, the Polypus of the Nose, the Cancer of the Scrotum, the Different Kinds of Ruptures, and the Mortification of the Toes and Feet.* Hawes, Clark, and Collins, London.

Rothman, K. 1976. Causes. *Am. J. Epidemiol.* **104:**587–592.

Rothman, K. J., and J. D. Boice. 1979. *Epidemiologic Analysis with a Programmable Calculator.* NIH Pub. No. 79-1649. U.S. Department of Health, Education, and Welfare, Washington, D.C.

Rothenberg, R. 1981. Morbidity study at a chemical dump—New York. *Morbid. Mortal. Weekly Rep.* **30:**293–294.

Rutstein, D. D., W. Berenberg, T. C. Chalmers, C. G. Child III, A. P. Fishman, and E. B. Perrin. 1976. Measuring the quality of medical care: A clinical method. *N. Engl. J. Med.* **294:**582–588.

Susser, M. 1973. *Causal Thinking in the Health Sciences. Concepts and Strategies of Epidemiology.* Oxford University Press, New York.

Susser, M. 1977. Judgment and causal inference: Criteria in epidemiologic studies. *Am. J. Epidemiol.* **105:**1–15.

Taylor, J. R., *et al.* 1976. Neurologic disorder induced by kepone: Preliminary report [Abstract] *Neurology* **26:**358.

U.S. Department of Health, Education, and Welfare. 1978. Protection of Human Subjects. *Fed. Regist.* **43:**56174–56198.

Waxweiler, R. J., and M. K. Haring. 1981. Comparison of proportionate mortality ratios based on company records and standardized mortality ratios based on completed cohort follow-up. Pp. 379–390 in R. Peto and M. Schneiderman, eds. *Banbury Report 9: Quantification of Occupational Cancer.* Cold Spring Harbor Laboratory, Cold Spring Harbor, New York.

7

Comprehensive Evaluation of Human Data

David G. Hoel and Philip J. Landrigan

Several techniques can be used to perform a comprehensive evaluation of data from multiple studies. Assessment of the quality of information provided by each individual study is, of course, a prerequisite to using the collective approaches; however, it is the simultaneous evaluation of data from multiple, often apparently conflicting studies that constitutes the core of epidemiological risk assessment and forms the basis for the establishment of dose–response relationships between chemical exposure and human disease.

Criteria currently used for the comprehensive evaluation of data from clinical and epidemiological studies evolved in large measure in the 1950s and 1960s during the controversy over the association between cigarette-smoking and cancer of the lung. The following paragraphs contain some brief comments on the nature of the reasoning and on the criteria developed for the assessment of causality in epidemiological studies.

THE NATURE OF EPIDEMIOLOGICAL REASONING

Physicians and epidemiologists have traditionally approached the establishment of causality through a process of inductive reasoning (Kleinbaum *et al.*, 1982; Susser, 1973, 1977). That is to say, they begin with scattered bits of data and develop a hypothesis, which attempts to explain the development of a disease

DAVID G. HOEL • Division of Biometry and Risk Assessment, National Institute of Environmental Health Sciences, Research Triangle Park, North Carolina 27709. PHILIP J. LANDRIGAN • Division of Environmental and Occupational Medicine, Mt. Sinai Medical Center, New York, New York 10029.

by attributing its origin to one or more causes or risk factors. In the simplest situation, a single cause is postulated, i.e., a cause that is sufficient in itself to produce the disease in an exposed person (Rothman, 1976). In other, more complex instances, a hypothesized cause may require the concomitant action of other risk factors to produce an effect, or it may be unable to exert its effect unless certain preexisting conditions are met. For example, toxicologists suggest that certain substances, such as phorbol esters, act as tumor promoters in animals already predisposed to the development of cancer as a result of their previous exposure to a tumor inducer (Weinstein, 1981).

It is important to realize that the establishment of a causal hypothesis is frequently a tenuous process (Susser, 1973). It depends upon imagination, insight, and the ability to reach beyond the data that are immediately at hand. Once a hypothesis has been proposed, it is the work of the epidemiologist to test it and to determine whether it provides an adequate explanation of the disease process. The techniques for the testing of epidemiological hypotheses were reviewed in detail in Chapter 6.

ESTABLISHING CAUSALITY: QUALITATIVE ASPECTS OF DATA EVALUATION

The observational nature of epidemiology has fostered the development of a set of reasonably well-standardized criteria for examining data pertaining to cause and effect (Hill, 1962; Susser, 1973). The use of these criteria serves to reduce uncertainty and imprecision in data analysis and thus "to carry our understanding beyond the level of intuition" (Susser, 1973, p. 162).

The first step in the systematic evaluation of data from a series of human studies is to evaluate individually the results of each separate report. The strengths and weaknesses of each study must be considered along with potential for the existence of bias (Gehlbach, 1982).

Studies that have been reported to show negative results, that is, studies reporting an apparent absence of evidence for a hypothesized causal relationship between exposure and effect, are of particular interest (Hernberg, 1980). Such studies should be analyzed for obvious flaws, such as dilution (the inclusion of unexposed people in an allegedly exposed group of persons), misclassification (Copeland *et al.,* 1977), omissions, or premature examination of subjects for diseases that may have long induction-latency periods. In addition, the statistical power of each negative study should be assessed. The statistical power of a study is the probability that the study will be able to demonstrate the presence of an effect, such as excessive disease or mortality, in a population (Beaumont and Beslow, 1981). A formal statement on statistical power can be found in a study of leukemia mortality among naval shipyard workers exposed to ionizing radiation (Rinskey *et al.,* 1981a). The authors of the study reported explicitly that their work had a 99% power, or probability, of being able to detect a threefold excess of leukemia mortality, had such an excess been present in the study population at the time of the investigation. The absence of an explicit statement regarding statistical power weakens the impact of a negative study.

An excellent summation of the approaches used to evaluate clinical and epidemiological studies has been provided by the International Agency for Research on Cancer (IARC) (1982):

> An analytical study that shows a positive association between an agent and an effect may be interpreted as implying causality to a greater or lesser extent, if the following criteria are met: (a) there is no identifiable positive bias; (b) the possibility of positive confounding has been considered; (c) the association is unlikely to be due to chance alone; (d) the association is strong; and (e) there is a dose–response relationship (International Agency for Research on Cancer, 1982, p. 18).

The IARC stated that negative epidemiological studies, i.e., studies showing no association between the agent and cancer, must be assessed according to these criteria. Moreover, the power of the study must be considered along with the calculation of upper confidence limits on the relative risk. The agency also pointed out that the results of a study apply only to those dose levels and lengths of exposure observed in the study. All these issues determine the degree of confidence one has in the study's negative findings.

After each report has been assessed, one can begin a comprehensive literature review in which published studies describing the health effects due to a particular agent are considered collectively. Guidelines for this exercise were developed by Hill (1965), who recommended that the following six factors be considered in determining whether an observed association represents causality: its strength, consistency, temporal relationship, biological gradient, specificity, and biological plausibility.

The Strength of the Association

The strength of an association between exposure and effect is measured in terms of relative risk. The stronger a measured relative risk, the greater the likelihood that an observed association is causal in nature. Monson (1980), for example, has suggested that relative risk in the range 1.5–3 (or 0.4–0.7) provides evidence of a moderate degree of association. Relative risks above and below this range are correspondingly classified as strong or weak, respectively.

In evaluating relative risks, it is important to note the actual numbers of observed and expected cases. If the expected number is very small, i.e., fewer than two or three cases, then even a very high relative risk should be evaluated with caution (Hernberg, 1980).

It is also necessary to be cognizant of the "healthy worker effect" in evaluating relative risk. This term describes the underestimation of mortality for a working population that typically results from a selection process in which only those members of a population free of obvious health defects and chronic debilitating disease become involved in active work. As a result of the healthy worker effect, the overall relative risk of a working population tends to be below that of the general population, especially in the youngest age groups and in the early years of employment.

Finally, the possible influences of such confounding factors as cigarette-smoking and other occupational exposures need to be considered in any assessment of relative risk in an occupational study. For example, calculations based on the relative risk of lung cancer associated with smoking and on the fraction of smokers in typical working populations indicate that relative risks for lung cancer above 2.0 can seldom be explained by confounding with smoking (Axelson, 1980).

The Consistency of the Association

The case for causal influences is strengthened by repetition of findings "by different persons, in different places, circumstances and times" (Hill, 1965). The reproducibility of findings constitutes one of the strongest arguments for the existence of causality. This is especially true when the studies are conducted under differing conditions.

The Temporal Relationship between Cause and Effect

In making an inference of causality, it is most important to determine that exposure preceded illness. When latency periods are involved, exposures must have occurred early enough to have produced an effect by the time of the study. To make such a judgment, it is reasonable to use the criterion that exposure must have begun at least during the first half of the mean latency period (Hernberg, 1980).

The Biological Gradient of the Association

If a factor is causally related to a disease, the risk of developing the disease should be related positively to the extent and severity of exposure to the factor (Weiss, 1981). Although the criterion of dose response is extremely useful in assessing causality, a confounding factor may be so intermingled with an exposure on some occasions that its effect cannot be separated, even by application of this criterion.

Relative to dose response is the concept that very strong evidence for causality is provided when a change in exposure brings about a change in disease frequency (Hernberg, 1980). The decrease in risk of lung cancer that follows cessation of smoking is a good example of this sort of evidence (Doll and Hill, 1956).

The Specificity of the Association

The term specificity refers to the notion that if a particular exposure is associated with only one disease and vice versa, there exists strong evidence for causality. Examples of such one-to-one correspondence abound in the study of infectious diseases, but are less common in noninfectious diseases, which tend to be of multifactorial origin. Reasonable examples are mesothelioma in persons exposed to asbestos and bladder cancer in persons exposed to certain aromatic amines such as benzidine. Although the presence of specificity seems to imply causality, its absence does not exclude it (Fralick, 1983).

The Biological Plausibility of the Association and Its Concordance with Generally Known Facts about the Disease under Study

Hill (1962, 1965) stated strongly that a proposed causal relationship should not seriously conflict with knowledge of the biology and pathophysiology of a disease under study. Although this criterion is not absolute, there is reason to believe that an epidemiological inference of causality is strengthened by similar findings in experimental studies and by the demonstration of consistency between epidemiological results and biological mechanisms. Exposure to ionizing radiation, for example, causes increased incidence of numerous cancers not only in many, but also in a variety, of animal species. It has also been demonstrated that the ability of ionizing radiation to react with and to alter the structure of chromosomes helps to explain its carcinogenic potential.

QUANTITATIVE ASPECTS OF DATA EVALUATION

The quantitative analysis of epidemiological data is often considered to be the crucial component of quantitative risk assessment. This is the process in which the risk of disease in a population exposed to a toxic agent is related quantitatively to the intensity and duration of exposure (International Agency for Research on Cancer, 1982). In a study of risk assessment, a committee of the National Research Council divided the process into four major steps (National Research Council, 1983). The first is referred to as *hazard identification*—the process of determining whether exposure to an agent can cause an increase of disease. This is basically what we have been discussing under the category of qualitative aspects of data evaluation. Given that a qualitative association has been established, there is often the need to make a quantitative determination of the potential health effects of the agent. This process is defined by the NRC study as *risk characterization*, and it is performed by combining the results of the second and third steps in the risk assessment process, which are *dose–response assessment* and *exposure assessment.*

Dose–response assessment is the characterization of the relationship between the dose of the agent received and the incidence of an adverse health effect in the exposed population. It should take into account the various aspects of the exposure, including intensity and temporal patterns, and should include evaluation of host factors, such as sex and lifestyle.

Exposure assessment is the process of determining the intensity, frequency, and duration of human exposure to the agent under study. It must include all routes of exposure and the size, nature, and classes of the exposed populations.

For *risk characterization,* which combines the dose–response and exposure assessments, models must be used so that study results can be compared quantitatively. It is often necessary to project quantitatively the risks found in an epidemiological study to other populations with different exposure patterns and host factors.

Quantitative estimates can often be misleading, because the degree of uncertainty in the estimates is often not clearly stated. Uncertainty in risk estimates can sometimes be expressed as statistical variability associated with the use of a specific mathematical model. The biological uncertainty associated with the model choice is often nearly impossible to express quantitatively, and as such is usually simply ignored. Unfortunately, the errors associated with an incorrect model choice often make up the major component of the overall source of error in the risk quantification process.

In determining carcinogenesis, epidemiological studies most often depend upon data on past exposures, which are almost always poorly understood. These are often studies of exposure in the workplace, where industrial hygiene has continually improved over time, thereby reducing the exposure of workers. Reconstructing historical exposure patterns is especially difficult, but is crucial to the quantitative risk estimation process. Often, the best that can be done is to provide upper and lower bounds on the exposures at various work sites. However, if these bounds are too broad, such as several orders of magnitude, the quantitative risk estimates will not provide much information beyond the qualitative results of the study.

Dose–response models are also crucial to risk estimation. Linear, nonlinear, and threshold models for dose-rate effects are well understood and have been much debated. The appropriate selection of a model generally cannot be determined from epidemiological studies in themselves, but must depend upon the study of experimental systems in the laboratory.

Less appreciated is the effect of temporal exposure patterns on risk. In chemical carcinogenesis, for example, whether the material is an initiator or promoter has considerable impact on risk after the cessation of exposure (Day and Brown, 1980; Hoel, 1985). That is, the risk associated with a promoter rapidly diminishes when the exposure is curtailed. Since occupational cohorts are often the key study groups in chemical carcinogenesis, these temporal issues are critical. This is especially true in those work forces whose employment histories are remarkably variable.

Epidemiological studies can, at least theoretically, be used individually for a quantitative risk estimation. These individual risk estimates can then be compared or combined quantitatively to produce an overall estimate of risk. Variabilities associated with the individual estimates provide guidance in the relative weight one assigns to a risk estimate from a particular study. For those studies that produce a qualitatively negative result, an upper bound on the risk can be calculated. This upper bound can in turn be compared with the quantitative results from studies that produce a positive effect. By means of this process, one may discover that there is reasonable consistency between studies. This is more meaningful than, for example, a simplistic statement that there were x positive studies and y negative studies.

The International Agency for Research on Cancer (1982) study on benzene and benzidine provides two excellent examples of quantitative analyses of epidemiological studies. For the benzidine analysis, the IARC working group identified only one study, that of Zavon *et al.* (1973), as providing sufficient information for a quantitative analysis of bladder cancer risk from occupational exposure to benzidine. In this study, 13 of 25 exposed men developed bladder tumors after an average latency of 13.5 years. It was estimated that the cumulative incidence rates were 25 and 75% after 15 and 20 years of exposure, respectively.

The issue of model selection becomes critical when one considers extrapolating these results of Zavon and colleagues to an average working lifetime of 45 years. For example, if one reduced the risk linearly by assuming an exposure of 1/100 so that the cumulative incidence would be 0.25% after 15 years, then a projected lifetime (45-year) incidence would be either 0.75 or 60%, depending upon whether one chooses a linear model or an exponential model for time effects. Work-site exposure measurements varied by two or three orders of magnitude. From this example, we can appreciate the difficulty in providing precision in any quantitative risk estimate for bladder cancer and benzidine exposure.

The second IARC example was concerned with leukemia deaths and benzene exposure. The main body of data used for quantification is attributed to Rinskey *et al.* (1981b), who studied seven leukemia deaths among 748 rubber hydrochloride workers. This study again illustrates the issue of temporal effects, in that the relative risk of leukemia was estimated to be 5.6. If the short-term employees are not included, then among those workers with at least 5 years of exposure, the relative risk becomes 21. No data are available to indicate whether high-dose-rate to low-dose-rate effects are linear or nonlinear. Using reasonably good exposure data, the IARC working group calculated lower bounds to leukemia risk for working lifetime benzene exposures. These risk estimates were found to be consistent with the quantitative results of several other epidemiological studies also concerned with benzene.

The IARC working group found that its risk estimation for benzene and benzidine was more feasible than expected. Its members were impressed that a large amount of quantitative information could be obtained from the key published studies. Their general conclusion, however, concerning this effort was stated as follows:

> First, the review of the literature has underscored the relative inadequacy of most published reports for quantitative risk estimation. Too few reports give sufficient detail on methods of exposure measurement, on variations in individual exposure over time, or on epidemiological methods. Clearly, it will be necessary in future epidemiological studies to collect the data which will permit more confident and complete estimation of quantitative risk. No longer can merely quantitative assessment of the carcinogenic potential of chemical agents be considered adequate (International Agency for Research on Cancer, 1982, p. 397).

Clearly, the IARC working group recognized the value of quantitative methods in the evaluation of data on humans. The two examples of benzene and benzidine illustrate the need for the collection of appropriate data in studies directed toward quantification of risk as well as the importance of various model assumptions.

A very interesting and sophisticated example of the use of quantitative models in epidemiology was reported by Pike and Henderson (1981). They were concerned with lung cancer risk resulting from exposure to polycyclic aromatic hydrocarbons (PAH) in air pollution. In their approach, they used benzo(*a*)pyrene (BP) as a surrogate indicator of PAH, both from an exposure basis and as an indicator of carcinogenic potency. As a model for cancer incidence they applied the multistage model and used data from lung cancer and cigarette exposure studies. The multistage model for cancer reflects the importance of both exposure duration and the rate of exposure to the carcinogen. Specifically, they assumed the incidence rate to be proportional to the daily rate of cigarette consumption and to the 4.5 power of smoking duration.

To determine the effect of BP levels on lung cancer, Pike and Henderson used the results of an epidemiological investigation by Doll *et al.* (1972), who studied carbonization workers in several British gasworks. Applying the multistage model to these data, they came to the conclusion that an air pollution BP level of 15 ng/m^3 is equivalent to a dose rate of one U.S. cigarette per day. The quantitative results derived from the Doll study corresponded very closely to those obtained by Lloyd (1971) in a study of coke-oven workers in the United States. The implication of this model is that the median value for urban air concentration (6 ng/m^3) should result in a relative incidence of lung cancer twice that found in rural areas (0.4 ng/m^3), which is approximately the concentration found by Haenszel and Täuber (1964) from a sample of nonsmokers in the United States during the late 1950s. A second quantitative conclusion reached by the authors was that a BP level of 1 ng/m^3 gave a lifetime lung cancer risk of slightly more than 1/1500. In this work, Pike and Henderson brought together several epidemiological studies with some very reasonable mathematical arguments to produce quantitative estimates of lung cancer risk from air pollution.

CONCLUSIONS

Despite the observational nature of epidemiology, it appears that reasonable criteria have been developed for the qualitative establishment of causality in a study. Moving from qualitative to quantitative assessment is a less certain process. Criteria are developing with regard to the quantitative analysis of individual studies as reflected through power calculations, dose–response determinations, and model refinements. The use of quantitative evaluation has two great benefits: first, it provides a more satisfying method than a purely qualitative approach for comparing and contrasting individual studies on the same agent. Second, there is a growing need for quantitative risk estimates of adverse health effects, and this need can best be met through the careful combination of epidemiological studies with mathematical models. To ensure the future development of the quantitative analysis of human studies, it is hoped that epidemiologists will follow the caveat of the IARC working group in its plea for the collection of specific types of data necessary for the application of quantitative methods. Finally, it is necessary that the quantitative methods themselves continue to be developed and refined. Much of this work will depend upon experimental results obtained from the laboratory.

REFERENCES

Axelson, O. 1980. Aspects of confounding and effect modification in the assessment of occupational cancer risk. *J. Toxicol. Environ. Health* **6**:1127–1131.

Beaumont, J. J., and N. J. Breslow. 1981. Power consideration in epidemiologic studies of vinyl chloride workers. *Am. J. Epidemiol.* **114**:725–734.

Copeland, K. T., H. Checkoway, A. J. McMichael, and R. H. Holbrook. 1977. Bias due to misclassification in the estimation of relative risk. *Am. J. Epidemiol.* **105**:488–494.

Day, N. E., and C. C. Brown. 1980. Multistage models and primary prevention of cancer. *J. Natl. Cancer Inst.* **64**:977–989.

Doll, R., and A. B. Hill. 1956. Lung cancer and other causes of death in relation to smoking—A second report on the mortality of British doctors. *Br. Med. J.* **2**:1071–1077.

Doll, R., M. P. Vessey, R. W. R. Beasley, A. R. Buckley, E. C. Fear, R. E. W. Fisher, E. J. Gammon, W. Gunn, G. O. Hughes, K. Lee, and B. Norman-Smith. 1972. Mortality of gasworkers: Final report of a prospective study. *Br. J. Ind. Med.* **29**:394–406.

Fralick, R. A. 1983. The role of epidemiology in defining an occupational carcinogen. *Occup. Health Ontario* **4**:2–16.

Gehlbach, S. H. 1982. *Interpreting the Medical Literature: A Clinician's Guide.* D.C. Health and Company, Lexington, Massachusetts.

Haenszel, W., and K. E. Täuber. 1964. Lung cancer mortality as related to residence and smoking histories. II. White females. *J. Natl. Cancer Instl.* **32**:803–838.

Hernberg, S. 1980. Evaluation of epidemiologic studies in assessing the long-term effects of occupational noxious agents. *Scand. J. Work Environ. Health* **6**:163–169.

Hill, A. B. 1962. *Statistical Methods in Clinical and Preventive Medicine.* Oxford University Press, New York.

Hill, A. B. 1965. The environment and disease. Association or causation? *Proc. R. Soc. Med.* **58**:295–300.

Hoel, D. G. 1985. The impact of occupational exposure patterns on quantitative risk estimation. In *Banbury Report 19, Risk Quantitation and Regulatory Policy.* Cold Spring Harbor Laboratory, Cold Spring Harbor, New York.

International Agency for Research on Cancer. 1982. *Some Industrial Chemicals and Dyestuffs.* IARC Monograph on the Evaluation of Carcinogenic Risk of Chemicals to Humans, Volume 29. Lyon.

Kleinbaum, D. G., L. L. Kupper, and H. Morganstern. 1982. Epidemiologic research. Principles and quantitative methods. Chapter 2 in *Fundamentals of Epidemiologic Research.* Lifetime Learning, Belmont, California.

Lloyd, J. W. 1971. Long-term mortality of steelworkers. V. Respiratory cancer in coke plant workers. *J. Occup. Med.* **13**:53–68.

Monson, R. R. 1980. *Occupational Epidemiology.* CRC Press, Boca Raton, Florida 219 pp.

National Research Council. 1983. *Risk Assessment in the Federal Government: Managing the Process.* National Academy Press, Washington, D.C.

Pike, M. C., and B. E. Henderson. 1981. Epidemiology of polycyclic hydrocarbons: Quantifying the cancer risk from cigarette smoking and air pollution. Pp. 317–334 in H. V. Gelboin and P. O. P. Ts'o, eds. *Polycyclic Hydrocarbons and Cancer,* Volume 3. Academic Press, New York.

Rinskey, R. A., R. D. Zumwalde, R. J. Waxweiler, W. E. Murray, Jr., P. J. Bierbaum, P. J. Landrigan, M. Terpilak, and C. Cox. 1981a. Cancer mortality at a naval nuclear shipyard. *Lancet* **1**:231–235.

Rinskey, R. A., R. J. Young, and A. B. Smith. 1981b. Leukemia in benzene workers. *Am. J. Ind. Med.* **2**:217–245.

Rothman, K. 1976. Causes. *Am. J. Epidemiol.* **104**:587–592.

Susser, M. 1973. *Causal Thinking in the Health Sciences: Concepts and Strategies of Epidemiology.* Oxford University Press, New York.

Susser, M. 1977. Judgment and causal inference: Criteria in epidemiologic studies. *Am. J. Epidemiol.* **105**:1–15.

Weinstein, I. B. 1981. Current concepts and controversies in chemical carcinogenesis. *J. Supramol. Struct. Cell. Biochem.* **17**:99–120.

Weiss, N. S. 1981. Inferring causal relationships: Elaboration of the criterion of "dose-response." *Am. J. Epidemiol.* **113**:487–490.

Zavon, M. R., W. Hoegg, and E. Bingham. 1973. Benzidine exposure as a cause of bladder tumors. *Arch. Environ. Health* **27**:1–7.

III

Interpretation of *in Vivo* Experimental Data for Evaluation of Hazards to Humans

8

Selection of Animal Models for Data Interpretation

Robert E. Menzer

Humans are exposed to chemicals in a variety of settings and through a variety of routes as they proceed through daily routines of work, recreation, relaxation, and maintenance of bodily condition. Exposure to chemicals in the modern world is unavoidable, and indeed modern society would not exist in its present form without the chemicals upon which we have come to depend for the maintenance of health, the provision of food, the production and maintenance of clothing, and the provision of the myriad of consumer products that maintain the comfortable lifestyles to which we have become accustomed. We must realize, however, that the chemicals that are responsible for our high standards of living also have the capacity to exert effects beyond those that are desired and anticipated. It is these effects that concern toxicologists as they seek to assess the benefits of chemicals versus the risks of undesirable side effects inherent in their use.

Human contacts with chemicals run the full gamut from single, short-term exposures to constant exposures over a very long period, perhaps as long as an entire lifetime or even beyond into succeeding generations. Routes of exposure include intentional ingestion of chemicals in the form of drugs, unintentional ingestion of food additives and pesticide residues, and inhalation of air pollutants of a great variety of chemical types, from nicotine and tars in cigarette smoke to the variety of gases produced by industrial and automotive emissions. Consideration must also be given to dermal exposure to the chemicals in cosmetics, the many chemicals used in the production of totally synthetic fibers and fibers from both animal and plant origin used to make and maintain our clothing, and atmospheric chemicals, which are absorbed systemically. Occasionally, humans are

ROBERT E. MENZER • Department of Entomology, University of Maryland, College Park, Maryland 20742-5575.

exposed by direct injection of chemicals into various parts of the body to maintain and improve health or of drugs of abuse.

The benefits and risks of chemicals can only rarely be assessed directly in humans. Toxicologists must depend on information obtained indirectly from other species, a process that is accompanied by problems of extrapolating the results from the test animals to the subjects of primary concern—humans. Generally speaking, the more data available on the widest possible variety of organisms, the better will be the confidence of the evaluator that the necessary extrapolations led to valid conclusions about risk. In practical terms, however, the accumulated data on a single compound from a variety of test organisms will seldom lead to definitive, clear-cut information about the effects of exposure of humans to that chemical. Hence, the problem confronting toxicologists is how to provide reasonable evaluations of human risks from information on exposures of a variety of laboratory animals.

The first and most important consideration is the nature and condition of the animal model, which serves as the principal tool of the toxicologist. The quality of the data obtained and the accuracy and value of the decisions based on them are directly related to the validity of the selection of the animal model and the condition of the animals themselves. Healthy, uniform animals are normally desired for toxicity testing and research unless there is a specified, known pathological condition required by the nature of the experiment. Much effort must be devoted to the production of the animals required for experimental research.

The most commonly used laboratory mammals are mice, rats, hamsters, gerbils, guinea pigs, rabbits, cats, dogs, and certain nonhuman primates. Frequently used birds include chickens, quail, pigeons, and turkeys. Certain amphibians, reptiles, and fish can also provide information related to their unique physiological adaptations to their living environments (Arrington, 1978). Data from more esoteric species occasionally shed light on specific unusual chemicals for unusual pathological conditions. For example, armadillos have been useful in the study of leprosy. Statistics on the quantities of laboratory animals used in the United States are presented in Table 8-1. Although such numbers are nearly impossible to obtain with any accuracy, these estimates serve to illustrate the magnitude of animal use in research (Arrington, 1978). Biological data for commonly used laboratory animals are summarized in Table 8-2.

Toxicologists have generally agreed on certain principles to govern experimental research designed to gather the data needed to predict the degree of safety of chemical substances to humans. The codification of these principles by Weil (1972) is the informally accepted standard for the selection of animal models and the conduct of experiments. One principle deals with the selection of animal models:

Use, wherever practical or possible, one or more species whose biological response to the test compound is qualitatively and/or quantitatively as similar as possible to that of humans. For this, metabolism, absorption, excretion, storage, and other physiological effects might be considered.

Other principles specify that multiple dose levels should be administered so

TABLE 8-1. Annual Use of Laboratory Animals[a]

Animal	Number used (thousands)
Mice	30,000
Rats	10,000
Hamsters	500
Guinea pigs	600
Gerbils	40
Rabbits	500
Cats	100
Dogs	200
Nonhuman primates	50
Chickens	1,500
Pigeons	30
Turkeys	60
Quail	80
Reptiles	100
Amphibians	1,500
Fish	300

[a] From Arrington (1978).

that the dose–response relationship will be established, that a threshold level can therefore be set by the use of a proper safety factor and competent scientific judgment, and that the statistical techniques used to establish the significance of data are valid only when animals have been mathematically randomized among dosed and control groups. Another principle advocated by Weil (1972) emphasizes the importance of route of administration. That is, *effects obtained in test animals by one route of administration are not a priori applicable to effects by another route of administration. The routes of administration selected for test animals should therefore be the same as those to which humans will be exposed. Thus, for example, food additives for humans should be tested by admixture of the material in the diet of animals.*

The various regulatory agencies responsible for the assessment of chemical safety to humans have generally used guidelines for data requirements that embody these principles. Although details may vary, the extent and quantity of data needed have been obtained within the parameters established by the acceptance of these principles by toxicologists.

In a provocative treatise, Kaiser (1980) makes the point that invertebrates are the future source of animal models for the experimental toxicologist. Invertebrates comprise approximately 90% of the species inhabiting the earth. Thus, they represent an important animal resource as well as an important component of the total environment for consideration of the benefits and risks of chemicals. The disadvantage of using invertebrates, of course, is their dissimilarity to humans. Despite this disadvantage, however, many biochemical and physiological advances have been made by using invertebrate models for research. Among

TABLE 8-2. Biological and Physiological Data for Laboratory Animals[a]

Animal	Mature weight (g)		Birth weight (g)	Weaning weight (g)	Weaning age (days)	Average litter size	Average life span (years)	Daily food (g)	Daily water (ml)	Daily urine (ml)	Rectal temperature (°C)	Heart rate (beats/min)[b]	Respiratory rate (breaths/min)[b]	Blood pressure (mg Hg)	
	Male	Female												Systolic	Diastolic
Mouse	20–40	18–35	1–1.5	10–12	18–21	10–12	1–2	4–5	4–7	1–2	36.5	600	163	113	81
Rat	300–450	250–330	5–6	35–45	21–23	9–11	2–3	12–15	25–30	11–15	37.5	380	90	116	90
Hamster	90–125	95–140	2	35–40	21–23	6–8	2–3	10–12	8–10	—	37.4	380–412	74	—	—
Gerbil	80–90	70–80	2.5–3.0	16–20	22–24	5	2–4	7–9	4	—	37	—	—	—	—
Guinea pig	900–1,500	700–1,300	75–100	200–250	14–21	3	6	40–50	80–100	—	38.5	230–280	82–90	77	47
New Zealand rabbit	4,000–5,000	4,500–5,500	64	1,600	42–56	8	6–7	130–170	300	—	39.2	250	53	110	80
Cat	2,000–4,000	2,000–4,000	125	800	42–56	4	13–17	150–200	300	100	38.6	130	25–35	155	100
Beagle dog	13,000–18,000	13,000–16,000	30–400	2,000–3,000	42–56	4–6	13–17	300–500	350	—	38.8	80	20	148	100
Rhesus monkey	9,000–13,000	5,000–11,000	500–700	900	90–180	1	15	150–250	450	250	38.8	193–260	40–55	159	127

[a] From Arrington (1978).
[b] Average or range of averages.

the advantages of using invertebrates for predictive toxicology, the following have been listed by Kaiser (1980):

1. Increased experimental scope. The inclusion of factors not present in vertebrates may prove useful in the illumination of questions about vertebrates. The ability of tissues or even whole organisms to regenerate lost or damaged parts is, for example, not present in vertebrates to the extent that it exists in the coelenterata. The study of this process may contribute to our understanding of the regenerative process in mammals.
2. Rapid reproduction and numerous offspring of invertebrates. Relative toxicity and other factors can be compared using large numbers of individuals in bioassays.
3. Flexibility afforded by use of short-lived species. Experiments can be conducted in a shorter time with a large, genetically homogeneous population.
4. Frequent occurrence of special characteristics in invertebrates limited to a small number of species. Because of their large populations and wide range of variability, special features appear more often than in mammals. Low-frequency mutations, for example, may be more readily detected in *Drosophila,* which has extremely high reproductive capability, than in a mammalian population.
5. Particular problems may be studied more simply in an invertebrate than in a vertebrate. For example, it is easier to study ameboid movement in ameba than in humans, where it is present in the white blood cells.

Although one would not propose replacing mammalian animal models completely by invertebrates, they may be a valuable adjunct to research conducted with higher animals and may assist in the use and interpretation of the data available from such studies.

Many writers (e.g., Dixon, 1976; Stara and Kello, 1979) have identified and discussed the problem of extrapolating to humans the results of research using homogeneous, standardized populations of animals. As expressed by Rall (1978), "The human population is different. The mouse doesn't smoke or breathe hydrocarbons or sulfur oxides from fossil fuels, doesn't drink, doesn't take medicine, doesn't eat bacon or smoked salmon, but man does."

The main purpose of toxicity testing using laboratory animals is to provide a data base that can be used to assess the risk or hazard of an adverse health effect associated with an exposure situation in which the chemical agent, the subject, and the exposure conditions are defined (Klaassen, 1986). In the past, these data have frequently been collected by using only laboratory animals under standard laboratory conditions without varying parameters of age, sex, genetic composition, nutrition, and disease states. Thus, extrapolations from laboratory animals to humans have not been made with comparable conditions between the test animals and the potential subjects of exposure. Furthermore, there have been occasional attempts to extrapolate from one mode of chemical exposure to another, thus further compounding difficulties of extrapolation. Finally, most experiments with chemicals use the pure test compound, without interference from formulation chemicals or the possibility of interactions with other chemicals to which a subject will be simultaneously exposed. Thus, risk–benefit analysis for a chemical agent is frequently confounded by violations of all three defined conditions in the establishment of the data base. The importance of considering all aspects of the animal models used in risk assessment is illustrated in the following sections of this chapter.

HEALTHY VERSUS DISEASED ANIMALS

Although it is an overstatement to say that toxicity is never tested in diseased animals (Menzel, 1979), it is certainly true that there are very few published reports of the testing of drugs or other chemicals in animals with preexisting disease conditions, except for circumstances where drugs are being specifically screened for their therapeutic effect on a particular disease state. Clearly, it would be a herculean task to include the parameter of disease state in all testing protocols for the toxicology of chemicals. However, the present rudimentary knowledge of the effect of disease states on toxicity demands at least some investigation of the possible exacerbation of toxicological effects by this variable.

One of the few documented examples of the positive interaction between disease and chemical toxicity is given by the testing of chemicals in alloxan-induced diabetic rats. A synergistic interaction between diabetic nephrotoxicity and both sodium arsenate toxicity and chronic mercury poisoning has been demonstrated in rats (Schiller *et al.,* 1978, 1981; Uvarenko, 1978). Effects at the molecular level demonstrated in Schiller's arsenate study included changes in the urinary levels of a number of enzymes; and at the cellular level in Uvarenko's mercury study, there were changes in both alpha and beta cells of Langerhans islets. Other investigators, such as Cook and Past (1979) and Wolfe and Schnell (1979), have shown that diabetes in the male rat affects hepatic drug metabolism. Thus, in diabetics one would expect alterations in the toxicity of any chemical that depends on microsomal enzyme alteration for its activity, or that is detoxified by this system. There are many other examples in which alloxan-induced diabetes has been observed to alter the effects of various drugs in laboratory animals (e.g., Essman, 1978; Hoult and Moore, 1980; Macleod and McNeill, 1981; Marshall *et al.,* 1978; Nakama *et al.,* 1978).

The monumental task of evaluating all possible combinations of chemical toxicant exposures and disease states in humans may be minimized by careful analysis of the possible biochemical and physiological effects of disease that might cause great changes in the toxicity of chemicals in humans. Particularly when a chemical's toxicity depends on a metabolic change in the organism or a physiological interaction with certain kinds of macromolecules or cells, specific test results of the chemical in an animal model that possesses the appropriate enzymes, macromolecules, or cells should be considered. This suggestion, of course, introduces the additional difficulty that for many human disease conditions specific comparable animal models do not exist. An extensive compilation and analysis of spontaneous animal models of human disease has been presented by Andrew *et al.* (1979).

YOUNG, MATURE, OR OLD ANIMALS

It has been evident for many years that immature and mature animals respond differently to chemicals. Recently, it has been demonstrated, however, that there are also differences between mature and aged individuals. In both instances, variations in toxicity seem to be related mostly to changes in the hepatic detoxification processes. Thus, earlier protocols, which included both mature and immature animals, now may also need to consider senescing animals in order to encompass the full range of possible variations in effects over the entire lifespan of the animal.

The importance of the age consideration in toxicological evaluation is illustrated by the cases of nitrate poisoning in children in the United States. Methemoglobinemia in humans is caused by high levels of nitrite or, indirectly, nitrate, which interferes with the oxygen transport system of the blood. Poisoning of infants from nitrate in well water was first reported in 1944. Cases numbering in the thousands have now been reported, mostly in rural areas, mostly involving poisonings in infants. Although the precise mechanism for high infant susceptibility to nitrate poisoning is not known, it has been deduced that susceptibility of infants to upset stomachs and achlorhydria results in increased stomach acidity, which allows nitrate-reducing organisms to enter and reduce nitrates to nitrites. Immature enzyme systems may also be of importance. The enzymatic capacity of infant erythrocytes to reduce methemoglobin to hemoglobin appears to be less than that of adults. Furthermore, fetal hemoglobin is oxidized by nitrite to methemoglobin at a rate twice as rapid as adult hemoglobin. Thus, several clinical, physiological, and metabolic factors seen to predispose infants to development of methemoglobinemia and acute nitrate poisoning (Ridder and Oehme, 1974). In general, children are known to be 3–5 times more sensitive to nitrates than are adults (Krasovskii, 1976).

The principal xenobiotic metabolizing mechanism in most animals is the hepatic cytochrome P450 monooxygenase system. In humans, monooxygenase activity may be detected in liver preparations by the sixth week of gestation and increases to a level that is approximately 20–40% of comparative adult values by midgestation. Laboratory animals have very low levels of hepatic monooxygenase activity by midgestation, but after birth the levels increase substantially and rapidly (Neims *et al.*, 1976). During postnatal maturation, it is not unusual to observe levels of hepatic monooxygenase activity that exceed subsequent adult values by 2–3 times. After puberty, activity toward some substrates decreases. Perhaps most difficult for the interpretability of data is the fact that the development of hepatic monooxygenase activity for different substrates varies within the maturation periods of different animal species. Thus, one can neither easily extrapolate from one species to another nor draw conclusions for one chemical based on another used as a substrate in *in vitro* assays of hepatic cytochrome P450 monooxygenase.

At the opposite end of the spectrum, the toxicity of chemical agents in aged animals must also be considered. One pertinent illustration is *itai-itai* disease, an occupational or environmental disease that was prevalent in Japan during the first half of the 20th century. Its etiology is still not certain, but it appears to have been at least associated with dietary exposure to cadmium and with low calcium and vitamin D deficiency. It was characterized by osteomalacia, proteinuria, and glycosuria and struck only elderly women who had borne many children, indicating an association with the stress of pregnancy and lactation (Fassett, 1975).

The high incidence of *itai-itai* disease in certain areas of Japan stimulated health authorities to undertake extensive epidemiological studies of victims. They have demonstrated that the disease was definitely environmental, not hereditary, and that those affected seem to have been located in areas where the daily intake of cadmium exceeded 300 μg. The source of the cadmium was apparently mine drainage, leading to contamination of paddy soils where rice was cultivated. Although the association of cadmium with *itai-itai* disease is strongly presumed, it is not the single causative agent, since other factors seem also to have been necessary for its development in aged women (Friberg *et al.,* 1974). The disease now seems to have disappeared (Fassett, 1975).

There is very little information on the effect of aging on the response of organisms to toxicants. Certainly, the history of *itai-itai* disease indicates that considerable research activity is needed to determine the effect of aging on the biochemistry and physiology of organisms, especially humans and laboratory animals. Highest priority for studies in this area should be given to the hepatic cytochrome P450 monooxygenase system, since it is so important to the development of toxicity and detoxification processes.

NUTRITIONAL STATUS OF ANIMALS

Nearly every known nutrient or nutritional conditioning has been shown to be capable of modifying the tolerance of animals to the biological activity of foreign compounds (Campbell, 1976). Thus, data should not be extrapolated indiscriminately from the well-nourished laboratory animal to the malnourished human without knowledge of how various nutritional deficiencies might alter toxic responses. Unfortunately, in the usual toxicological testing little effort is made to control the nutritional status of laboratory animals beyond that required for normal growth and development. Standard, commercially available laboratory chows provide a balanced diet designed to rear and maintain animals in optimal conditions of health. Therefore, few animals reared for experimental use, including tests to satisfy protocols for registration or approval of drugs or other chemicals, are intentionally rendered deficient or even marginal in vitamins or minerals or other conditions of malnutrition known to be limiting in human populations. This situation is surprising in view of the literally hundreds of studies demonstrating marked effects of nutritional factors on the responses of animals to toxic agents of all kinds. Particularly in the areas of vitamin deficiency, protein malnutrition, and trace metal and mineral imbalances, there are many reports that the response of animals to toxicants has been altered, in some cases by several orders of magnitude.

Particularly striking is the evidence for enhanced susceptiblity to chemical carcinogenesis in animals deficient in dietary vitamins. For example, vitamin A deficiency disposes rats to cancer of the respiratory system, bladder, and colon (Sporn *et al.,* 1976). Retinoids (synthetic vitamin A analogs) have been shown to prevent chemical carcinogenesis in some animal models. They are believed to act in this way because of their essential role in controlling normal differentiation of many epithelial tissues. Thiamine deficiency in the rat markedly stimulates the metabolism and excretion of dimethylnitrosamine, thus altering the sensitivity of the rat to an acutely toxic dose of the compound and, presumably, to its carcinogenic effects as well (Ruchirawat *et al.,* 1978).

Deficiencies of vitamins A and E, panthothenic acid, riboflavin, and pteroylglutamic acid have been shown to result in teratogenic effects in rats (Johnson, 1965). Both thalidomide and tolbutamide are more teratogenic in riboflavin-deficient rats (Rogers *et al.,* 1965). Presumably, testing of other chemicals for teratogenic effects would show similar influences of vitamin deficiencies in rats and other animal models.

Vitamin A deficiency produced by omission of the vitamin from the diets of rats considerably increased the toxicity of hexachlorophene administered daily at 50 or 75 mg/kg (Hanig *et al.,* 1977). On the other hand, equivocal results were obtained by two groups of investigators with respect to the relationship of vitamin A deficiency to polychlorinated biphenyl (PCB) toxicity. In one case young rats fed a vitamin A-deficient diet with 0.1% PCB showed growth retardation, rough coat, and closed eyes. Supplementation of the diet with vitamin A reversed the symptoms (Innami *et al.,* 1975). However, Narbonne (1979) found that intestinal absorption of PCB was unaffected by vitamin A deficiency. This discrepancy demonstrates the need to understand the mechanisms and effects of nutritional deficiencies so that the influence of such states on the toxicity of xenobiotics can be anticipated and predicted. In one such study, Siddik *et al.* (1980) demonstrated the effects of vitamin A deficiency on liver glutathione transferase activities. The investigators showed that glutathione *S*-aryl, *S*-aralkyl, *S*-alkyl, and *S*-epoxide transferase in liver cytosol and glutathione *S*-aralkyl and *S*-alkyl transferase in kidney were significantly increased because of the deficiency. Pulmonary glutathione transferases were unaffected. Furthermore, they showed that the increases were only in the V_{max}, not in the K_m, of any of the enzymes. The relationship between these results and enhanced susceptibility to chemical carcinogens in vitamin A deficiency was discussed in this report.

Some investigators have expressed the hypothesis that certain nutritional deficiencies, notably of protein, should retard tumor growth (Syrotuck and Worthington, 1977). These workers fed mice isoleucine-, leucine-, or phenylalanine–tyrosine–deficient diets, but found that tumor growth was not restricted. In some cases deficiencies enhanced tumor growth.

Azaserine-induced pancreatic carcinogenesis in the rat is influenced by dietary modification. A number of different dietary regimens were fed to rats, and the effect on pancreatic adenomas and carcinomas was evaluated. The number of tumors was reduced in rats on high-protein diets (Roebuck *et al.,* 1981). In another study raw soya flour was shown to sensitize the rat pancreas to the action of azaserine (McGuinness *et al.,* 1981). A similar approach was used to assess the effect of protein-deficient diets on alloxan-induced diabetes. Low-protein diets reduced the diabetogenicity of alloxan (Young and Dixit, 1980).

Increased protein in the diet of rats leads to increased elimination of dieldrin, and there is a clear relationship between dieldrin toxicity in rats and the protein content of the diet; mortality from dieldrin decreases as the protein content of the diet increases (Oshiba, 1977). Similar effects have been observed for other organochlorine compounds. Investigations with the organophosphorus insecticide parathion have shown that protein malnutrition has an additive effect with parathion on lipid metabolism in the rat (Chakravarty and Ghosh, 1980). Presumably, toxicity of parathion under these conditions would also be altered.

Trace metal and mineral nutrition has many diverse effects on the toxicity of metals and other xenobiotics in animals. Iron deficiency, a condition widely distributed among the human population, has been extensively studied and found to increase the absorption of other metals. In iron-deficient rats the body burden of administered lead or cadmium was increased (Ragan, 1977, 1978a), and in another study they were observed to absorb approximately 2.5 times as much nickel as controls (Ragan, 1978b). Similarly, insufficient dietary zinc or copper also increased the absorption and toxicity of orally ingested lead and cadmium (Petering, 1978). The chronic oral toxicity of cadmium and lead under these conditions may be due as much to alterations of zinc, copper, and iron nutrition and metabolism as it is to the direct cellular inhibitory action of the metals themselves.

There is a sex-related difference in the ability of selenium to alter hepatic drug metabolism and drug response in the rat. Sodium selenite administered intraperitoneally in doses of 2.4 mg/kg prolonged the duration of pentobarbital-induced hypnosis in males, but not in females. Similar treatment led to significant inhibition of ethylmorphine demethylase activity by hepatic microsomal enzymes (Schnell and Early, 1981). Prior administration of sodium selenite, however, blocked the cadmium-induced prolongation of hexobarbital-induced hypnosis and the inhibition of ethylmorphine or aniline metabolism by hepatic microsomes (Early and Schnell, 1981).

Malnutrition of whatever sort almost invariably leads to a reduced capacity of the animal to metabolize and, thus, in many cases, to detoxify xenobiotics. Hence, the effect of malnutrition is to increase the toxicity of such exogeneous chemicals. It is now well known that nutritional status affects the hepatic microsomal drug-metabolizing enzyme system. It is also true that extensive or prolonged metabolism of drugs and xenobiotics may make high nutritional demands and lead to dietary deficiencies, thereby rendering the animal more susceptible to effects from other drugs or chemicals (Ioannides and Parke, 1979; Parke 1978).

Clearly, it is important to consider the interaction of nutritional state and xenobiotic metabolism in animals.

The human population of the United States is known to have a marginal intake of bioavailable iron, riboflavin, and folic acid. Certain groups within the population may also have low intakes of calcium and other trace minerals. There are also considerable geographic, age, sex, and racial variations in nutrition in the United States. All these parameters are even more variable in other parts of the world. With respect to the testing and evaluation of drugs and xenobiotics, there is a definite need to model these marginal nutritional difficulties of certain segments of the population, such as the neonate, the growing child, pregnant women, and the aged (Menzel, 1979).

Genetic Homogeneity versus Heterogeneity

Most toxicological testing and research is conducted using standard animal models—either inbred or outbred strains. Occasionally, wild populations will be sampled, sometimes cultured, and used for experimental purposes, but generally the genetic variability thus introduced is considered to be undesirable. Although there is some controversy among investigators on the relative merits of using inbred versus outbred animals, most agree that standard laboratory strains of animals are preferable to the use of wild populations. Unfortunately, in too many instances those responsible for conducting toxicological evaluations of drugs and other chemicals do not give sufficient consideration to the genetic characteristics of the animals they use.

To illustrate the magnitude of the variability that may be introduced into the experimental situation by using different strains of animals, consider the case of the variation in hexobarbital-induced sleep time in different strains of mice (Table 8-3). In addition to the variation between strains that is evident from these data, it is also apparent that there is considerably more variability within the non-inbred strain, as measured by the large standard deviation. That is the result that

TABLE 8-3. Strain Differences in Duration of Action of Hexobarbital in Mice[a]

Strain[b]	Mean sleep time ± SD (min)
A/NL	48 ± 4
BALB/cAnN	41 ± 2
C57L/HeN	33 ± 3
C3HfB/HeN	22 ± 3
SWR/HeN	18 ± 4
Swiss (non-inbred)	43 ± 15

[a] From Goldstein *et al.* (1974)
[b] Male mice, 70–80 days old, given an intraperitoneal injection of 125 mg/kg.

would be expected when one is assessing a trait under genetic control (Goldstein *et al.*, 1974).

Other reports of studies on drugs and other xenobiotics indicate that there are considerable strain differences in responses of animals to chemicals (Fox and Reid, 1980; Tucker *et al.*, 1980; Webster *et al.*, 1980). The important conclusion to be drawn from such studies is that there is no one optimum strain to use in the assessment of toxicity of a chemical using almost any measurement parameter. All effects, including lethality, cytotoxicity, and teratogenesis, seem to vary among strains in a largely unpredictable way. Although it would be desirable to be able to predict toxic responses on the basis of the genetic characteristics of a strain of laboratory animals, there is no one strain that is either the most sensitive or the most resistant to the range of toxicants tested. Therefore, it is not possible to select one strain that would be representative of the entire range of animal models for routine use in screening tests.

Of particular importance in toxicology are strain differences in the animal models used in carcinogenicity screening. In most toxicity screening, the relatively genetically variable outbred stocks of animals are used (Festing, 1980). A better approach would be to test several inbred strains in a factorial experimental design. The range of genetic variables thus involved in the test would be broader, but also more under experimental control within the conditions of the test. Interpretation of results in such experiments would be easier than in the currently used tests. This is particularly important in carcinogenicity assays because of the significant variability of carcinoma incidence among strains (Madewell, 1981; Martin *et al.*, 1974; Reuber, 1976). Different strains or species of animals differ in their genetic susceptibility to carcinogens for a variety of reasons: mutation rates, elimination of mutant cells, growth control, metabolism of potential mutagens, and the number of mutational steps necessary for malignancy (Strong, 1978). The same may be said for groups within human populations. In order to establish acceptable exposure levels for environmental carcinogens, one must consider the existence of these groups and the resulting appearance of heterogeneity in the population. Thus, testing with a variety of different species and strains of known genetic composition will facilitate the extrapolation of animal test results to humans.

Sex Differences in Animal Responses

The characteristic perhaps best known to lead to unpredictable variation in animal models for toxicity testing is the sex of the animal. Sex differences in responses of animals to toxic chemicals have been most extensively reported in rats; however, there are also many examples of toxicity varying as much as an order of magnitude between males and females of the same species or strain in other animals. Furthermore, these differences are observed in all types of tests, from the short-term determination of acute LD_{50} to long-term testing for carcinogenicity.

One of the most striking examples of sex variation in response to toxicants is the acute toxicity of organophosphorus insecticides to rats. Table 8-4 lists the

TABLE 8-4. Acute Toxicity of Organophosphorus Insecticides to Adult Rats[a]

Compound	Oral LD_{50} (mg/kg)		Dermal LD_{50} (mg/kg)	
	Males	Females	Males	Females
Chlorthion	880	980	1500–4500	4100
Dichlorvos	80	56	107	75
Dioxathion	43	23	235	63
Demeton	6.2	2.5	14	8.2
Diazinon	108	76	900	455
Dicapthon	400	330	790	1250
Trichlorphon	630	560	>2000	>2000
EPN	36	7.7	230	25
Azinphosmethyl	13	11	220	220
Malathion	1375	1000	>4444	>4444
Methyl parathion	14	24	67	67
Parathion	13	3.6	21	6.8
Mevinphos	6.1	3.7	4.7	4.2
Ronnel	1250	2630	—	—
Schradan	9.1	42	15	44
Phorate	2.3	1.1	6.2	2.5
Carbophenothion	30	10	54	27

[a] From Gaines (1960).

oral and dermal LD_{50} values for a number of these compounds (Gaines, 1960). It is evident from these data that in some instances there is very little difference in the responses of males and females, whereas the difference in others may be as great as tenfold. In general, female rats were more susceptible to these compounds than males. However, schradan and ronnel orally and schradan dermally appear to be appreciably more toxic to male rats than to females. There is also some indication that orally administered methyl parathion is more toxic to males than to females. A number of investigators have demonstrated that the differences are related to metabolism of the compounds involved, but the relationships are neither simple nor direct.

There are many other examples of sex differences in the acute toxicity of drugs and chemicals (Benke and Murphy, 1974; Clausing *et al.*, 1980; Irving, 1975; Stevens *et al.*, 1978; Whitehouse and Ecobichon, 1974), but there are also many studies in which no differences were noted between males and females (McCollister *et al.*, 1974; Sasaki *et al.*, 1976; Tauchi *et al.*, 1976). There seem to be no patterns that could be used to predict which chemicals will show sex differences in toxicity or whether differences, when observed, will favor males or females.

Acute toxicity is not the only toxic response to involve sex differences. Pargyline and tryptophan potentiated hypothermia in female but not in male rats (Biegon *et al.*, 1979). Organ levels of chlorprothixene and nortriptyline were higher in female than in male rats when total amines were measured in brain, liver, kidneys, and lungs (Dell *et al.,* 1976). Serum and liver concentrations of organic fluorine were greater in male than in female rats following subchronic feeding studies with ammonium perfluorooctanoate, a commerical surfactant (Griffith and Long, 1980). In long-term toxicity testing, sex differences have been observed in the carcinogenicity of some chemicals in animal models. Female rats were more sensitive to tumor induction by diethylnitrosamine than were males (Deml *et al.,* 1981). Sex differences were not observed in mice, but female rats were less susceptible than males to tumors caused by aflatoxin B_1 (Degen and Neumann, 1981). The authors demonstrated that differences in ability to metabolize the compound accounted for strain differences as well as the sex differences.

Sex differences in response of animal models to chemicals lie not only in differential metabolism, but also in the storage and excretion of foreign compounds. Differences in metabolism are well known to be associated with sex differences, in that the principal metabolic capability of the principal metabolizing organ, the liver, is associated with endocrine metabolism, which is crucial to sexual and reproductive cycles. Studies in which endocrine balance has been altered by injection of sex hormones or by castration of males have unequivocably demonstrated this relationship (Gurtoo and Motycka, 1976; Muraoka and Itoh, 1980). The response of an organism to a toxicant can be markedly altered by manipulating the endocrine balance, thereby altering metabolism of the foreign compound. Inducers and inhibitors of metabolism have a similar effect for the same reason: toxicity is altered by alteration of metabolism (Beatty *et al.,* 1978). Similarly, sex differences in storage in various organs (Dell *et al.,* 1976; Griffith and Long, 1980; Muraoka and Itoh, 1980) and in the rate of excretion of the foreign compound and its metabolites (Dayton and Inturrisi, 1976) have also been observed.

Species Differences in Metabolism and Physiology

Differences of age, sex, genetic composition, state of health, and nutritional state ultimately influence the ability of the animal to activate or detoxify a chemical substance, thereby determining whether the activity of the material will be to the benefit of the individual or will cause it harm. Likewise, differences between species are frequently the result of differences in their ability to metabolize a xenobiotic substance.

Much has been written about species differences in metabolism of drugs and other foreign compounds. Different species may have entirely different pathways for metabolizing a given substance. Even when the same pathways of metabolism exist in different species, the rates may vary markedly. No consistent rationale for some of these differences has emerged (Goldstein *et al.,* 1974). The differences may be directly genetic in origin, as for the acetylation mechanism of aromatic amines or hydrazines, where certain species are "rapid acetylators" and others are "slow acetylators" (Weber, 1978), or they may be only indirectly genetic, where differences in metabolism result from the toxicity or therapeutic activity of a drug or chemical (Williams, 1978).

Williams (1978) has succinctly summarized species variations in the pathways of drug metabolism using as examples the metabolism of the amphetamines, the arylacetic acids, and the phenols, three important groups of compounds to which both humans and animal models are frequently exposed. He noted that the amphetamines may be attacked metabolically at two sites—the ring and the side chain. In the rat, the principal attack is *p*-hydroxylation, whereas guinea pigs and rabbits extensively degrade the side chain. In the metabolism of the amphetamines, unfortunately, neither the rat nor the guinea pig is like the human. Phenylacetic acid metabolism has been studied in 25 species, and in most of them metabolism occurs by conjugation with glutamine. On the other hand, in the Syrian hamster it is also hydroxylated in the *meta* or *para* positions. With naphthylacetic acid, however, the glucuronide conjugate is formed. With this compound, there is considerable species variation in conjugation with glucuronic acid versus amino acids, including both glutamine and taurine. With phenols, it is well established that both glucuronides and sulfates are formed by many different species; however, some species are defective in their ability to perform such conjugations. The domestic cat, for example, cannot make glucuronides, and other members of the cat family are similar. In contrast, swine cannot make sulfates, but form glucuronides from phenols. Rats use both routes almost equally well with simple phenols. These examples illustrate the unpredictability of metabolism among different species with respect to different compounds. Extrapolating metabolism results from animal models to humans is extremely risky in the absence of some comparative results from humans, perhaps from *in vitro* experiments with human tissues.

The basic differences in physiology among different species (Table 8-2) also result in differences in storage, disposition, and excretion of xenobiotics. Simply noting the ratios of metabolic rates as measured by food and water consumption compared with body size illustrates and perhaps partially explains the differences that exist among species with respect to metabolism of xenobiotics. This relationship has led to the proposition that for comparisons of toxicity among species, a more useful measure is the weight of toxicant per unit of body surface area rather than per unit body weight (Dixon, 1976; National Research Council, 1980). Better correlations among species seem to be based on surface area than on weight.

Animal Models for Extrapolation to Humans

The professional charged with evaluating the risk of xenobiotics to humans is often faced with data bases composed of studies having two or more species of test animals and, perhaps, data from several strains of some species. Difficulties arise in selecting the appropriate models and data for extrapolation to humans. The problem must be faced at two levels: (1) When designing test protocols before data have been generated, or even before animal models have been selected, what principles should guide the selection of animal models? (2) After data have been generated on the toxicity of a xenobiotic, often from many sources, what principles should guide the determination of which data will be used from which animal models in the assessment of risk to humans?

This chapter illustrates that the responses of animals to chemicals are influenced, sometimes markedly, by some characteristics of the animal model that may be manipulated and by others that must be at least recognized. It should be obvious that careful attention must be paid to sex, age, state of health, and nutritional state and that care must be exercised when selecting the appropriate model on the basis of the genetic composition of the animal population. Clearly, as demonstrated in the foregoing, a critical consideration is the ability of the animal model to metabolize the xenobiotic in question in a manner similar to that of humans. It has become apparent that this single factor, more than any other, influences the ultimate determination of whether a chemical will be therapeutic or toxic, will be stored or excreted, or will cause acute effect or long-term chronic disease in animals, including humans.

Regulatory agencies responsible for assessing the risk of chemicals to humans generally use guidelines that suggest testing protocols to those responsible for gathering toxicological data. These protocols have, in general, followed at least the framework of the "Weil principles" discussed earlier (Weil, 1972). In light of the foregoing, the first principle, dealing with the selection of animal models, should be reinforced and given greater prominence. It is, unfortuantely, the one of the four principles that is the most frequently ignored in the gathering of toxicological data, since there are usually no data from humans. In fact, it should be the one that governs the selection of the animal models to which the other three principles will be applied.

If the first Weil principle is properly applied and data are then gathered on the basis of the other three, risk assessment can proceed with some confidence that the extrapolation to humans will be based on sound scientific principles incorporating the current state of knowledge in toxicology. When accumulated data are deficient in some way, perhaps where information on the similarity of metabolism and disposition of the xenobiotic between humans and animal models is not available, one must evaluate all the evidence, giving increased weight to any data that include information about metabolic similarity between the animal model and humans.

When assessing the risk of a xenobiotic, a critical consideration is the potential exposure of humans. Toxicological data on experimental animal models must be applied to the potential exposure mechanisms and situations that are likely. For example, data obtained from oral or dietary administration of chemicals to animals have far less applicability to a situation in which humans are exposed by the inhalation route. For a discussion of considerations of interroute extrapolation, the reader is referred to Chapter 12. Furthermore, whenever possible, one should not apply data obtained on young, healthy, well-nourished, inbred animals to a human population that is aged, diseased, malnourished, and diverse in its genetic composition. Ideally, the data to be evaluated to establish potential risk will have been obtained from animal models that closely approximate the human condition.

Unfortunately, one must frequently use incomplete or inadequate data to assess human risk from a xenobiotic. In such cases, great care must be taken to avoid overinterpretation or overextrapolation. The key to the potential utility of data that do not fit the specific human situation is information about the metabolism of the compound in the animal model. Much greater confidence can be attained if one knows that the animal model metabolizes the xenobiotic in a manner similar to that in the exposed human population. This knowledge, along with the application of appropriate safety factors, will permit reasonable and the most accurate conclusions to be drawn about a chemical. Conclusions about risk reached in the absence of metabolic information will not have the same confidence level as those obtained from tests that were properly and completely performed.

It is hoped that future protocols for the assessment of risk to humans from chemicals will require active consideration of all the factors discussed in this chapter. Then there will be a higher degree of confidence that the data generated and the resulting judgments made are highly accurate.

REFERENCES

Andrews, E. J., B. C. Ward, and N. H. Altman, eds. 1979. *Spontaneous Animal Models of Human Disease,* 2 vols. Academic Press, New York. 322 and 324 pp.

Arrington, L. R. 1978. *Introductory Laboratory Animal Science.* Interstate, Danville, Illinois. 211 pp.

Beatty, P. W., W. K. Vaughn, and R. A. Neal. 1978. Effect of alteration of rat hepatic mixed-function oxidase (MFO) activity on the toxicity of 2,3,7,8-tetrachlorodibenzo-*p*-dioxin (TCDD). *Toxicol. Appl. Pharmacol.* **45:**513–519.

Benke, G. M., and S. D. Murphy. 1974. The influence of age and sex on the toxicity and multiple pathways of metabolism of methyl parathion and parathion in rats. *Toxicol. Appl. Pharmacol.* **29:**125.

Biegon, A., M. Segal, and D. Samuel. 1979. Sex differences in behavioral and thermal responses to pargyline and trytophan. *Psychopharmacology* (Berlin) **61:**77–80.

Campbell, T. C. 1976. Modern concepts in nutritional status and foreign compound toxicity. Pp. 11–31 in M. A. Mehlman, R. E. Shapiro, and H. Blumenthal, eds. *Advances in Modern Toxicology,* Volume 1, part 1. Hemisphere, Washington, D.C.

Chakravarty, I., and A. Ghosh. 1980. Effect of parathion under different nutritional conditions on brain lipids. *IRCS Med. Sci. Libr. Compend.* **8:**696.

Clausing, P., R. Bieleke, G. G. Grun, and H. Beitz. 1980. Acute toxicity of mercurial seed-dressing agents for rats and Japanese quail with special reference to coergistic effects. *Arch. Exp. Veterinaermed.* **34:**383–388.

Cook, D. E., and M. R. Past. 1979. Drug metabolism in diabetic isolated perfused rat liver. *Res. Commun. Chem. Pathol. Pharmacol.* **24:**389–392.

Dayton, H. E., and C. E. Inturrisi. 1976. The urinary excretion profiles of naltrexone in man, monkey, rabbit, and rat. *Drug Metab. Dispos.* **4:**474–478.

Degen, G. H., and H. G. Neumann. 1981. Differences in aflatoxin B1 susceptibility of rat and mouse are correlated with the capability *in vitro* to inactivate aflatoxin B1 epoxide. *Carcinogenesis* (London) **2:**299–306.

Dell, H. D., W. Fassbender, R. Kamp, and D. Lorenz. 1976. Age and sex specificity of organ distribution and metabolism of chlorprothixene and nortriptyline in the rat. *Arzneim. Forsch.* **26:**1098–1100.

Deml, E., D. Oesterle, T. Wolff, and H. Greim. 1981. Age-, sex-, and strain-dependent differences in the induction of enzyme-altered islands in rat liver by diethylnitrosamine. *J. Cancer Res. Clin. Oncol.* **100:**125–134.

Dixon, R. L. 1976. Problems in extrapolating toxicity data for laboratory animals to man. *Environ. Health Perspect.* **13:**43–50.

Early, J. L., Jr., and R. C. Schnell. 1981. Selenium antagonism of cadmium-induced inhibition of hepatic drug metabolism in the male rat. *Toxicol. Appl. Pharmacol.* **58:**57–66.

Essman, W. B. 1978. Morphine action in myocardial metabolopathies. Pp. 681–701 in M. W. Adler,

L. Manara, and R. Samanin, eds. *Factors Affecting the Action of Narcotics.* Monographs of the Mario Negri Institute for Pharmacological Research. Raven Press, New York.

Fassett, D. W. 1975. Cadmium: Biological effects and occurrence in the environment. *Annu. Rev. Pharmacol.* **15**:425–435.

Festing, M. F. 1980. The choice of animals in toxicological screening: Inbred strains and the factorial design of experiments. *Acta Zool. Pathol. Antverp.* **75**:117–131.

Fox, P. K., and O. J. Reid. 1980. The importance of animal strain in a mouse model assessing drug cytotoxicity. Pp. 73–80 in A. Spiegel, S. Erichsen, and H. A. Solleveld, eds. *Animal Quality and Models in Biomedical Research.* 7th Symposium of the International Council for Laboratory Animal Science, Utrecht, 1979. Fischer, New York.

Friberg, L., M. Piscator, G. F. Nordberg, and T. Kjellstrom. 1974. *Cadmium in the Environment.* CRC Press, Cleveland, Ohio. 248 pp.

Gaines, T. B. 1960. The acute toxicity of pesticides to rats. *Toxicol. Appl. Pharmacol.* **2**:88–89.

Goldstein, A., L. Aronow, and S. M. Kalman. 1974. *Principles of Drug Action: The Basis of Pharmacology,* 2nd ed. Wiley, New York. 854 pp.

Griffith, F. D., and J. E. Long. 1980. Animal toxicity studies with ammonium perfluoro-octanoate. *Am. Ind. Hyg. Assoc. J.* **41**:576–583.

Gurtoo, H. L., and L. Motycka. 1976. Effect of sex difference on the *in vitro* and *in vivo* metabolism of aflatoxin B1 by the rat. *Cancer Res.* **36**:4663–4671.

Hanig, J. P., J. M. Morrisson, Jr., A. G. Barr, and S. Krop. 1977. Effects of vitamin A on toxicity of hexachlorophene in the rat. *Food Cosmet. Toxicol.* **15**:35–38.

Hoult, J. R., and P. K. Moore. 1980. Prostaglandin synthesis and inactivation in kidneys and lungs of rats with experimental diabetes. *Clin. Sci.* **58**:63–66.

Innami, S., A. Nakamura, K. Kato, M. Miyazaki, S. Nagayama, and E. Nishide. 1975. Polychlorinated biphenyl (PCB) toxicity and nutrition. IV. PCB toxicity and vitamin A. *Fukuoka Igaku Zasshi* **66**:579–584.

Ionnides, C., and D. V. Parke. 1979. Effect of diet on the metabolism and toxicology of drugs. *J. Hum. Nutr.* **33**:357–366.

Irving, C. C. 1975. Comparative toxicity of *N*-hydroxy-2-acetylaminofluorene in several strains of rats. *Cancer Res.* **35**:2959–2961.

Johnson, E. 1965. Nutritional factors in mammalian teratology. Pp. 113–124 in *Teratology: Principles and Techniques.* University of Chicago Press, Chicago.

Kaiser, H. E. 1980. *Species-Specific Potential of Invertebrates for Toxicological Research.* University Park Press, Baltimore, Maryland. 224 pp.

Klaassen, C. D. 1986. Principles of toxicology. Pp. 11–32 in C. D. Klassen, M. O. Amdur, and J. Doull, eds. *Casarett and Doull's Toxicology: The Basic Science of Poisons,* 3rd ed., Macmillan, New York.

Krasovskii, G. N. 1976. Extrapolation of experimental data from animals to man. *Environ. Health Perspect.* **13**:51–58.

MacLeod, K. M., and J. H. McNeill. 1981. The effects of alloxan on streptozotocin on responses of rat vascular smooth muscle to vasoactive agents. *Proc. West. Pharmacol. Soc.* **24**:69–71.

Madewell, B. R. 1981. Neoplasms in domestic animals: A review of experimental and spontaneous carcinogenesis. *Yale J. Biol. Med.* **54**:111–125.

Marshall, J. F., M. I. Friedman, and T. G. Heffner. 1978. Reduced anorectic and locomotor stimulant action of amphetamine in experimental diabetes mellitus: Relation to brain catecholamines. *Cent. Mech. Anorectic Drugs* **1978**:111–125.

Martin, M. S., F. Martin, E. Justrabo, R. Michiels, H. Bastien, and S. Knobel. 1974. Susceptibility of inbred rats to gastric and duodenal carcinomas induced by *N*-methyl-*N'*-nitro-*N*-nitrosoguanidine. *J. Natl. Cancer Inst.* **53**:837–840.

McCollister, S. B., R. J. Kociba, C. G. Humiston, D. D. McCollister, and P. J. Gehring. 1974. Studies of the acute and long-term oral toxicity of chlorpyrifos (0.0-diethyl-0 (3,5,6-trichloro-2-pyridyl) phosphorothioate). *Food Cosmet. Toxicol.* **12**:45–61.

McGuinness, E. E., R. G. Morgan, D. A. Levison, D. Hopwood, and K. G. Wormsley. 1981. Interaction of azaserine and raw soya flour on the rat pancreas. *Scand. J. Gastroenterol.* **16**:49–56.

Menzel, D. B. 1979. From animals to man, the grand extrapolation of environmental toxicology. Pp.

1–14 in S. D. Lee and J. B. Mudd, eds. *Assessing Toxic Effects of Environmental Pollutants.* Ann Arbor Science, Ann Arbor, Michigan.

Muraoka, Y., and F. Itoh. 1980. Sex difference of mercuric chloride-induced renal tubular necrosis in rats—From the aspects of sex differences in renal mercury concentration and sulfhydryl levels. *J. Toxicol. Sci.* **5**:203–214.

Nakama, K., T. Okamura, K. Okamoto, Y. Fujie, K. Fujie, and S. Takeuchi. 1978. Studies on electroretinogram in alloxan rats and Mongolian gerbils *(Meriones unguiculatas)* and influences of long-term tolbutamide administration. *Rinsho Yakuri* **9**:277–287.

Narbonne, J. R. 1979. Polychlorinated biphenyl intoxication and nutrition. III. Effect of vitamin A deficiency on PCB storage in the rat. *Toxicol. Eur. Res.* **2**:41–45.

National Research Council. 1980. *Drinking Water and Health,* Volume 3. Report of the Safe Drinking Water Committee. National Academy Press, Washington, D.C. 415 pp.

Neims, A. H., M. Warner, P. M. Loughnan, and J. V. Aranda. 1976. Developmental aspects of the hepatic cytochrome P450 monooxygenase system. *Annu. Rev. Pharmacol. Toxicol.* **16**:427–445.

Oshiba, K. 1977. Nutrition and the toxicity of organic chlorides. *Paediatrician* **6**:34–44.

Parke, D. V. 1978. Effects of nutrition and enzyme induction in toxicology. *World Rev. Nutr. Diet* **29**:96–114.

Petering, H. G. 1978. Some observations on the interaction of zinc, copper, and iron metabolism in lead and cadmium toxicity. *Environ. Health Perspect.* **25**:141–145.

Ragan, H. A. 1977. Effects of iron deficiency on the absorption and distribution of lead and cadmium in rats. *J. Lab. Clin. Med.* **90**:700–706.

Ragan, H. A. 1978a. Hematologic effects of chronic lead and cadmium ingestion in iron-deficient rats. Pp. 1.11–1.13 in *Annual Report for 1977,* Part 1, *Biomedical Sciences.* Battelle Memorial Institute, Pacific Northwest Laboratories, U.S. Department of Energy, Assistant Secretary for the Environment.

Ragan, H. A. 1978b. Effects of iron deficiency on absorption of nickel. Pp. 1.9–1.10 in *Annual Report for 1977,* Part 1, *Biomedical Sciences.* Battelle Memorial Institute, Pacific Northwest Laboratories, U.S. Department of Energy, Assistant Secretary for the Environment.

Rall, D. P. 1978. Thresholds? *Environ. Health Perspect.* **22**:163–165.

Reuber, M. D. 1976. Various degrees of susceptibility of different stocks of rats to *N*-2-fluorenyldiacetamide hepatic carcinogenesis. *J. Natl. Cancer Inst.* **57**:111–114.

Ridder, W. E., and F. W. Oehme. 1974. Nitrates as an environmental, animal, and human hazard. *Clin. Toxicol.* **7**:145–159.

Roebuck, B. D., J. D. Yager, Jr., and D. S. Longnecker. 1981. Dietary modulation of azaserine-induced pancreatic carcinogenesis in the rat. *Cancer Res.* **41**:888–893.

Rogers, L. L., A. J. Lloyd, and A. C. Fowler. 1965. Teratogenic effects of drugs combined with a mild nutritional deficiency. *Fed. Proc.* **24**:443.

Ruchirawat, M., W. Mahathanatrakul, Y. Sinrashatanant, and D. Kittikool. 1978. Effects of thiamine deficiency on the metabolism and acute toxicity of dimethylnitrosamine in the rat. *Biochem. Pharmacol.* **27**:1782–1786.

Sasaki, K., M. Saitoh, C. Tooyama, and G. Takayanagi. 1976. The antiinflammatory action and the acute toxicity of the combination of betamethasome with *d*-chlorpheniramine malate. *Tokyo Yakka Daigaku Kenkyu Nempo* **23**:79–91.

Schiller, C. M., R. Walden, and B. A. Fowler. 1981. Interaction between arsenic and alloxan-induced diabetes. Effects on rat urinary enzyme levels. *Biochem. Pharmacol.* **30**:168–170.

Schiller, C. M., R. Walden, T. E. Kee, W. H. Curley, G. M. Gafford, R. J. Shirkey, G. W. Lucier, and B. A. Fowler. 1978. Development of a model to investigate the interaction between alloxan induced diabetes and exposure to arsenic elevated levels of rat urinary enzymes. *Fed. Proc.* **37**:505.

Schnell, R. C., and J. L. Early. 1981. Sex related differences in selenium induced alterations in drug action in the rat. *Res. Commun. Chem. Pathol. Pharmacol.* **32**:561–564.

Siddik, Z. H., E. G. Mimnaugh, M. A. Trush, and T. E. Gram. 1980. The effect of vitamin A deficiency on hepatic, renal and pulmonary glutathione *S*-transferase activities in the rat. *Biochem. J.* **188**:889–893.

Sporn, M. B., N. M. Dunlop, D. L. Newton, and J. M. Smith. 1976. Prevention of chemical carcinogenesis by vitamin A and its synthetic analogs (retinoids). *Fed. Proc.* **35**:1332–1338.
Stara, J. F., and D. Kello. 1979. Relationship of long-term animal studies to human disease. Pp. 43–76 in S. D. Lee and J. B. Mudd, eds. *Assessing Toxic Effects of Environmental Pollutants.* Ann Arbor Science, Ann Arbor, Michigan.
Stevens, J. T., J. D. Farmer, and L. C. Dipasquale. 1978. The acute inhalation toxicity of technical captan and folpet. *Toxicol. Appl. Pharmacol.* **45**:320.
Strong, L. C. 1978. Genetically susceptible subgroups. *Environ. Health Perspect.* **22**:139–140.
Syrotuck, J. A., and B. S. Worthington. 1977. Nutritional effects on syngeneic tumor immunity and carcinogenesis in mice. Part I. Selected essential amino-acids. *Fed. Proc.* **36**:1163.
Tauchi, K., H. Kawanishi, N. Igarashi, Y. Maeda, Y. Maeyama, K. Ebino, K. Suzuki, and T. Imamichi. 1979. Studies on toxicity of cefadroxil (S-578). I. Acute toxicity. *Jpn. J. Antibiot.* **32**:1230–1232.
Tucker, S. P., D. P. Lovell, A. A. Seawright, and V. J. Cunningham. 1980. Variation in the hepatotoxic effects of carbon disulfide between different strains of rat. *Arch. Toxicol.* **45**:287–296.
Uvarenko, A. R. 1978. Effect of chronic mercury poisoning on the structure and function of the island apparatus of the pancreas of animals. *Endokrinologiya* **8**:33–36.
Weber, W. 1978. Genetic variability and extrapolation from animals to man: Some perspectives on susceptibility to chemical carcinogenesis from aromatic amines. *Environ. Health Perspect.* **22**:141–144.
Webster, W. S., D. A. Walsh, A. H. Lipson, and S. E. McEwen. 1980. Teratogenesis after acute alcohol exposure in inbred and outbred mice. *Neurobehav. Toxicol.* **2**:227–234.
Weil, C. S. 1972. Guidelines for experiments to predict the degree of safety of a material for man. *Toxicol. Appl. Pharmacol.* **21**:194–199.
Whitehouse, L., and D. J. Ecobichon. 1974. Species variations in the activation and degradation of parathion by mammalian hepatic enzymes. *Toxicol. Appl. Pharmacol.* **25**:473.
Williams, R. T. 1978. Species variations in the pathways of drug metabolism. *Environ. Health Perspect.* **22**:133–138.
Wolfe, G. W., and R. C. Schnell. 1979. Influence of hormonal factors on daily variations in hepatic drug metabolism in male rats. *J. Interdiscip. Cycle Res.* **10**:173–183.
Young, J. K., and P. K. Dixit. 1980. Lack of diabetogenic effect of alloxan in protein calorie malnourished rats. *J. Nutr.* **110**:703–709.

9

Toxicokinetics

Edward R. Garrett

INTRODUCTION

Two major phenomena underlie the biological responses of an organism to a drug or a toxicant. One is the action of the substance on the organism, which results in pharmacological, chemotherapeutic, psychotropic, pharmacodynamic, or toxic effects. The other is the action of the organism on the substance to impede its penetration or absorption, and to dilute, deplete, transform, and excrete it, and thus diminish its availability at the site of action. These factors modify the time course of the active substance in the body. The elucidation of the quantitative factors that affect the time course of toxic substances is the function of toxicokinetics; whereas the elucidation of those factors that affect the time course of drugs is the function of pharmacokinetics (Garrett, 1971a).

The pharmacological, chemotherapeutic, or toxic effects of a substance are manifested through the interaction of its molecules with specific and specialized sites in the body. The extent and duration of action depend on how much of the compound reaches these receptor sites and how long it remains there. These receptor sites reside in one or more kinetically defined compartments called biophases. The extent and duration of interaction between receptor sites and various invading molecules depend on their concentration in this biophase.

The time course of substance concentration in the biophase is a function of the amount delivered, the rate of delivery, the mode of administration, and the properties of the organism.

Kinetics is a vital discipline in modern toxicology. It provides information on the time that the toxic agent and its metabolites persist in the body after

EDWARD R. GARRETT • The Beehive, College of Pharmacy, J. Hillis Miller Health Center, University of Florida, Gainesville, Florida 32610-0494.

administration of a dose. It also permits prediction of the accumulation and steady-state concentrations of toxic agents in the compartments of the body on the basis of the known pharmacokinetic rate constants, volumes of distribution, and estimated exposure to these toxic agents. Furthermore, it enables investigators to predict the elimination of such agents from the body when ingestion or other types of exposure has diminished or ceased. The magnitude of toxicity as a function of rates of ingestion (or exposure) and elimination can be predicted from the correlation of toxicant concentrations in the various compartments and observed toxic action.

BASIC CONCEPTS OF TOXICOKINETICS

When a systemically acting compound is delivered to the circulation, it is presumed to equilibrate instantaneously in a volume of the central compartment. The model shown in Fig. 9-1 can be based on the postulation of first-order transfers within, and first-order elimination from, the body. The term "first-order" means that the rates at which amounts leave a compartment are proportional to the amount or concentration in that compartment.

The apparent volume of distribution in the central compartment V_c can be calculated as

$$V_c = D/p_0 \tag{1}$$

where D is the dose and p_0 in plasma (or blood or plasma water) is the estimated concentration of an intravenously administered dose at zero time. This volume varies with the substance and its properties. The compartment may contain the plasma and its associated fluids and tissues (e.g., protein, erythrocytes, lymph, and extracellular water). Subsequently, the compound can diffuse relatively rapidly into "shallow" compartments or relatively slowly into "deep" compartments. A hydraulic analogy to the model in Fig. 9-1 is presented in Fig. 9-2, and typical time curves for amounts in these generalized compartments are given in Fig. 9-3.

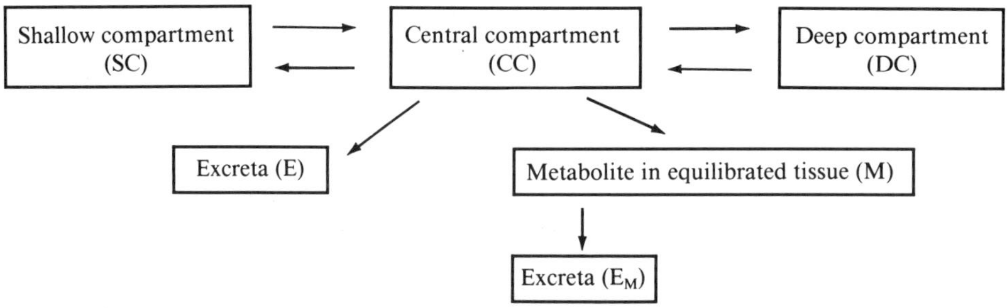

FIGURE 9-1. A general model for the distribution and elimination of a toxicant administered to the systemic circulation in the central compartment.

When a bolus dose is administered, the amount of compound in the shallow compartment increases relatively rapidly and, within a short period, the ratio of the compound present in the shallow compartment to that in the central compartment becomes relatively constant. Semilogarithmic plots of amounts in each of these compartments against time would show curves that eventually become parallel.

The amount of toxicant in the deep compartment rises slowly and continually, even when the greatest portion of the substance is diminishing in the body, plasma, and central or shallow compartments. Thus, if the biophase for toxic action lies in the central compartment, toxic action will parallel plasma levels. If it lies in the peripheral shallow compartment, action will increase quickly until it parallels plasma levels. If it lies in a deep compartment, toxic action may never parallel plasma or body levels of the toxicant, and the activity will peak much later than its availability in the systemic circulation (Fig. 9-3).

The apparent volume of distribution of the central compartment V_c of a toxicant will be greater the larger the extent of the toxicant's instantaneous distribution into its associated tissues. This volume will also depend on the degree of the toxicant's protein binding and erythrocyte partitioning as well as whether the substance's concentrations are monitored in (and thus, referenced to), the blood, plasma, or plasma water (Garrett and Lambert, 1973). The best physiological reference for this volume is frequently the extracellular water of the body where the toxicant concentration is measured in the plasma water. The same volume estimates can be obtained when the plasma concentrations are corrected for the toxicant's plasma protein binding, or when the blood concentrations are corrected for both protein binding and erythrocyte partitioning. The apparent volume of distribution referenced to plasma water concentrations of the toxicant may exceed the physiological value if the drug is highly fat-soluble or rapidly bound to skeletal tissues. In the extreme case of almost complete plasma protein binding, the V_c referenced to total plasma concentration may be equal to the actual volume of the plasma. Thus, volumes of distribution provide insight into the toxicant's destination in the body.

In Fig. 9-3, phase *a*, which is commonly called the distribution phase, corresponds to a rapid but not instantaneous loss of substance into peripheral tissues. These tissues generally consist of intracellular water and other spaces, which need time to equilibrate with the toxicant concentration in the central compartment. The concentration gradients that permit diffusion are due to the soluble toxicant in plasma water and generally are not directly related to the amounts of compound bound to plasma protein or distributed into erythrocytes. In many cases, the concentration gradients are established only by the nonionized form of the compound in the plasma water, which can vary with the pH of the plasma and be dependent on the dissociation constant of the particular chemical. Measured blood levels of drugs such as the alkyl barbiturates, which have a dissociation constant (pK_a) at the pH of the plasma, are particularly susceptible to mild acidosis or alkalosis. If the apparent volume of distribution of the peripheral tissues V_T is relatively large, the equilibrated tissue or plasma concentrations of nonionized drug may not effectively change, although the total plasma concentration

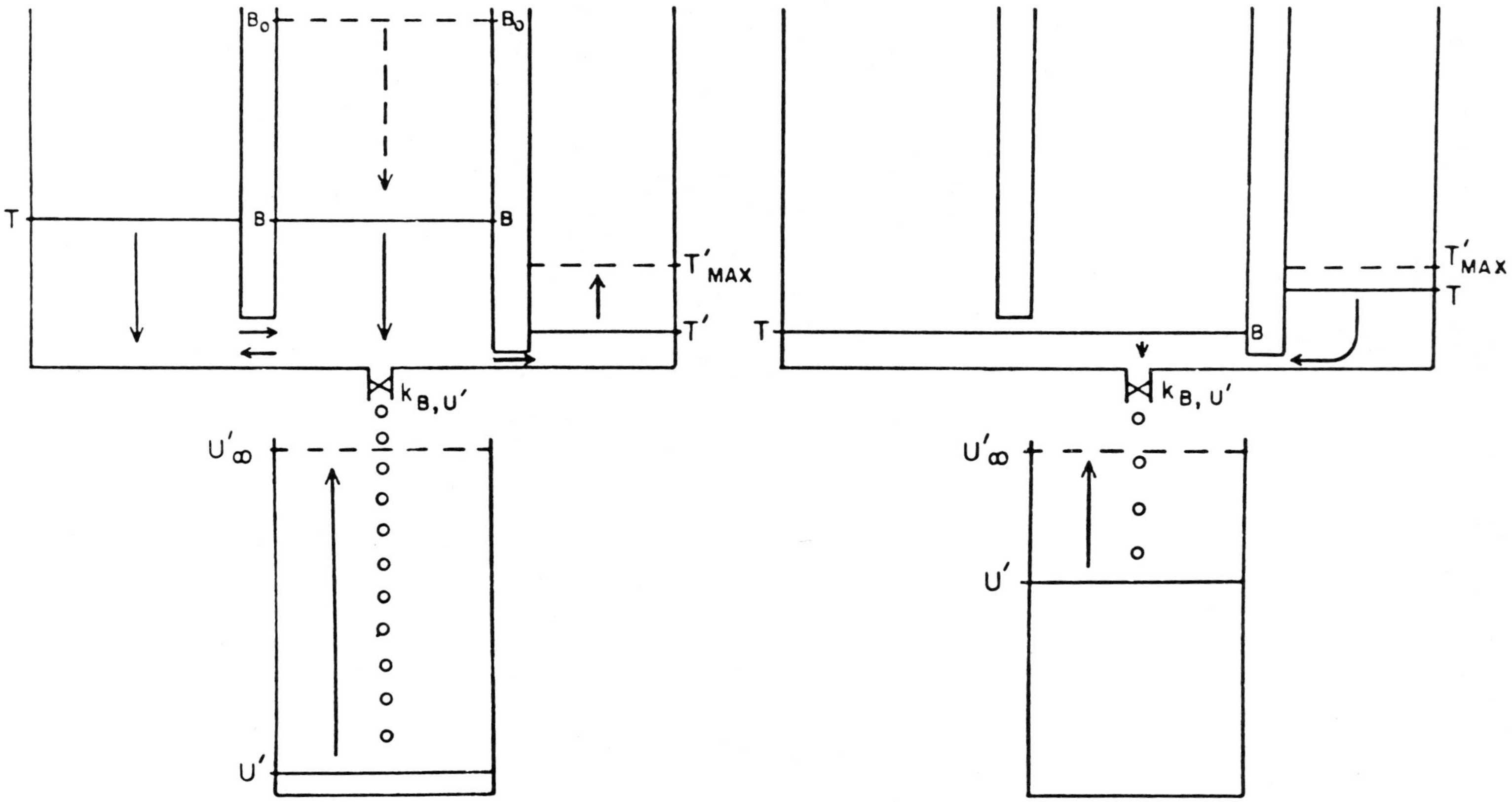

B₀
B₀
T
B
B
T'MAX
T'
kB, U'
U'∞
U'
T'MAX
T
T
B
kB, U'
U'∞
U'

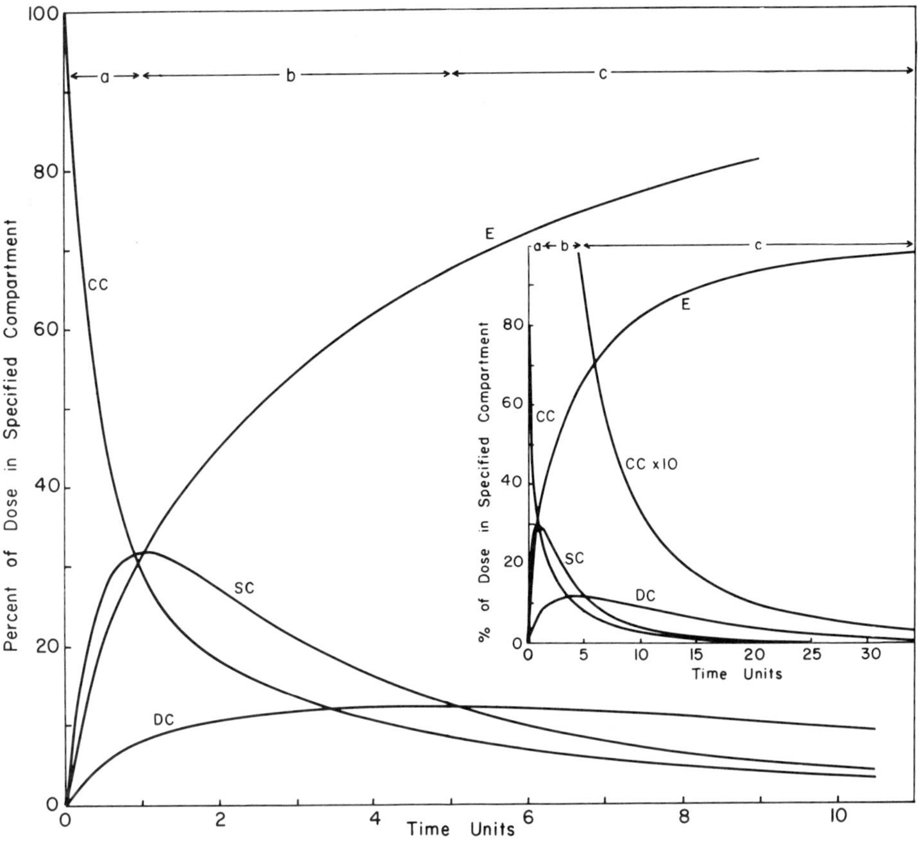

FIGURE 9-3. Analog computer-generated curves representing the time course of an intravenously administered toxicant bolus in the instantaneously equilibrated central compartment (CC), which rapidly equilibrates with a shallow compartment (SC) and slowly approaches a pseudo-steady-state equilibration with a deep compartment (DC). It is eliminated (E) by metabolism and excretion. Phase a is largely assignable to the rapid equilibration of CC and SC and the slow transfer to DC; phase b, to the elimination (E) and the slow transfer to DC; and phase c, to the rate-determining return of toxicant from the deep compartment (DC), which maintains the pseudo-steady-state rate of elimination (E) and slowly approaches a reasonably constant ratio between CC and DC. The approximate durations of the phases are given at the top of the figure. The inset covers a longer period and plots the contents of the central compartment on an expanded scale. [From Garrett (1977).]

FIGURE 9-2. Hydraulic analogy of the three-compartment body model. B_0 is the level corresponding to the original amount in the blood compartment before any equilibration or loss. B is the amount of compound at any time. In both halves of the figure, it represents levels when B has equilibrated with the shallow or rapidly equilibrating compartment T. The levels of compound in the deep or less rapidly equilibrating tissue compartment are represented by T'; T'_{max} is the maximum value achieved in this compartment for such a dosage (B_0). The sizes of the channels between the blood (B) and tissue compartments (T and T') represent the magnitudes of the first-order transfer rates. U' represents the level of compound eliminated and metabolized; U'_∞ is the level that would correspond to the amount of compound eliminated or metabolized at infinite time. The left half of the figure represents a time after equilibration of the blood and rapidly equilibrating tissues (T), while the slowly equilibrating compartment tends to increase in amount. In the right half of the figure the subsequent slow release of compound from the slowly equilibrating compartment (T') to the blood compartment (B) will modify and prolong the overall release rate of compound into U'. [From Garrett (1971a).]

may vary significantly with minor changes in blood pH (Garrett *et al.*, 1974).

With most compounds, but not all, the plasma concentration equilibrates with these peripheral tissue levels after the cessation of phase a and throughout phase b (Fig. 9-3). Thus, this equilibrated plasma concentration can be the criterion for, and proportional to, the effective concentration of a substance at a site of its action in the equilibrated tissues. With some substances, the effective amount may be that amount in a deep compartment. The time course of the plasma level could only reflect this amount during the terminal period of phase c (Fig. 9-3).

PARAMETERS OF FIRST-ORDER TOXICOKINETICS

Mathematical Descriptions of Time Course

If there are first-order transfers within, and first-order transfers from, the body, the plasma p kinetics with time t can be described by a linear sum of exponentials,

$$\begin{aligned} p/D &= (a/D)e^{-\alpha t} + (b/D)e^{-\beta t} + (c/D)e^{-\gamma t} + \cdots + (z/D)e^{-\omega t} \\ &= \sum (a_i/D)e^{-\lambda_i t} \end{aligned} \quad (2)$$

where the coefficients $a/D = a/D, b/D, \ldots, z/D$ are in concentrations per unit of dose D, and the rate constants $\lambda_i = \alpha, \beta, \gamma, \ldots, \omega$ are in reciprocal time units. These parameters do not change with dose for first-order toxicokinetics. They can be obtained from experimental data by "feathering" exercises—the method of residuals (Riggs, 1970). The number of exponentials in Eq. (2) necessary to describe the plasma level–time curve is equal to the kinetically observable compartments in the body.

An appropriate example for first-order transferences is the three-compartment body model:

$$T_{SC} \underset{k_{TP}}{\overset{k_{PT}}{\leftrightharpoons}} P_{CC} \underset{k_{T'P}}{\overset{k_{PT'}}{\rightleftharpoons}} T'_{DC}$$

$$\downarrow k_{PE}$$

$$E$$

Scheme I

Here, the toxicant undergoes first-order elimination on bolus administration into the systemic circulation and the amount P_{CC} is instantaneously equilibrated in the central compartment associated with plasma. This material undergoes relatively rapid equilibration with an amount T_{SC} in a tissue (shallow compartment) and a slower, reversible transfer with an amount T_{DC} in a tissue (deep compartment). The agent is eliminated (to E) by first-order processes (Figs. 9-3 and 9-4).

The sum of exponentials that can describe the plasma concentration per unit

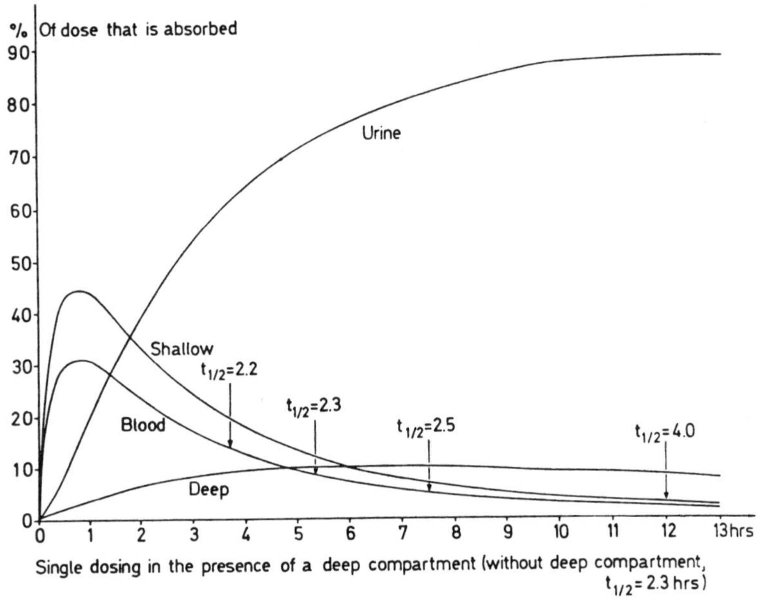

FIGURE 9-4. Amounts of toxicant in the blood, shallow compartment, deep compartment, and urine after first-order absorption from the gastrointestinal tract, as generated by the analog computer with first-order transferences among compartments. The shallow compartment is a rapidly equilibrating tissue, whereas the deep compartment is a slowly equilibrating tissue. [From Garrett (1971a).]

of dose p/D at any time t is

$$p/D = (a/D)e^{-\alpha t} + (b/D)e^{-\beta t} + (c/D)e^{-\gamma t} \tag{3}$$

The value of p_0/D at the time t = 0 of administration of the dose D is given by

$$p_0/D = a/D + b/D + c/D = \sum_i a_i/D \tag{4}$$

The area under the plasma concentration–time curve AUC at any time t is

$$\begin{aligned} AUC &= (a/\alpha)(1 - e^{-\alpha t}) + (b/\beta)(1 - e^{-\beta t}) + (c/\gamma)(1 - e^{-\gamma t}) \\ &= \sum (a_i/\lambda_i)(1 - e^{-\lambda_i t}) \end{aligned} \tag{5}$$

The total area under the plasma concentration–time curve AUC_∞ when $t = \infty$ is

$$AUC_\infty = a/\alpha + b/\beta + c/\gamma = \sum a_i/\lambda_i \tag{6}$$

In the general case of first-order toxicokinetic processes, the microscopic first-order rate constants in reciprocal time units k_j (e.g., k_{PT}, k_{TP}, kPT', $k_{T'P}$, and k_{PE} in the three-compartment body model of Scheme I) are the proportionality constants between rate of egress from the ith compartment to the jth and the concentration of the ith compartment C_i. Thus, $-dA_i/dt = k_{ij}C_i$, where $C_i = A_i/V_i$, i.e., the amount ($A_i = P, T$, etc.) divided by the apparent volume of distribution V_i of that compartment. These microscopic rate constants can be readily calculated from the determined parameters a_i and λ_i (Gibaldi and Perrier, 1975) of the linear sum of exponentials that best fit the experimentally determined plasma concentration–time data.

Accumulation to Steady State on Chronic Dosing

When kinetic processes are first order, so that plasma concentrations can be characterized by a sum of exponentials, there are simple equations to predict the accumulation and the ultimate steady-state plasma concentrations of toxic agents. An example of accumulation on repetitive dosing with first-order absorption is given in Fig. 9-5.

Respective accumulation factors (Garrett, 1978)

$$\frac{1 - e^{-n\lambda_i\tau}}{1 - e^{-\lambda_i\tau}}$$

may be inserted as multipliers before each exponential to calculate the plasma concentration p_n at any time t after a repetitive dose D is given at intervals of τ time units for n number of times. For the example from Eq. (3), when the drug is administered n times as an i.v. bolus at τ intervals we have

$$\begin{aligned} p_n &= a\frac{1 - e^{-n\alpha\tau}}{1 - e^{-\alpha\tau}}e^{-\alpha t} + b\frac{1 - e^{-n\beta\tau}}{1 - e^{-\beta\tau}}e^{-\beta t} + c\frac{1 - e^{-n\gamma\tau}}{1 - e^{-\gamma\tau}}e^{-\gamma t} \\ &= \sum a_i \frac{1 - e^{-n\lambda_i\tau}}{1 - e^{-\lambda_i\tau}}e^{-\lambda_i t} \end{aligned} \tag{7}$$

In this case, the maximum plasma concentration $(p_n)_{max}$ within the nth dosing interval occur when $t = 0$, and thus $e^{-n\lambda_i\tau} = 1$, in Eq. (7), so that

$$\begin{aligned} (p_n)_{max} &= a\frac{1 - e^{-n\alpha\tau}}{1 - e^{-\alpha\tau}} + b\frac{1 - e^{-n\beta\tau}}{1 - e^{-\beta\tau}} + c\frac{1 - e^{-n\gamma\tau}}{1 - e^{-\gamma\tau}} \\ &= \sum a_i \frac{1 - e^{-n\lambda_i\tau}}{1 - e^{-\lambda_i\tau}} \end{aligned} \tag{8}$$

The minimum plasma level $(p_n)_{min}$ after the nth dose occurs just before the subsequent dosing, and thus $t = \tau$ and $e^{-\lambda_i t} = e^{-\lambda_i\tau}$ in Eq. (7), and

$$\begin{aligned} (p_n)_{min} &= a\frac{1 - e^{-n\alpha\tau}}{1 - e^{-\alpha\tau}}e^{-\alpha\tau} + b\frac{1 - e^{-n\beta\tau}}{1 - e^{-\beta\tau}}e^{-\beta\tau} + c\frac{1 - e^{-n\gamma\tau}}{1 - e^{-\gamma\tau}}e^{-\gamma\tau} \\ &= \sum a_i \frac{1 - e^{-n\lambda_i\tau}}{1 - e^{-\lambda_i\tau}}e^{-\lambda_i\tau} \end{aligned} \tag{9}$$

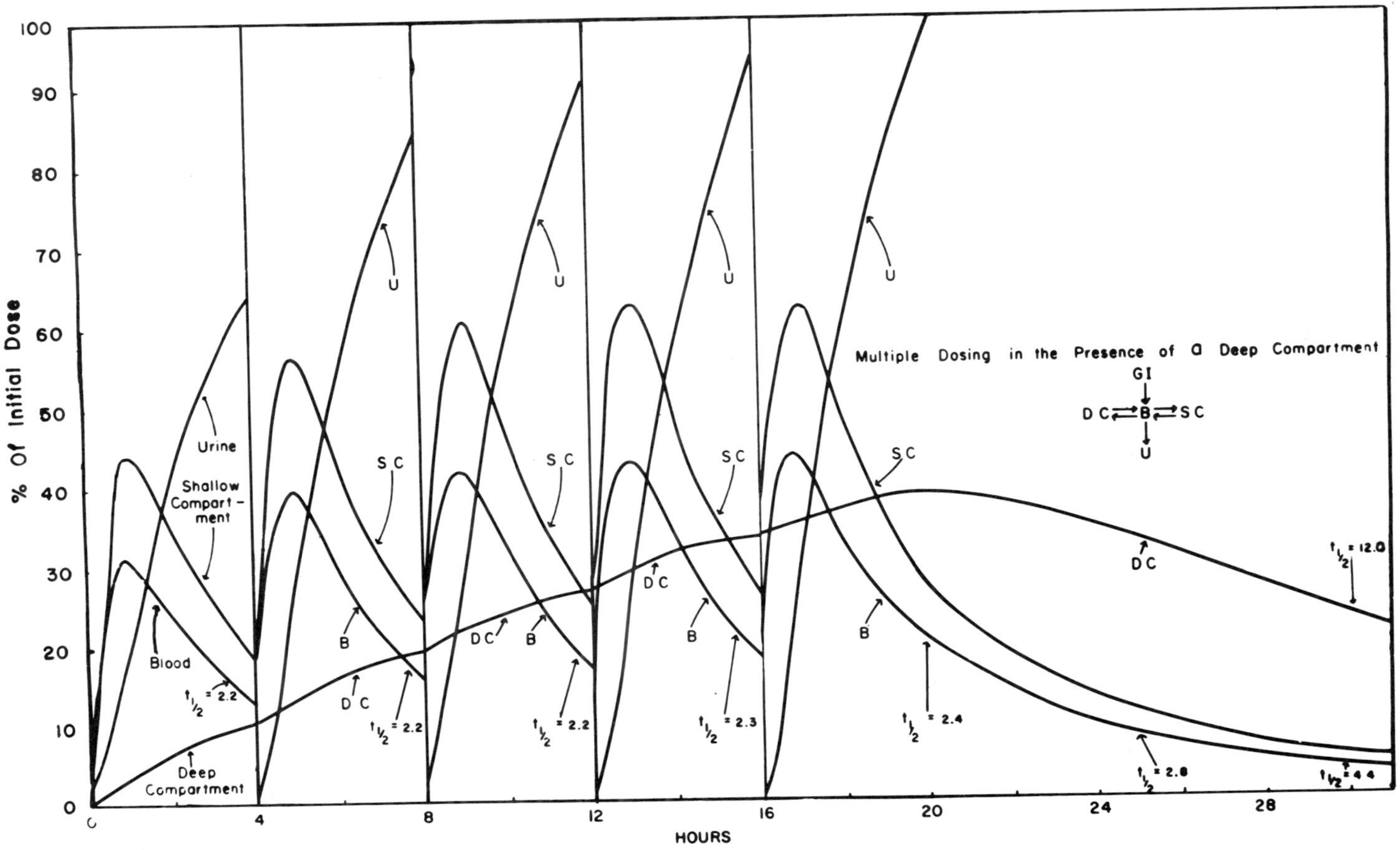

FIGURE 9-5. Curves showing the amounts of toxicant in blood (B), shallow compartment (SC), deep compartment (DC), and urine (U) after first-order absorption of toxicant from the gastrointestinal tract as generated by the analog computer on *repetitive dosing*. The data for the urine (U) are corrected to the zero amount line at each dosing interval for convenience of representation. [From Garrett (1971a).]

After a large number of doses, repetitive oscillations in plasma level–time curves will occur within the dosing intervals, such as shown in Fig. 9-5 for the case of first-order absorption and steady-state conditions will be achieved. Thus, when $n = \infty$, $e^{-n\lambda_i\tau} = e^{-\infty} = 0$, and the accumulation factor is simplified to $1/(1 - e^{-\lambda_i\tau})$. Thus, the plasma concentration p_∞ at any time t after each intravenous bolus dose will be

$$p_\infty = \frac{ae^{-\alpha t}}{1 - e^{-\alpha\tau}} + \frac{be^{-\beta t}}{1 - e^{-\beta\tau}} + \frac{ce^{-\gamma t}}{1 - e^{-\gamma\tau}} = \sum \frac{a_i e^{-\lambda_i t}}{1 - e^{-\lambda_i\tau}} \qquad (10)$$

The repetitive maximum plasma level $(p_\infty)_{\max}$ after accumulation from a bolus dose would occur when $t = 0$ in Eq. (10), so that the exponentials reduce to unity, e.g., $e^{-\alpha t} = 1$, and thus

$$(p_\infty)_{\max} = \frac{a}{1 - e^{-\alpha\tau}} + \frac{b}{1 - e^{-\beta\tau}} + \frac{c}{1 - e^{-\gamma\tau}} = \sum \frac{a_i}{1 - e^{-\lambda_i\tau}} \qquad (11)$$

The repetitive minimum plasma concentration $(p_\infty)_{\min}$ after accumulation would be reached at the end of the dosing interval τ, so that $t = \tau$ in Eq. (10) and

$$(p_\infty)_{\min} = \frac{ae^{-\alpha\tau}}{1 - e^{-\alpha\tau}} + \frac{be^{-\beta\tau}}{1 - e^{-\beta\tau}} + \frac{ce^{-\gamma\tau}}{1 - e^{-\gamma\tau}} = \sum \frac{a_i e^{-\lambda_i\tau}}{1 - e^{-\lambda_i\tau}} \qquad (12)$$

The plasma concentration at any time after a terminal (or nth) dose is given can be calculated from Eq. (7). The plasma concentration at any time after a terminal dose when steady state has been achieved can be calculated from Eq. (10).

Half-Lives and Their Limitations

With most substances, only the second phase (b) of slower plasma level decay can be observed (Fig. 9-3). This commonly called disposition phase is due primarily to the renal and metabolic clearances of the compound from the body. It has been characterized by a "half-life" concept, defined by the time $t_{1/2}$ it takes for a plasma concentration to fall to one-half its value. The equilibration of the compound in the plasma and the peripheral tissues may be relatively fast compared to the processes of elimination from the body, and the clearances may be constant and independent of plasma level. Under these conditions and with a two-compartment body model, a plot of the natural logarithms of plasma concentrations against time is usually a straight line for phase b with a slope β. The half-life concept is valid for estimating how long the administered dose will remain in the body when absorption and tissue equilibrium are relatively rapid. This simple concept has been a cornerstone of operational clinical pharmacology and toxicology. The half-life can be calculated from the slope β as

$$t_{1/2} = 0.693/\beta \qquad (13)$$

Even if there are only first-order transfers, a single half-life cannot characterize a substance's disposition when there are deep compartments. The change in the rate-limiting processes from renal, biliary, or metabolic elimination to release from deep compartments—the change from phase *b* to phase *c* in Figs. 9-3 and 9-4—is accompanied by steadily decreasing apparent half-lives from $0.693/\beta$ to $0.693/\gamma$ with time. Examples are β-methyldigoxin (Hinderling and Garrett, 1977) and tetrahydrocannabinol (Garrett and Hunt, 1977).

Clearances

A clearance CL is usually given in ml/min. Hypothetically, it is that volume (ml) of circulating fluid *completely* cleared in a specific period (min), although it is actually the flow rate Q through a clearing or extracting organ multiplied by the extraction efficiency ϵ, the portion of the total flow that is extracted by the organ (Garrett, 1978).

Figure 9-6 illustrates the general concept of clearance for an amount of compound F in the reservoir of equilibrated fluids of an apparent volume of distribution V_f to give a concentration p. The clearance of F is

$$CL = \epsilon Q \tag{14}$$

Clearance can also be defined as the proportionality constant λV between rate of elimination dE/dt in amount per time unit and the concentration p in amount/volume of the circulating fluid. Thus, when the rate of elimination is constructed as proportional to the amount P in a volume V of circulating fluid,

$$dE/dt = \lambda P = (\lambda V)p = CL \times p \tag{15}$$

and clearance ($CL = \lambda V$ in ml/min) can be defined as the product of an overall rate constant λ in min^{-1} (the terminal rate constant in a multicompartmental model) and the overall apparent volume of distribution V_d in ml. The λ is β for two compartments, and γ for the three compartments characterized by Eq. (3).

Clearance can be estimated from the quotient of the amount of compound ΔE excreted (or transformed) per time interval Δt divided by the plasma concentration $p_{t_{\text{mid}}}$ at the midtime of the collection interval:

$$CL = (\Delta E/\Delta t)p_{t_{\text{mid}}} = \lambda V \tag{16}$$

For renal clearance CL_{ren}, ΔE is the amount of compound excreted in a specific time. For metabolic clearance, ΔE is the amount of compound transformed.

The constant clearances of first-order eliminations could be estimated from the slopes obtained from plots of the rates of amounts eliminated against the plasma levels at the midtime of the collection interval.

Equation (15) can be integrated for the constant clearances of a first-order

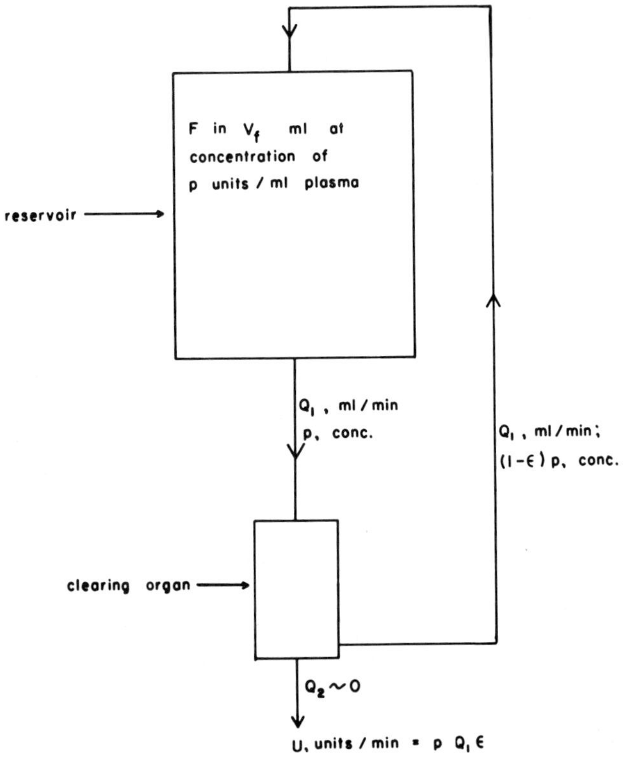

FIGURE 9-6. Schematic diagram demonstrating the general concept of clearance by an organ. The Q values represent flow rates, usually given in ml/min; the ϵ values represent the efficiencies of the process by which the toxicant is transported. It is presumed that negligible volume is eliminated from the cycling system. [From Garrett (1978).]

process between the limits of 0 and time t, to give

$$E = CL \int_0^t p\, dt = CL \times AUC \tag{17}$$

where the integral is the area under the plasma concentration–time curve AUC to time t. Thus, the clearance can be calculated from the quotient of the cumulative amount E eliminated (or transformed) at a time t and the AUC calculated [Eq. (5)] or estimated by the trapezoidal rule to that time:

$$CL = E/AUC = E_\infty/AUC_\infty \tag{18}$$

Similarly, the clearance can be estimated from the total amount transformed E_∞ and the total area under the curve AUC_∞.

The constant clearances of first-order eliminations could be estimated from the slopes of plots of cumulative amounts eliminated E against the area under the plasma level–time curve to that time. The total clearance CL_{tot}, which is the sum of the separate clearances, e.g., renal CL_{ren} and nonrenal CL_{nonren}, can be calculated similarly, where dose D is the total toxicant eliminated and transformed at infinite time:

$$CL_{\text{tot}} = \frac{E_{\infty}^{\text{ren}} + E_{\infty}^{\text{nonren}}}{AUC_{\infty}} = \frac{D}{AUC_{\infty}} \tag{19}$$

Rearrangement of this expression to

$$AUC_{\infty} = D/CL_{\text{tot}} = D/\lambda_n V_d \tag{20}$$

shows that the total area under the curve for first-order elimination processes with an overall elimination rate constant λ_n is proportional to the administered dose.

It follows that the ratio of these areas AUC_{∞} for equivalent oral and intravenous doses, when the total clearance $\lambda_n V_d$ is constant, is a measure of the absorbed dose that reaches the systemic circulation unchanged.

Apparent Volumes of Distribution

The amounts of a toxicant in the body A are related to the assayed concentration of the toxicant in the circulating fluids of the body, such as plasma, by a proportionality constant. This constant has the units (e.g., ml) of an apparent volume of distribution V_d,

$$A = V_d p \tag{21}$$

On repetitive dosing or infusion, the toxicant will equilibrate among all exchangeable compartments of the body and achieve steady-state levels. The overall apparent steady-state volume of distribution V_d^{ss} relates the amount and plasma concentration under such conditions [Eq. (21)] and is the sum of the apparent volumes of the separate compartments:

$$V_d^{\text{ss}} = V_c + V_{T\text{SC}} + V_{T'\text{DC}} \tag{22}$$

where the respective volumes V_c, $V_{T\text{SC}}$, and $V_{T'\text{DC}}$ are of the central compartment, the shallow compartment, and the deep compartment. From detailed compartmental analysis, such as that reported by Gibaldi and Perrier (1975), one can derive the forward (k_{PT} and $k_{PT'}$) and backward (k_{TP} and $k_{T'P}$) first-order rate constants so that the apparent volumes of the peripheral tissues can be calculated at steady state as follows:

$$V_T = (k_{PT}/k_{TP})V_c \tag{23}$$

$$V_{T'} = (k_{PT'}/k_{T'P})V_c \tag{24}$$

After administration of acute doses, a pseudo-steady-state equilibration can be established among compartments. For first-order transfers, the result is a terminal linear semilogarithmic plot of plasma concentration against time with an overall elimination rate constant λ_n from the slope of this terminal line.

Under these circumstances, the ratio of the amount of toxicant in one compartment to that in another remains constant throughout the terminal phase, even though the compartments are not in true equilibrium. The pseudo-steady-state overall apparent volume of distribution V_d^{pss}, which relates the amount of toxicant in the body to that in the plasma [Eq. (21)] during the pseudo steady state of exponentially decreasing elimination, can be estimated from the following equation:

$$V_d^{pss} = CL_{tot}/\lambda_n = (D/AUC_\infty)\lambda_n = CL_{tot}(t_{1/2})_n/0.693 \tag{25}$$

where the total clearance is calculated from the total area under the curve [Eqs. (6), (16), and (19)].

If there is instantaneous equilibration of the administered toxicant among all the tissues of the body, the system reduces to the one-compartment body model. There is only one rate constant, λ, or half-life, $t_{1/2}$, over the entire time course of monoexponentially decreasing plasma concentrations, and the apparent overall pseudo-steady-state volume of distribution and the apparent volume of distribution of the central compartment, $V_c = D/p_0$, are synonymous.

Saturable Processes, Varying Clearances, and Dose-Dependent Toxicokinetics

The parameters a_i/D, λ_i, CL, and $t_{1/2}$ of the toxicokinetics [Eqs. (2)–(6) and (14)–(19)] may not always be invariant with dose. The apparent "half-life" of a toxicant in the body at a given time after administration is not always independent of plasma concentration. The renal processes of tubular secretion and reabsorption may not show constant clearances; the clearance of toxicants from circulating fluids may vary with the concentration of fluids and, thus, the dose (Garrett, 1977, 1978).

If the toxicant is eliminated primarily through metabolic processes, metabolizing enzymes can be limited and saturable. Since the rate of metabolism may be proportional to the amount of substrate–enzyme complex formed, the rate at which the toxicant is removed from the body by a saturable metabolism approaches constancy at high plasma concentrations. Thus, when this rate [Eq. (15)] is no longer proportional to plasma concentration, the metabolic clearance of a toxicant, and therefore the "half-life," can change with dose. Since species have different enzyme capacities, it is not reasonable to assume that they would have the same "half-life" or enzyme saturabilities for the same dose. Saturable metabolisms can permit accelerating accumulation of toxicants in the body on chronic dosing. The net result is that toxicities may not be strictly proportional to dose. An initial high dose may then not produce evidence of toxicity, which could be manifested only with subsequent smaller maintenance doses. Some compounds in this category are diphenylhydantoin, ethanol (Wagner, 1973), and the salicylates (Levy, 1965), whose apparent half-lives in humans vary with dose.

If the loss of toxicant by elimination is not a first-order process, the kinetics becomes dependent on dose or on plasma concentration. Thus, multiplication of a dose by a specific factor does not mean that the amount of toxicant in a given compartment at any given time is multiplied by the same factor. The rates at which an amount P in the central compartment is eliminated by a saturable process (Garrett, 1978; Garrett *et al.*, 1974) after bolus administration can be expressed for the one-compartment body model as follows:

$$dP/dt = dE/dt = k_{max}KP/(1 + KP) = k_{max}KVp/(1 + KVp) \tag{26}$$

where the equilibrium constant K, a measure of the degree of saturability of the process, can vary from a small number to infinity, and k_{max} is the rate at complete saturation, i.e., when $K \rightarrow \infty$. The clearance CL of such a saturable process is generally not constant, but varies as a function of plasma concentration p, since

$$CL = (dE/dt)/p = k_{max}KV/(1 + KVp) \tag{27}$$

unless $KV_p \ll 1$. Thus, a clearance plot of $\Delta E/\Delta t$ against p will not be linear, as in Eq. (15). When the toxicant is eliminated from the body by one saturable process, a semilogarithmic plot of plasma concentration against time cannot be linear (Figs. 9-7A and 9-8B), even in the pseudo steady state, when equilibrations of a toxicant among all the compartments of the body are instantaneous. At the lowest concentrations, when $KV_p \ll 1$, the functional dependence approaches the first-order process of elimination,

$$\lim_{p \to 0}(-dP/dt) = \lim_{p \to 0} dE/dt = k_{max}KVp \tag{28}$$

and since Eqs. (15) and (28) are equivalent at low concentrations, $k_{max}K$ is the overall elimination rate constant λ. The clearance, $k_{max}KV$, tends to a constant value with lowered concentrations and the terminal slope of the semilogarithmic plot of plasma concentration against time (Fig. 9-7A) approaches the constancy represented by $\lambda = k_e = k_{max}K$. A linear plot of plasma concentration against time (Figs. 9-7B and 9-8a) approaches linearity at the high plasma concentrations, since the elimination rate tends to approach the constancy of k_{max} when $KV_p \gg 1$ in Eq. (26) and in the following equation:

$$\lim_{p \to \infty}(-dP/dt) = \lim_{p \to \infty}(dE/dt) = k_{max} \tag{29}$$

There is no explicit solution for plasma concentration as a function of time for such a process (Wagner, 1973), since

$$p_0 - p - \frac{1}{KV} \ln \frac{p}{p_0} = \frac{k_{max}}{V} t \tag{30}$$

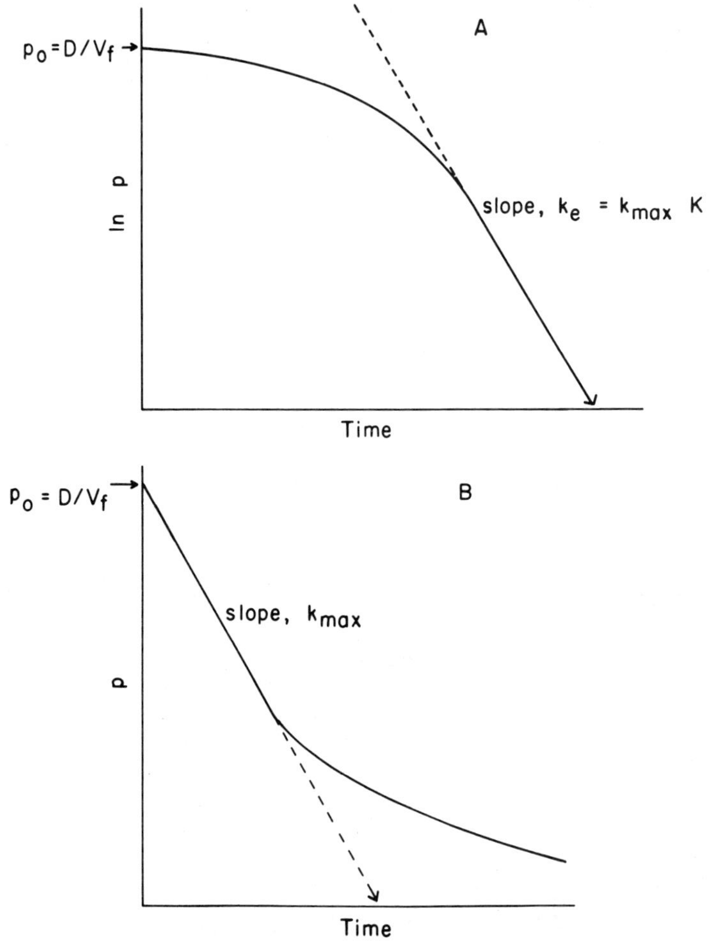

FIGURE 9-7. Typical (A) semilogarithmic and (B) linear plots of plasma levels against time in the one-compartment body model for a toxicant that undergoes a saturable elimination process. (A) The terminal slope of the logarithmic plot against time estimates $\lambda = k_{max}K$, the apparent first-order rate constant for elimination at low plasma concentrations. (B) The initial slope of the linear plot against time estimates k_{max}, the zeroth-order rate for elimination at high plasma concentration. [From Garrett (1978).]

FIGURE 9-8. Examples of (A) linear and (B) semilogarithmic plots of plasma levels against time for a drug that undergoes a saturable elimination process. The drug is amobarbital administered intravenously to a dog at the specified levels. The designated values of $[A_p]_2$ are the intercept values for the disposition phases and are the p_0 values considered in the text for the one-compartment body model. Top: A plot of the initial distribution data for the 5 mg/kg dose on an expanded time scale used to estimate the slope α and intercept ($[A_p]_1$) of the distribution phase by the method of residuals. [From Garrett *et al.* (1974).]

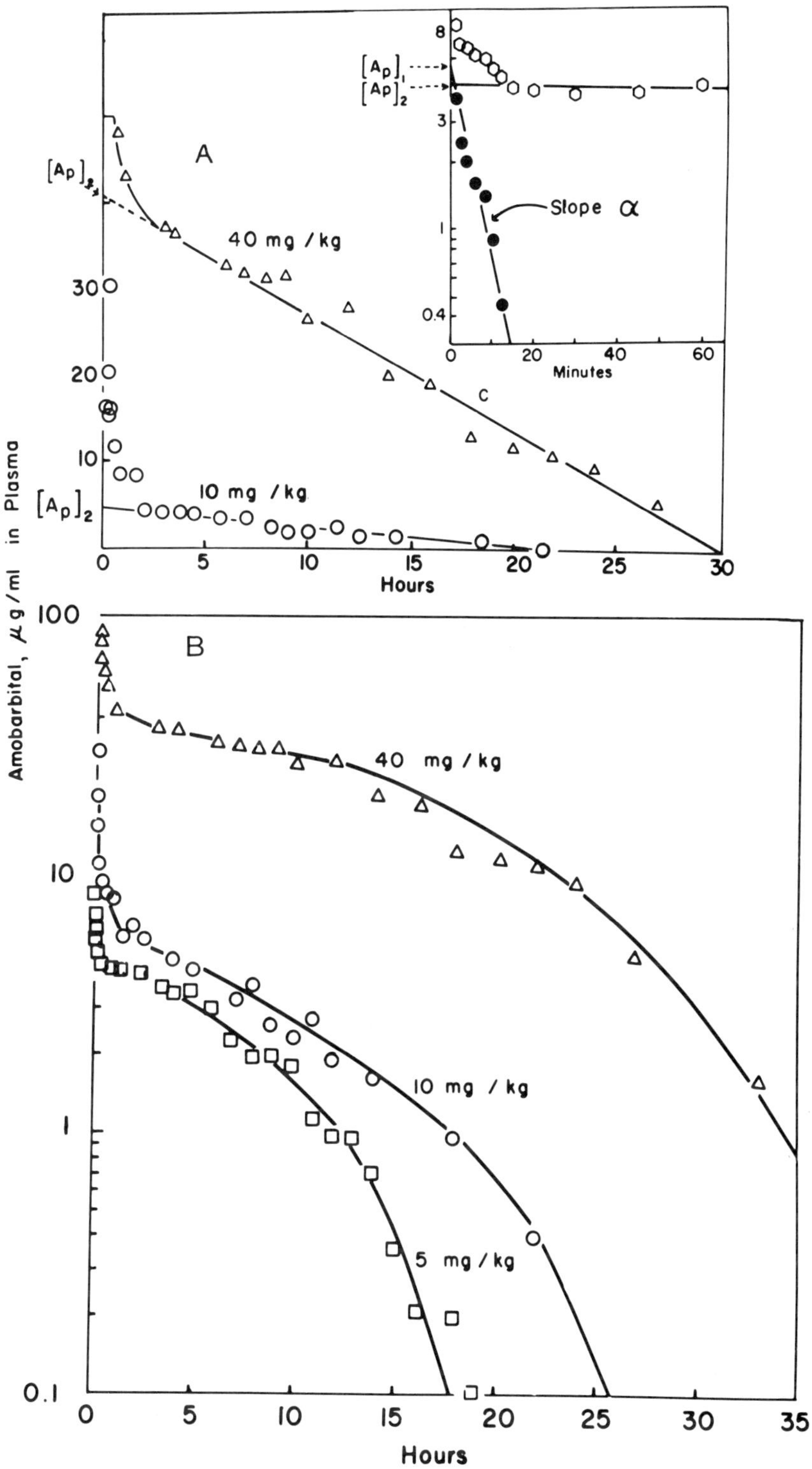
A
[Ap]1
[Ap]2
Slope α
Minutes
40 mg / kg
10 mg / kg
C
Hours
Amobarbital, μg/ml in Plasma
B
40 mg / kg
10 mg / kg
5 mg / kg
Hours

In such a saturable elimination process, the total area under the toxicant plasma concentration–time curve AUC_∞ to infinite time is *not* directly proportional to the intravenously administered dose, as was shown for first-order elimination in Eq. (20). The AUC is a complex function of dose for saturable elimination from the one-compartment body model after intravenous bolus administration of a dose D (Garrett, 1971a; Garrett *et al.*, 1974; Wagner, 1973),

$$AUC = \int_0^\infty p\,dt = \frac{\gamma D^2}{2k_{\max}V} + \frac{\gamma D}{k_{\max}KV} \tag{31}$$

where γ is the fraction of dose absorbed unchanged.

Thus, in contrast to the consequences of Eq. (19) for first-order elimination, where $CL_{\text{tot}} = \gamma D/AUC$, the total clearance for saturable elimination of an intravenous bolus dose is given by

$$CL_{\text{tot}} = \frac{\gamma D}{AUC}\,\frac{K\gamma D + 2}{2} \tag{32}$$

The expression for total clearance also depends on the rates at which nonbolus doses are administered.

The dependence of plasma concentration on time for concomitant saturable and first-order processes is more complex, even if restricted to the one-compartment body model with intravenous dosing of a bolus. The steady-state disposition phase for the loss of plasma levels over time is a linear sum of saturable and first-order processes:

$$-dp/dt = k_{\max}Kp/(1 + KVp) + \lambda' p \tag{33}$$

where λ' is the rate constant for the concomitant first-order process. Thus, on integration, we have the following complex equation:

$$\frac{k_{\max}K + \lambda' - \lambda' KV}{\lambda' KV}\log\frac{k_{\max}K + \lambda' + \lambda' KVp}{k_{\max}K + \lambda' + \lambda' KVp_0} + \log\frac{p}{p_0} = -\frac{k_{\max}K + \lambda'}{2.303}\,t \tag{34}$$

CORRELATION OF TOXICOKINETICS AND ACTION

Simultaneous monitoring of blood concentrations, amounts excreted, and pharmacological or toxic action over time can provide insight into the properties and nature of the biophase compartment in contact with the receptor sites (Garrett *et al.*, 1967).

There may be agents whose half-life of pharmacodynamic or toxic action is much shorter than the classical plasma half-life of first-order kinetics. It may be considered, for example, that there is no blood–brain barrier to thiopental, i.e., that the amount of the drug in the central compartment and the amount in the brain are in relatively instantaneous equilibration. Brain concentrations of this agent are diminished primarily by losses through the central compartment to the tissues of the shallow compartment, which equilibrate relatively rapidly (Mark, 1963).

Although the central compartment may be the site for action, the biological effect may be a consequence of a time-dependent process. A plasma level of drug maintained for long periods may be pertinent for antibacterial, anticancer, or toxic action (Garrett and Hinderling, 1976; Garrett *et al.*, 1967).

If the time course of toxicity parallels the time course of a toxicant's elimination from the shallow compartment, a tissue component of this compartment may be the biophase, and the plasma level would be a proper determinant of toxic action upon both acute and chronic administration.

When the appearance of a toxic action and its maximum effect occur later than would be expected from the plasma level data (*CC*, Figure 9-3) and from the amount of toxicant in the shallow compartment (*SC*, Fig. 9-3), there is a definite implication that the biophase is in a deep compartment. The quantification of toxicity would follow the time course given as *DC* in Fig. 9-3. Such a deep compartment may not be analytically discernible, since insignificant amounts of a toxicant in that deep compartment may be responsible for the toxic action. The actual blood concentrations after acute administration would have *no* direct relationship to the activity of the toxicant, although the magnitude of the activity at any time would still be dose-related, mediated by the time course of the toxicant's existence in the observable compartments. In general, there would be no strict proportionality between toxicity and plasma concentration after acute administration until the rate-determining step becomes the rate of elimination of the toxicant from this deep compartment. Only if the deep compartment can be equilibrated with the plasma upon chronic administration would the plasma concentration be indicative of toxic action at all times.

Figure 9-4 shows the delay before the amounts of an administered toxicant in the deep compartment reach their maximum and that there is no apparent proportionality with the amounts in the shallow and central compartments associated with the blood. Furthermore, it demonstrates that the apparent half-life of the compound in the blood changes with time even when all kinetic processes are first order. The rate-limiting step of removal from the body changes from the renal or metabolic clearances to the limited rate of return from the deep compartment, as shown in the hydraulic analogy of Fig. 9-2. This can explain why many substances dribble out of the body for a much longer period than can be anticipated from given estimates of half-lives.

The presence of this deep compartment may be confirmed by evidence of slowed and almost constant terminal rates of elimination or metabolism of a toxicant, which can be obtained by monitoring the rate at which the toxicant and metabolite appear in the urine or excreta. After chronic dosing, when the body contains a deep compartment with sufficient capacity, large amounts of such a toxicant will be accumulated. The elimination of significant amounts, even after termination of chronic administration, will be slow and prolonged (Fig. 9-5).

Consideration of the embryo or fetus as a deep compartment can demonstrate why a substance that is harmless to the pregnant woman can be toxic to the prenatal organism. Such a substance would tend to persist in the fetus and accumulate there upon chronic administration. The immature organs would be incapable of metabolizing even such low concentrations or of transferring a toxicant or its metabolites efficiently to the maternal body for elimination.

Many drugs, such as tetrahydrocannabinol (Garrett and Hunt, 1977) or β-methyldigoxin (Hinderling and Garrett, 1977), are readily metabolized or eliminated from the central compartment. Their long apparent "half-life" is attributable largely to their slow rate of return from such deep compartments. In fact, the prolonged effects from long-term administration of tetrahydrocannabinol shown by electroencephalogram readings may well be consequences of this accumulation process and of lengthy drug retention rather than of any irreversible brain damage.

Typical physiological references for such deep compartments include the eye. After oral administration of acetazolamide, a carbonic anhydrase inhibitor, the maximum decrease in the eye's outflow pressure occurred long after the maximum in plasma concentration was reached (Lehmann *et al.*, 1970).

A pus pocket or a tubercular lesion may also be considered as a deep compartment (Garrett and Hinderling, 1973, 1976). Fat for polychlorinated aromatic hydrocarbons and bone for heavy metal ions are other possible deep compartments (Garrett *et al.*, 1962, 1963).

A toxicokinetically observed deep compartment is not necessarily the same as, or equivalent to, the biophase, which may be in a deep compartment. There must be parallelism between the time course of biological activity and the amounts of toxicants analyzed in such a compartment as evidence for such an equivalence.

Delayed onset and delayed attainment of maximum activity of an acutely administered compound with respect to its observable time courses in the central and shallow compartments may implicate its metabolites as the active or coactive species. This can be ascertained by comparing the time courses of the activity and the metabolites after administration of the compound. Correlation of the kinetics of the administered metabolite and the time course of toxicity is needed.

The action of a toxicant on its receptor sites may not produce an immediately discernible effect. The resultant toxicity may be mediated by a biochemical process and delayed. Only after sufficient time has elapsed may certain critical components be depleted or accumulated to those levels that permit toxicity. The action of sulfonamide on microorganisms is a case in point. The inhibition of the production of *p*-aminobenzoic acid, a folic acid precursor, was not manifested until the metabolic store of *p*-aminobenzoic acid was depleted by subsequent division of the microorganisms. The bacteriostatic action of sulfonamide was not demonstrated until five generations had passed after it was introduced (Garrett, 1971b).

Occupancy Receptor Site Theory

A simple model that can be used to rationalize the pharmacodynamic, biological, or toxic action of an agent is based on the presumption that response is proportional to the numbers of receptor sites that are occupied (Garrett *et al.*, 1967). These receptor sites are in equilibrium with the molecules of the toxicant in its biophase. When the sites are limited and completely occupied, the maximum biological response is achieved.

If the biological response is alteration or an inhibition of a necessary biological process, the degree of inhibition I can be expressed as a function of the concentration C of the toxic agent in the biophase:

$$I = aC/(1 + aC) \tag{35}$$

where a is a proportionality constant (Garrett, 1977).

Thus, at high concentrations, $C \to \infty$, the degree of inhibition of the process is total, $I = 1$, and the maximal toxic response is achieved. At very low concentrations, $aC \ll 1$, the degree of inhibition is proportional to the fraction of occupied receptor sites and to the biophasic concentration:

$$I = aC \tag{36}$$

Other Aspects of Receptor Site Models

Quantitative theories of the action of drugs and other biologically active agents are based on the law of mass action (Ariens, 1964; Clark, 1933; Garrett, 1977; Garrett *et al.*, 1967). It is assumed that the drug or agent reacts with pertinent receptors and that the biological response is proportional to, or is some function of, the extent of this reaction or combination. The maximum effect in this "occupation" theory is attained when a maximum number of receptors is occupied. This does not necessarily mean that all receptors must be occupied for the maximum effect to occur, nor does it necessarily mean that the biological effect is linearly related to degree of occupancy (Paton, 1961). There are two reasons that a biological effect may diminish with dose or biophase concentration: either the receptor sites are saturable and the degree of occupancy does not increase, or there is a nonlinear functional dependence of biological response on degree of occupancy.

In an alternative model for action on receptor sites, the effect of an agent is proportional to the rate of its reaction with the receptors (Gibaldi *et al.*, 1972; Paton and Waud, 1964). In this "rate" theory, the agent–receptor reaction will attain a constant rate with a concomitant constant biological effect when an equilibrium between the agent and receptor site is established. As the agent–receptor complex subsequently dissociates due to excretion or degradation of the agent, the biological effect will decay.

The main difference between the time courses of biological response in the "occupation" and "rate" theories occurs at an early stage. In the former case, the effect increases with time with increased occupancy of sites. The subsequent decay occurs upon dissociation from the receptor sites caused by displacement of the equilibrium when the agent diffuses away from the biophase as a result of its excretion, degradation, or metabolism. In the latter case, if the agent is administered directly into the biophase (i.e., the compartment containing the effective receptor sites), the maximum effect occurs immediately upon administration and then fades to an equilibrium value due to a decreased number of unoccupied sites (Paton, 1961; Paton and Waud, 1964). Subsequent decay is also due to removal of the agent from the biophase by various elimination processes.

This "fading" with time to an equilibrium value can only be observed in those cases in which the agent has immediate access to receptor sites; it becomes less and less observable when diffusion of an agent to a biophase becomes more and more rate-determining. The presence of receptor sites in a peripheral tissue or deep compartment would permit diffusion-controlled entry of an agent into the biophase. Mackay (1966), who reviewed and compared the "occupation" and "rate" theories, concluded that there is no strong reason for discarding the "occupation" theory.

TOXICOKINETICS AND TOXICOLOGY

Toxic responses at the high doses administered in standard toxicity studies are not necessarily extrapolatable to low doses. It can be argued that all toxicokinetic processes of metabolism or elimination are saturable at sufficiently high doses, but that this is not necessarily true at low doses. Thus, the homeostatic repair processes of the body may be overwhelmed only at the high doses and not at the lower ones. Elimination rates at high doses may not vary with concentrations in the body at high doses, although they are functions of concentration at low doses. Thus, the duration of toxicant action in the biophase is relatively much greater at the higher doses than in normal dose ranges. Thus, measurements of toxicity at high doses may be of uncertain predictive value at the lower doses unless the kinetics of the toxicant is well known and the saturable toxicokinetics has been well delineated.

Experimental Methods to Determine if Toxicokinetics Is Dose-Independent

The kinetics of a toxic agent should be delineated in animals as a function of dose to determine if there are first-order transfers within, and first-order eliminations from, the body (Garrett, 1970). The doses should range from levels at which toxicity is well manifested down to the lower levels for which toxicity is to be predicted. The simplest method to challenge for dose-dependent toxicokinetics is to monitor plasma and excreta concentrations of specifically assayed toxicants and metabolites at various administered doses, preferably in the same animal. If the quotient of each concentration and dose (e.g., p/D or E/D) gives superimposable curves over time for all doses, it can be concluded that the toxicokinetics of the agent is first-order, linear, and dose-independent. The parameters of curve fitting [Eqs. (2)–(4)] would be coincident, and the clearances [Eqs. (14), (16), (18), and (19)] would be constant for all doses within the time course for any dose administered.

The accumulation of toxicant in the body can be readily predicted if dose-independent toxicokinetics exists. The simplest procedure that can be used to determine this is to plot the expected time courses of plasma concentrations for each individual dose, and then to add the specific plasma concentrations that would have resulted if each dose had been administered separately. The arithmetic sum constructs the plasma concentration–time curves that would result after repetitive administration. Alternatively, the cumulative levels could be calculated as shown in Eqs. (7)–(12).

The calculation of the time course of a toxicant in the body when saturable or dose-dependent toxicokinetics exists is more complex. Even the simplest example of one saturable elimination process in a one-compartment body model does not have an equation that provides an explicit solution for determining plasma concentrations over time [Eq. (30)].

When p/D or E/D plots against time for various doses cannot be superimposed, toxicokinetics is dose dependent and there are many possible causes. The saturation of plasma protein-binding sites could result in increased rates of elimination or metabolism, since such rates depend on the concentration of free, unbound toxicant in plasma, which is relatively greater at the higher doses (compare curves B and A in Fig. 9-9). The saturation of insufficient quantities of metabolizing enzymes would result in decreased rates of metabolism and subsequent loss of toxicant blood concentrations at higher doses (compare curves C and A in Fig. 9-9). Alternative explanations for such a phenomenon could be toxicant-inhibited hepatic blood flow at the higher plasma concentrations or metabolite-inhibited enzymes. Substrate induction of metabolizing enzymes would result in increased metabolism rates with each sequential study (compare curves D and A in Fig. 9-9). An anomalous increase in toxicant blood level subsequent to the distributive toxicokinetic phase may be assigned to an enterohe-

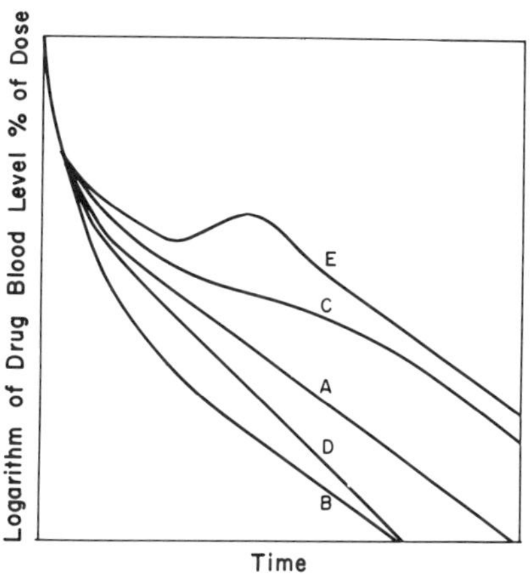

FIGURE 9-9. Anomalous deviations from the simple model of toxicant distribution into tissues with subsequent first-order elimination. T ⇄ B → E when the logarithm of the blood level of the drug is plotted against time as a percent of dose. (A) If the model holds, log of the blood level as percent of dose curves versus time are superimposable for all doses. (B) Curve for higher doses when blood protein shows saturable binding. (C) Curve for higher doses when metabolizing enzymes are saturable. (D) Curve for same or different dose after enzyme induction. (E) Form of curves when there is an enterohepatic shunt. [From Garrett (1970).]

patic shunt—the biliary excretion of a toxicant or its conjugate, where the latter is split in the gastrointestinal tract, where the toxicant, reintroduced into the gastrointestinal tract, is reabsorbed at later times.

Other dose-related toxicokinetic phenomena may involve changes in diffusivity, permeability, renal clearance (saturable tubular reabsorption or excretion), biliary clearance, or size of distributive compartments as functions of dose. There are possible diurnal variations in, or substrate-modified effects on, urinary pH that could modify renal tubular reabsorption.

Methods for testing these possibilities experimentally have been outlined (Garrett, 1970, 1971a, 1978).

Calculation of Amounts Excreted, Metabolized, and Retained by the Body

The amounts of toxicant metabolized or eliminated can be readily calculated for dose-independent toxicokinetics from the obtained constant clearances [Eqs. (16), (18), and (19)] by Eq. (17). For dose-dependent toxicokinetics, the clearances are more complex functions [Eq. (27)].

If the clearance of a metabolite CL_M has reached constancy, the plasma concentration–time course of the metabolite concentration p_m can be calculated (Garrett and Jackson, 1980) as follows

$$p_m = \frac{CL_{A\to M} \cdot AUC_A - CL_M \cdot AUC_M}{V_M} \tag{37}$$

where AUC_A and AUC_M are the respective areas under the plasma concentration–time curves of the toxicant A and metabolite M up to the time that plasma concentration p_m has been reached, and $CL_{A\to M}$ is the constant clearance of toxicant to that metabolite. The V_M is the overall apparent volume of distribution of the metabolite based on the presumption that upon formation it is readily equilibrated among all the body's tissues, i.e., the one-compartment body model is operative for the metabolite.

The parameters $CL_{A\to M}$ and V_M can be experimentally determined. When a formed metabolite is excreted solely into the urine, $CL_M = CL_M^{\text{ren}}$ and the cumulative amount of metabolite excreted in the urine is either $U_M = CL_M^{\text{ren}}(AUC_M)$ from Eq. (17) or the measured amount. The modified Eq. (37) can be rearranged as follows:

$$\frac{U_M}{AUC_A} = CL_{A\to M} - V_M \frac{p_M}{AUC_A} = \frac{CL(AUC_M)}{AUC_A} \tag{38}$$

Thus, a plot of the quotient U_M/AUC_A at a time t against the quotient p_m/AUC_A should give a straight line where the negative of the slope $-V_M$ is the overall volume of distribution of the metabolite and the intercept is the clearance $CL_{A\to M}$ of the drug A to the metabolite M.

The determined apparent volumes of distribution can relate amounts in the central compartment [Eq. (1)] or in the body after acute [Eq. (25)] or chronic [Eq. (22)] administration [Eq. (21)].

Some Caveats for Toxicokinetics

A primary function of toxicokinetics in toxicological appraisals is to estimate how long a toxic agent and its possibly toxic metabolites persist in the body, based on the reasonable assumption that toxicity relates to this persistence. The kinetics of a toxic substance can be truly delineated only if specific and sensitive assays are available to monitor the time courses of the toxicant and its metabolites in the circulating fluids, tissues, and excreta of the intact animal.

A frequently used procedure is to administer a radiolabeled dose of a toxicant and to monitor the total radioactivity over time in the tissues and excreta. This can provide important information concerning the time that total elimination of the toxicant and its derived substances occurs for a given dose. However, it does not provide any information about the separate time courses of the toxicant and its metabolites. If the toxicokinetic activity of radiolabeled substances is to be monitored, the toxicant and its metabolites should be separated before counting, despite the relatively small increase in cost. In such procedures as thin-layer chromatography and high-pressure liquid chromatography, the separated fractions can be radioactively monitored. Even if the separated metabolites have not been structurally characterized, the quantitative monitoring of the chromatographic separations can permit delineation of the toxicokinetics of each component and provide quantitative expressions for transformations and eliminations (Garrett, 1970).

Anomalous blood concentration–time curves will result when total radioactivity of the toxicant and its metabolites is monitored and when there are significant circulating levels of metabolites. Examples are given in Fig. 9-10 for the semilogarithmic curves of total radioactivity against time, where rapid equilibrations among the tissues of distribution are assumed and the toxicant and metabolite eliminations are first order. Curve A is representative of the true curve for

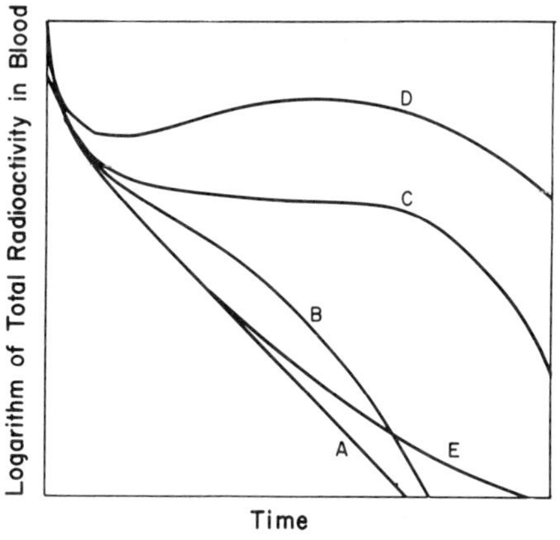

FIGURE 9-10. Typical curves for the monitoring of total radioactivity concentrations of the blood with time when circulating levels of a metabolite exist. (A) Toxicant is excreted unchanged or a formed metabolite is immediately excreted. (B) A metabolite with similar volumes of distribution as the toxicant is formed. (C, D) Metabolites are less apt to distribute into tissues or are highly protein-bound. (E) Metabolites are preferentially distributed into tissues and are slowly eliminated from the body. [From Garrett (1970).]

the toxicant alone and also represents a situation where no metabolites are formed, or when they are excreted as rapidly as they are produced. If a radiolabeled metabolite is formed with the same volume of distribution among blood and equilibrating tissues as that of the toxicant, curve B will be observed. Curves of types C and D are anticipated when the metabolites are less apt to distribute in the equilibrating tissues or are preferentially bound to protein. Curve E may result when a major metabolite is preferentially distributed into the equilibrating tissues, which act as a deep compartment, or when it is preferentially trapped in an intermediary compartment and has a slower rate of elimination than does the toxicant.

The renal clearance of a toxicant, as monitored by its appearance in the urine, usually depends on the concentration of the free or unbound toxicant in the plasma (Garrett, 1971a). A plot of this rate of appearance in the urine should be linearly related to the toxicant's concentration in the plasma if the extent of protein binding does not change with concentration. This would also be true if the rate of radioactivity increase in the urine dR_U/dt is plotted against the concentration of radioactivity in the blood r when the toxicant is excreted unchanged (Fig. 9-11, curve A). A similar plot is to be expected when a formed radioactive metabolite is rapidly equilibrated among the body's tissues and has similar apparent volumes of distribution and similar rates of glomerular filtration. However, these are unusual circumstances. Such plots will normally be nonlinear. If we consider equivalent rates of glomerular filtration and rapid distribution among the equilibrating tissues, curve B of Fig. 9-11 would result when the apparent volume of distribution of the metabolite exceeds that of the toxicant, and curve C would result when it is less. When monitoring the time courses of a radioactive dose, nonlinearities in such plots are useful indications of formed metabolites.

A BASIC TOXICOKINETIC MODEL FOR TOXICOLOGY

In contrast to pharmacology, toxicology deals with the effects of and responses due to large doses of substance, which can more readily overwhelm the

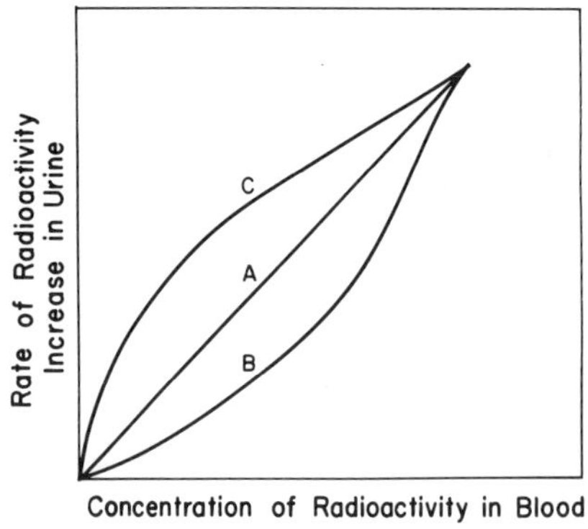

FIGURE 9-11. Possible relationships between rates of excretion of radioactivity and concentrations in blood with time. (A) No metabolites formed, or they have the same volume of distribution as the toxicant. (B) Metabolites have greater volumes of distribution than toxicant. (C) Metabolites have smaller volumes of distribution than toxicant. [From Garrett (1970).]

homeostatic capacity of the body to adjust to or rid itself of the insults of foreign compounds. Toxicokinetics is concerned with saturable excretions, metabolic pathways, and receptor sites.

There is a need for a generalized model capable of quantifying the time course and dose dependences of toxicity. There is also a need for a model that incorporates the concept of threshold concentrations or doses for toxicity to be manifested. Such a model should be able to accommodate potential saturation of elimination and receptor loci and partial or total reversibility of receptor-site binding. Such a model should permit interpretation of toxicological data for predictive purposes. Observed nonlinearities in dose–response correlations could enable investigators to predict toxicities at the lower chronic doses that cannot be efficiently monitored by classical toxicological technology; there is an economic and logistic limit to the numbers of animals that can be subjected to the low stresses needed, for example, to determine the LD_{01}. The apparent linearization of data by the existing probit or logit transformations of toxicological data is no guarantee that these models are capable of such prediction. Nevertheless, regulatory agencies need such predictive models to bridge the gap between the low levels of exposure to atmospheric, marine, or food toxicants and the knowledge of responses at the high doses obtained in the laboratory.

A model of toxicity has been developed (Garrett, 1977) that has capabilities for analysis and prediction. Its fundamental premise is that the response rate (the number of organisms dying or exhibiting a specific toxicity per unit of time) is proportional to the number of unaffected organisms and to the degree of alteration of a critical metabolic or biological activity beyond what can be compensated or repaired by existing or inducible homeostatic mechanisms. The identity of the critical metabolic or biological activity need not be known; it is necessary only that key processes exist. The degree of alteration can be expressed as a function of the concentration of the toxicant in the biophase, which in turn can be expressed as a function of administered dose (e.g., concentration in an equilibrating atmosphere) over time. The model has been developed for homogeneous populations in which individual susceptibilities to toxicants are equal and do not vary within the time of the study. The spectrum of susceptibilities in a heterogeneous population would necessitate the introduction of statistical considerations.

The simplest possible form of the model considers toxicities after a distribution steady state is reached. This is generally applicable to populations of organisms suspended in a fluid containing constant concentrations of toxic agents that rapidly equilibrate with the distribution volumes of the organisms. Examples of such populations are marine organisms in fresh or ocean waters and bacteria in nutrient media. These simple models are also valid for chronic oral or intravenous doses where the toxicant accumulates in the organisms to steady-state concentrations at relatively rapid rates.

In a more general case, an induction period will occur in a chronic study since a measurable time must elapse before the amount of toxicant in the biophase can attain its steady-state concentration. The monitoring of toxicities may show increasing responses over time under these circumstances. Zero responses would be observed until the time-dependent concentration of the toxicant in the biophase accumulated in excess of the threshold amount.

It is important to distinguish whether an apparent time lag in toxicity manifestation is caused by the induction phase, which can be attributed either to time-dependent accumulation in the biophase, or to the existence of a toxicity threshold.

When the saturable processes of elimination are overwhelmed by rates of administration or absorption, concentrations in the body can accumulate without bound. Under such circumstances, toxicities can accelerate with time.

Janků and Farghalli (1972) have proposed that the mortality rate is proportional to the difference between toxicant concentration and the maximum concentration that can be tolerated. A more fundamental restatement of this thesis would be that the number of organisms or animals dying or exhibiting specific toxicities per unit of time is proportional not only to the net degree of complete inhibition I (or of alteration) of the biological or metabolic action necessary for organism survival (or resistance to toxicity), but also to the number of organisms N. Thus

$$-dN/dt = k_D(\Delta I)N = k_D(I - I_m)N \tag{39}$$

when $I > I_m$. The net degree of inhibition ΔI is a finite value when the degree of inhibition I exceeds a value I_m below which the repair or homeostatic capacity of an individual permits the biological processes to continue at a level that maintains survival. If the degree of inhibition is less than I_m, the death or toxicity rate $-dN/dt$ is zero. The implicit premise is that the repair capacity in every individual organism is the same. In the ensuing discussion, resistance to toxicity can be equated with survival, and manifestation of toxicity can be equated with death.

The postulates imply that the larger the net degree of inhibition, the greater the probability that an individual death will occur in a shorter time. The inhibitory mechanism eventually results in the critical depletion of some biologically necessary product or process. Thus, death results ultimately in all organisms when I consistently exceeds I_m of a substance—a condition that will occur upon chronic administration.

It follows (Garrett, 1977) that when $I > I_m$,

$$\int_{N_0}^{N} \frac{dN}{N} = -\int_0^t k_D(I - I_m)\, dt \tag{40}$$

where N_0 is the number of individuals alive at time zero and N is the number of individuals alive at time t. Thus

$$\ln f_A = \ln (N/N_0) = -\int_0^t k_D(I - I_m)\, dt \tag{41}$$

and the fraction f_A of organisms that remain alive at any time t is

$$f_A = N/N_0 = \exp\left[-\int_0^t k_D(I - I_m)\, dt\right] \tag{42}$$

where all ultimately die if $I > I_m$ and $f = 0$ at infinite time. Also, $f_A = 1$ for all values of $I \leq I_m$ for all times.

DEDUCTION OF TOXICOKINETICS WITH BIOPHASIC STEADY-STATE CONCENTRATIONS OF TOXICANT

Chronic administration of a given dose of toxicant can maintain a constant concentration of the toxicant in its target biophase provided that the elimination process is not saturable. Thus, the net degree of action is kept constant and Eq. (42) reduces to

$$f_A = \exp\left[-k_D(I - I_m)\int_0^t dt\right] \tag{43}$$

which, upon integration, becomes

$$\log N = -[k_D(I - I_m)/2.303]t + \log N_0 \tag{44a}$$

or

$$\log f_A = -[k_D(I - I_m)/2.303]t \tag{44b}$$

which is consistent with a linear plot of the logarithm of the number alive, log N, or of the logarithm of the fraction alive, f_A, against time when a constant amount of toxic agent in the animal results in a constant degree of inhibition I.

Thus, the slope of such a plot is

$$\text{slope} = -k_D(I - I_m)/2.303 = -k_{app}/2.303 \tag{45}$$

where k_{app} is inversely proportional to the time $t_{50\%}$ of 50% fatality (when $N/N_0 = 0.5$). The k_{app} is also directly proportional to the negative logarithm or cologarithm of the fraction of organisms that survived, $-\log f_A = -\log(N/N_0)$, at any given duration of dosing t_D:

$$k_{app} = -\frac{2.303\log(N/N_0)}{t_D} = \frac{0.693}{t_{50\%}} \tag{46}$$

The degree to which the necessary biological process is inhibited should be related to a concentration C of the toxic agent in the biophase. In an occupied receptor site model with a finite number of receptor sites, the degree of inhibition can be expressed as a function of this concentration [see Eq. (35)].

First-Order Toxicokinetics with Constant Biophasic Toxicant Concentrations

The concentration in the biophase, equilibrated with all the tissues in the body, is proportional to the dose of the toxicant administered chronically when the toxicant distribution and elimination rates adhere to first-order kinetics.

In general, the rate of change dA/dt of an amount A of toxicant in the body is equal to the difference in the rate of administration $dD/dt = \text{Dose}/\Delta t_D$ and its rate of elimination dE/dt:

$$\frac{dA}{dt} = \frac{dD}{dt} - \frac{dE}{dt} \tag{47}$$

The rate of change for amount A of a chronically administered toxicant in the body at steady state is zero. Thus,

$$\frac{dD}{dt} = \frac{D}{\Delta t_D} = \frac{dE}{dt} \tag{48}$$

where D is the dose administered at intervals of time Δt_D. If the rate of elimination is first order, then

$$\frac{D}{\Delta t_D} = \frac{dE}{dt} = \lambda A = \lambda VC \tag{49}$$

where V is the apparent volume of distribution containing the concentration C, which can be formulated as follows:

$$C = \frac{D}{\lambda V \Delta t_D} = \frac{a'}{a} D \tag{50}$$

where the plasma concentration is directly proportional to the dose D and $a' = a/\lambda V \Delta t_D$. Substitution of this value into Eq. (35) for saturable receptor sites and consideration of Eq. (45) results in

$$-2.303 \cdot \text{slope} = k_{\text{app}} = \frac{a' k_D D}{1 + a'D} - \frac{a' k_D D_m}{1 + a' D_m} \tag{51}$$

where D_m is the dose that gives the steady-state concentration C_m in the biophase that results in the inhibition I_m that does not cause death. The $k_{\text{app}} = k_D(I - I_m)$

is defined as the apparent mortality (or toxicity) rate constant. Typical plots for various doses are given in Fig. 9-12.

At high doses, the maximum apparent mortality rate constant is

$$\lim_{D\to\infty} k_{\text{app}} = (k_{\text{app}})_\infty = k_D - \frac{a' k_D D_m}{1 + a' D_m} \tag{52}$$

These principles have been confirmed in studies with the antibacterials aminosidine (Garrett and Lewis, 1975) and sulfisoxazole (Garrett and Wright, 1967). These compounds are only bacteriostatic below certain minimum concentrations $C_{\text{min}} = D_m/V$ in a volume of nutrient fluid. Above a minimum bactericidal concentration there is an invariant ultimate first-order death rate for microorganisms characterized by a maximum apparent mortality rate constant $(k_{\text{app}})_\infty$, as in Eq. (52). In this particular *in vitro* system, $\lambda \Delta t_D = 1$.

This example of microorganism toxicity (Fig. 9-13) with saturable receptor sites shows the aminosidine is bacteriostatic below a minimum value and that the organisms demonstrate an invariant ultimate first-order death rate for organisms inhibited by these agents. There appears to be a C_m value below which there is no biocidal activity; otherwise, viable and total counts would not be coincident in the subbactericidal concentration range.

It can be shown from Eqs. (51) and (52) that

$$\frac{1}{(k_{\text{app}})_\infty - k_{\text{app}}} = \frac{1}{k_D} + \frac{a'D}{k_D} \tag{53}$$

and a plot of the reciprocal of the difference between the maximum mortality rate constant $(k_{\text{app}})_\infty$ and the constant k_{app} against the chronic dose for which it was

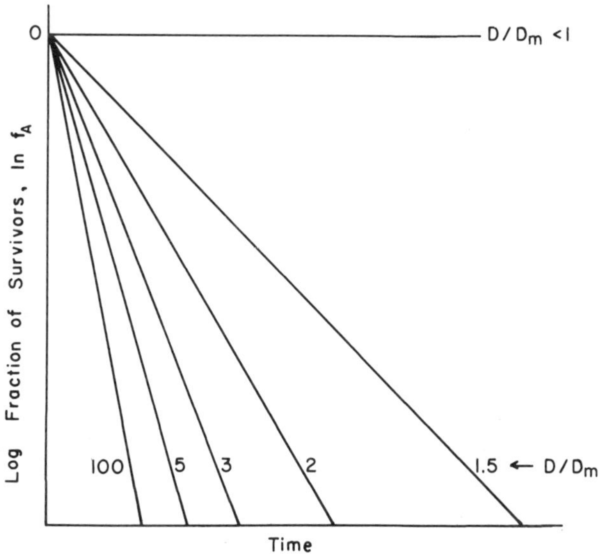

FIGURE 9-12. Typical plots of the logarithm of the survivors $\ln f_A = \ln (N/N_0)$ as a function of time in a homogeneous population of N_0 organisms when the ratio of a chronic dose D to the chronic dose that does not cause death or toxicity D_m is varied. The elimination rate constant λ is first order and the inhibitory process that results in death can be saturable. The slopes of such plots approach a constant $[k_D = (k_{\text{app}})_\infty]$ with increasing dose unless the inhibitory process is not saturable. When $D < D_m$ there is no toxicity and the slope k_{app} is zero. [From Garrett (1977).]

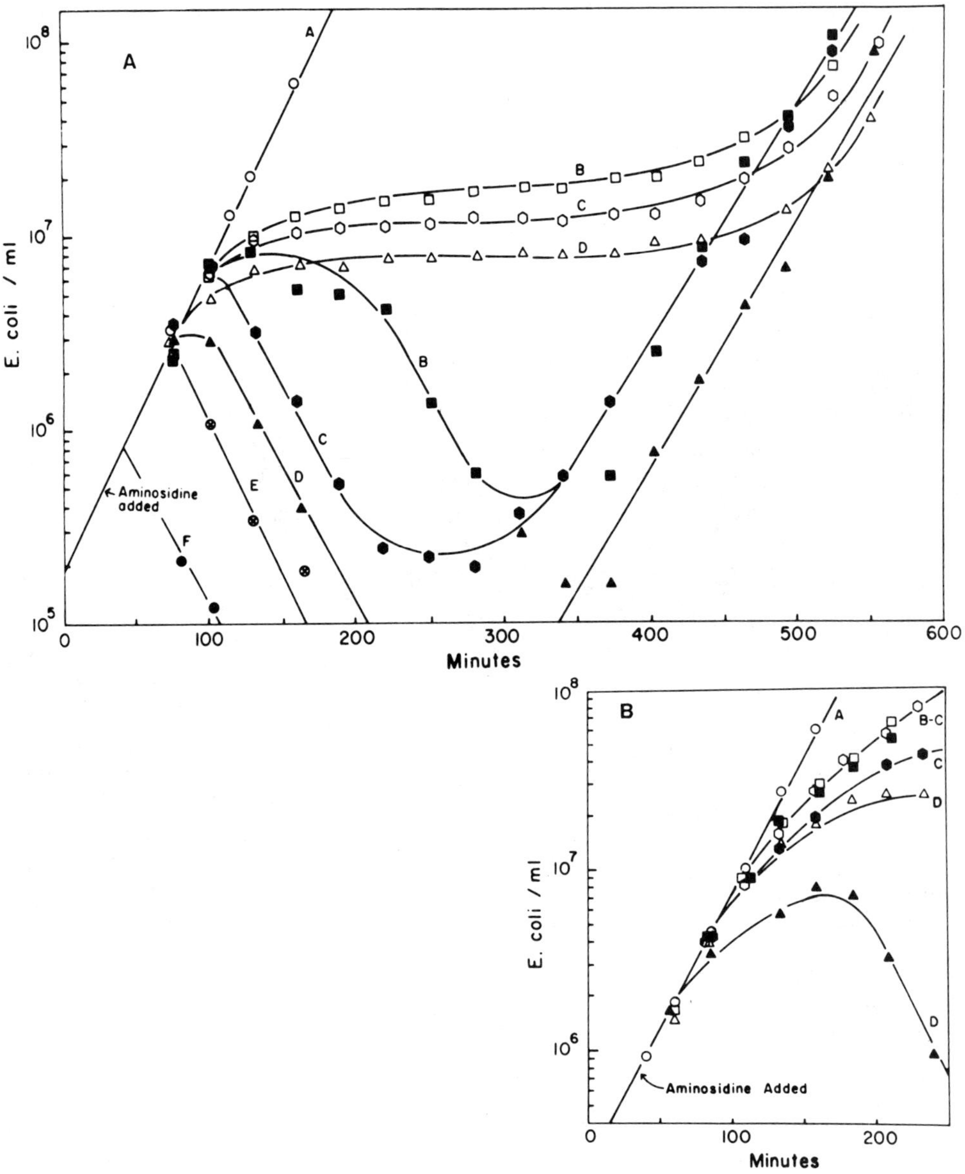

FIGURE 9-13. Drug-affected generation and death rates of *E. coli* in the presence of aminosidine sulfate at pH 7.00 in peptone broth USP. Open symbols are total counts, and solid symbols are viable counts. The curves are labeled with the drug concentration in micrograms per milliliter. [From Garrett and Lewis (1975).]

observed gives a slope of a'/k_D and an intercept of $1/k_D$, from which the values of k_D and $a'D$ can be calculated. Rearrangement of Eq. (52) then permits one to estimate the minimum toxic dose as follows:

$$D_m = \frac{k_D - (k_{\text{app}})_\infty}{a'(k_{\text{app}})_\infty} \tag{54}$$

If D_m is very small or negligible, k_D approximates $(k_{\text{app}})_\infty$ in Eq. (54). Under these circumstances, Eq. (51) can be rearranged as follows:

$$\frac{1}{k_{\text{app}}} = \frac{1}{a'k_D}\frac{1}{D} + \frac{1}{k_D} \tag{55}$$

where $k_D = (k_{\text{app}})_\infty$ when D becomes large. A plot of the reciprocal of the mortality rate constant k_{app} against the reciprocal of the chronic dose gives a slope of $1/a'k_D$ and an intercept of $1/(k_{\text{app}})_\infty$, from which the values of a' and $(k_{\text{app}})_\infty = k_D$ can be calculated when $D_m = 0$.

If the receptor sites are nonsaturable, the degree of inhibition is directly proportional to the chronic dose and the mortality rate constant is a linear function of dose:

$$k_{\text{app}} = a'k_D(D - D_m) \tag{56}$$

The apparent linear dependence of the kill rate of bactericidal concentrations of penicillin, kanamycin, and rifampin (Garrett and Won, 1973) at their studied concentrations implies that Eq. (56) holds in those cases (Figs. 9-14 and 9-15).

The action of bactericidal agents on microorganisms provides excellent examples of achieved steady-state toxicant concentrations in the biophase. The action of various concentrations of antibiotics in the nutrient medium of *Escherichia coli* is demonstrated in Fig. 9-14A for penicillin G and in Fig. 9-14B for kanamycin. The curves are similar for rifampin (Garrett and Won, 1973). The significant decreases in viable *E. coli* per millimeter as a function of time, compared to the total numbers obtained by the Coulter counter, show that penicillin, kanamycin, and rifampin kill organisms. After the drugs were applied, there were definite lag periods before the decreasing slopes of logarithms of viable numbers plotted against time eventually became linear. These lag times were due to the fact that drug partitioning from the medium into and through the bacterial membrane was not instantaneous. The durations of these intervals were apparent functions of drug concentration.

The apparent first-order generation rate constants k'_{app} were obtained from the slopes of the later linear portions of the semilogarithmic plots, as shown in Fig. 9-14. This reflected the toxicant-affected steady state achieved prior to the emergence of resistant organisms and/or inactivation of the toxicant—facts that can explain the terminal increase in microbial generation for penicillin. Kanamycin and rifampin were not inactivated in their cultures, but resistant organisms did develop in their presence.

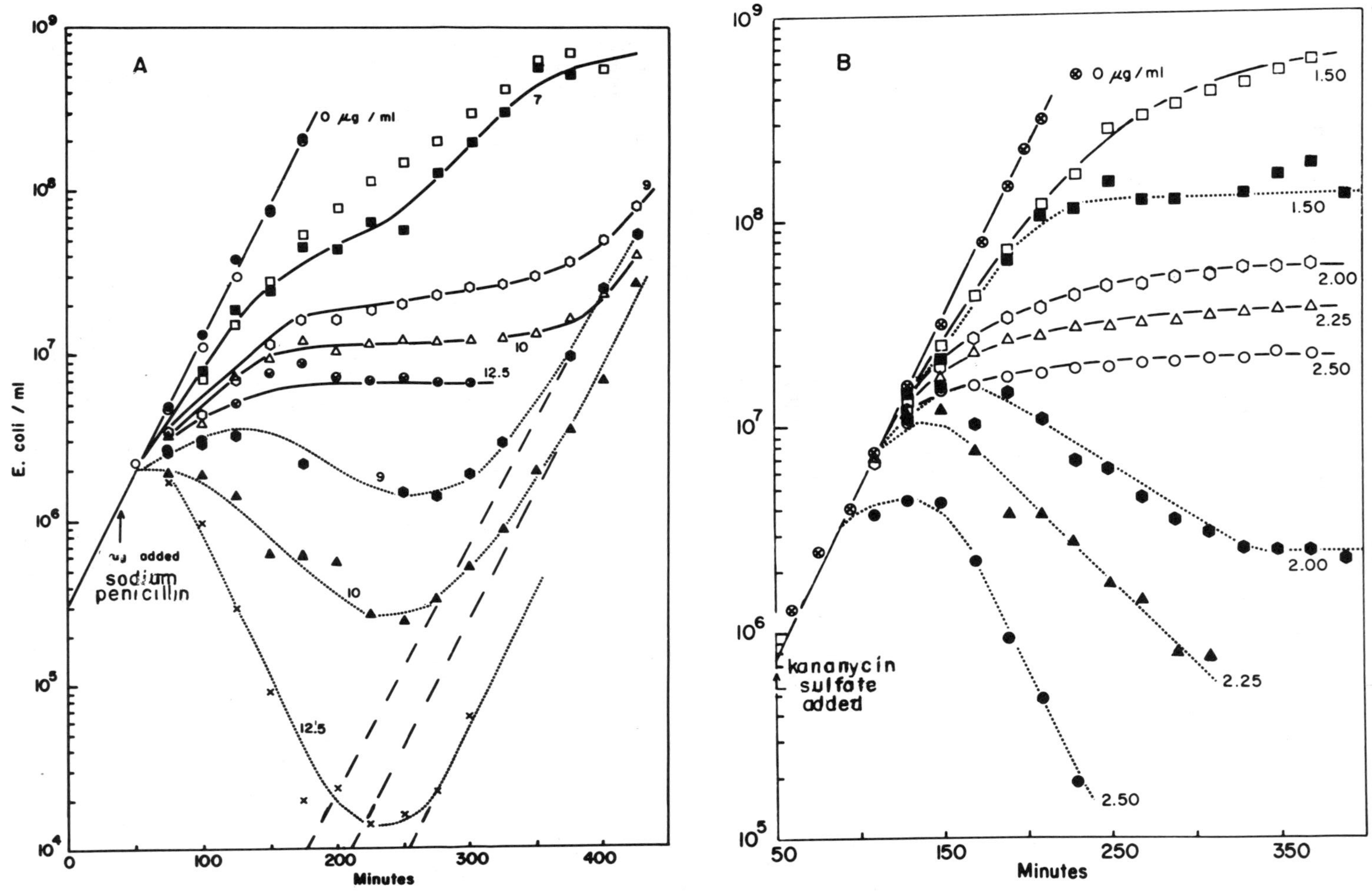
A
B
E. coli / ml
Minutes
Minutes
0 μg / ml
7
9
10
12.5
sodium penicillin
added
kananycin sulfate added
0 μg/ml
1.50
2.00
2.25
2.50

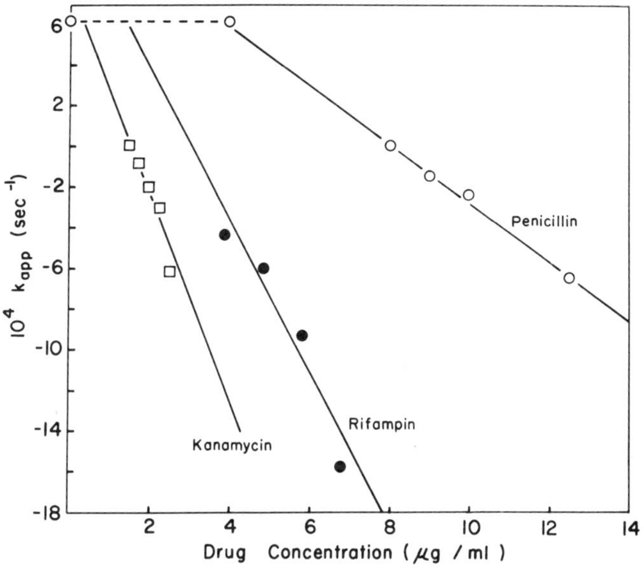

FIGURE 9-15. Dependence of the apparent *E. coli* generation rate constant k'_{app} on sodium penicillin C, kanamycin sulfate, and rifampin concentrations when the curves are constructed from two separate experiments. [From Garrett and Won (1973).]

The apparent generation rate constant $k'_{app} = k_0 - k_{app}$ can be obtained from the slope of the logarithm of the numbers of viable organisms plotted against time, where k_{app} is the mortality rate constant defined in Eq. (56) and k_0 is the generation rate constant in the absence of the drug.

The plot of these estimated k'_{app} values against drug concentrations demonstrated reasonable linear dependences for all three antibiotics, thereby confirming the apparent nonsaturation of receptor sites in the concentration ranges studies. The fact that the plots show infinite values of concentration C when $k'_{app} = k_0$ implies the existence of a C_m—a concentration of antibiotic that has no toxic effect on the organism.

All the above arguments are based on the premise that there is biological identity among individuals and relatively constant rates of chronic dosing. When biological variability in elimination rates and in the minimum toxic dose D_m is considered, typical log survivor–time curves on chronic dosing may not be completely linear, except at the higher doses. Typical curves for log survivors against time are given in Fig. 9-16 for such variable populations. Curve a represents the dose below the minimum toxic dose D_m for all animals. In curve b, the D_m for only some of the animals may be exceeded and the others survive. In curve c, at some higher dose, all animals die eventually, but those with the higher D_m will

←

FIGURE 9-14. Typical generation curves of *E. coli* in the absence and presence of various concentrations of (A) sodium penicillin G and (B) kanamycin sulfate, obtained by (—) total and (- -) viable counts. The curves are labeled with the drug concentrations in micrograms per milliliter. (· · ·) The logarithmic growth of the emergent resistant organisms. [From Garrett and Won (1973).]

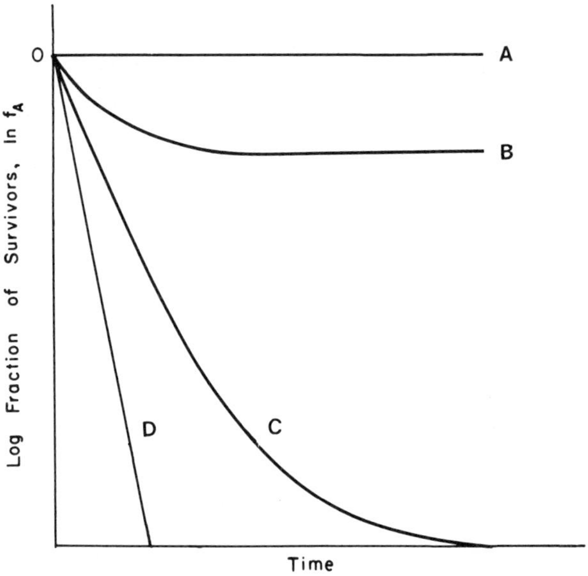

FIGURE 9-16. Typical plots of the fraction of survivors ln f_A = ln (N/N_0) as a function of time in a heterogeneous population of N_0 organisms when the ratio of a chronic dose D to the chronic dose that does not cause death in all animals D_m is varied. The elimination rate constant can be first order and the receptor sites can be saturable. (A) Doses below the minimum toxic dose D_m for all animals. (B) Doses that exceed the D_m for some animals but not for all. (C) Doses that exceed the D_m for all animals. (D) The highest doses where D_m/D approaches zero. [From Garrett (1977).]

survive the longest and will have the lowest mortality rate constant, as demonstrated by the lower terminal negative slope of the semilogarithmic plot. When the D_m is negligible with respect to the highest dose, the log survivor–time plot may approach linearity, as in curve d.

Experimental data of this type are given in Fig. 9-17.

Saturable Toxicokinetics with Constant Biophasic Toxicant Concentrations

The above discussion is based on the assumption that the concentration of toxicant in the biophase is proportional to the chronic dose. This implies non-

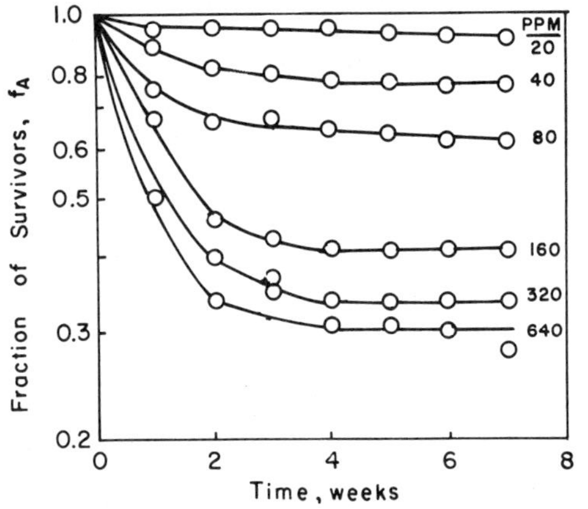

FIGURE 9-17. Semilogarithmic plot of surviving insects *(Tribolium castaneum)* against time when the population was maintained in successive petri dishes coated with 1.0 ml of acetone containing different concentrations of dichlofenthion shown in parts per million (ppm). Data from Tammes *et al.* (1970). [From Garrett (1977).]

saturable toxicokinetics. If the elimination processes are saturable, the concentration of chronically administered toxicant in the biophase (that is proportional to the amount of toxicant in the body and the plasma concentration) *increases* with dose.

If the rate of elimination proceeds by a saturable process and its maximum rate is not exceeded by the rate of administration, then at steady state, from Eqs. (26), (47), and (48), it follows that

$$\frac{D}{\Delta t_D} = \frac{dE}{dt} = \frac{k_{\max} KA}{1 + KA} \tag{57}$$

provided that the rate of dosing $D/\Delta t_D$ does not exceed $k_{\max}$, the maximum possible excretion rate at high values of A in the body.

This can also be formulated in terms of plasma concentration $C = A/V$, where V is the apparent volume of distribution. Thus,

$$\frac{D}{\Delta t_D} = \frac{k_{\max} KVC}{1 + KVC} \tag{58}$$

Rearrangement of this expression relates plasma concentration and its proportional biophase concentration to the chronically administered dose D:

$$C = \frac{D}{k_{\max} KV\Delta t_D - KVD} = \frac{a'D}{a(b' - D)} \tag{59}$$

and the ratio of the plasma concentration to the dose C/D would increase with increasing dose D, where the constants are pooled so that $a' = a/KV$ and $b' = k_{\max}\Delta t_D$.

Substitution of this value for C in Eq. (35) for saturable receptor sites and subsequent substitution in Eq. (45) results in the following rearrangement:

$$k_{\text{app}} = \frac{k_D a' D}{b' + D(a' - 1)} - \frac{k_D a' D_m}{b' + D_m(a' - 1)} \tag{60}$$

where D_m and the final term are constants.

The mortality rate constant per unit of dose k_{app}/D decreases upon chronic administration when the receptor sites are saturable and the factor a' is smaller than unity in Eq. (60). This reflects the fact that the receptor sites are relatively more quickly saturable than are the elimination processes (Fig. 9-18d, curve A) and there is a maximum mortality rate constant $(k_{\text{app}})_\infty$ achieved at infinite doses:

$$(k_{\text{app}})_\infty = \frac{a' k_D}{a' - 1} - \frac{k_D a' D_m}{b' + D_m(a' - 1)} \tag{61}$$

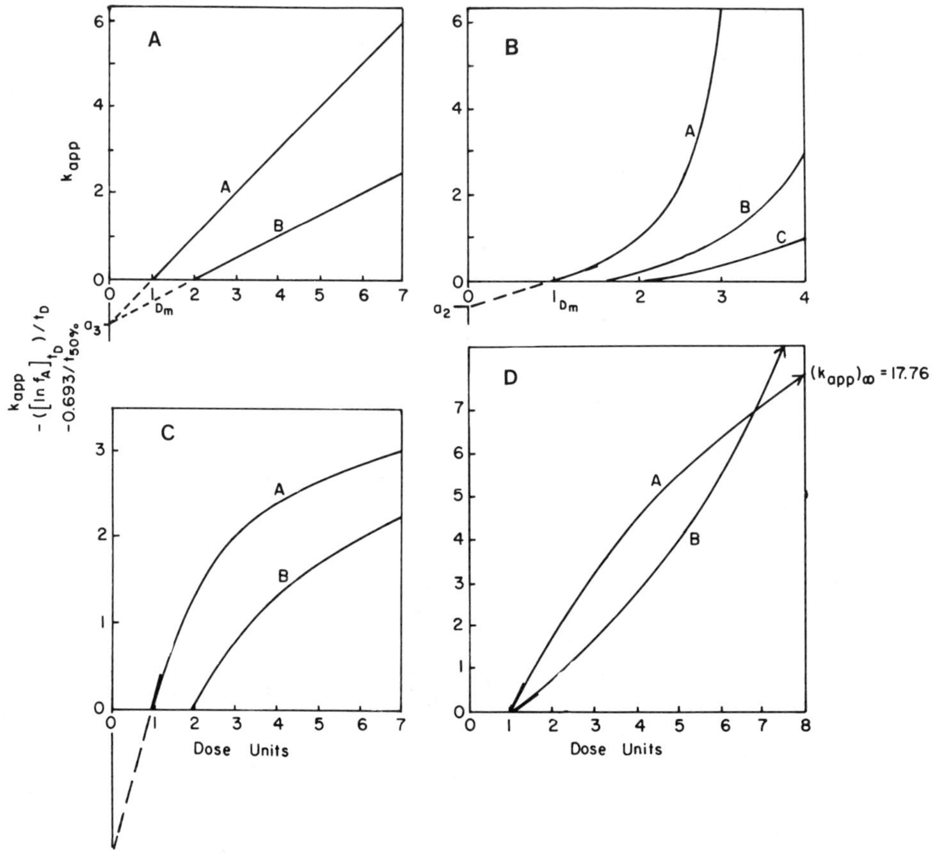

FIGURE 9-18. Examples showing the dependencies of the mortality rate constant k_{app} (hr^{-1}) on a dose D chronically administered at time intervals Δt_D (hr) for various combinations of saturable or nonsaturable receptor sites and elimination processes, provided that steady-state concentrations in biological fluids are achieved and the dosing schedule does not overwhelm the elimination processes. (A) Nonsaturable receptors and eliminations. (B) Nonsaturable receptors and saturable eliminations. (C) Saturable receptors and nonsaturable eliminations. (D) Saturable receptors and eliminations. [From Garrett (1977).]

The mortality rate constant per unit of dose k_{app}/D increases when the factor a' is larger than unity in Eq. (60). This reflects the fact that the receptor sites are relatively less quickly saturable than are the elimination processes (Fig. 9-18d, curve B) and that the dosing eventually overwhelms the elimination process at a dose $D' = b'/(1 - a')$. Above this dose D' the plots of $\ln f_A$ against time would not be linear as in Fig. 9-12, but would have increasing negative slopes over time, so that no valid estimate of a constant k_{app} can be obtained.

For saturable elimination toxicokinetics with nonsaturable toxic receptor sites, Eq. (59) is substituted into Eq. (35) with $aC \ll 1$, and this modified expression is substituted into Eq. (45):

$$k_{\text{app}} = \frac{k_D a' D}{b' - D} - \frac{k_D a' D_m}{b' - D_m} \tag{62}$$

This results in plots (Fig. 9-18b) where the mortality rate constant per unit dose increases. Above a dose $D'' = b'$, the plots of ln f_A against time would not be linear as in Fig. 9-12, but would have increasing negative slopes with time, so that no valid estimate of a constant k_{app} can be obtained.

Data on the effect of *N*-ethylmaleimide on the response of synchronized HeLa cells to x rays (Han *et al.*, 1976) show reasonable linearity for semilogarithmic plots of f_A against time at low concentrations (0.5–1.0 μM) in the culture for as long as 50 min. After that time, the controls (0.0 μM) demonstrated toxicity, so that further data could not be ascribed to toxicant action alone (Fig. 9-19). This indicated a quasi-steady state in which the organism's elimination processes reasonably balanced the rate of absorption into the organism. Thus, the estimated linear slopes k_{app} of ln f_A plotted against time are reasonably related to toxicant concentration C at these low toxicant concentrations (Fig. 9-19B). However, any approximations of k_{app} at higher (2.0 and 5.0 μM) concentrations show no such linear relationship with concentration, but instead one similar to that shown in Fig. 9-18b. This indicates that elimination processes can be saturated at the higher toxicant concentrations, so that toxicities accelerate with increasing concentrations or doses. Since the extrapolated curves of Fig. 9-19a (0.0–1.0 μM) may go through the origin, a significant threshold of C_m or t_m cannot be stated.

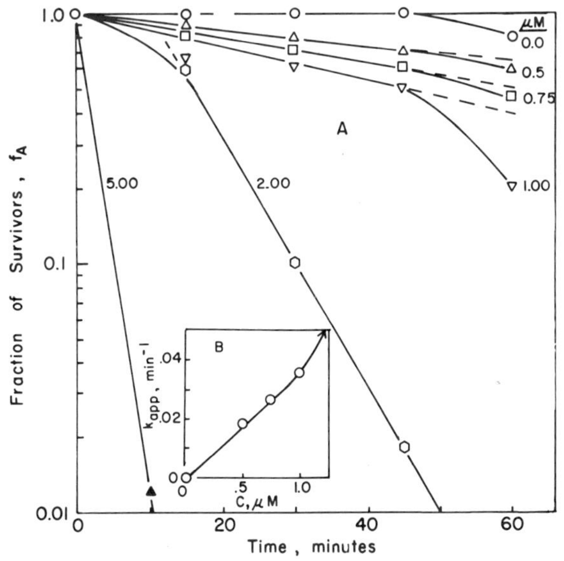

FIGURE 9-19. (A) Semilogarithmic plots of synchronized HeLa cells that survive exposure to x rays when subjected to labeled concentrations of *N*-ethylmaleimide in phosphate-buffered saline. (- -) Estimates of linear relationships based on data from Han *et al.* (1976). (B) Plots of slopes k_{app} of ln f_A against time for the *lower* toxicant concentrations in accordance with the presumptions of Eq. (56) for steady-state conditions with nonsaturable receptor sites, first-order eliminations, and no threshold. Note the indications of nonlinearity at the higher (2 and 5 μM) concentrations.

Summary of Methods and Limitations When Biophasic Steady-State Concentrations Exist

Examples of dependences of k_{app} values, the negative slopes of the semilogarithmic plots of survivors from a homogeneous population against time (Fig. 9-12), on dose for various combinations of saturable or nonsaturable receptor sites and elimination processes are given in Fig. 9-18. Other experimentally determined parameters can be substituted for k_{app} [Eq. (46)] in such plots. They are the reciprocal of the time that 50% fatality results in the population or the quotient of the logarithm or cologarithm of surviving organisms and the time of the organism's chronic exposure to the toxicant, i.e., $-(\ln f_A)/t_D$.

If there were no threshold value D_m for toxicity, all such plots would pass through the origin. The presence of extrapolatable values for finite doses D_m when $k_{app} = 0$ provides acceptable evidence for the existence of a threshold dose for toxicity in the test group of organisms.

Linear plots show no saturabilities; nonlinear plots show a saturability. Slopes that increase with dose show receptor-site saturability in excess of elimination processes, whereas decreasing slopes indicate the converse. Plots of k_{app} versus dose are not possible when logarithms of survivors plotted against time (from which k_{app} values are to be obtained) are nonlinear, never become linear even at longer times, and show increasing negative slopes over time for a given chronic dose. This observation leads to the conclusion that elimination processes are much more readily saturable than receptor sites at such doses.

To deduce the kinetic dependences, sufficient points are required to define a curve showing the number or fraction of organisms unaffected by an agent's toxicity with time. Acceptance of the hypothesis of steady-state concentrations in the biophase demands only two points to define the straight lines (Fig. 9-12) and yield proper k_{app} values. More points would be needed if accumulation occurred over time to provide assurance of linearity when steady-state conditions are reached or if the kinetic model is to be challenged and not assumed.

If the logarithm of surviving organisms plotted against the time of chronic dosing or the duration of contact with a known concentration of toxicant for a wide range of doses and concentrations is linear for more than five or six apparent half-lives (Fig. 9-12), it is reasonable to presume (1) that the population of organisms is homogeneous in its toxicokinetics and in its susceptibility to the administered toxicant, (2) that steady-state and reasonably constant levels of drug in the equilibrated compartments are readily achieved, and (3) that there is no significant lag between the toxic insult imposed on the receptor sites and the manifestation of death or toxicity. A corollary to the last presumption is that there is no significant accumulation of bodily produced toxins and that over time there is no significant depletion of physiologically vital metabolites. Oscillations in such plots indicate that the interval between repetitive doses is too long to achieve constant body levels of the toxicant or that there have been chronobiological perturbations of the time course of toxicity.

Continuously decreasing negative slopes of such plots over time (Figs. 9-16 and 9-17) that approach constant slopes or finite, constant values at lower doses are indicative of a heterogeneous population of organisms with different absorption efficiencies, susceptibilities, or thresholds. Acceleration of elimination processes by enzyme induction or an adaptive physiological mechanism over time cannot be excluded as alternative explanations.

Continuously increasing negative slopes of such plots over time that approach a constant slope are indicative of a rate-determining accumulation process in the biophase until either a steady-state concentration is reached or the receptor sites are saturated.

TOXICOKINETICS ON CHRONIC TOXICANT ADMINISTRATION WHEN TISSUES ARE NOT EQUILIBRATED INSTANTANEOUSLY

The previous discussion was based on the assumption that constant biophasic concentrations of toxicants in an organism are achieved instantaneously and maintained over time. This is unrealistic. After introducing an organism into an environment containing fixed toxicant concentrations or after administering repetitive doses, it should take time to accumulate toxicant to steady-state biophasic concentrations.

First-Order Toxicokinetics upon Chronic Toxicant Administration

The rate at which toxicant concentrations change with time in equilibrated body fluids (the one-compartment body model of apparent volume of distribution V) when the rate of chronic administration is $dD/dt = \text{Dose}/\Delta t_D$ and elimination rates are first order [Eq. (15)], can be constructed as follows:

$$\frac{dC}{dt} = \frac{D}{\Delta t_D V} - \lambda C \tag{63}$$

where λ is the apparent overall first-order rate constant for elimination. This expression is readily integrated to

$$C = \frac{D}{\Delta t_D V\lambda}(1 - e^{-\lambda t}) \tag{64}$$

Nonsaturable Receptor Site with No Threshold Toxic Dose

When receptor sites are nonsaturable and there is no threshold biophasic concentration, $C_m = D_m/V = 0$, for toxicity, the following equation (Garrett, 1977) is applicable:

$$\ln f_A = -a'k_D Dt - (a'k_D D/\lambda)(e^{-\lambda t} - 1) \tag{65}$$

where $a' = a/\lambda V \Delta t_D$. As time increases and steady-state concentrations can be achieved, $e^{-\lambda t} \rightarrow 0$ and

$$\ln f_A \rightarrow -a'k_D Dt + a'k_D D/\lambda = a'k_D D(t + 1/\lambda) \tag{66}$$

Ultimately, $1/\lambda$ is small relative to time t, and

$$\ln f_A \to -a'k_D Dt = -k_{app}t \tag{67}$$

so that the terminal linear slope $-k_{app} = -a'k_D D$ of a $\ln f_A$ versus t plot for a specific chronically administered dose D is proportional to dose and k_{app}/D is a constant, $a'k_D$.

Typical plots of data conforming to the premises of Eq. (65), $C_m = 0$, are given in Fig. 9-20A. The dashed lines are extrapolations of the anticipated terminal linearity of the semilogarithmic plots from which $\ln f_{A,extrap}$ values can be estimated at a given time t.

Subtraction of an actual $\ln f_A$ value from an extrapolated value $\ln f_{A,extrap}$ [Eqs. (65) and (67)] yields

$$\ln f_{A,extrap} - \ln f_A = (a'k_D D/\lambda)e^{-\lambda t} \tag{68}$$

so that semilogarithmic plot of this difference in the logarithms for a given dose D gives a straight line of constant slope, permitting the estimation of the overall elimination rate constant (insert A′ in Fig. 9-20A):

$$\ln(\ln f_{A,extrap} - \ln f_A) = \ln(a'k_D D/\lambda) - \lambda t \tag{69}$$

The antilogarithms of the intercepts of such plots would give straight lines of slope $a'k_D/\lambda$ when plotted against the dose.

Nonsaturable Receptor Sites with a Threshold Toxic Dose

When a threshold biophasic concentration C_m for toxicity exists, the following equation (Garrett, 1977) is applicable:

$$\ln f_A = (a'k_D D - \beta C_m)(t - t_m) - (a'k_D D/\lambda)(e^{-\lambda t} - e^{-\lambda t_m}) \tag{70}$$

where $\beta = a'k_D$ and t_m is the time when the biophasic concentration has achieved a C_m value and toxicity can result.

As time increases and steady-state concentrations can be achieved, $e^{-\lambda t} \to 0$ and

$$\ln f_A \to -(a'k_D D - \beta C_m)(t - t_m) + (a'k_D D/\lambda)e^{-\lambda t_m} \tag{71}$$

Ultimately, the second factor is negligible with respect to the first and

$$\ln f_A \to -(a'k_D D - \beta C_m)(t - t_m) = -k_{app}t + k_{app}t_m \tag{72}$$

Thus, the terminal linear slope $-k_{app} = -(a'k_D D - \beta C_m)$ of a plot of $\ln f_A$ versus t for a specific chronic dose D is linearly related to dose. However, in contrast to the case where $C_m = 0$, k_{app}/D is not constant and the intercept of a plot of k_{app}

versus D is not zero, but has a finite value of βC_m indicating the existence of a threshold value for toxicity.

Typical plots of data conforming to the premise of Eq. (69), i.e., that C_m is finite, are given in Fig. 9-20B. Equations (68) and (69) are also valid in this case, and constant first-order eliminations with an overall elimination rate constant λ can be substantiated by appropriate plots (e.g., Fig. 9-20A′) in accordance with these equations.

Examples of Nonsaturable Receptor Sites and First-Order Eliminations

The pattern for the action of chronically administered *p,p′*-DDT [2,2-bis(*p*-chlorophenyl)-1,1,1-trichloroethane] in the diet of fingerling chinook salmon *(Oncorhynchus tshawytscha)* (Buhler *et al.*, 1969) is typical of that expected for nonsaturable receptor sites and nonsaturable elimination toxicokinetics, as can be seen by comparing Figs. 9-20 and 9-21. It takes time for the accumulation of the toxicant in the organism to approach linearity in the semilogarithmic plot of f_A against time, especially in the studies conducted at 37.5- and 150-ppm doses (Fig. 9-21). This final linearity indicates the attainment of steady-state concentrations in the fingerlings. The ratios of the negatives of the terminal slopes (k_{app} = 0.123 day^{-1} at 37.5 ppm, 0.439 day^{-1} at 150 ppm, and 1.8 day^{-1} at 600 ppm) of ln f_A against time to dose are reasonably constant for the three doses, i.e., k_{app}/D = 3×10^{-3} day^{-1} ppm^{-1}. In accordance with Eq. (56), therefore, the threshold dose is not significantly different from zero and is less than 9 ppm. The pattern was similar for the survival of fingerling coho salmon *(Oncorhynchus kisutch)*, which had significant mortalities at only two studied concentrations, 100 and 400 ppm. However, the facts that 5% of the population showed resistance at the 100-ppm level and no significant number of deaths from the drug occurred within 95 days at 25 ppm indicated a significant toxicity threshold in this species. This has been explained (Buhler *et al.*, 1969) not by a higher volume of distribution, but because of *higher* lipid content in the coho than in the chinook salmon studied by that group.

Goldberg *et al.* (1964) studied the effect of ethylene glycol dimethyl ether inhalation on avoidance behavior. The reasonably linear plots of the terminal data of ln f_A versus time provided estimates of k_{app} values (Fig. 9-22A) that were linearly related to dose (Fig. 9-22B) in accordance with Eq. (67). This indicated nonsaturability of receptor sites and no significant threshold, D_m or C_m, for toxicity. Thus, the induction periods in Fig. 9-22A can be assigned to time-dependent accumulations up to steady-state concentrations in the organism.

Examples with Saturable Receptor Sites

When the receptor sites are saturable and elimination processes are first order, the shape of a semilogarithmic plot of f_A against time for a given chronic dose is similar to those for nonsaturable receptor sites (Fig. 9-20). The major difference would be that the terminal slopes of such plots would approach a constant

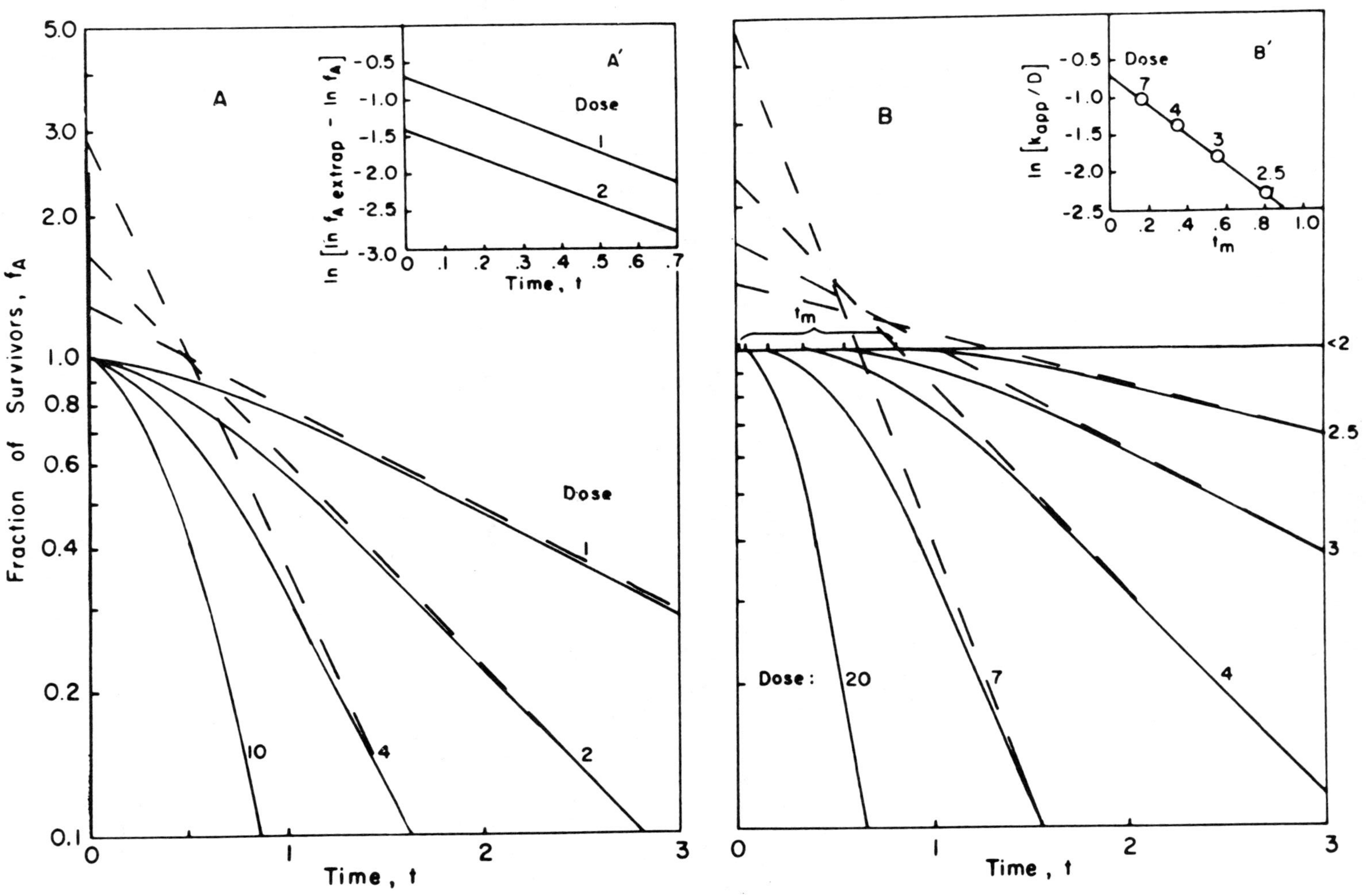

A
A′
B
B′
Dose
Dose:
1
2
4
10
20
7
3
2.5
<2
t_m
Fraction of Survivors, f_A
Time, t
$\ln[\ln f_A \text{ extrap} - \ln f_A]$
$\ln[k_{app}/D]$
5.0
3.0
2.0
1.0
0.8
0.6
0.4
0.2
0.1
0
1
2
3
-0.5
-1.0
-1.5
-2.0
-2.5
-3.0
.1
.2
.3
.4
.5
.6
.7
.8
1.0

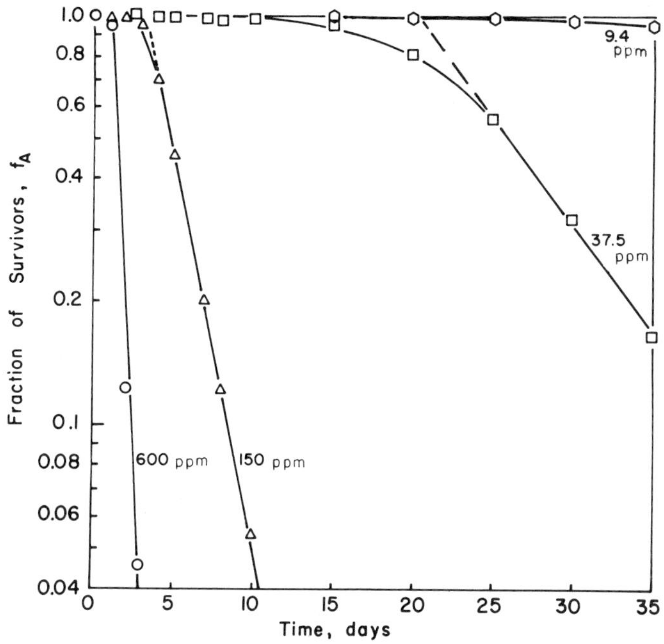

FIGURE 9-21. Semilogarithmic plots of surviving Chinook salmon fingerlings against time on chronic oral feeding of *p,p'*-DDT in their daily diet at the labeled ppm values. Data from Buhler *et al.* (1969). (- -) The estimated linearities of the terminal data. [From Garrett (1977).]

value $-k_D$ with increasing doses [Eq. (52)], in contrast to the dose-dependent slopes with nonsaturable receptor sites in accordance with Eqs. (65) and (70).

Newts *(Traicha granulosa)* immersed in various concentrations (v/v) of dimethylsulfoxide (DMSO) (Lappenbusch and Willis, 1971) and oil (Lappenbusch and Ward, 1973) had increasing values of t_m with decreasing concentrations. A t_m value is the time until a biophasic concentration has achieved its threshold toxic concentration to produce significant toxicity (Fig. 9-23). Subsequently, the mortality rates were approximately equivalent for all concentrations above 2% DMSO and 5% oil, indicating that the toxicity receptor sites were readily saturated. Apparently, the homeostatic processes of the organisms were able to compensate for insults of $\leq$2% DMSO and 1% oil and a C_m was manifested. A preliminary assumption that the t_m is inversely proportional to concentration is not inconsistent.

FIGURE 9-20. Semilogarithmic plots of survivors against time in accordance with Eq. (65) for chronic toxicant administration, first-order eliminations, and nonsaturable receptor sites for various doses. (A) $C_m = 0$, (B) $C_m \neq 0$, where the t_m values are for a constant C_m. Extrapolations of the linear terminal data in accordance with Eq. (67) for (A) and Eq. (72) for (B). Inset A': Logarithm of the differences between the extrapolated data, $\ln f_{A,\text{extrap}}$, and the actual data, $\ln f_A$, against time for several doses of (A) in accordance with Eq. (69) to confirm the theoretical slope of $-\lambda = -2$. The ratios of the terminal slopes and dose for the data in (A) for $C_m = 0$ were invariant for all doses, $-S/D = 2$. [From Garrett (1977).]

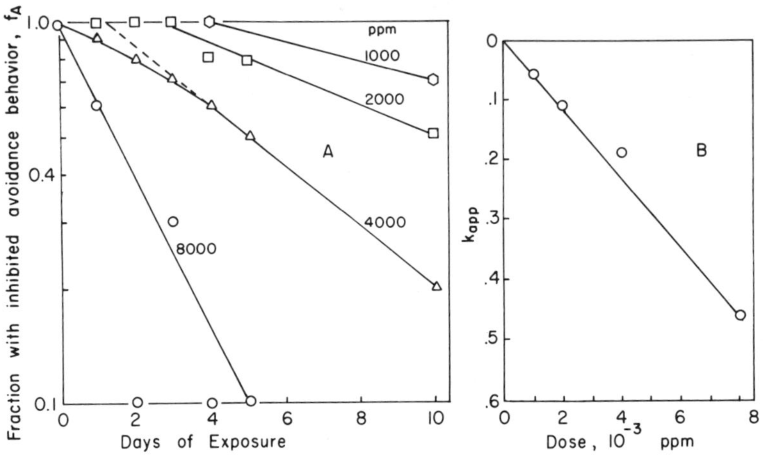

FIGURE 9-22. (A) Semilogarithmic plots of animals f_A with inhibited avoidance against days of exposure to the stated ppm of ethylene glycol dimethyl ether in the atmosphere. Data from Goldberg *et al.* (1964). (B) Dependence of slope k_{app} of $\ln f_A$ versus time on ppm of ethylene glycol dimethyl ether in atmosphere. [From Garrett (1977).]

The survival of green sunfish *(Lepomis cyanellus)* exposed to sodium arsenate in aerated well water was monitored by Sorenson (1976). The semilogarithmically plotted data in Fig. 9-24 demonstrate the presence of saturated toxicity receptor sites, since the curves at 500 and 1000 ppm demonstrate similar slopes and k_{app} values. The decreasing slope over time at 100 ppm indicates that there is variation in the susceptibility of the fish. This is readily explained by the fact that toxicity depends on the size of the organism. Sorenson (1976) also reported that a spectrum of sizes existed in the population. The time lag t_m before toxicity was

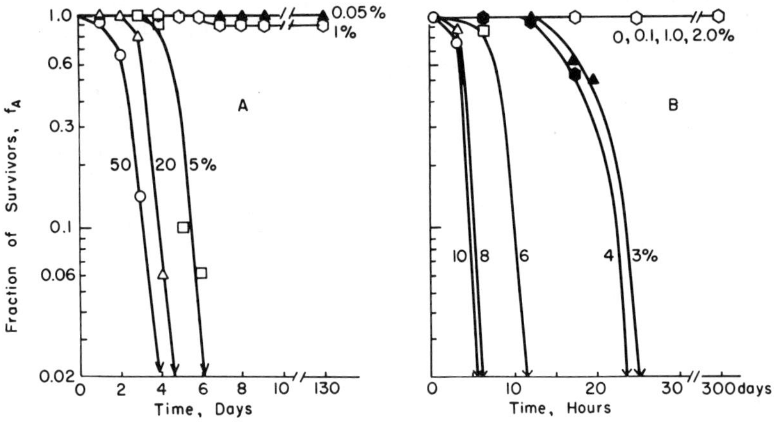

FIGURE 9-23. Semilogarithmic plots of newts that survived immersion in water containing the variously labeled percentages of (A) dimethylsulfoxide and (B) nondetergent motor oil. Data from Lappenbusch and Willis (1971) and Lappenbusch and Ward (1973), respectively. [From Garrett (1977).]

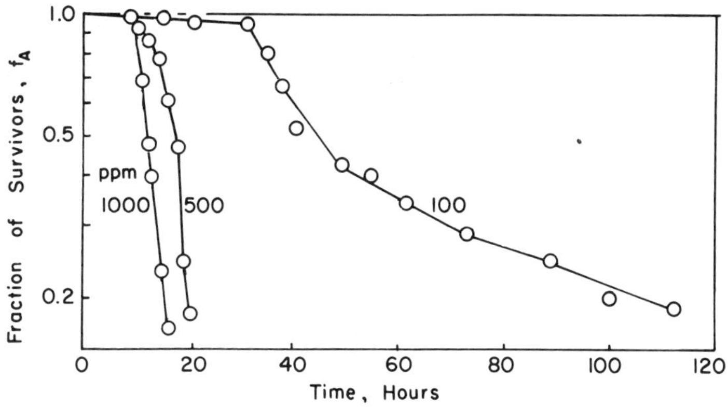

FIGURE 9-24. Semilogarithmic plot of survivors f_A against time for green sunfish exposed to various ppm of sodium arsenate in aerated well water at 20°C. Data from Sorenson (1976). [From Garrett (1977).]

manifested decreased with dose, indicating increased rate of accumulation to the threshold concentration C_m in the biophase with increased dose.

Studies on the survival of quail *(Colinus virginianus virginianus)* fed diets containing the insecticide guthion (*O,O*-dimethyl *S*-4 OXO-1,2,3-benzotriazin-3[4H]-y) methyl phosphorodethioate) (Gough *et al.*, 1967) clearly showed saturable toxicity receptor sites. This was demonstrated by the parallelism of the terminal data for 1620 and 4800 ppm on semilogarithmic plots of the survivors against time (Fig. 9-25). The initial nonlinearity of these plots can be assigned to the accumulation to steady-state concentrations. The lower toxicity in the Japanese quail *(Coturnix coturnix japonica)* than in the bobwhites *(Colinus virgini-*

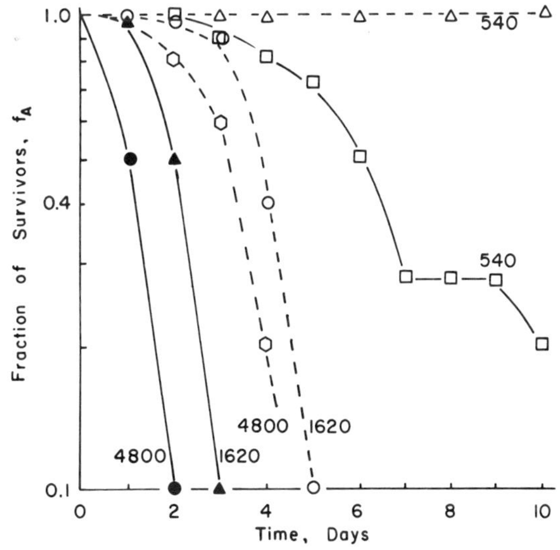

FIGURE 9-25. Semilogarithmic plots of survivors against time for Japanese (dashed curves) and bobwhite (solid curves) quail chronically fed the labeled ppm amounts of guthion (*O,O*-dimethyl *S*-4-OXO-1,2,3-benzotriazin-3[4H]-y) methyl phosphorodethioate. Data from Gough *et al.* (1967). [From Garrett (1977).]

anus) has been attributed to differences in metabolic rates or absorption efficiencies (Gough *et al.,* 1967). The similarities in terminal slopes of the plots reflect the similarities in the toxic nature of the receptor sites of both species.

Zeroth-Order and Saturable Pharmacokinetics for Chronic Exposure

In contrast to the time course of concentration on first-order elimination [Eq. (64)], the time course of concentration on zeroth-order or constant-rate elimination from equilibrated body fluids can be formulated as follows:

$$C = \frac{D}{V\Delta t_D} - \frac{k_{\max}}{V}t \tag{73}$$

where $k_{\max}$ is the constant rate of toxicant loss from the body.

When receptor sites are nonsaturable, equations of the following forms (Garrett, 1977) are applicable:

$$\ln f_A = -(\alpha D - k_0)(t^2 - t_m^2) + ak_DC_m(t - t_m) \tag{74}$$

and

$$\frac{\ln f_A}{t - t_m} = -(\alpha D - k_0)(t + t_m) + ak_DC_m \tag{75}$$

where

$$\alpha = \frac{ak_D}{2V\Delta t_D}, \qquad k_0 = \frac{k_{\max}ak_D}{2V}$$

At time increases, t_m is relatively small with respect to t, and

$$\ln f_A \rightarrow -(\alpha D - k_0)t^2 + ak_DC_mt = -k'_{app}t^2 + ak_0C_mt \tag{76}$$

and

$$\frac{\ln f_A}{t - t_m} \rightarrow -(\alpha D - k_0)t + ak_DC_m = -k'_{app}t + ak_DC_m \tag{77}$$

where $k'_{app} = (\alpha D - k_0)$.

When there is no threshold value for toxicity and D_m and C_m are zero, $t_m = 0$, and

$$\ln f_A = -(\alpha D - k_0)t^2 = -k'_{app}t^2 \tag{78}$$

and semilogarithmic plots of survivors against the square of the time should be linear (Fig. 9-26C). The plots of the slopes against dose should also be linear, and

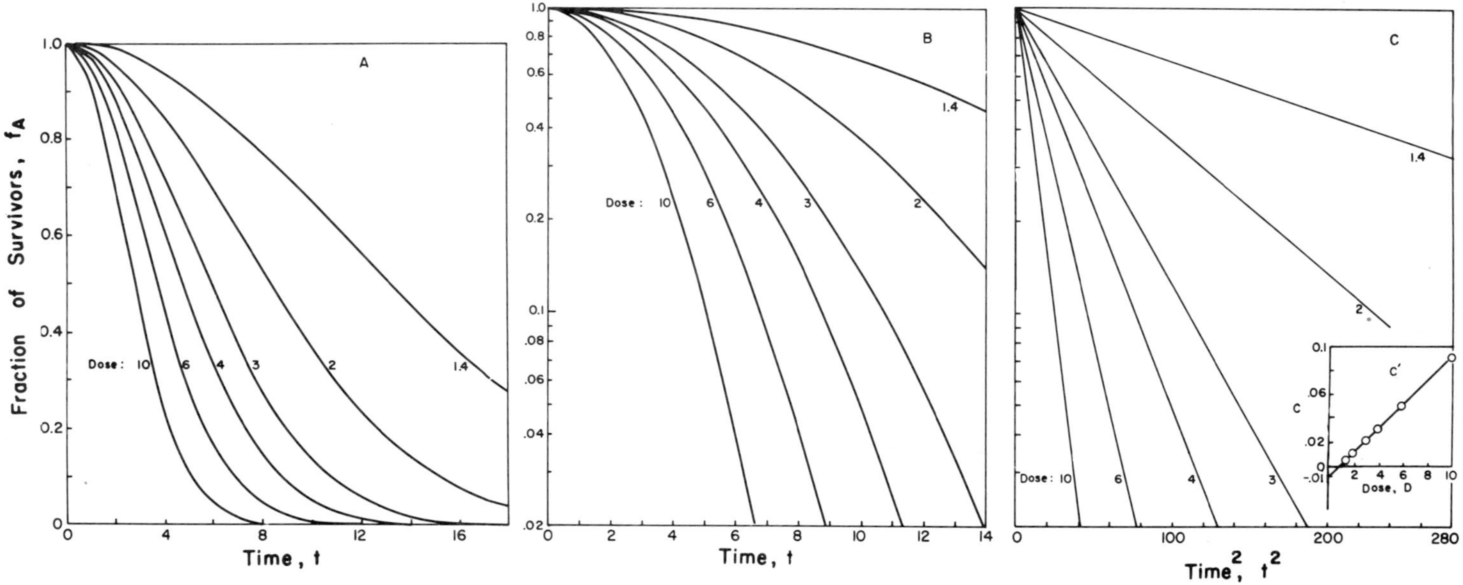

FIGURE 9-26. Time dependences of survivor functions (A) f_A and (B, C) $\ln f_A$ for chronic rates of drug administration, zeroth-order elimination, and nonsaturable receptor sites when C_m and t_m are zero in accordance with Eq. (78). Inset C′: The linear dependence of the slope $-k'_{app}$ of $\ln f_A$ versus t^2 in part C on dose D. The ratio of the intercept to the slope is $-k_{max}$. [From Garrett (1977).]

the ratio of the intercept to the negative of the slope should be k_{max}, the constant rate of drug loss from the body (inset C′ in Fig. 26C). It should be noted that ln f_A versus t plots never asymptotically approach linearity (Fig. 26B).

The presence of a threshold dose for toxicity D_m giving rise to a C_m or t_m does not permit linear plots of ln f_A versus t^2 until the time t of chronic dosing increases greater than t_m in Eq. (72) (Fig. 9-27C). Typical plots of functions of f_A against time for this model are given in Fig. 9-27.

If t_m can be estimated for the different doses (Figs. 27A–27C), plots of (ln f_A)/ $(t - t_m)$ against $t - t_m$ or t in accordance with Eq. (77) should be linear (Fig. 9-27C) for a given chronic dose. The k'_{app} values obtained from the slopes, when plotted against dose, should provide estimates of k_{max} (inset C′ in Fig. 27C).

The determination of t_m may be difficult experimentally. However, if ln f_A versus t plots show perpetually increasing negative slopes with time (Fig. 9-27B) and ln f_A versus t^2 plots indicate asymptotic terminal constant slopes k'_{app} (Fig. 9-27D), it is possible to select a t_m that linearizes the data for all doses in the (ln f_A)/ $(t - t_m)$ versus t plots (Fig. 9-27C). Such values should show a linear dependence of $1/t_m$ values on dose (Fig. 9-27A′).

The major difference in such plots when the receptor sites are potentially saturable (Fig. 9-28) is that the plot of logarithm of survivors against time is linear at the higher doses with slopes relatively insensitive to dose. When the affinities of the drug to receptor sites are moderate, such plots would generally not be linear but would have appearances similar to the plots in Figs. 9-26B and 27B—slopes increasing over time, eventually approaching the constant slope obtained at the highest doses. Saturable receptor sites would not permit linear plots of ln f_A against t^2.

Equations for saturable elimination toxicokinetics are obviously more complex (Garrett, 1977). However, the shapes of the plots for various functions of f_A are very similar to those for zeroth-order elimination (Fig. 9-28). Plots of ln f_A versus t^2 become reasonably linear for nonsaturable receptor sites when the toxicokinetics has the proper degree of saturability, especially at the higher doses. The plots of the slopes k'_{app} against dose will approach linearity only at the high doses. Plots of ln f_A versus t are also linear at higher doses and at greater times of dosing, and the slopes approach constancy with increasing dose when receptor sites are saturable (Figs. 9-23, 9-24, and 9-28).

Examples of Nonsaturable Receptor Sites and Possible Saturable Eliminations

Harbison *et al.* (1976) studied the time dependence of fractions f_A of control pregnancies per total breeding population of rabbits. They administered weekly 200- and 325-mg doses of trimethyl phosphate per kg of body weight. Estimated slopes k_{app} of linear plots of ln f_A against time were 0.154 and 0.512 week^{-1}, respectively (Fig. 9-29). Although the data were scattered, the ratios of these k_{app} values to dose were 7.7×10^{-4} week^{-1} per mg/kg at 200 mg/kg and 16×10^{-4} week^{-1} per mg/kg at 325 mg/kg, indicating that the toxic receptor sites were not readily saturable with this dosage regimen. There is a possibility of saturation elimination toxicokinetics since the ratio increased with dose.

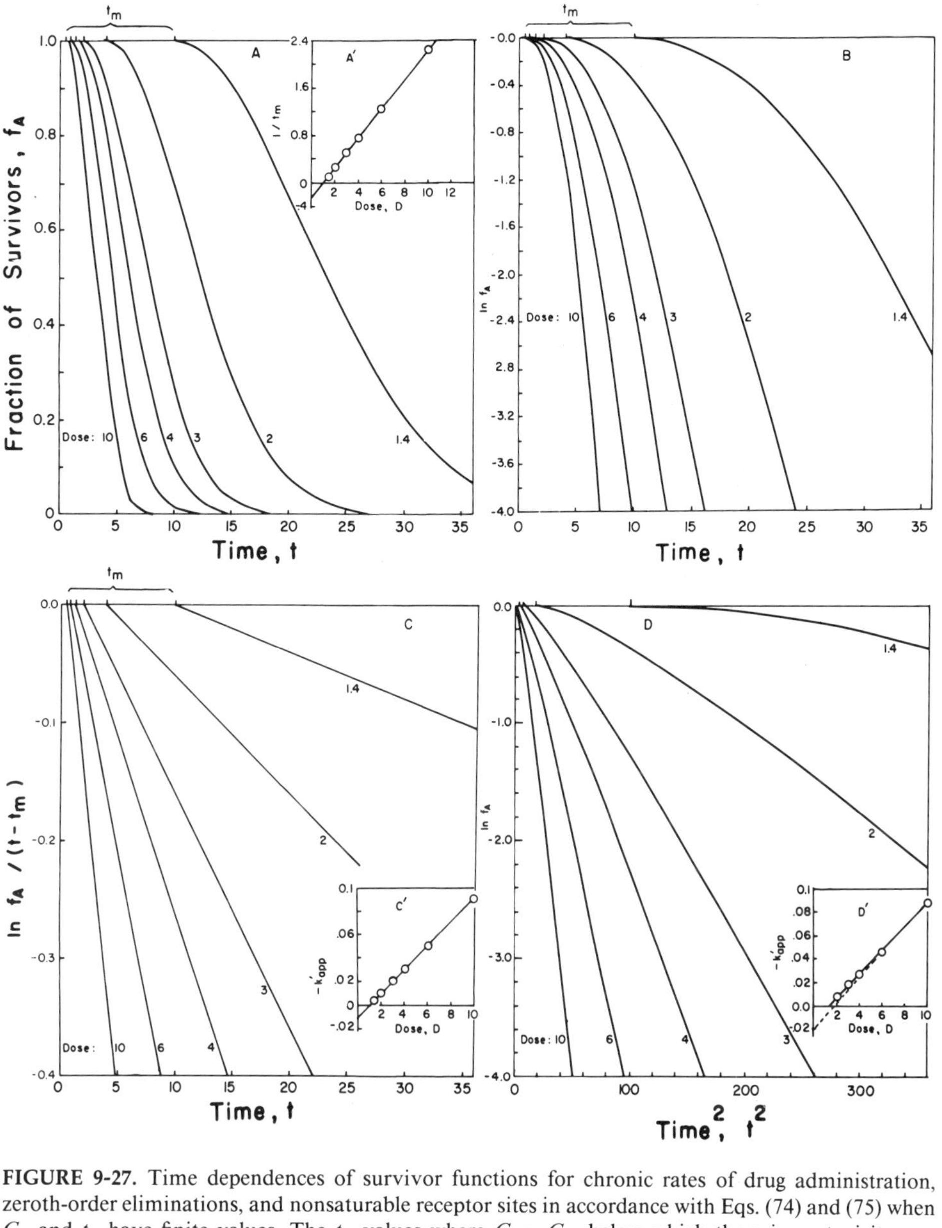

FIGURE 9-27. Time dependences of survivor functions for chronic rates of drug administration, zeroth-order eliminations, and nonsaturable receptor sites in accordance with Eqs. (74) and (75) when C_m and t_m have finite values. The t_m values where $C = C_m$, below which there is no toxicity, are designated for the various doses. Inset A′: The linear dependence of $1/t_m$ on dose in accordance with the ratio of the intercept to the slope of $-k_{max}$. Inset C′: The linear dependence of the slope $-k'_{app}$ of $(\ln f_A)/(t - t_m)$ versus t in part C on dose D in accordance with Eq. (77), where the ratio of the intercept to the slope is $-k_{max}$. (D) Curves plotted as $\ln f_A$ versus t^2, which should be linear when C_m and t_m = 0. Under these conditions, however, they curve significantly at the lower values of the dose D, but approach linearity at high doses where t_m is small and C_m becomes negligible relative to the drug concentration in the body. Inset D′: The dependence of the slopes of the terminal data for the various curves on dose; the ratio of the intercept to the slope from the higher dose values of the dashed line is $-k_{max}$. [From Garrett (1977).]

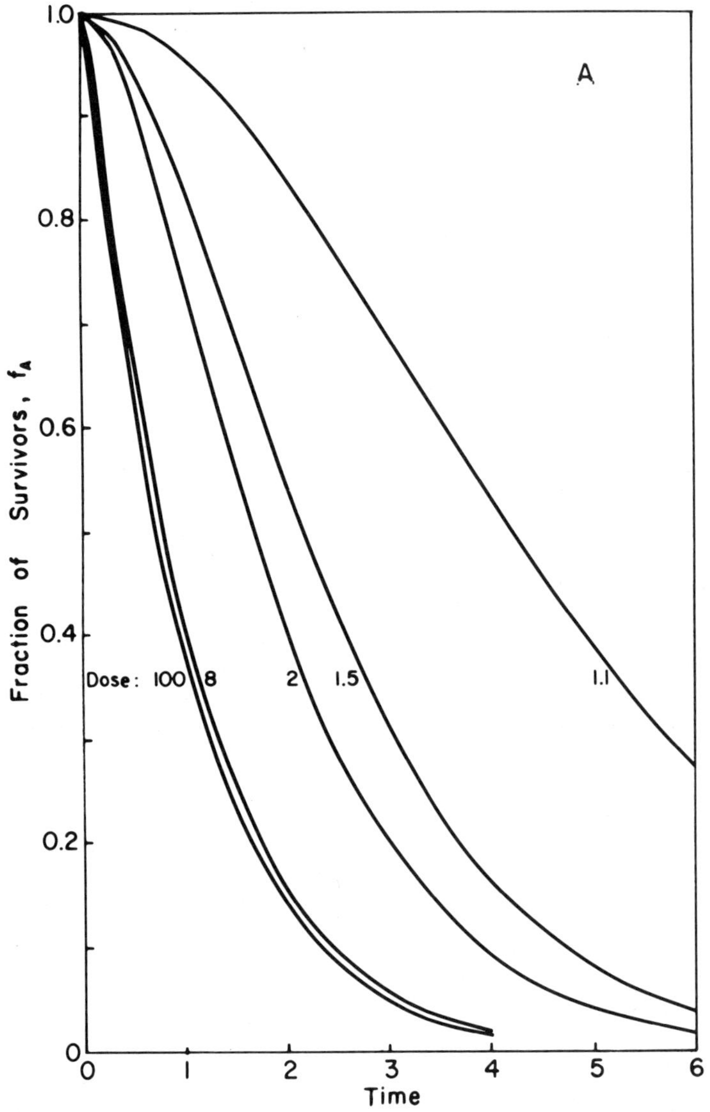

FIGURE 9-28. Time dependences of survivor functions (A) f_A and (B) ln f_A for chronic rates of drug administration, zeroth-order elimination, and saturable receptor sites for various doses. The slopes at high doses and at terminal times approach the same values. [From Garrett (1977).]

In a study on the effect of dieldrin on the red flour beetle *(Tribolium castaneum),* Tammes *et al.* (1970) found a different pattern from the effect of dichlofenthion (Fig. 9-17). In addition to the indication of a resistant population at low doses, e.g., <0.8 ppm (Fig. 9-30A), the increasing negative slopes of the semilogarithmic plots of f_A against time for the higher doses may indicate significant rate-determining accumulation of drug in the organisms on chronic exposure to diel-

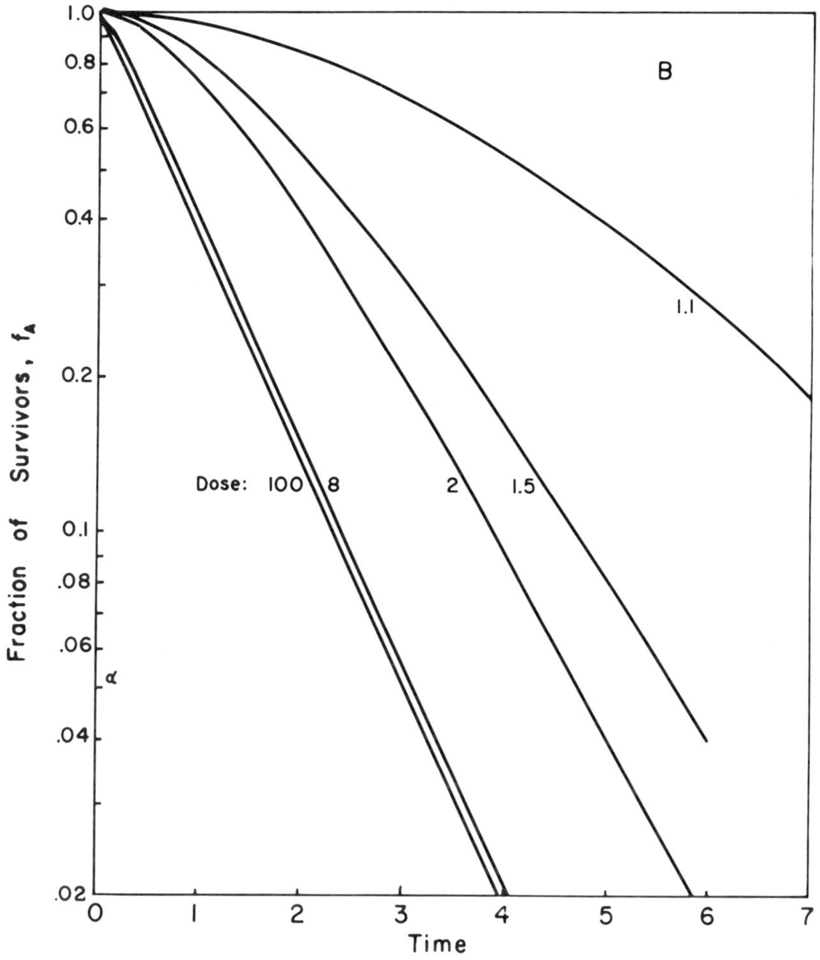

FIGURE 9-28. *(continued)*

drin. The shapes of the curves at these higher doses (Fig. 9-30A) (especially at 1.0 ppm, which has significant numbers of values) are similar to those given in Fig. 9-20B for chronic drug administration and first-order elimination, with an apparent C_m or t_m greater than zero. The data at 1.0 ppm were plotted (Fig. 9-31A) in accordance with Eq. (69), which is based on the presumption that there were nonsaturable receptor sites. The first-order rate constant $\lambda = 0.56$ week^{-1} ($t_{1/2} = 1.24$ weeks) for the elimination of dieldrin was estimated from the slope of this plot. This was not inconsistent with the $\lambda = 0.72$ week^{-1} ($t_{1/2} = 0.97$ week) obtained from the semilogarithmic plot of the amount of dieldrin in *T. castaneum* after an acute dosing (Tammes *et al.*, 1970).

The fact that it takes weeks to achieve a possible linearity in the semilogarithmic plots may cast doubt on the validity of this hypothesis of accumulation to a steady-state concentration in the organism as a cause of the curvature shown

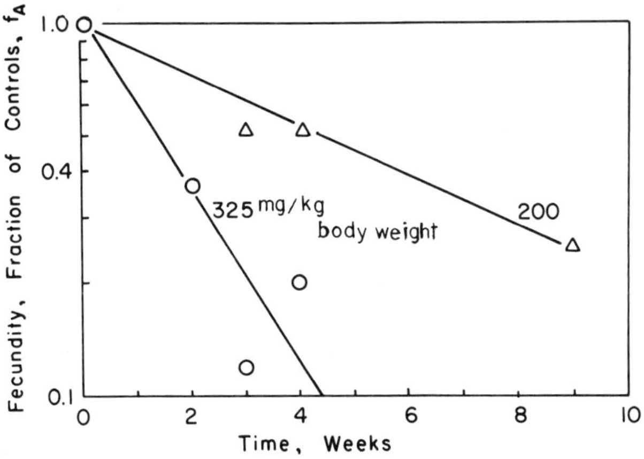

FIGURE 9-29. Semilogarithmic plots of pregnancies in a control population of rabbits given chronic mg/kg doses of labeled trimethylphosphate at weekly intervals. Data from Harbison *et al.* (1976). [From Garrett (1977).]

in Fig. 9-30A. The shape of the curve at 1 ppm in that figure is similar to those in Figs. 9-26B, 9-27B, and 9-28B for zeroth-order or saturable eliminations and nonsaturable receptor sites upon chronic dosing. A semilogarithmic plot of f_A against t^2 is excellently linear for the dieldrin data at the 1.0-ppm concentration (Fig. 9-30B). The slopes $-k'_{app}$ estimated for such ln f_A versus t^2 plots at 1.0 ppm and for the few data points of 10 and 100 ppm in Fig. 9-30B consistently increase with concentration in accordance with Eqs. (74) and (76). This indicates that

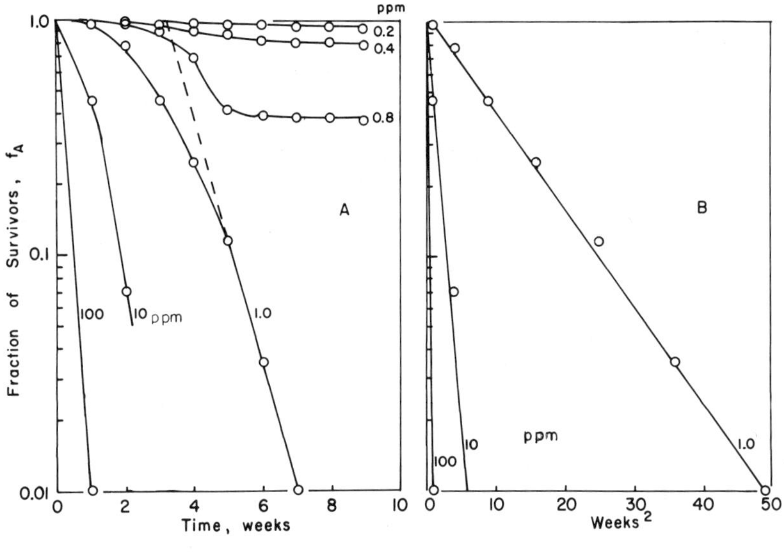

FIGURE 9-30. Semilogarithmic plots against (A) time and (B) time^2 of surviving insects *(Tribolium castaneum)* in a population maintained in successive petri dishes coated with 1.0 ml of acetone containing different concentrations of dieldrin, shown in parts per million (ppm). (- -) A linear extrapolation of the terminal points. Data from Tammes *et al.* (1970). [From Garrett (1977).]

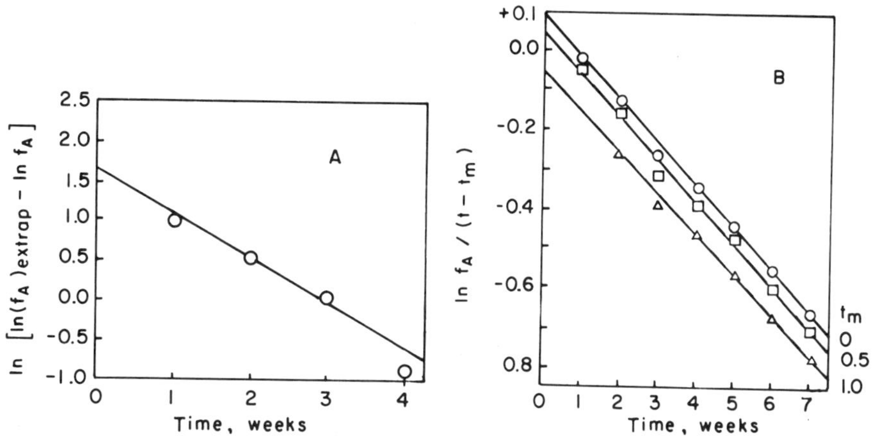

FIGURE 9-31. Graphic treatments of the data in Fig. 9-30 for 1.0 ppn of dieldrin in acetone. (A) Plot of ln [ln $(f_A)_{extrap}$ $-$ln f_A] against time in accordance with Eq. (69), based on the presumption that dieldrin accumulates in the organism, reaching a steady-state concentration, and that there is first-order elimination, where the ln $(f_A)_{extrap}$ values were obtained from extrapolation of an apparent terminal linear segment of the plot in Fig. 9-30A. (B) Plot of (ln f_A)/(t − t_m) against time for various assumed t_m values in accordance with Eq. (75) based on the presumption that there are saturable eliminations and nonsaturable receptor sites with chronic administration. [From Garrett (1977).]

there is neither potential for saturable receptor sites nor saturable elimination toxicokinetics. The plot for 1.0 ppm in Fig. 9-30B apparently intersects the value ln $f_A = 0, f_A = 1$, at a time between 0 and 1.0 week. This indicates a probable C_m and t_m, a minimum biophase concentration for manifestation of toxicity and its time of achievement.

The data for concentrations less than 1 ppm (Fig. 9-30) indicate the presence of resistant populations after 5 weeks of exposure. The larger t_m values indicated at 0.8, 0.4, and 0.2 ppm show an approximate dependence on these concentrations C of $1/t_m = 1.5C$.

Another study that provides fragmented information but permits preliminary conclusions concerning the toxicokinetic model and nature of toxic receptor sites is that of Kaplan *et al.* (1973), who studied the ability of rats to maintain their balance on a rod rotating at 8 rpm over an electrode floor as a measure of resistance to acrylamide-induced neuropathy. The rats were given only one 40 mg/kg intraperitoneal dose. The data (Fig. 9-32) demonstrated a delay in onset followed by a rapidly accelerating toxicity with time. Neither the semilogarithmic plots of the proportion of rats f_A maintaining balance against time or time^2 are completely linear (Fig. 9-32A). The data are best rationalized by the presumption that there is a t_m of 14 days to achieve the minimum biophase concentration necessary for toxicity and that there are saturable toxicokinetics and nonsaturable receptor sites. This explanation is supported by the linearity of the plot against time in accordance with Eq. (77), which shows that (ln f_A)/(t − t_m) = 0 at 14 days (Fig. 9-32C).

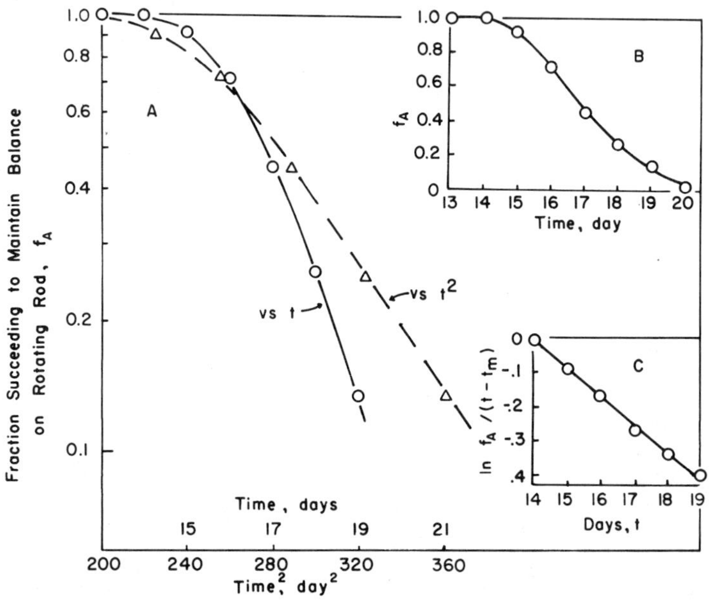

FIGURE 9-32. Plots of various balance functions of rats f_A given 40-mg chronic intraperitoneal doses of acrylamide/kg. The criterion was an animal's success in maintaining its balance on a rotating rod above an electrode floor. Data from Kaplan *et al.* (1973) for normal and deficient diets, where there were no significant differences among responses for all diets. (A) Semilogarithmic plots of f_A against (–○–) time and (--△--) time². (B) Linear plots of f_A against time t. (C) Plot of $(\ln f_A)/(t - t_m)$ in accordance with Eq. (77), when t_m = 14 days. [From Garrett (1977).]

Summary of Methods and Limitations When Accumulation Occurs on Chronic Toxicant Administration

As stated previously, investigators studying the manifestation of toxicity resulting from the chronic administration of a substance must consider the toxicant's accumulation in the organism. No toxic responses would be observed until accumulation in the biophase exceeds the threshold toxic concentration. Also, toxicities may show increasing rates with time. Induction periods for manifestation of toxicity can be ascribed to both phenomena.

Accumulation of toxicants in the body on chronic dosing would always increase initial rates of loss of survivors with time, as demonstrated by the increasing negative slopes of plots of f_A and $\ln f_A$ against time in Figs. 9-20 and 9-26 to 9-28.

If semilogarithmic plots of f_A versus t eventually become linear and the slopes k_{app} are proportional to the chronically administered dose, first-order eliminations can be concluded. There would be no significant slope and the k_{app} values would be negligible at the lower doses if there were a threshold toxicity. Constant slopes of the linearly plotted data at the higher doses would be evidence for saturable receptor sites, regardless of whether the toxicokinetics was saturable or not.

Evidence for saturable toxicokinetics with nonsaturable receptor sites would be ln f_A versus t plots that never become linear and have increasing negative slopes over time. If the threshold toxic dose is not significant or high, plots of ln f_A versus t^2 would approach linearity asymptotically. In the presence of a significant threshold biphasic toxic concentration, where t_m is the time that must be exceeded on chronic dosing for toxicity to be manifested, a plot of $(\ln f_A)/(t - t_m)$ versus t is linear. Plots of this function for variously hypothesized t_m values give the best linearity for the most valid estimation of t_m when the toxicokinetics is saturable.

Alternative explanations for such nonlinear semilogarithmic plots of f_A with increasing negative slopes over time could include a time-dependent accumulation of toxins when the toxicant exposed to the organism inhibits their removal, a time-dependent depletion of physiologically vital anabolites or metabolites when the toxicant inhibits their removal, a time-dependent depletion of physiologically vital anabolites or metabolites when the toxicant inhibits their production as with the anticoagulants, or a lag between the initial insult and death or manifestation of toxicity, e.g., when there is renal failure or biochemically mediated processes.

Chronic toxicity has been frequently studied in many living organisms and animals. The results can be helpful in determining the underlying toxicokinetic processes and the nature of toxic receptor sites if the times of death or toxic manifestations were monitored as functions of the chronic dose.

THE TIME COURSE OF TOXICITY ON ACUTE ADMINISTRATION OF A TOXICANT

The concentration of toxicant in the biophase always changes with time on acute dosing, in contrast to chronic dosing. Equations (39)–(42) are still valid, but the net degree of alteration or inhibition $\Delta I = I - I_m$ of a bodily process that would result in a manifestation of toxicity changes with time. Thus, these equations cannot be readily integrated unless ΔI can be expressed as a function of time. Then, the fraction of organisms manifesting toxicities is

$$f_D = 1 - f_A = 1 - \exp\left[-\int_0^t k_D(I - I_m)\,dt\right] \quad \text{for} \quad I > I_m \qquad (79)$$

This f_D may also be looked upon as the probability of manifesting toxicity, e.g., death.

All organisms will manifest toxicity if I is always maintained above the threshold I_m necessary to manifest toxicity. Thus, $f_D = 1$ at infinite time when ΔI is maintained positive, and all organisms would manifest toxicity. Obviously, if a positive value for $I - I_m$ were not maintained, the probability of toxicity f_D as a consequence of the insult is lessened and the probability of living has been determined at time t (if $I \leq I_m$). In the simplest case when the toxicity is not mediated by a subsequent time-dependent result of deterioration of an impaired biochemical process, death is an immediate consequence of the time integral in Eq. (79) and the fraction alive, surviving, or manifesting no toxicity f_A at that time t would remain alive.

Under the conditions of nonsaturable toxic receptor sites, the degree of inhibition I can be postulated as proportional to the drug concentrations C' in the biophasic compartment:

$$I = aC' \tag{80}$$
$$I_m = aC'_m \tag{81}$$

and

$$f_A = \exp\left[-\int_0^t ak_D(C' - C'_m)\,dt\right] \tag{82}$$

where C'_m is the biophase concentration below which no toxicity occurs. In contrast to the chronic toxicity studies, the concentration C' is always variable and varies between zero at infinite time and a maximum value C'_{max} related to the dose and rate of administration.

If the biophasic receptor sites are saturable and in instantaneous equilibration with the body fluids and tissues, the degree of inhibition is characterized by Eq. (35) rather than Eq. (80) and

$$\begin{aligned} f_A &= \exp\left[-\int_0^t k_D(I - I_m)\,dt\right] \\ &= \exp\left\{-\int_0^t ak_D[C'/(1 + aC') - C_m/(1 + aC_m)]dt\right\} \end{aligned} \tag{83}$$

The fundamental premise of this development is that the insult that ultimately results in death or toxicity is a consequence of the time duration and magnitude of a net degree of alteration or inhibition ΔI of a critical metabolic or biological activity. This net is in excess of what can be compensated by existing or inducible homeostatic mechanisms. The totality of the insult can be expressed as the time integral of this net degree of inhibition ΔI for an acute dose. This ΔI is related to the toxicant concentration in the biophase [Eqs. (35) and (80)], which is a function of the administered dose and the time course, i.e., the toxicokinetics, of the resultant concentrations of toxicant in the biophase. The model has been developed here for homogeneous populations, where susceptibilities to toxicants are equal. A heterogeneous population would demand statistical considerations to relate the spectrum of susceptibilities. The model has been developed here on the premise of bolus or extremely fast administration to an organism, where the biophase instantaneously equilibrates with other tissue and fluid volumes of distribution. The basic premise is that the probability that an animal will die (or manifest toxicity) has been decided at the time when the biophasic concentration finally has decreased below its threshold toxicity level C_m by the processes of toxicant elimination and metabolism.

This does not mean that the agent-induced death or toxicity must have occurred by this time, merely that the probability has been fixed. There can be a time lag mediated by biochemical or physiological processes between the insult and the manifested toxicity. There can be a time lag mediated by biochemical or physiological processes between the insult and the manifested toxicity. The immediate effect could be the irreversible or noncompensatory injury to a process that results in a time-dependent depletion of an anabolite necessary for the maintenance of life or permits the time-dependent accumulation of toxins that ultimately leads to death.

DOSE DEPENDENCES OF ACUTE TOXICITY WITH NONSATURABLE TOXIC RECEPTOR SITES

First-Order Elimination Processes

An agent administered as a bolus to the body may instantaneously equilibrate with the tissues and fluids of the body and thus with the biophase. Then, the concentration C' in the biophase at any time is equal to the concentration C in the apparent volume of distribution V of the body. With first-order elimination processes, the concentration decreases exponentially as in

$$C = C_0 e^{-\lambda(t-t0)} \tag{84}$$

where λ is the overall first-order elimination rate constant and C_0 is the concentration of toxicant in the body at time $t_0 = 0$. This value can be substituted into Eq. (82) and

$$-\ln f_A = \int_0^{tCm} ak_D C_0 e^{-\lambda t}\, dt - \int_0^{tCm} ak_D C_m\, dt \tag{85}$$

which gives the cologarithm of the fraction surviving until that time $t = t_{C_m}$ when the concentration C falls below the minimum toxic concentration C_m. Hereafter, the use of the term "survive" can be equated to nonmanifestation of toxicity. All organisms who have survived at that time continue to survive.

Appropriate integration gives

$$-\ln f_A = \frac{ak_D}{\lambda} C_0[1 - \exp(-\lambda_{tCm})] - ak_D C_m t_{C_m} \tag{86}$$

Rearrangement of the logarithmic transformation of Eq. (84) for $t_0 = 0$ permits an explicit definition of

$$t_{C_m} = \frac{1}{\lambda} \ln \frac{C_0}{C_m} = \frac{1}{\lambda} \ln \frac{D}{D_m} \tag{87}$$

where $C_0 = D/V$ and $C_m = D_m/V$.

Substitution of this value into Eq. (84) gives

$$\ln f_A = -\frac{ak_D}{\lambda V}(D - D_m \ln D) + \frac{ak_D}{\lambda V}(D_m - D_m \ln D_m) \qquad (88)$$

Thus, a plot of the logarithm of fraction of survivors against dose D after a dose interval of no effect is generally curved downward (Fig. 9-33), i.e., the slope becomes more negative with increasing dose (see Fig. 9-34A, curve c, for nonsaturable receptor sites and first-order toxicokinetics, NS-1). The semilogarithmic plot of fraction of survivors against dose conforming to Eq. (88) should approach linearity as D becomes much greater than D_m. If there were no threshold toxic dose, $D_m = 0$, toxicity would occur at all doses and the semilogarithmic plot of the fraction against dose always would be linear and the intercept at $D = 0$ would correspond to $f_A = 1$ (curve c in Fig. 9-35A for NS-1) (Garrett, 1977).

Examples with Possible Nonsaturable Toxic Receptor Sites and First-Order Eliminations

Jusko (1972) has applied toxicokinetic modeling to teratogens. Linear plots of the logarithm of the fraction of survivors $\ln f_A$, where f_A is actually the fraction of nonresorbed rat embryos or nonaffected fetuses against dose, were linear for cyclophosphamide (i.p.) and 6-aminonicotinamide (intra-aminiotic) on mouse embryos and thalidomide (i.p.) on rabbit embryos (Fig. 9-36). This is consistent with the acute toxicity model for homogeneous populations, where there is no threshold value for toxicity, i.e., C_m or $D_m = 0$, and neither elimination rates nor receptor sites are saturable [Eq. (88) and curved in Fig. 9-35A for NS-1]. Semilogarithmic plots of f_A against dose for hydroxyurea (i.p.) and 5-fluorouracil (i.p.),

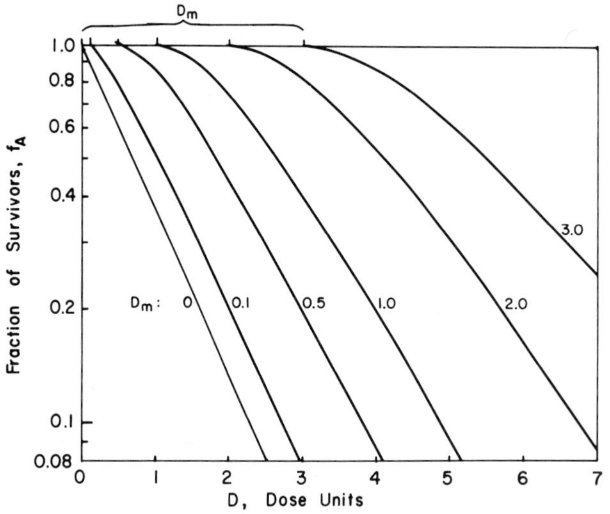

FIGURE 9-33. Semilogarithmic plots of fraction of survivors against dose D for the variously labeled threshold doses D_m. The curves were constructed in accordance with Eq. (88) for acute dosing of a bolus to a one-compartment body model organism with nonsaturable toxicity receptor sites and first-order elimination. The value of $ak_D/\lambda V$ was taken as unity. [From Garrett (1977).]

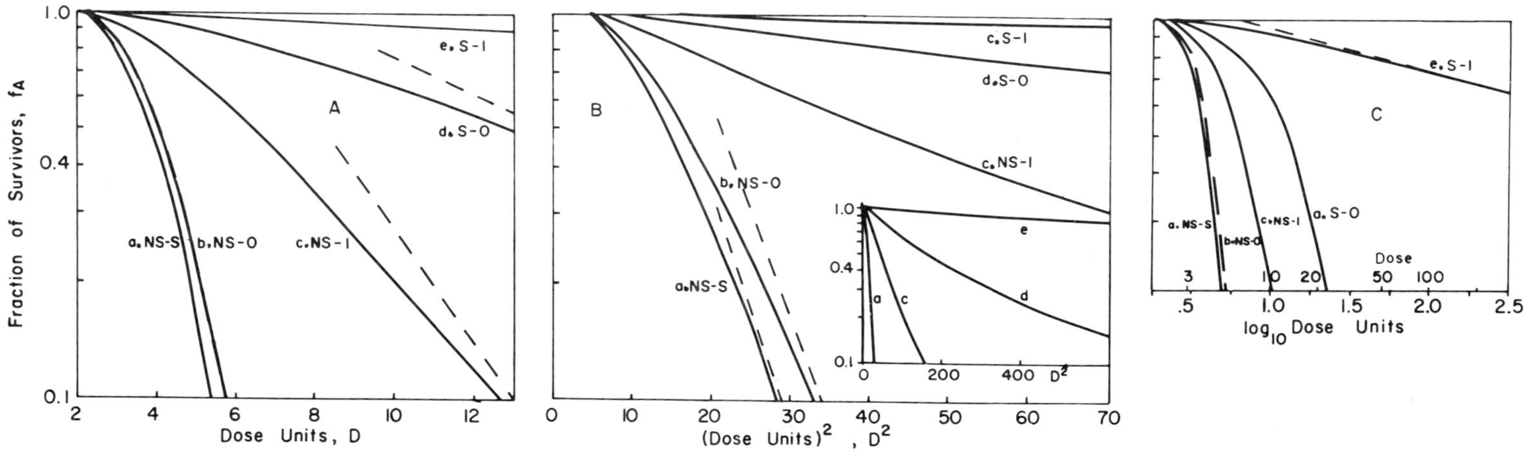

FIGURE 9-34. Comparisons of various graphical treatments of data from representative models of acute toxicity on bolus administration to the one-compartment body model with a threshold toxicity level taken as $D_m = 2$. The representative models were (a) NS-S, nonsaturable receptor sites and saturable elimination process with $K = 1$; (b) NS-0, nonsaturable receptor sites and zeroth-order elimination processes with $K = 0$; (c) NS-1, nonsaturable receptor sites and first-order elimination processes; (d) S-0, saturable receptor sites and zeroth-order elimination processes; (e) S-1 saturable receptor sites and first-order elimination processes. The curves were constructed for k_D/k_{max} and k_D/λ values of 1/3 and a and V values of 1.00. (- -) The model that conforms to, or approaches, linearity for the plotted function dependences. The semilogarithmic plots of fraction of survivors f_A are given against (A) dose D; (B) D^2; (C) $\log_{10} D$. (D) Fraction dead against dose. Fraction of survivors f_A against dose. (F) Fraction f_A against $\log_{10}$ dose. (G) The quotient ($\ln f_A$)/dose against dose. [From Garrett (1977).]

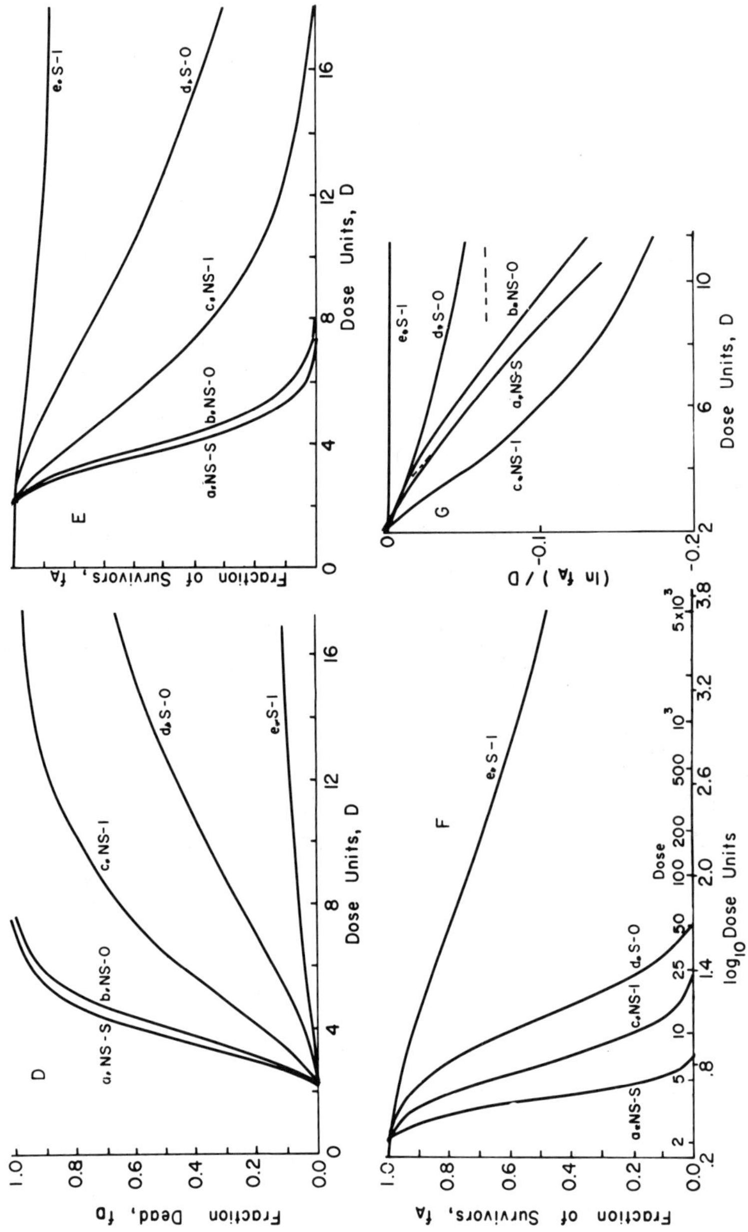

FIGURE 9-34. *(continued)*

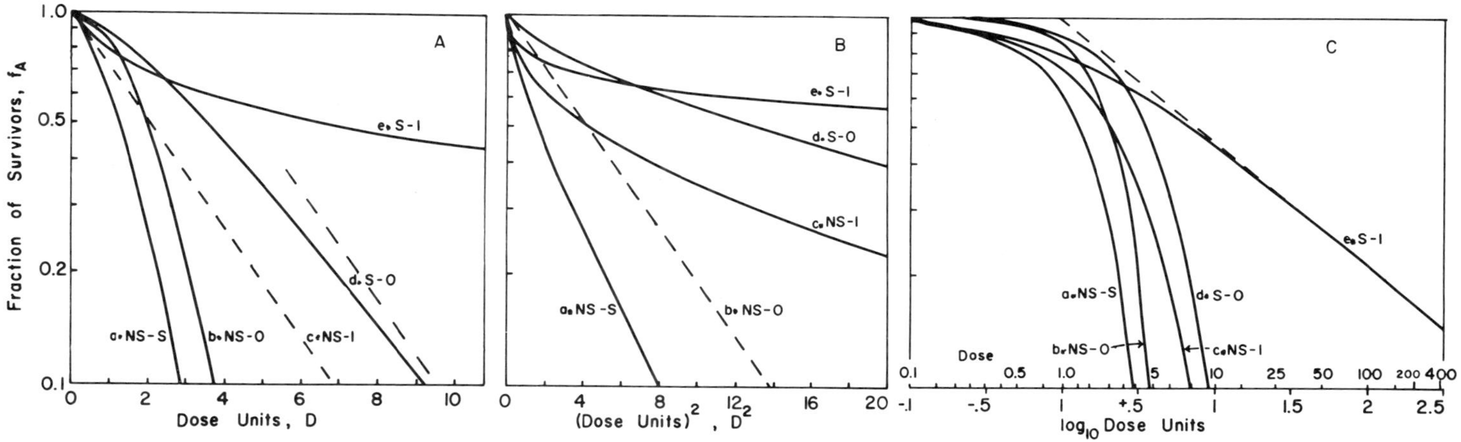

FIGURE 9-35. Comparison of various graphical treatments of data from representative models of acute toxicity on bolus administration to the one-compartment body model with no threshold toxic level, $D_m = 0$. The representative models were (a) NS-S, nonsaturable receptor sites and saturable elimination processes; (b) NS-0, nonsaturable receptor sites and zeroth-order elimination processes; (c) NS-1, nonsaturable receptor sites and first-order elimination processes; (d) S-0, saturable receptor sites and zeroth-order or constant rates of elimination; (e) S-1, saturable receptor sites and first-order elimination. The curves were constructed for k_D/k_{max} and k_D/λ values of 1/2 and a, K, and V values of 1.00. (- -) The model that conforms to, or approaches, linearity for the plotted functional dependences. The semilogarithmic plots of fraction of survivors f_A against (A) dose D; (B) D^2; (C) $\log_{10} D$. (D) Fraction dead against dose. (E) Fraction of survivors f_A against dose. (F) Fraction f_A against $\log_{10}$ dose. (G) The quotient ($\ln f_A$)/dose against dose. [From Garrett (1977).]

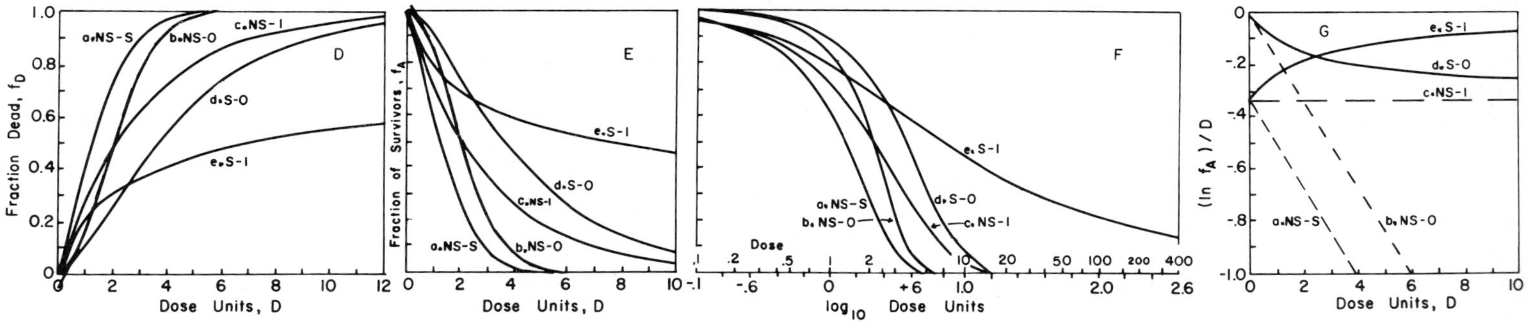

FIGURE 9-35. *(continued)*

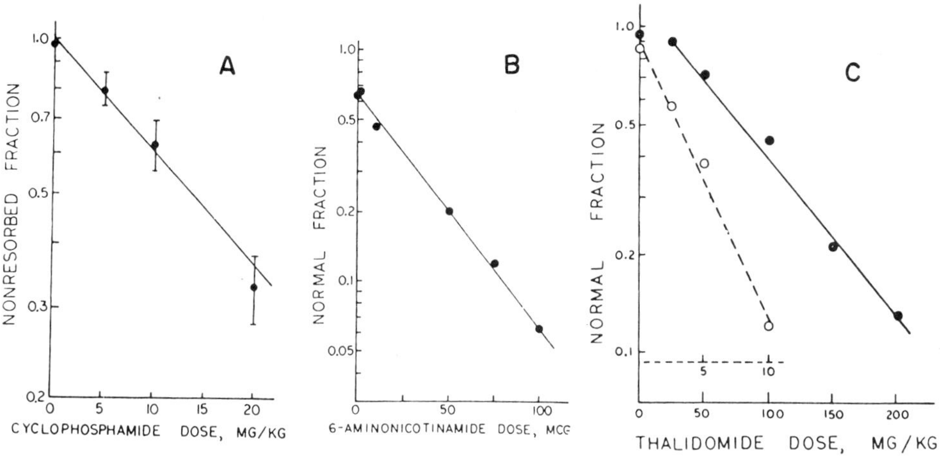

FIGURE 9-36. Semilogarithmic plots of fraction of survivors f_A against time with intercepts of f_A = 1 at time zero in accordance with Eq. (88) for D_m = 0. (A) Total effect of i.p. dose of cyclophosphamide on day-10 mouse embryos. Vertical lines show ±1 S.E. Data are from Gibson and Becker (1968). [From Jusko (1972).] (B) Total effect of intra-amniotic dose of 6-aminonicotinamide on day-15 rat embryos. Data from Turbow and Chamberlain (1968). [From Jusko (1972).] The intercept of less than 1.0 was assigned to the fact that intra-amniotic injection killed a constant fraction of embryos independent of dose. (C) Total effect of (○) i.p. and (●) p.o. doses of thalidomide on day 8–12 rabbit embryos. Data from Schumacher *et al.* (1968), showing the effect of slower rates of absorption. The i.p. doses are represented on the inserted abscissa. [From Jusko (1972).]

aspirin (p.o.), actinomycin D (i.p.), and reserpine (i.p.) were nonlinear (Fig. 9-37), indicating a conformity to Eq. (86) with $D_m > 0$ (curve c in Fig. 9-34A, for NS-1).

Saturable Elimination Processes

It has been shown (Garrett, 1977) for a saturable elimination process [e.g., Eq. (26) with nonsaturable receptor sites for a bolus dose in the one-compartment body model] that a plot of the logarithm of fraction of survivors against dose D conforms to an equation of the form

$$\ln f_A = -(a''D^2 + \beta''D + \gamma \ln D + \delta) \qquad (89)$$

At a given dose D, the coefficients are dependent on the values of k_{max} (the maximum rate of elimination of drug from the body), K (the association constant of drug with elimination sites), D_m (the threshold dose needed for toxicity, and V (the overall volume of distribution). An example of a semilogarithmic plot against dose for such conditions is given in Fig. 9-34A (NS-S, curve a) for a finite threshold toxicity for nonsaturable receptor sites and saturable toxicokinetics. The curve continuously curves downward with increasing dose.

When elimination rates become constant (the zero order of complete satu-

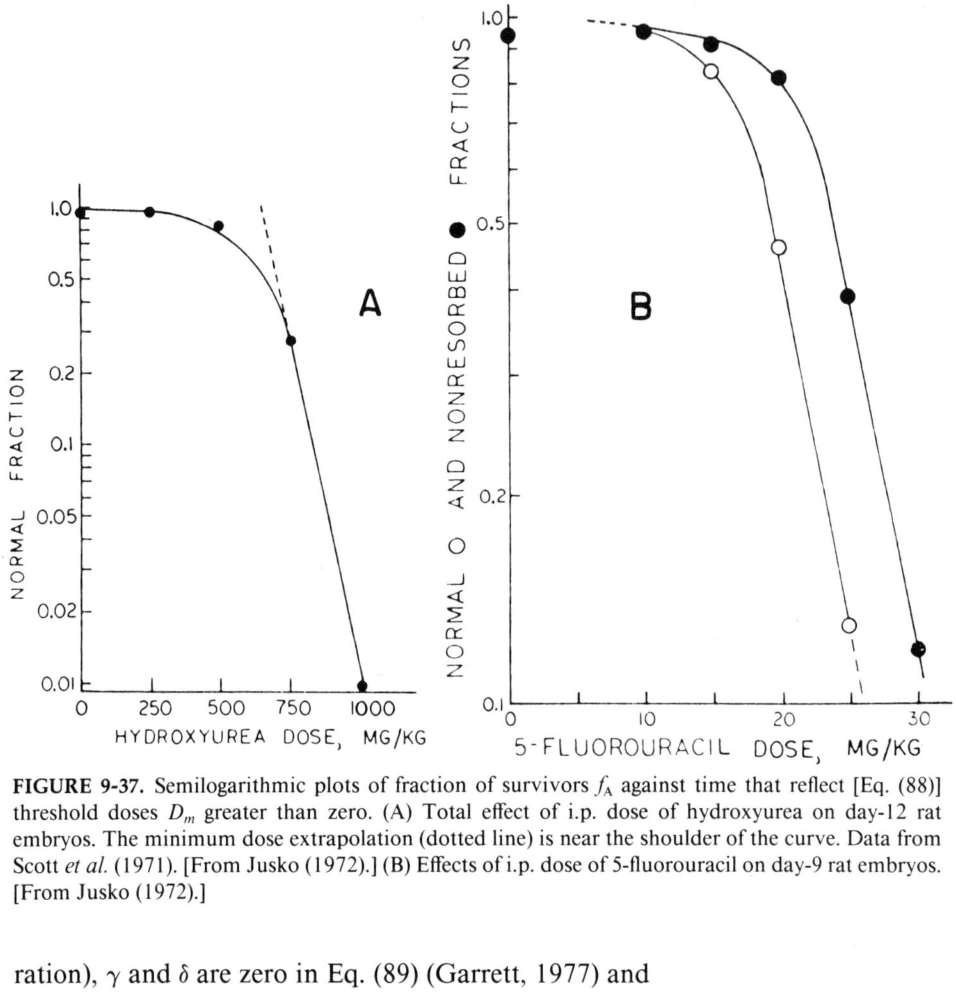

FIGURE 9-37. Semilogarithmic plots of fraction of survivors f_A against time that reflect [Eq. (88)] threshold doses D_m greater than zero. (A) Total effect of i.p. dose of hydroxyurea on day-12 rat embryos. The minimum dose extrapolation (dotted line) is near the shoulder of the curve. Data from Scott *et al.* (1971). [From Jusko (1972).] (B) Effects of i.p. dose of 5-fluorouracil on day-9 rat embryos. [From Jusko (1972).]

ration), γ and δ are zero in Eq. (89) (Garrett, 1977) and

$$\ln f_a = -(a''D^2 + \beta''D) \tag{90}$$

However, the plots for NS-0 (curve b in Fig. 9-34A) are similar in shape to those of NS-S (curve a in Fig. 9-34A).

An equation of the form of Eq. (90) is applicable when there is no threshold dose for toxicity, $D_m = 0$, for nonsaturable receptor sites and saturable elimination processes (NS-S, curve a in Fig. 9-35A). Terminal data at high doses approach linearity when $\ln f_A$ is plotted against D^2(NS-S, curve a in Fig. 9-35B) since, when the saturable elimination processes show constant rates of elimination (NS-0, curve b of Fig. 9-35B), β'' in Eq. (90) equals zero, it can be shown (Garrett, 1977) that

$$\ln f_A = -a''D^2 \tag{91}$$

In all cases of nonsaturable (NS) receptor sites and saturable eliminations, when D becomes very much greater than the threshold dose D_m and elimination rates approach constancy, the quotient $(\ln f_A)/D$ should become linear with dose at the higher doses (NS-S and NS-0, curves a and b in Fig. 9-34G). If there is no threshold toxicity, this quotient should be linear for all doses (NS-S and NS-0, curves a and b in Fig. 9-35G).

DOSE DEPENDENCES OF ACUTE TOXICITIES WITH SATURABLE TOXIC RECEPTOR SITES

The prior discussion was based on the postulate that the degree of alteration or inhibition of the biological receptor sites was directly proportional to the toxicant concentration in the biophase. If the biophasic receptor sites are saturable and in instantaneous equilibration with the toxicant in the body fluids and tissues, Eq. (83) is operational.

First-Order Elimination Processes

The presence of saturable receptor sites modifies the equation for logarithm of fraction of survivors (Garrett, 1977) to

$$\ln f_A = [(k_D/\lambda) \ln (V/a + D) + \beta \ln D + \kappa'] \tag{92}$$

where coefficients β and κ' depend on the values of D_m, V, and λ. If $D_m = 0$, $\beta = 0$.

Semilogarithmic plots of fractions of survivors against dose (Fig. 9-38 and curve e for S-1 in both Figs. 9-34A and 9-35A) differ from those with nonsaturable receptor sites (Fig. 9-33 for NS-1) in that they bend upward (show decreasing negative slopes) with dose. In contrast to the data in Fig. 9-33 and curve a for NS-1 in Figs. 9-34C and 9-35C, which show increasing negative slopes against the logarithm of dose, such plots for saturable receptor sites tend to become linear with increasing dose (curve e for S-1 in Figs. 9-34C and 9-35C).

Saturable Elimination Processes

With both saturable elimination processes and saturable toxicity receptor sites, the dose dependencies of toxicity depend on which is the most readily saturable. A particular case is when the pharmacokinetics is completely saturated and the rate of loss of toxicant from the body is constant and dose-independent. The concentration in the biophase is then proportional to time,

$$C' = C'_0 - k_{max}t \tag{93}$$

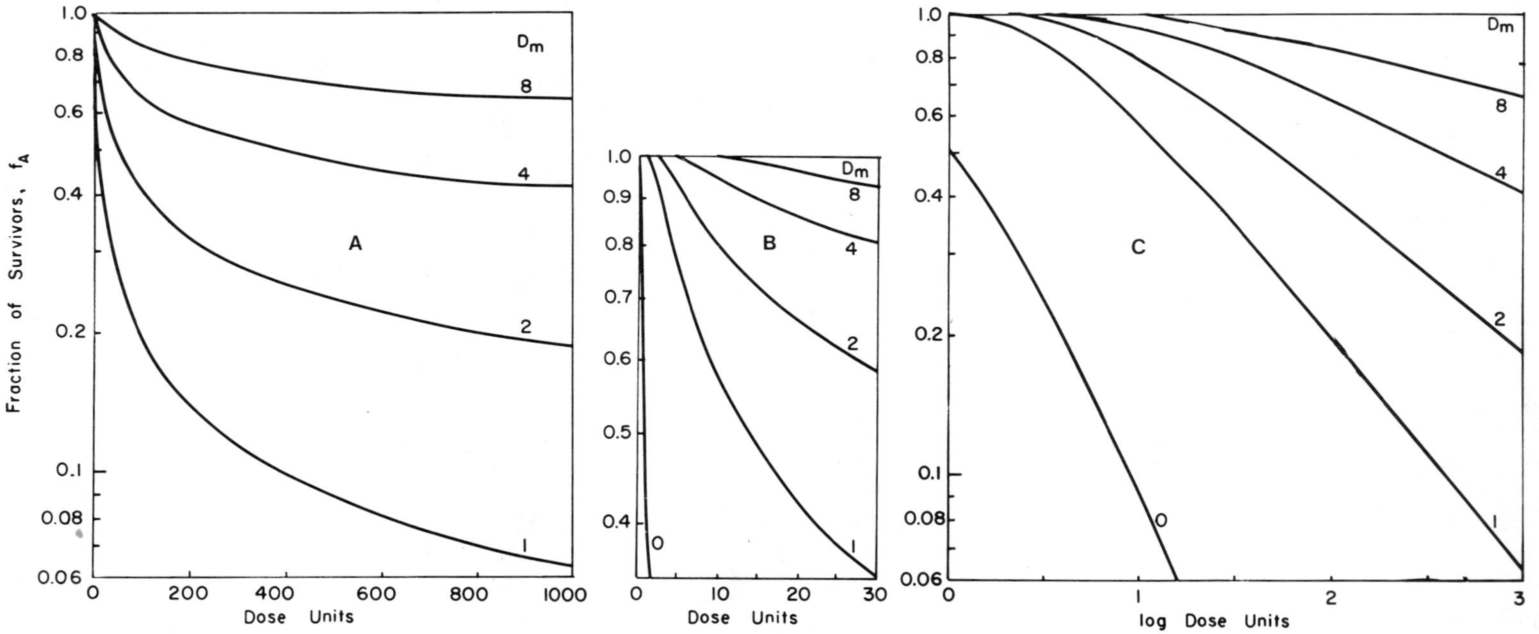

FIGURE 9-38. Semilogarithmic plots against (A, B) dose and (C) logarithm of acute bolus administration of dose units for the one-compartment body model with saturable receptor sites and first-order elimination. The curves are labeled with their respective threshold toxic doses D_m. [From Garrett (1977).]

and it can be shown (Garrett, 1977) that

$$\ln f_A = \beta'' D + \alpha' \ln(1 + \gamma D) + \kappa''' \quad (94)$$

where the coefficients β'' and κ''' for a given dose depend on values of D_m, V, and k_{max}. Such plots of the logarithm of fraction of survivors against dose ultimately approach linearity with increasing dose, since the second and third terms of Eq. (94) become less significant at the higher doses with increasing saturation of toxic receptor sites (S-0, curves d in Figs. 9-34A and 9-35A).

The simplest case is when $D_m = C_m = 0$, where

$$\ln f_A = -\frac{k_D}{k_{max} V} D + \frac{k_D}{a k_{max}} \ln\left(1 + \frac{aD}{V}\right) \quad (95)$$

which approaches $-(k_D/k_{max}D)/D$ at higher doses or with high toxic receptor site saturabilities, i.e., when a is high.

DEDUCTION OF PHARMACOKINETIC PROPERTIES OF A TOXICANT FROM ACUTE TOXICITY STUDIES

The possible identification of the toxicokinetic processes underlying toxicity as a function of dose may be deduced from the shape of the curves seen on appropriate plots of functions of the data (Figs. 9-34 and 9-35). Table 9-2 summarizes the relations between the shape of a plotted curve and the underlying nature of receptor sites and toxicokinetic process, i.e., combination of nonsaturable (NS) or saturable (S) receptor sites and kinetic order [zero (0), first (1), or saturable (S)] of the elimination process. These plotted curves are based on the mathematical analyses given in Garrett (1977).

Comparisons of Toxicokinetically and Statistically Based Models of Acute Toxicity

The toxicity responses considered here are quantal (all or nothing) responses. These responses also have been considered as functions of dose on the purely statistical premise of variability in test subjects. Finney (1964) has ably summarized these concepts with particular respect to probit analysis. Responses are presumed to result from a normal distribution of tolerance against dose or logarithm of dose. The latter is favored on the premise that it provides greater symmetry to the distribution. Although attempts have been made to rationalize this preference, Finney (1964) has concluded that its major justification is that it simplifies the analysis, i.e., it works.

These statistical hypotheses lead to plots of the fraction of organisms surviving against dose or logarithm of dose that are sigmoidal curves that resemble those in Figs. 34F and 35F. Such sigmoidal curves can be transformed into straight lines by mathematical manipulations, such as the probit transformation (Finney, 1964), where probits are linearly plotted against dose, log dose, or some power of the dose. The assumed linearity of the probit-transformed data has been used to permit prediction of toxicity at lower doses, even though Finney and others have warned against such a temptation. The LD_{50} is the dose at $D_{0.5}$, when 0.5 is the fraction of animals that have died or survived the toxic stress of the toxicant.

The data of Fig. 9-34 are plotted in Fig. 9-39 on probability paper against the logarithm of the dose. These plots of transformed data against the logarithm of the dose are reasonably linear for all possible toxicokinetic circumstances between the 10 and 90% mortality where toxicity data are normally obtained. This could be so notwithstanding the fact that a threshold dose for toxicity (or death) could be manifested by nonlinearity below 10% mortality.

The underlying postulates of the statistical model predict complete linearity in such plots and will always predict a finite probability of death or toxicity at low or any dose even though such a threshold dose could exist.

Examples of Applications

Studies with appropriate and adequate acute toxicity data in the literature are listed in Table 9-1. The fractions of organisms unaffected by the toxicants f_A are plotted semilogarithmically against dose for these studies in Fig. 9-40. Very few published acute toxicity studies were performed at more than six or seven doses or at doses with fraction of survivors f_A below 0.1. Unfortunately, popula-

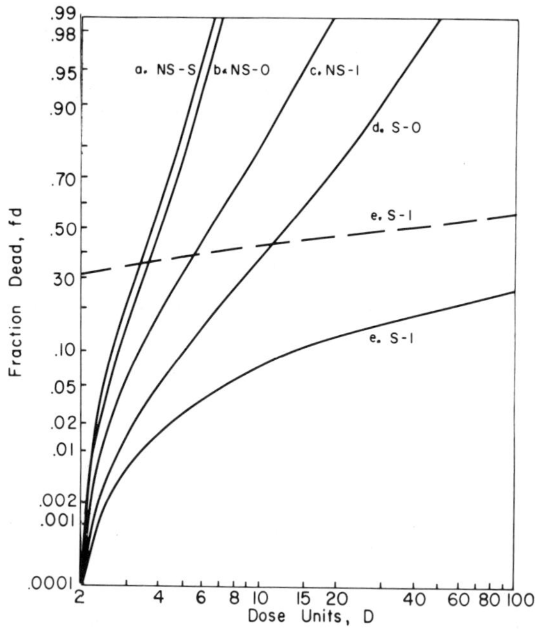

FIGURE 9-39. Plots of data for fraction dead f_D on probability paper from representative models of acute toxicity on bolus administration to the one-compartment body model for a finite threshold dose of $D_m = 2$. The parameters are the same as in Fig. 9-34. The representative models were: (a) NS-S, nonsaturable receptor sites and saturable eliminations; (b) NS-0, nonsaturabe receptor sites and zeroth-order eliminations; (c) NS-1, nonsaturable receptor sites and first-order elimination processes; (d) S-0, saturable receptor sites and zeroth-order eliminations; and (e) S-1, saturable receptor sites and first-order eliminations. (- -) For S-1 for the dose range of 100–10,000 units. [From Garrett (1977).]

tions of rats and mice at each acute dose were frequently limited to about ten individuals, which minimizes the reliability of an individual f_A estimate. Nevertheless, the data from the reasonably adequate studies listed in Table 9-1 do permit plots of transforms of the fractions of survivors f_A against functions of the acute dose (Figs. 9-40 to 9-41 and Table 9-2). The evaluation of these plots of the experimental data in light of the forms and patterns anticipated from various underlying toxicokinetic models (Table 9-2) is summarized in Table 9-3.

The greater number of studies (i.e., A–F, including D′, E′, and E″ in Table 9-1) demonstrated saturable elimination processes, nonsaturable elimination processes, and significant threshold doses necessary for the demonstration of toxicity (i.e., NS-S or NS-0 and $D_m > 0$). This is not unexpected. Toxicities are usually engendered and studied at high doses where elimination processes may be over-

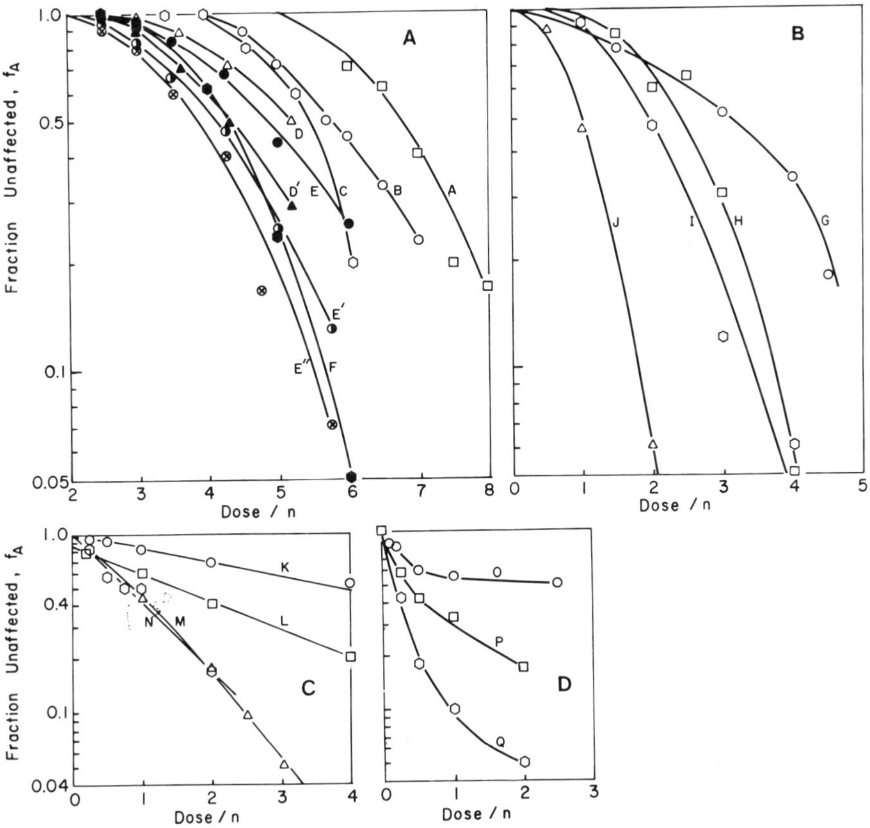

FIGURE 9-40. Semilogarithmic plots of unaffected fraction of organisms f_A against D/n, where D is the acute dose for the studies listed in Table 9-1. (A) Studies A–F with an apparent threshold dose for toxicity D_m. (B) Studies G–J with no apparent threshold dose for toxicity $D_m = 0$, and increasing negative slopes with dose. (C) Studies K–N with no apparent threshold dose for toxicity, giving apparent linear plots. (D) Studies O–Q with no apparent threshold dose for toxicity and decreasing negative slopes with dose. [From Garrett (1977).]

TABLE 9-1. Acute Toxicity Studies Analyzed for Consistency with Pharmacokinetic Receptor-Site Models

Study[a]	Agent	Mode of administration	Organism	Number of organisms per dose	Dose units D	Adjustment factors for plots n	Adjustment factors for plots p	Means used to monitor fraction unaffected f_A	Time monitored after administration
Figure 9-40a. Studies with an apparent threshold dose for toxicity D_m and saturable elimination processes with nonsaturable receptor sites (NS-S)									
A	Paraoxon (plus 10 μg PCB/fly)[b]	Topically	Fly, *Musca domesticus*	45	μg/fly	0.002	100	Survival	24 h
B	Paraoxon	Topically	Fly, *Musca domesticus*	45	μg/fly	0.01	20	Survival	24 h
C	Glycopyrrolate	IP	Mouse	10	mg/kg	20	0.1	Survival	72 h
D	Indomethacin	Oral	Rat	10	mg/kg	2	0.2	Nonperforation of intestine	72 h
D′	Indomethacin	Subcutaneous	Rat	10	mg/kg	2	0.2	Nonperforation of intestine	72 h
E	Ichthyotoxin of *Pyrmnesium parvum*	IP	Fish, *Gambusia affinis*	10	μg/fish	4	0.1	Survival	15 min
E′	Ichthyotoxin of *Pyrmnesium parvum*	IP	Fish, *Gambusia affinis*	10	μg/fish	4	0.1	Survival	30 min
E″	Ichthyotoxin of *Pyrmnesium parvum*	IP	Fish, *Gambusia affinis*	10	μg/fish	4	0.1	Survival	45 min[c]
F	Ethanol	IP	Bat, *Myotis lucifugus*	18–44	g/kg	1	0.5	Survival	24 h
Figure 9-40b. Studies with no apparent threshold dose for toxicity and increasing negative slopes of semilogarithmic plots verse dose									
G	Cyclohexylamine	IP	Mouse	10	mg/kg	100	0.005	Survival	2 h
H	Apholate (insecticide, alkylating agent)	Oral (gelatin capsule)	Single-comb white leghorn 11-day-old cockerels	20–40	mg/kg	100	0.01	Survival	3 weeks

I	Lucanthone	Immersion	*Neurospora crassa*	—	mM	0.02	30	Survival	
J	Dieldrin	Topically	Insect, *Tribolium castaneum*	—	ppm in applied spray to container	100	0.1	Survival	5 weeks
Figure 9-40c. Studies with no apparent threshold dose for toxicity D_m and linear semilogarithmic plots versus dose (NS-1)									
K	Methaqualone	IV	Cat	6–12	μg/kg	125	0.0025	Fraction of original blood pressure maintained	3 min
L	Fentanyl citrate	Subcutaneous	Mouse	10	mg/kg	50	0.1	Survival	1–2 min
M	Hycanthone	Topically (immersion)	*Neurospora crassa*	—	mM	0.1	10	Survival	—
N	Prednisolone (plus 60 mg/kg propionitrile 12 h later)	Subcutaneous	Rat	20	mg/rat	1	2	Survival	36 h
Figure 9-40d. Studies with no D_m and decreasing negative slopes of semilogarithmic plots versus dose									
O	Chlordane[d]	IP	Rat	6–8	mg/kg	10	0.2	Fraction of net wet weight of uteri maintained	4 h[d]
P	Dimethylselenide (plus 4 mg As/kg)	IP	Rat	12–14	mg Se/kg	10	1	Survival	8 h
Q	PCB[b] (plus 0.02 μg paraoxon per fly)	Topically	Fly, *Musca domesticus*	45	μ/fly	2	2	Survival	24 h

[a] References for cited study are: A) and B) Fuhremann and Lichtenstein, 1972; C) Franco *et al.*, 1970; D) and D′) Brodie *et al.*, 1970; E) E′) and E″) Bergman *et al.*, 1963; F) Greenwald *et al.*, 1968; G) Lee and Dixon, 1972; H) Sherman and Gerrick, 1966; I) Ong and de Serres, 1975; J) Tammes *et al.*, 1970; K) Quest *et al.*, 1974; L) Gardocki and Yelnosky, 1964; M) Ong and de Serres, 1975; N) Robert *et al.*, 1975; O) Welch *et al.*, 1971; P) Obermeyer *et al.* 1971; Q) Fuhremann and Lichtenstein, 1972.

[b] PCB is Aroclor® 1248 brand of polychlorinated biphenyls.

[c] No more organisms died after this time.

[d] Actually administered daily for 7 days every 24 h and 0.2 μg estrone was injected IP on day 8. The uteri were removed 4 h after estrone administration and the wet weight determined.

TABLE 9-2. Shapes and Patterns of Plots of Functions of Fractions of Population f_A Manifesting Toxicity against Functions of Acute Doses D for Saturable (S) and Nonsaturable (NS) Receptor Sites and Various Orders of Elimination Toxicokinetics

Process[a]	D_m[b]	Curve[b]	log f_A vs. D (Figs. 9-34A, 9-35A)	log f_A vs. D^2 (Figs. 9-34B, 9-35B)	log f_A vs. log D (Figs. 9-34C, 9-35C)	f_A vs. D (Figs. 9-34E, 9-35E)	f_A vs. log D (Figs. 9-34F, 9-35F)	(ln f_A)/D vs. D (Figs. 9-34G, 9-35G)
NS-1	0	c	Completely linear with negative slope	Slope becomes less negative with increasing dose	Slope becomes more negative with increasing dose[c]	Slope becomes less negative with increasing dose[a]		No change with dose, slope = 0
	>0	c	Slope becomes more negative with increasing dose[c]; curve approaches terminal linearity if D_m is relatively small	As above	As above	As D_m increases, so does the sigmoidal nature of the curve		Slope becomes less negative with increasing dose[d]; if D_m relatively small, curve approaches an asymptote with increasing dose
NS-S	0	a	Slope becomes more negative with increasing dose[c]	Slope becomes more[c] or less[d] negative (dependent on ease of saturability) with increasing dose; curve approaches terminal linearity	As above	Slope becomes less negative with increasing dose[d]	All curves have sigmoidal character with the appearance of much greater symmetry than when f_A is plotted against dose D	Completely linear with negative slope and intercept less than zero
	>0	a	As above	Slope becomes more[c] negative with increasing dose; curve approaches terminal linearity if D_m is relatively small	As above	As D_m increases, so does the sigmoidal nature of the curve		Slope becomes more negative with increasing dose[c]; curve approaches terminal linearity if D_m is relatively small

NS-0	0	b	As above	Completely linear with negative slope	As above	Curve has sigmoidal character	Completely linear with negative slope and intercept of zero
	>0	b	As above	Slope becomes more negative[c] with increasing dose; curve approaches terminal linearity if D_m is relatively small	As above	As above	Slope becomes more negative with increasing dose[c]; curve asymptotically approaches zero from a lower value
S-1	0	e	Slope becomes less negative with increasing dose[d,e]	Slope becomes less negative with increasing dose[d]	Slope becomes more negative with increasing dose[c]; curve approaches terminal linearity	Slope becomes less negative with increasing dose[d]	Positive slope decreases with increasing dose; curve asymptotically approaches zero from a lower value
	>0	e	As above	As above	As above	As D_m increases, so does the sigmoidal nature of the curve	Curve can initially decrease and then increase; subsequent positive slope then decreases with increasing dose
S-0	0	d	Slope becomes more negative with increasing dose[c] and approaches terminal linearity if D_m is relatively small	As above	Slope becomes more negative with increasing dose[c]	Curve has sigmoidal character	Slope becomes less negative with decreasing dose[d] and approaches an asymptote at higher doses
	>0	d	As above	As above	As above	As above	As above

[a] The first component of the labeled process (S or NS) refers to whether the receptor sites for toxic activity are saturable or nonsaturable, respectively. The second component (1, S, or 0) refers to whether the elimination process is first order, saturable or zeroth order, respectively.

[b] The designated curve is in Fig. 9-34 for $D_m > 0$ and Fig. 9-35 for $D_m = 0$, where D_m is the necessary threshold value of the toxicant dose that must be exceeded for toxicity to be manifested by any fraction f_A of the population.

[c] The decreasing curve bends downward.

[d] The decreasing curve bends upward as it decreases.

[e] The presence of heterogeneous or resistant populations would show the same curvature.

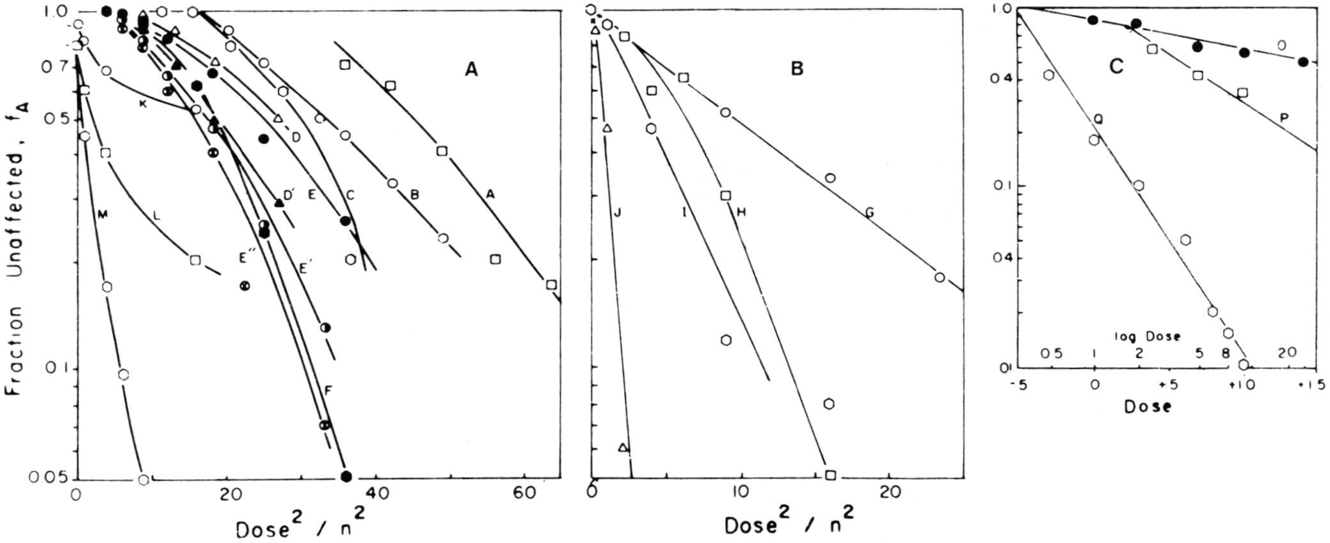

FIGURE 9-41. Semilogarithmic plots of unaffected fraction of organisms f_A against various functions of the acute dose D for the studies listed in Table 9-1. Those plotted against D^2/n^2 are (A) studies A–F with apparent threshold dose of toxicity that give nonlinear semilogarithmic plots against dose (Fig. 9-40a) and studies K–M with no apparent threshold dose for toxicity that gave linear plots when semilogarithmically plotted against dose (Fig. 9-40c) and (B) studies G–J with no apparent threshold dose for toxicity that give nonlinear semilogarithmic plots against dose with increasing negative slopes (Fig. 9-40b). (C) Studies O–Q plotted against logarithm of dose, log D, with no apparent threshold dose for toxicity, and which gave nonlinear semilogarithmic plots against dose with decreasing negative slopes (Fig. 9-40d). [From Garrett (1977).]

TABLE 9-3. Evaluation of Probable Toxicokinetic Models Underlying Various Plots of Fractions of Populations f_A Manifesting Toxicity against Functions of Acute Dose D

Study[a]	Estimated D_m[b]	$\log f_A$ vs. D	Processes possible[c]	$\log f_A$ vs. D^2	Processes possible[c]	$(\log f_A)/D$ vs. D	Processes possible[c]	Conclusion as to underlying model
A–F	>0	Slope becomes more negative with increasing dose (Fig. 9-40a)	NS-1 NS-S NS-0 S-0	Slope becomes more negative with increasing dose (Fig. 9-41a)	NS-S NS-0	Slope becomes more negative with increasing dose (Figs. 9-42a and 9-42b)	NS-S NS-0	Saturable or zeroth-order elimination processes with nonsaturable toxicity receptor sites and a threshold dose D_m that must be exceeded for toxicity manifestation
G–J	0	Slope becomes more negative with increasing dose (Fig. 9-40b)	NS-S NS-0 S-0	Linear with negative slope (Fig. 9-41b)	NS-0	Negative slope reasonably linear with intercept close to zero (Fig. 9-42c)	NS-0	Zeroth-order or constant-rate elimination processes with nonsaturable toxicity receptor sites with no apparent threshold dose D_m to manifest toxicity
K–N	0	Linear and with negative slope (Fig. 9-40c)	NS-1	Slope becomes less negative with increasing dose (Fig. 9-41a)	NS-1	Horizontal lines with zero slopes (except *M*) (Fig. 9-42c)	NS-1	Nonsaturable toxicity receptor sites and first-order elimination processes with no apparent threshold toxicity dose, $D_m = 0$
O–Q	0	Slope becomes less negative with increasing dose (Fig. 9-40d)	S-1	$\log f_A$ vs. $\log D$ terminal negative slope shows excellent linearity (Fig. 9-41c)	S-1	Positive slope decreasing with dose (Fig. 9-42d)	S-1	Saturable toxicity receptor sites and first-order elimination processes, probably $D_m = 0$, but could be $D_m > 0$

[a] Listed in Table I.
[b] Based on observation of negligible toxicity until a threshold dose D_m is reached.
[c] See Table II for patterns of plots for the stated models, i.e., first component of the classification is for saturable (S) or nonsaturable (NS) toxicity receptor sites; the second component is for the nature of the elimination process: 0 and 1 are for zeroth-order and first-order eliminations, respectively; S is for a saturable elimination process.

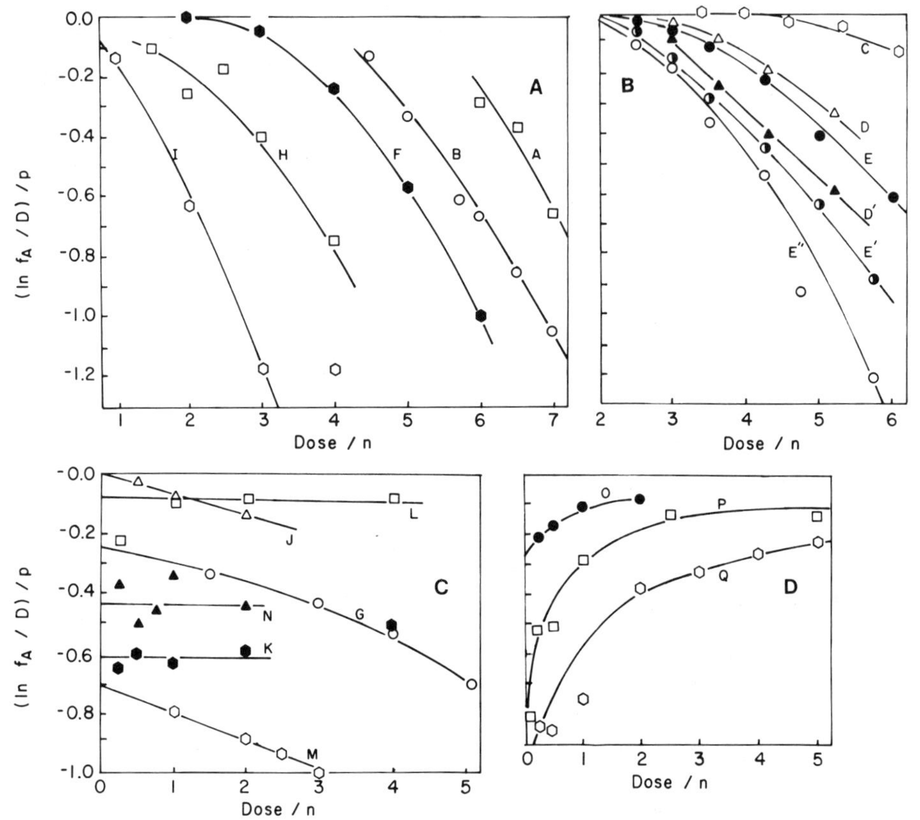

FIGURE 9-42. Plots of $(\ln f_A/D)/p$ against D/n, where D is the acute dose for the studies listed in Table 9-1. Studies with an apparent threshold dose for toxicity: (A) A, B, and F and (B) C–E″. Studies with no apparent threshold dose for toxicity and whose semilogarithmic plots of f_A (Fig. 9-40B) showed increasing negative slopes with dose: (A) H and I and (C) G and J. Studies with no apparent threshold dose whose semilogarithmic plots of f_A were apparently linear against dose (Fig. 9-40C): (C) K–N. Studies with no apparent threshold dose whose semilogarithmic plots of f_A (Fig. 9-40D) showed decreasing negative slopes with dose: (D) O–Q. [From Garrett (1977).]

whelmed. However, such elimination processes can relatively rapidly eliminate toxicant from the organisms at lower doses as compared to the higher doses. Thus, C_m in the biophase necessary for toxicity may not be readily exceeded at these low doses. Thus, even though the plots of toxicity on probability paper against the logarithm of the dose are reasonably linear under these conditions, prediction of a probability of toxicity at low doses *cannot* be effected by a simple linear extrapolation of the data plotted in Fig. 9-43A, since the toxicokinetic mod-

FIGURE 9-43. Plots of fraction of organisms unaffected f_A on probability paper against D/p, where D is the acute dose for the studies listed in Table 9-1. Studies with an apparent threshold dose for toxicity: (A) A–E″ and (B) F. Studies with no apparent threshold dose for toxicity and whose semilogarithmic plots of f_A (Fig. 9-40B) showed increasing negative slopes with dose: (B) G–J. (C) Studies (K–M) with linear semilogarithmic plots of f_A against dose (Fig. 9-40C) and studies (O–Q) with decreasing negative slopes with dose (Fig. 9-40D) [From Garrett (1977).]

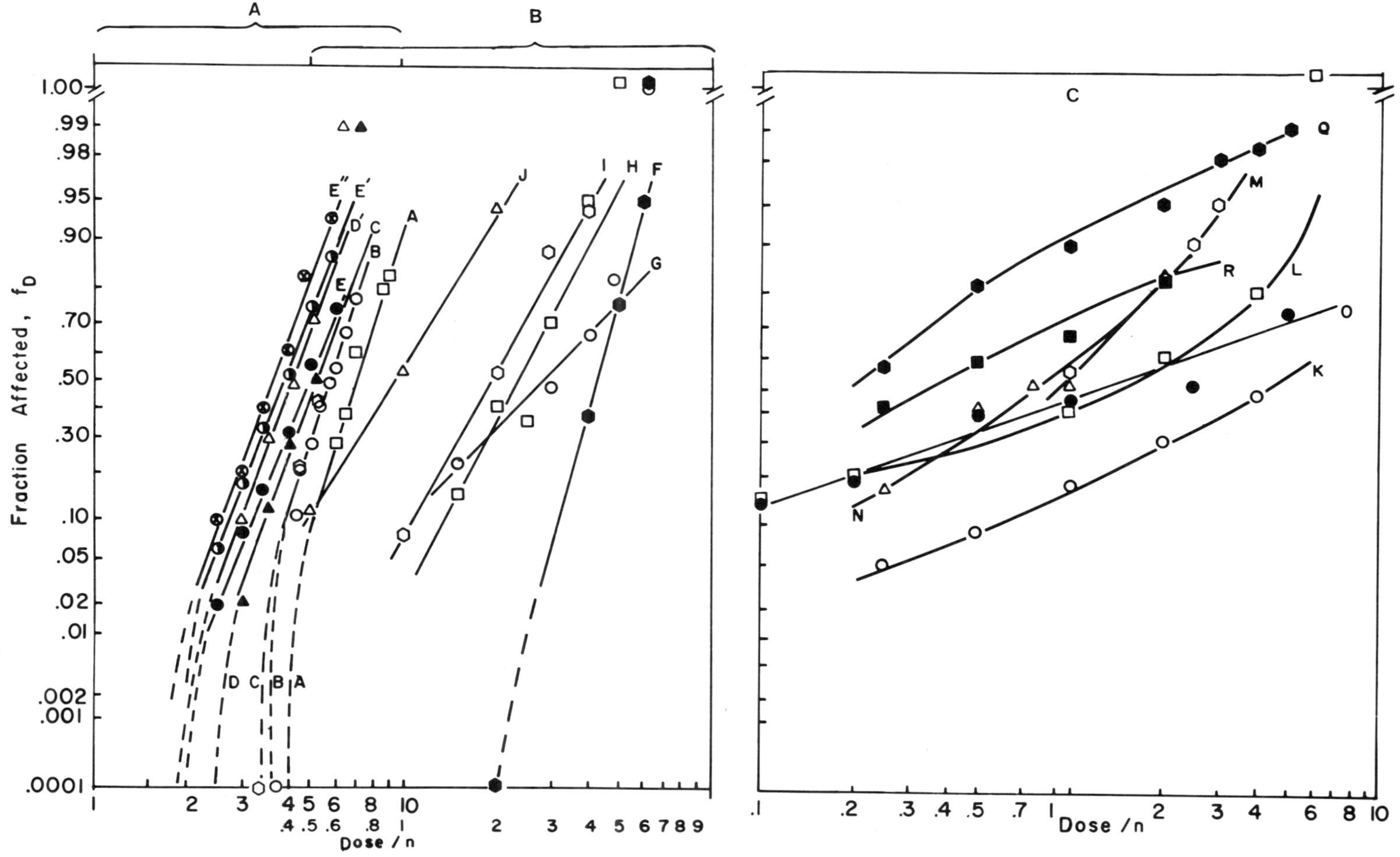

A
B
C
Fraction Affected, f_D
Dose /n
Dose /n

els predict that such plots for NS-S would be nonlinear at these low doses (Fig. 9-39). It should be noted that, as expected, the toxicity studies of ethanol in bats (F in Table 9-1) conform to the premises of saturable elimination processes and demonstrate nonsaturable receptor sites and a threshold dose for toxicity.

Studies G–J in Table 9-1 are apparent examples of saturable elimination processes and nonsaturable receptor sites with no threshold doses for toxicity (NS-S, NS-0, $D_m = 0$).

The studies on chronic poisoning of insects *(Tribolium castaneum)* by dieldrin were discussed previously, and it was concluded that there was saturable elimination of the toxicant above 0.9 ppm from the organism and no potential saturation of receptor sites. This analysis of the acute toxicity data (J in Table 9-1) for the action of dieldrin on the same organism at acute doses above 50 ppm results in the same conclusion.

Studies that conform to the premises for NS-1 would have linear semilogarithmic plots of unaffected fractions against dose (Fig. 9-35A). Studies K–N in Table 9-1 apparently fit the conditions for nonsaturable elimination processes and nonsaturable receptor sites with no threshold dose for toxicity. An intercept of the linear plots with $f_A < 1$ can be assigned (Jusko, 1972) to a procedural killing of constant fraction of organisms independent of dose. Plots of these studies on probability paper against the logarithm of the dose (Fig. 9-43C) are *not* linear, a pattern that would be predicted for such underlying conditions (Garrett, 1977).

The studies O–Q of Table 9-1 follow the patterns characteristic of saturable receptor sites and first-order eliminations (S-1).

The patterns of plots of such data (S-1) on probability paper against log dose should be reasonably linear for f_A values of 0.80 or less, provided that there is no significant threshold dose for toxicity. The pertinent plots for studies O–Q are given in Fig. 9-43C and their patterns are confirmatory.

CRITIQUE OF MODELS FOR TOXICITY

In classic toxicity studies, the mean or median lethal or toxic dose of an agent is estimated. It is assumed that the variation of toxicity in a population of organisms is due to a statistical distribution of tolerance. It follows from this statistical model that there is a finite probability of toxicity at any dose. Any predictive power for such a model must be based on this premise of toxicity at all doses and the assumption that transformed functions such as probits, when plotted against dose or logarithm of the dose, are linear. The presumption of a normal distribution of tolerances around the logarithm of the dose does not take into account the saturability of toxicokinetic processes and receptor sites at higher doses. The prediction of toxic doses in one species based on data obtained from another is frequently based on a simplified premise, usually that toxicities can be correlated on the basis of mg/kg of body weight. This premise has definite limitations: it assumes that the underlying sensitivities of all species are the same and that volumes of distribution, rates of elimination, and detoxification processes are merely proportional to weight. It ignores the variations in genetic patterns, metabolic pathways, and quantities of enzymes, and the fact that volumes of distribution of drugs and toxicants are *not* strictly proportional to weight.

The toxicokinetically based toxicity models developed in this chapter are based on the premise of receptor-site interaction with the toxicant. These models can be used to explain all the functional dependences of toxicity on dose and time of dosing. In addition, they permit the use of known metabolic pathways and volumes of distribution among the various species to permit prediction and correlation. They do not demand finite probability of toxicity at any dose.

Moreover, these models take into account the potential saturation of physiological processes at higher doses and permit tests to evaluate them. They do not assume linear processes and linear extrapolations to estimate toxicities at lower doses, but instead provide challenges and methods of evaluating the formal dependence of toxicity on dose.

Toxicity studies should be conducted primarily with homogeneous populations. Then the effects of inhomogeneities in the population on toxicities can be determined. When populations of diverse sizes and morphologies with different volumes of distribution are used, heterogeneous responses result and obscure the underlying model. Then, only statistical estimates are possible. The toxicity studies of the future must provide sufficient numbers of organisms dosed at adequately spaced intervals to permit the models to be challenged. Present chronic toxicity studies can be put to advantage by monitoring the times of death or toxic manifestation.

Such efforts would provide the information to challenge the theses presented here; such theses could determine the conditions under which high-dose toxicological studies were capable of predicting toxicities at low-level exposures.

An operational hypothesis could be that equivalent concentrations of an agonist in the biophase cause equivalent responses in various mammals. Knowledge of the relative volumes of distribution of two different mammals, the differences in their metabolic and elimination rates, and the differences in their toxicokinetic dependences on dose could serve as the criteria for utilizing the known toxicity in the one to predict that in the other.

ACKNOWLEDGMENT. This chapter is based in part on material from "The Pharmacokinetic Bases of Biological Response Quantification in Toxicology, Pharmacology, and Pharmacodynamics" in *Progress in Drug Research,* Volume 21, Birkhäuser, Basel, 1977.

REFERENCES

Ariens, E. J. 1964. *Molecular Pharmacology.* Academic Press, New York. 119 pp.

Bergmann, F., I. Parnas, and K. Reich. 1963. Observations on the mechanism of action and on the quantitative assay of Ichthyotoxin from *Prymnesium parvum* Carter. *Toxic. Appl. Pharmacol.* **5**:637–649.

Brodie, D. A., P. G. Cooke, B. J. Bauer, and G. E. Dagle, 1970. Indomethacin-induced intestinal lesions in the rat. *Toxic. Appl. Pharmacol.* **17**:615–624.

Buhler, E. R., M. E. Rasmusson, and W. E. Shanks. 1969. Chronic oral DDT toxicity in juvenile Coho and Chinook salmon. *Toxic. Appl. Pharmacol.* **14**:535–555.

Clark, A. J. 1933. *The Mode of Action of Drugs on Cells.* Arnold, London. 147 pp.

Finney, D. J. 1964. *Probit Analysis—A Statistical Treatment of the Sigmoid Response Curve.* 2nd ed. Cambridge University Press, England. 318 pp.

Franco, B. V., J. W. Ward, D. L. Gilbert, and G. Woodward. 1970. A condensed format for reporting toxicologic data—results of studies on glycopyrrolate. *Toxic. Appl. Pharmacol.* **17**:361–365.

Fuhremann, T. W., and E. P. Lichtenstein. 1972. Increase in the toxicity of organophosphorous insecticides to house flies due to polychlorinated biphenyl compounds. *Toxic. Appl. Pharmacol.* **22**:628–640.

Gardocki, J. F., and J. Yelnosky. 1964. A study of some of the pharmacological actions of fentanyl citrate. *Toxic Appl. Pharmacol.* **6**:48–62.

Garrett, E. R. 1970. The significance of properly designed pre-clinical studies in the clinical evaluation of new drugs. *Int. J. Clin. Pharmacol.* **4**:6–20.

Garrett, E. R. 1971a. Theoretical pharmacokinetics. Pp 27–40 in H. P. Kuemmerle. Urban and Schwarzenberg Verlag, Munich.

Garrett, E. R. 1971b. Drug action and assay by microbial kinetics. Pp. 271–352 in E. Jucker, ed. *Progress in Drug Research,* Volume 15. Berkhäuser, Basel.

Garrett, E. R. 1977. The pharmacokinetic bases of biological response quantification in toxicology, pharmacology and pharmacodynamics. Pp. 105–230 in E. Jucker, ed. *Progress in Drug Research,* Volume 21. Birkhäuser, Basel.

Garrett, E. R. 1978. Pharmacokinetics and clearances related to renal processes. *Int. J. Clin. Pharmacol.* **16**:155–172.

Garrett, E. R., A. J. Ågren, and H. J. Lambert. 1967. Pharmacokinetic analysis of receptor-site models in multicompartmental systems. *Int. J. Clin. Pharmacol.* **1**:1–14.

Garrett, E. R., J. Brès, K. Schnelle, and L. L. Rolf, Jr. 1974. Pharmacokinetics of saturably metabolized amobarbital. *J. Pharmacokin. Biopharm.* **2**:43–103.

Garrett, E. R., and P. H. Hinderling. 1973. Clinical relevance of pharmacokinetics in antibiotic treatment. *Med. Sci. Donat.* **15**:3–15.

Garrett, E. R., and C. A. Hunt. 1977. Pharmacokinetics of Δ^9-tetrahydrocannabinol in dogs. *J. Pharm. Sci.* **66**:395–407.

Garrett, E. R., and A. Jackson. 1980. Pharmacokinetics of morphine and its surrogates. III: Morphine and morphine-3-glucuronide in the dog as a function of dose. *J. Pharm. Sci.* **68**:753–771.

Garrett, E. R., R. L. Johnston, and E. J. Collins. 1962. Kinetics of steroid effects of Ca^{47} dynamics in dogs with the analog computer, I. *J. Pharm. Sci.* **51**:1050–1057.

Garrett, E. R., R. L. Johnston, and E. J. Collins. 1963. Kinetics of steroid effects on Ca^{47} dynamics in dogs with the analog computer. II. *J. Pharm. Sci.* **52**:668–678.

Garrett, E. R., and H. J. Lambert. 1973. Pharmacokinetics of trichloroethanol and metabolites and the interconversion among variously referenced pharmacokinetic parameters. *J. Pharm. Sci.* **62**:550–572.

Garrett, E. R., and J. K. Lewis. 1975. Kinetics and mechanisms of drug action on microorganisms. XXII: Effects of aminosidine with and without organism pretreatment with bacteriostatic agents. *J. Pharm. Sci.* **64**:1936–1940.

Garrett, E. R., and C. M. Won. 1973. Kinetics and mechanisms of drug action on microorganisms. XVII: The bactericidal effects of penicillin, kanamycin and rifampin with and without organism pretreatment with the bacteriostatic chloramphenicol, tetracycline, and novobiocin. *J. Pharm. Sci.* **62**:1666–1673.

Garrett, E. R., and O. K. Wright. 1967. Kinetics and mechanism of drug action on microorganisms. VII: Quantitative adherence of sulfonamide action on microbial growth to a receptor-site model. *J. Pharm. Sci.* **56**:1576–1585.

Gibaldi, M., G. Levy, and W. L. Hayton. 1972. Kinetics of the elimination and neuromuscular blocking effect of *d*-tubocurarine in man. *Anesthesiology* **36**:213–218.

Gibaldi, M. and D. Perrier. 1975. *Pharmacokinetics.* Marcel Dekker, New York. 329 pp.

Gibson, J. E., and B. A. Becker. 1968. The teratogenicity of cyclophosphamide in mice. *Cancer Res.* **28**:475–478.

Goldberg, M. E., H. E. Johnson, U. C. Pozzani, and H. J. Smyth, Jr. 1964. Effect of repeated inhalation of vapors of industrial solvents on animal behavior. *Am. Ind. Hyg. Assoc. J.* **25**:369–375.

Gough, B. J., L. A. Esuriex, and T. E. Shellenberger. 1967. A comparative toxicologic study of a phosphoradithioate in Japanese and bobwhite quail. *Toxic. Appl. Pharmacol.* **10**:12–19.

Greenwald, E. K., R. C. Martz, P. D. Harris, D. J. Brown, R. B. Forney, and F. W. Hughes. 1968. Ethyl alcohol toxicity in the bat *(Myotis lucifugus). Toxic Appl. Pharmacol.* **13**:358–362.

Han, A., W. K. Sinclair, and B. F. Kimler. 1976. The effect on *N*-ethylmaleimide on the response to x-rays of synchronized HeLa cells. *Radiat. Res.* **65**:337–350.

Harbison, R. D., D. Dwivedi, and M. A. Evans. 1976. A proposed mechanism for trimethylphosphate-induced sterility. *Toxicol. Appl. Pharmacol.* **35**:481–490.

Hinderling, P. H., and E. R. Garrett. 1977. Pharmacokinetics of β-methyldigoxin in healthy humans. I: Intravenous studies. *J. Pharm. Sci.* **66**:242–253.

Janků, I., and H. Farghalli. 1972. Relationships between elimination and acute toxicity of barbital during development. *Proc. Eur. Soc. Study Toxicity* **13**:289–293.

Jusko, W. J. 1972. Pharmacodynamic principles in chemical teratology: Dose effect relationships. *J. Pharmacol. Exp. Ther.* **183**:469–480.

Kaplan, M. L., S. D. Murphy, and F. H. Gilles. 1973. Modification of acrylamide neuropathy in rats by selected factors. *Toxic. Appl. Pharmacol.* **24**:564–579.

Lappenbusch, W. L., and D. L. Willis. 1971. Acute toxicologic effects of dimethyl sulfoxide on the rough skinned newt *(Traicha granulosa). Toxic. Appl. Pharmacol.* **18**:141–150.

Lappenbusch, W. L., and J. M. Ward. 1973. Synergistic bioeffects of oil and irradiation in an aquatic organism, *Traicha granulosa. Environ. Contam. Toxicol.* **9**:75–79.

Lee, I. P., and R. L. Dixon. 1972. Various factors affecting the lethality of cyclohexylamine. *Toxic. Appl. Pharmacol.* **22**:465–473.

Lehmann, B., E. Linner, and P. J. Wistrand. 1970. The pharmacokinetics of acetazalamide in relation to its use in the treatment of glaucoma and to its effects as an inhibitor of carbonic anhydrasses. Pp. 197–217 in G. Raspe, ed. *Advances in Biosciences,* Volume 5. Pergamon, Braunschweig.

Mackay, D. 1966. A general analysis of the receptor drug interactions. *J. Pharm. Pharmacol.* **18**:201–222.

Mark, L. C. 1963. Thiobarbiturates. Pp. 289–301 in E. M. Papper and R. J. Kitz, eds. *Uptake and Distribution of Anesthetic Agents.* McGraw-Hill, New York.

Obermeyer, B. P., I. S. Palmer, O. E. Olson, and A. W. Halverson. 1971. Toxicity of trimethylselenonium chloride in the rat with and without arsenates. *Toxic. Appl. Pharmacol.* **20**:135–146.

Ong, T., and F. J. de Serres. 1975. Mutagenic evaluation of antischistosomal drugs and their derivatives in *Neurospora crassa. J. Toxicol. Environ. Health* **1**(2):271–279.

Paton, W. D. M. 1961. A theory of drug action based on the rate of drug–receptor combination. *Proc. R. Soc. B* **154**:21–69.

Paton, W. D. M. and D. R. Waud. 1964. A quantitative investigation of the relationship between rate of access of a drug to receptor and the rate of onset or offset action. *Naunyn-Schmiedebergs Arch. Exp. Pathol. Pharmakol.* **248**:124–143.

Quest, J. A., G. S. Rowles, L. T. Mulligan, and P. P. Mathur. 1974. Mechanism of the hypotensive effect of IV methaqualone in the cat. *Toxic. Appl. Pharmacol.* **29**:420–433.

Riggs, D. S. 1970. *The Mathematical Approach to Physiological Problems.* MIT Press, Cambridge. 445 pp.

Robert, A., J. E. Nezamis, and C. Lancaster. 1975. Duodenal ulcers produced in rats by proprionitrile: Factors inhibiting and aggravating such ulcers. *Toxic Appl. Pharmacol.* **31**: 201–207.

Schumacher, H., D. A. Blake, J. M. Gurian, and J. R. Gillette. 1968. A comparison of the teratogenic activity of thalidomide in rabbits and rats. *J. Pharmacol. Exp. Ther.* **160**:189–200.

Scott, W. J., E. J. Ritter, and J. G. Wilson. 1971. DNA synthesis inhibition of cell death associated with hydroxyurea teratogenesis in rat embryos. *Dev. Biol.* **6**:306–315.

Sherman, M., and R. B. Gerrick. 1966. Acute and subacute toxicity of aphalate to the chick and Japanese quail. *Toxic. Appl. Pharmacol.* **9**:279–292.

Sorenson, E. M. B. 1976. Toxicity and accumulation of arsenic in green sunfish *Lepomis cyanellus* exposed to arsenate in water. *Bull. Environ. Contam. Toxicol.* **15**:756–761.

Tammes, P. M. L., F. E. Loosjes, and R. Wijnen. 1970. Time response experiments with single dose and chronic treatment of dieldrin on *Tribolium castaneum. Bull. Environ. Contam. Toxicol.* **5**:397–398.

Turbow, M. J., and J. G. Chamberlain. 1968. Direct effects of 6-aminonicotinamide on the developing rat embryo *in vitro* and *in vivo. Teratology* **1**:103–108.

Wagner, J. G., 1973. Properties of the Michaelis–Menten equation and its integrated form which are useful in pharmacokinetics. *J. Pharmacokin. Biopharm.* **1**:103–121.

Welch, R. M., W. Levin, R. Kuntzman, M. Jacobson, and A. H. Conney. 1971. Effect of halogenated hydrocarbon, insecticide on the metabolism and uterotropic action of estrogens in rat and mice. *Toxic. Appl. Pharmacol.* **19**:234–246.

10

Approaches to Intraspecies Dose Extrapolation

Charles C. Brown

In recent years, as the serious long-range health hazards of environmental toxicants have become recognized, the need has arisen for quantitative estimates of their effects on humans exposed to low levels. Often inherent in this estimation procedure is the necessity to extrapolate evidence observed under one set of conditions in one population group or biological system to arrive at an estimate of the effects expected in the population of interest under another set of conditions.

The quantitative assessment of risk to human health from exposure to toxic agents has been approached by relating the exposure level to measures of the chemical's heath risk, based on either epidemiological or clinical data on humans or on experimental data on animals or other biological systems. Unfortunately, there are often serious limitations with both approaches. Since human populations cannot be used as experimental subjects to investigate deleterious effects on health, the observational data from such sources are often incomplete and not of the most desirable form and substance. Attendant with epidemiological studies are difficulties in the accurate measurement of individual exposure patterns and the control of factors that may modify or confound the quantitative measures of health risk. Moreover, long delays often occur between exposure and the occurrence of a measurable effect. Such delays can range up to decades, as seen in many cases of carcinogenesis associated with occupational exposure to certain agents, such as asbestos-induced lung cancer.

It is necessary to test the potentially deleterious effects of chemical compounds in laboratory animals. To extrapolate results from animal studies to

CHARLES C. BROWN • National Cancer Institute, National Institutes of Health, Bethesda, Maryland 20892-4200.

humans, much care should be given to the design and conduct of these studies, since many factors may influence their results. These factors include the dosage and frequency of exposure, route of administration, species, strain, sex, and age of the animal, duration of the study, and various other modifying factors deemed important for the specific agent and effect being studied.

In order to estimate risk, both observational and experimental information on dose–response relations in animals and humans must be based on exposure levels higher than those for which the risk estimation is to be made. Some consideration has been given to the possibility of conducting extremely large experiments at very low dose levels. Use of large numbers of experimental subjects is necessary to reduce the statistical error so that very small effects can be adequately quantified. However, as Schneiderman *et al.* (1975) remarked, "Purely logistical problems might guarantee failure." Therefore, to obtain reliably measurable effects, the observational or experimental information must be based on levels of exposure high enough to show positive results. Since large segments of the human population are often exposed to much lower levels, these high-exposure-level data must be extrapolated to lower levels of exposure. Subsequent sections of this chapter describe the statistical methods used for this "high-to-low-dose" extrapolation and indicate the uncertainties necessarily attached to the estimates made with these methodologies.

High-to-low dose extrapolation is conceptually straightforward. The probability of a toxic response is modeled by the dose–response function $P(D)$—some quantitative or qualitative measure of the toxic response when exposed to D units of the toxic agent. A general mathematical model is chosen to describe this functional relationship, its unknown parameters are estimated from the available data, and this estimated dose–response function $P(D)$ is then used either to estimate the response measure at a particular low dose of interest or to estimate the dose level corresponding to a desired low level of response, i.e., a virtually safe dose (VSD).

Many mathematical models have been proposed for high-to-low-dose extrapolation. Those currently in use are described in the next section of this chapter. One of the major difficulties inherent in this process is that the estimates of risk at low doses and, correspondingly, the estimates of VSDs for low response levels are highly dependent upon the mathematical form assumed for the underlying dose–response relations.

MATHEMATICAL MODELS OF DOSE–RESPONSE RELATIONSHIP

To estimate the effects expected outside the range of experimental or observational data, a mathematical model relating dose (i.e., level of exposure to the toxic agent) to response (i.e., a quantitative measure of the deleterious effect produced) is necessary. In general terms, a dose–response relation is a relationship between a measurable physical, chemical, or biological stimulus and the response of living matter, measured in terms of the reaction produced over some range of the degree or level of the stimulus.

The reactions to any one stimulus might be multiple in nature, e.g., loss of weight, decrease in organ function, or even death. Each reaction may have its own unique relationship to the level of the stimulus. In addition, any specific reaction may be measured in terms of the magnitude of the effect produced *(quantitative response),* whether or not a specific effect is produced *(quantal response),* or the time required to produce a specific effect *(time to response).* These responses may be acute reactions, sometimes occurring within minutes of the stimulus, or they may be long-delayed effects, such as cancer, which may not appear clinically until most of the subject's normal life span has elapsed. Other responses may not even appear in the exposed subject, but may become manifest in some later progeny.

The level of the stimulus (dose level) may also be measured in different ways. For example, because subjects may be exposed to a toxicant in their environment through the air breathed, the food eaten, or some other external source of exposure, the dose level may be quantified by concentration in the air or food or by the quantity of the substance reaching the target receptor, some internal organ, or other tissue. Concentrations in air or food may be regarded as the environmental, or "external," exposures, whereas the latter quantifications may be termed the "internal" exposure levels. Due to a subject's biochemical and physiological internal mechanisms, the dose–response relation may be quite different for the internal and external measures of dose. Since the following material is applicable to dose measured on any scale, no distinction is made between these two general bases of measurement.

Quantal or Dichotomous Response Models

This section describes some of the more commonly used quantal, all-or-nothing response models. Quantitative measurement of a response is almost always preferable; however, certain responses, such as death or diagnosis of a particular disease, permit no gradation and can be expressed only as occurring or not.

Tolerance Distribution Models

When the response is quantal, its occurrence for any particular subject will depend upon the level of the stimulus. Under constant environmental conditions, it is commonly assumed that there is a certain dose level below which the subject will not respond in a specified manner and above which the subject will respond with certainty. This level is referred to as the subject's tolerance. Because of biological variability among exposed individuals in the population, their tolerance levels will also vary. When the measurable response is quantitative, the dose level required to provide a response of a constant magnitude will similarly vary among individuals. For quantal responses, it is therefore natural to consider the frequency distribution of tolerances over the population studied. If D represents the level of a particular stimulus, or dose, then the frequency distribution of tolerances may be mathematically expressed as

$$dP = f(D)\, dD \tag{1}$$

which represents the proportion of subjects whose tolerances lie between D and $D + dD$, where dD is small. If all subjects in the population are exposed to a dose D_0, then all subjects with tolerances less than or equal to D_0 will respond, and the responsive proportion $P(D_0)$ of the total population is

$$P(D_0) = \int_0^{D_0} f(D)\, dD \tag{2}$$

Assuming that all subjects in the population will respond to a sufficiently high dose level, then

$$P(\infty) = \int_0^{\infty} f(D)\, dD = 1 \tag{3}$$

Figure 10-1 shows a hypothetical tolerance frequency distribution $f(D)\, dD$ along with its corresponding cumulative distribution $P(D)$ as defined in Eq. (2). Thus, when the response is quantal in nature, the function $P(D)$ can be thought of as representing the dose–response relation, either for the population as a whole or for a randomly selected subject. The notion that a tolerance distribution, or dose–response function, could be determined solely from consideration of the statistical characteristics of a study population was introduced independently by Gaddum (1933) and Bliss (1934).

This dose–response function depends upon the frequency distribution function $f(D)$, which may be any nonnegative function that satisfies Eq. (3). Thus, this class of tolerance distribution dose–response models is very large. For all the models in this chapter, it is assumed that $P(0) = 0$, i.e., no responders at a zero dose. The incorporation of a spontaneous, or background, response rate caused by factors other than the toxicant in question will be described in a later section.

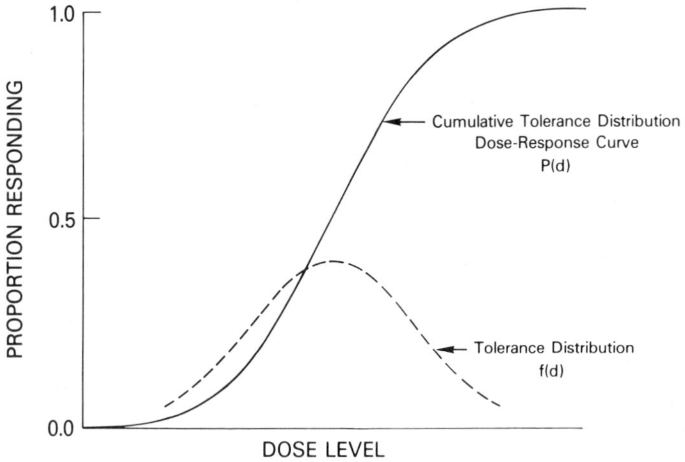

FIGURE 10-1. Relationship between tolerance distribution and dose–response curve.

The results of toxicity tests have often shown that the proportion of responders monotonically increases with dose and exhibits a sigmoid relationship with some function of the exposure level. This observation led to the development of the normal, or probit, model, for the tolerance frequency distribution,

$$f(D;\mu,\sigma) = (2\pi\sigma^2)^{-1/2} \exp\left\{-\frac{1}{2}\left[\frac{x(D) - \mu}{\sigma}\right]^2\right\}, \qquad \sigma > 0 \tag{4}$$

while the dose–response function is given by the cumulative normal probability,

$$P(D;\mu,\sigma) = \Phi\{[x(D) - \mu]/\sigma\} \tag{5}$$

where μ and σ^2 represent the mean and variance of the distribution of the transformed tolerances $x(D) = x$. Some transformations commonly used in practice are

$$x = \log_{10}(D)$$

and, more generally,

$$x = D^{\lambda}$$

where $\lambda \leq 1$. The use of these transformations is somewhat arbitrary. Finney (1949) proposed the use of any transformation that allows this normal model to provide an adequate fit to the data under consideration. The more familiar form of this log-normal model is the linear relationship,

$$y = \Phi^{-1}[P(D;\, \mu,\, \phi)] = (x - \mu)/\sigma = a + bx \tag{6}$$

where $a = -\mu/\sigma$, $b = 1/\sigma$, and x is the logarithm of dose [$x = \log_{10}(D)$]. The slope of this linear relationship with the transformed dose level x can be interpreted as the reciprocal of the tolerance distribution standard deviation. Thus, small variability in this distribution will produce a large slope, and large variability, a shallow slope. This method was put into its modern form by Bliss (1935). Finney (1971) has reported a brief history of its development.

This dose–response model was originally proposed as a method to solve problems of biological assay (i.e., assessment of the potency of toxicants, drugs, and other biological stimuli) and has been used primarily for dose–response interpolation (i.e., estimation within the range of observable response rates) rather than for dose–response extrapolation (i.e., estimation outside the range of observable rates). Mantel and Bryan (1961) and Mantel *et al.* (1975) later proposed its use, with suitable modification, to extrapolate experimentally induced effects observed at "high" dose levels to those expected at "low" levels. Their modification assumed a conservatively shallower slope than that observed in the experimental animal study. Their reasons for this modification were twofold: (1) to guard conservatively against the possibility that the true dose–response relation in the "low-dose" region might be different from that observed in the "high dose" region and (2) to compensate for the possibility that inbred strains of laboratory animals are more likely to show steeper dose–response relationships than the heterogeneous human population to which the extrapolation is to apply. This assumed conservative slope is a key feature of the Mantel–Bryan methodology, though its choice is arbitrary. For the purpose of extrapolation, the slope selected is not meant to represent the "true" slope in the low-dose region, but to represent a conservatively shallow slope regardless of the true dose–response relation in this region. Therefore, the Mantel–Bryan method was not designed to provide necessarily valid estimates of low-dose risk, but to provide "conservative" estimates of this risk. However, the "conservative" nature of this extrapolation methodology has been questioned by many authors, including Schneiderman (1974) and Hoel *et al.* (1975).

Other mathematical models or tolerance distributions that produce a sigmoid appearance of their corresponding dose–response functions have been suggested. The most commonly used is the log-logistic function

$$P(D;a,b) = \{1 + \exp[a + b\log_{10}(D)]\}^{-1}, \qquad b < 0 \tag{7}$$

which, like the log-normal model, is sigmoid and symmetric at approximately the 50% response level, but approaches the extremes, 0 and 100% response, more slowly than does the log normal. The logistic function, derived from chemical kinetic theory, was proposed as a dose–response model by Worcester and Wilson (1943) and by Berkson (1944). The log-logistic and log-normal functions are so similar in appearance that discrimination between them is nearly impossible. Because other proposed models are not widely accepted, they will not be discussed in this chapter.

These tolerance distribution models may also be used when the measurable response is quantitative. When studied over a range of dose levels, these responses also frequently show a sigmoid relationship with dose, either increasing or decreasing with dose, depending upon the measure of the effect of the toxic agent. If it is assumed that the effect decreases with dose, e.g., a decrease in the weight of some target organ, quantitative response measure Y at dose D can be given by

$$E(Y) = \theta[1 - P(D] \tag{8}$$

where $P(D)$ is a dose–response function and θ is an unknown "normal" value of Y when the subject is not exposed to the toxic agent. Thus, as the dose increases, $P(D)$ will also increase, and the expected value of the quantitative response will decrease from its "normal" value of θ. This procedure is also applicable to effects that increase with dose. The statistical estimation of the unknown θ and dose–response $P(D)$ will be based on a different procedure than that for quantal responses. Finney (1952) discusses the relevant methodology.

For quantitative measures of response, the extrapolated measure of low-dose risk is often difficult to interpret. If the "normal" value of this quantitative measure is θ_0, then how does one determine the risk for an extrapolated measure of, for example, $0.99\theta_0$—an average 1% reduction in the "normal" value? This problem is discussed in a report by the Safe Drinking Water Committee (National Research Council, 1980b). Much research on this subject remains to be done.

Models Derived from Mechanistic Assumptions

A number of proposed dose–response models have been based on assumptions regarding the mechanism of action of the toxic agent upon its target site. The "hit theory" for interaction between radiation particles and susceptible biological targets has produced a general class of these models (Turner, 1975). This theory is also applicable to the action of chemical toxicants upon their target sites. In general, the hit theory rests upon a number of postulates: (1) the organism has some number M of "critical targets"; (2) the organism responds if m or more of these critical targets are "destroyed"; (3) a critical target is destroyed if it is "hit" by k or more toxic particles; and (4) the probability P of a hit is proportional to the dose level of the toxic agent, i.e., Prob(hit) $= \lambda D$, $\lambda > 0$.

Some commonly used special cases of this general theory are the single-hit model:

$$P(D; \lambda) = 1 - \exp(-\lambda D) \tag{9}$$

where the subjects respond if a single critical target is destroyed by a single hit, and the multihit model:

$$P(D; \lambda, k) = \int_0^{\lambda D} \frac{x^{k-1} \exp(-x)}{\Gamma(k)} dx \tag{10}$$

where $\Gamma(k)$ denotes the gamma function and the subject responds if a single critical target is destroyed by k hits. This multihit model, also referred to as the gamma model, may also be interpreted as a tolerance distribution model in which the tolerance distribution is gamma with parameters λ and k. For a discussion of the single-hit model as applied to the high-to-low-dose extrapolation problem, see Hoel *et al.* (1975) and the BEIR report (National Research Council, 1980). The Scientific Committee, Food Safety Council (1978) and Rai and Van Ryzin (1979, 1981) have also discussed the application of the multihit model for dose extrapolation.

Other mechanistic models derived from quantitative theories of carcinogenesis include the multistage carcinogenesis model (Armitage and Doll, 1954, 1961; Crump *et al.*, 1976), which assumes that a single cell can generate a malignant tumor only after it has undergone a certain number of heritable changes:

$$P(D; \lambda_1, \ldots, \lambda_k) = 1 - \exp[-\lambda_1 D + \lambda_2 D^2 + \cdots + \lambda_k D^k)] \tag{11}$$
$$\lambda_i \geq 0, \quad i = 1, \ldots, k$$

The use of this model for extrapolation has been described by Brown (1978a) and Guess and Crump (1976, 1978).

The multistage theory of Nordling (1953) and the multicell carcinogenesis theory of Fisher and Holloman (1951) lead to a dose–response function with extrapolation characteristics similar to those of the multihit model,

$$P(D; \lambda, k) = 1 - \exp(-\lambda D^k), \qquad \lambda, k > 0 \tag{12}$$

This model has also been termed the Weibull model. Van Ryzin (1980) has discussed its application.

For radiation carcinogenesis, Burch (1960) and Marshall and Groer (1977) introduced the concept of dose-dependent cell death to analyze the dose–response relation of radiation-induced cancer. The Marshall and Groer dose–response function can be written as

$$P(D; \lambda) = 1 - (1 + \lambda D) \exp(-\lambda D), \qquad \lambda > 0 \tag{13}$$

which is mathematically equivalent to a two-hit model.

Both the single-hit model and the multistage model (when $\lambda_1 > 0$) become approximately linear at low dose levels. This low-dose linearity is an important aspect of "conservative" extrapolation models. Many researchers believe that the true dose–response relation at low exposure levels is convex, i.e., it may have some degree of upward curvature. Therefore, linearity represents an "upper bound" to the general class of convex functions, thereby providing conservative extrapolated risk estimates at low doses ("conservative" in the sense that they produce estimated risks higher than those of other convex functions). High-to-low-dose extrapolation methods using the simple linear model $P(D; \lambda) = \lambda D$ outside the observational or experimental range have also been proposed as "conservative" extrapolation procedures. Gaylor and Kodell (1980) described a method in which any model is used to estimate the response rate at the lowest experimental dose followed by linear extrapolation from that point to lower dose levels. Van Ryzin (1980) has suggested a similar methodology in which the dose corresponding to a risk of 10^{-2} is first estimated by a dose–response model, then linear extrapolation is used from that point. Both approaches seek to provide "conservative" extrapolations that do not depend strongly upon the dose–response model selected.

Pharmacokinetic Models

In the pharmacokinetic hypotheses concerning toxicity from foreign chemicals, biological effects are regarded as manifestations of biochemical interactions between the foreign substances (or substances derived from them) and components of the body. Actual mechanisms of toxicity are many and varied, and the kinetics that relate the concentration and exposure duration of the toxic substance at its site of action with its effect depends upon the mechanism.

A critical problem in the application of pharmacokinetic principles to risk extrapolation is the potential change in metabolism or other biochemical reactions as external exposure levels of the toxic agent decrease (Hoel *et al.* 1983). Linear pharmacokinetic models are often used; however, there are numerous examples of nonlinear behavior in the dose range studied, and these nonlinear kinetics pose significant problems for quantitative extrapolation from "high" to "low" doses if the kinetic parameters are not measured.

A simplified pharmacokinetic reaction scheme suggested by Gillette (1976) is given as follows:

$$\begin{array}{ccccccc} & r_0 & & r_{AB} & & r_{BC} & \\ & \longrightarrow & A & \longrightarrow & B & \longrightarrow & C \\ & & \downarrow r_A & & \downarrow r_B & & \end{array}$$

This simple steady-state model assumes that exposure to the toxic substance A occurs at a constant rate r_0. This substance is eliminated through excretion, detoxification, and other processes, at a rate r_A and is converted into its reactive toxic form B at a rate r_{AB}. This reactive intermediate is eliminated at a rate r_B; its rate of irreversible reaction at the site of toxicity is represented by r_{BC}. Each of these biochemical and physiological reactions may follow either linear or nonlinear kinetics.

Linear kinetics assumes that the reaction rate is proportional to the concentration C of the substance, $r = kC$; whereas nonlinear kinetics is most often described in the form of a Michaelis-Menten expression (Riggs, 1963), $r = aC/(b + C)$, where C represents the concentration of the substance. For low concentrations, $r \approx (a/b)C$, i.e., linear kinetics, whereas for high concentrations, $r \approx a$, independent of the concentration C, often referred to as "saturable" kinetics.

Clearly, if all the processes in the above diagram are linear, then the rate of concentration of the toxic substance at its site of action r_{BC} will be proportional to the external exposure rate r_0. However, saturation phenomena will produce different results, depending upon the steps where they act. If the elimination pathways corresponding to r_A or r_B are saturable, then r_{BC} will increase more rapidly with dose than linear kinetics would suggest. If the reaction pathways corresponding to r_{AB} or r_{BC} are saturable, then r_{BC} will increase less rapidly with dose. These simplified pharmacokinetic models may provide more realistic explanations of observed nonlinear dose–response relationships than other dose–response models.

These pharmacokinetic models involving nonlinear kinetics of the Michaelis–Menten form have the important extrapolation characteristic of being linear at low dose levels. This low-dose linearity contrasts with the low-dose nonlinearity of the multihit and Weibull models. Each model—pharmacokinetic, multihit, and Weibull—has the desirable ability to describe either convex (upward curvature) or concave (downward curvature) dose–response relationships. Other models, such as the log-normal or multistage models, are not consistent with concave relationships. However, the pharmacokinetic model differs from the multihit and Weibull models in that it does not assume that the nonlinear behavior observed at high dose levels will necessarily correspond to the same nonlinear behavior at low dose levels.

Gehring and Blau (1977) and Gehring *et al.* (1977) have discussed this simplified pharmacokinetic model and its extension to more complex reactions with respect to extrapolation of carcinogenic risk from high to low doses. Gehring *et al.* (1978) applied pharmacokinetic principles to the dose–response relation of hepatic angiosarcomas in rats exposed to different concentrations of atmospheric vinyl chloride over a period of 12 months. The results of their study are shown in Fig. 10-2. Since the metabolic activation of vinyl chloride appears to be a saturable process, the observed relationship between response, as measured by the proportion of rats with hepatic angiosarcomas, and dose, as measured by the external atmospheric exposure level of vinyl chloride, is clearly nonlinear, showing a leveling out at the highest exposure levels, which cannot be explained by a number of the previously discussed dose–response models (e.g., the log-normal and multistage models), but is consistent with a multihit model with $k < 1$ "hits" or a Weibull model with $k < 1$ "stages," both of questionable meaning. However, if dose is measured in terms of the amount of vinyl chloride metabolized, then the dose–response relation becomes much more linear, and most models provide an adequate fit to the data (Anderson *et al.,* 1980).

A variant of the reaction scheme shown in the previous diagram was considered by Cornfield (1977) to support the possibility of a threshold level for a carcinogenic substance:

$$C + S \overset{k}{\rightleftharpoons} X$$

$$C + T \xrightarrow{k^*} Y$$

In this scheme, the toxic substance C combines with a free substrate S to form an activated toxicant X in a reversible manner. C also reacts with a deactivator T to form a toxin-deactivator Y in an irreversible reaction. The probability of a toxic

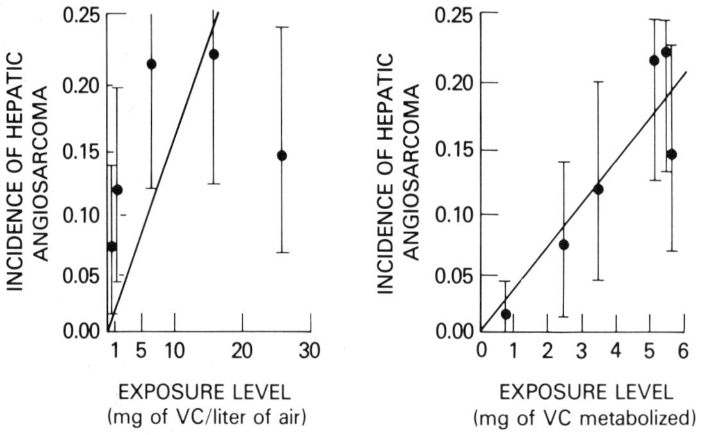

FIGURE 10-2. Dose–response relation of vinyl chloride-induced tumors in rats.

response is assumed to be proportional to the amount of the activated toxin X. However, in subsequent correspondence Brown *et al.* (1978) noted several limitations of this model: it assumes a single acute exposure rather than chronic exposure, it ignores variations in threshold levels among the exposed individuals, and it relies on the questionable assumption of an irreversible deactivator reaction. Cornfield (1977) also examined a chain of such reactions (assuming all processes are reversible) and concluded that the resulting dose–response relation would be shaped like a hockey stick, with the low-dose region being flat or nearly so, and the high-dose region showing a steep dose response, in agreement with the Gehring and Blau (1977) kinetic model.

Time-to-Occurrence Models

The preceding section was concerned primarily with quantal response models. For many experimental or observational sources of toxicity data, however, the time to occurrence of the toxic response may provide a useful additional piece of information. These data may take one of two forms: the elapsed time between initiation of exposure and occurrence of response or the total length of time a subject was observed without showing the desired response. This information may be of benefit in two ways. First, it should add information to define the dose–response relation more clearly. This is especially true for experiments in which most dose levels produce close to a 100% response. In such cases, there is little information on quantal dose–response relationships, but times to response may indicate a clear relationship to dose. Second, mathematical models of the relationship between dose and time to response will provide estimates of risk for any time subsequent to initiation of exposure. In addition, these models could be important by allowing investigators to adjust for competing risks from two or more toxic responses produced by the agent under consideration. A dose–response relationship for a late-occurring response may be obscured by a different, earlier toxic response, but a proper time-to-response analysis will provide a valid estimate of the dose–response relationship.

In general, the mathematical time-to-occurrence model consists of two parts. The first is the mathematical form for the probability distribution of the response-time random variable T (assuming not all responses occur at precisely the same time), i.e.,

$$\mathrm{Prob}(T \leq t) = F(t; \lambda_1, \ldots, \lambda_k) \tag{14}$$

where $\lambda_1, \ldots, \lambda_k$ are parameters describing the shape and scale of the cumulative distribution function F. The probability density function corresponding to this cumulative distribution is denoted by $f(t) = dF(t)/dt$. The general mathematical form of F is assumed for each dose level, and at least one of the parameters λ_i, $i = 1, \ldots, k$, is a function of dose. It is also assumed that all subjects eventually respond, i.e., $F(\infty) = 1$, but due to competing risks, such as death without the desired response, the subject may be removed from observation before the response can occur.

The second assumption concerns the relationship between the dose level D

and the parameters in Eq. (14). A general empirical relationship proposed by Busvine (1938) is

$$\lambda = \alpha D^m \tag{15a}$$

or

$$\log \lambda = \log \alpha + m \log D \tag{15b}$$

When the parameter λ represents the mean, or median time to occurrence, this linear relationship between the logarithm of λ and the logarithm of dose has often been observed in toxicity experiments.

Many mathematical time-to-occurrence models have been proposed, and all have a direct correspondence with quantal dose–response models (Chand and Hoel, 1974). One of the first models, proposed by Druckrey (1967), assumes that the probability distribution of the response times is log-normal:

$$f(t;\mu,\sigma) = (2\pi\sigma^2 t)^{-1/2} \exp\left[-\frac{1}{2}\left(\frac{\log t - \log \mu}{\sigma}\right)^2\right], \qquad \sigma > 0 \tag{16}$$

where μ is the median time to response and σ is the standard deviation of the log response times. The value of μ is assumed to be related to the dose level through Eq. (15), where σ is assumed to be independent of dose. This log-normal time-to-occurrence model corresponds to the log-normal quantal response model when response is defined as occurrence at or before a specified time. Albert and Altshuler (1973) examined this model as applied to a number of experimental and observational data sets of the time to occurrence of various cancers and found excellent agreement between the model and the data.

The Weibull time-to-occurrence model is derived from the multistage theory of carcinogenesis for continuous lifetime carcinogenic exposure:

$$f(t; \lambda, k) = \lambda k t^{k-1} \exp(-\lambda t^k) \tag{17}$$
$$F(t; \lambda, k) = \int_0^t f(x; \lambda, k)\, dx = 1 - \exp(-\lambda t^k)$$

where λ is assumed to be a function of the dose level and k is independent of dose. This model, which corresponds to the Weibull quantal response model, has been studied by Peto *et al.* (1972), among others, who fit both the log-normal and Weibull models to a number of carcinogenesis data sets and concluded that the Weibull generally provided a better fit to the data.

Hartley and Sielken (1977) proposed a general class of time-to-occurrence models, which take the following form:

$$\begin{aligned} &f(t; D, \theta_1, \ldots, \theta_k, \lambda_1, \ldots, \lambda_l) \\ &\quad = g(D; \theta_1, \ldots, \theta_k)h'(t; \lambda_1, \ldots, \lambda_l) \\ &\quad \exp[-g(D; \theta_1, \ldots, \theta_k)h(t; \lambda_1, \ldots, \lambda_l)] \end{aligned} \tag{18}$$

in which $g(D; \theta_1, \ldots, \theta_k)$ is a function only of dose, $h(t; \lambda_1, \ldots, \lambda_l)$ is a function only of time, and h' denotes the derivative of h with respect to time t. Hartley and Sielken called this the "general product model" and showed that the Weibull model is a special case: $h(t; \lambda) = t^\lambda$. Crump (1977, 1979) has also studied methods for estimating the parameters in this general model.

When dealing with responses measured in terms of time to occurrence, some care should be given to the correct interpretations derived from these models. In many instances, especially for carcinogenesis data, the actual time to the response of interest, e.g., time to tumor occurrence, may not be directly measurable. In experimental animal studies, the time to occurrence of internal tumors, as opposed to those that are palpable or directly observable, is commonly observable only upon death or sacrifice of the animal. The correct response endpoint in this situation is open to question. Death caused by the tumor or death with discovery of a particular tumor at necropsy are two possible response endpoints. The former requires cause-of-death information, which is often not available or difficult to determine; the latter is more commonly used, but this approach may be confounded by competing causes of death attributable to the toxic substance (Peto *et al.*, 1980). As an example, consider an experimental animal carcinogenesis study that produces no carcinogenic effect, but does shorten the lives of the exposed animals because of some general toxicity. Therefore, slow-growing, nonlethal tumors will be discovered earlier in the exposed animals and an incorrect dose–response analysis may erroneously estimate a carcinogenic dose response. The results from a hypothetical study are given in Table 10-1.

In this hypothetical study, 100 animals were tested in each of two groups: one composed of unexposed, control animals and one of exposed, treated animals. The tumorigenic response is assumed to be *nonlethal,* unrelated to exposure, and with a prevalence that increases with time. In Table 10-1, note the increase from 10 to 50% prevalence rate between time periods 1 and 5. This example also assumes that exposure to the treatment is toxic and causes an earlier death of the exposed animals (90% of the exposed animals and only 45% of the unexposed animals die within three time periods). Comparing the prevalence rates between the two groups for each time period clearly shows no effect of the exposure. However, two improper comparisons (improper due to the bias induced by the toxicity associated with exposure) lead to opposite conclusions: a

TABLE 10-1. Hypothetical Results from a Dose–Response Carcinogenesis Experiment

	Results for given experimental group[a]	
Period	Unexposed	Exposed
1	1/10	4/40
2	3/15	6/30
3	6/20	6/20
4	10/25	4/10
5	15/30	0/0

[a]Number of responders/number of deaths during period.

simple comparison of the total number of responders (35 in the unexposed group and 20 in the exposed group) indicates a significant *decrease* in tumor occurrence associated with exposure, whereas a time-to-tumor-occurrence analysis based on the incorrect assumption that the tumors caused the death of the animals indicates a significant *increase* in tumor occurrence, i.e., a shorter latency period. Both conclusions are incorrectly biased by the differential competing risk from other toxicity induced by the exposure. This example provides clear evidence that one should be careful in the analysis and interpretation of time-to-occurrence toxicity data. Hartley and Sielken (1977) provide a general method for including both lethal and nonlethal response endpoints into their general product model, which can also be used for any time-to-occurrence model.

Adjustments for Natural Responsiveness

The mathematical dose–response models described in the preceding sections have assumed that responses of the subjects are due solely to the applied stimuli. However, many toxicity experiments and observational studies show clear evidence that responses can occur even at a zero dose. Thus, any mathematical dose–response function should properly allow for this natural, or "background," responsiveness.

Two methods have been proposed to incorporate the possibility of response due to factors other than the stimulus in question. The first is commonly known as Abbott's corrrection, which is based on the assumption of an independent action between the stimulus and the background (Abbott, 1925). If the probability of response in the absence of any stimulus is denoted by P_0, then the overall response probability at dose level D, assuming independent actions, becomes

$$P(D) = P_0 + (1 - P_0)P^*(D) \tag{19}$$

where $P^*(D)$ represents the dose-induced probability of response. The second method assumes that the dose acts in an additive manner with the background environment, producing the following overall dose–response model (Albert and Altshuler, 1973):

$$P(D) = P^*(D + D_0) \tag{20}$$

where D_0 represents some unknown background level of the stimulus (or other stimuli that produce the response in a mechanistically dose-additive manner).

It is often difficult to discriminate between the independent and additivity assumptions on the basis of dose–response data. Figure 10-3 compares the dose–response relationships of the two assumptions shown in Eqs. (19) and (20), where $P^*(D)$ is a log-logistic model. Clearly, a large set of data would be required to determine the proper manner in which to incorporate background response. To describe the dose–response relation in this observable response range, this figure shows that this assumption is not an important issue, since both adequately describe the data. However, for purposes of low-dose extrapolation, this assump-

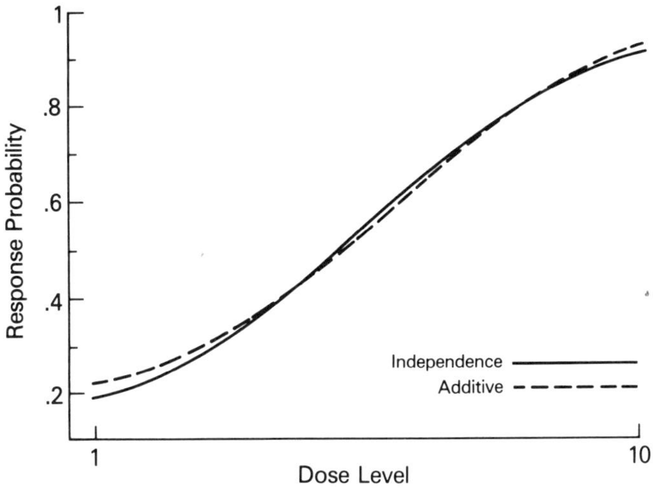

FIGURE 10-3. Comparison of dose–response relations assuming independent and additive backgrounds.

tion can have important consequences. Crump *et al.* (1976) have shown mathematically that no matter what dose–response model $P^*(D)$ is used, the additivity assumption will lead to a linear dose–response in the low-dose region. This will not necessarily be true for the independent action assumption. (Note that both assumptions lead to identical mathematical models for overall response rates when the assumed dose-induced model is either the single-hit or multistage model.) Hoel (1980) compares low-dose-risk extrapolations based on the two assumptions applied to a log-normal dose–response model. His results are given in Table 10-2, which clearly shows the low-dose linearity of the additive assumption and the substantial difference between the additive and independent assumptions at low dose levels. Hoel also examined models that incorporate a mixture of independent and additive background responses and found that low-

TABLE 10-2. Excess Risk $P(D) - P(0)$ for Log-Normal Dose–Response Model, Assuming Independent and Additive Background[a]

	Result for given type of background	
Dose	Independent	Additive
10^{0}	4.0×10^{-1}	4.0×10^{-1}
10^{-1}	1.5×10^{-2}	5.2×10^{-2}
10^{-2}	1.6×10^{-5}	5.2×10^{-3}
10^{-3}	3.8×10^{-10}	5.1×10^{-4}
10^{-4}	1.8×10^{-16}	5.1×10^{-5}

[a]From Hoel (1980). $P(0) = 0.1$; log-normal model slope = 2.

dose linearity prevails, except when the background mechanism is *totally independent* of the dose-induced mechanism.

Crump *et al.* (1976) argued in a general manner that the dose–response relation should be linear at low doses for carcinogens that affect the cancer processes by influencing cellular mechanisms that produce "background" cancers in the absence of the carcinogenic stimulation. They concluded that the additive assumption is proper for carcinogenic responses. However, Mantel (1978) stated that Crump *et al.* had not scientifically established the validity of this hypothesis and suggested that other processes, such as cellular repair, may affect low-dose linearity. No such theoretical distinction between the independent and additive hypotheses has been made for other toxic responses.

Thresholds and Heterogeneity of Response Rates

The existence of a threshold is an important consideration in the evaluation of risk to low levels of environmental toxicants. In this section, the term "threshold" for a particular toxic response is defined as the critical level of exposure below which the response attributable to the specific agent is impossible. If there are thresholds, and if they can be quantified, then truly safe levels of a toxic agent can be established.

The existence of thresholds may depend upon the type of toxic effect produced—either a reversible or irreversible effect. Freese (1973) discussed the possibility of thresholds for general toxic effects and more specific teratogenic, mutagenic, and carcinogenic effects. He suggested that many toxic agents inhibit cellular reactions in a reversible manner and that a true threshold may exist if the inhibited reactions do not normally limit the rate of cell metabolism or an organ's function until a certain critical level is attained. However, he believes that thresholds for irreversible mutagenic effects are less likely, since the heritable effect upon a single cell may produce untoward effects if the mutated cell replicates.

In discussing thresholds for carcinogenesis, Rall (1978) and Brown (1976) argued against the existence of a single threshold. They believe that thresholds, if they exist, are likely to vary among members of the population at risk and may be modified by other environmental agents. Mantel *et al.* (1961) and Brown (1976) have shown mathematically that such population heterogeneity induces an increasing convexity in the population dose–response relationship at low dose levels and that these variable-threshold models are difficult to distinguish from nonthreshold convex models. Therefore, when individual thresholds do vary within the population, extrapolation of an observed dose–response relation to estimate a population threshold level will at best estimate the average threshold of the population at risk. The estimate will have little practical utility, since many subjects in the population will have their threshold below this value.

In many experimental animal studies and human clinical studies, evidence for the existence of a threshold is usually presented in the form of an observed dose–response relationship. Either the existence of exposure levels not leading to an increase over unexposed controls in the occurrence of a toxic response or a linear extrapolation to low exposure levels, which apparently would result in no excess toxic response, is cited as indication of the existence of some threshold below which no response is possible. This type of evidence, however, is far from compelling. For example, the finding of a "no-effect" dose level is highly dependent upon the number of subjects exposed to the toxicant at each dose level. Such "no-effect" levels should more properly be called "no-*observed*-effect" levels. Table 10-3 shows the dependence of the determination of a no-observed-effect level upon the sample size of the observational or experimental data. This hypothetical example assumes that the study was conducted at three dose levels and the probabilities of observing a toxic response for a single subject are $P(D_1) = 0.01$, $P(D_2) = 0.05$, and $P(D_3) = 0.10$. The table also lists the probabilities of determining various no-observed-effect levels from such a study for different numbers of subjects examined at each dose level.

When the study sample size is small, e.g., $N = 5$, then the most likely no-observed-effect level is equal to the highest dose level D_3, i.e., no responders at any of the three dose levels. However, as the sample size increases, the most likely no-observed-effect level decreases.

The linear extrapolation method is also laden with uncertainty. Observed dose–response data can be plotted in a number of ways: linear dose scale against linear response rate scale, some power of dose against the same or different power of response rates, or logarithmic scales. Although any of these scales can be used to display the dose response, extrapolation to lower dose values is justified only if there are theoretical reasons for a particular scaling. In a review of 151 observed carcinogenesis dose–response curves, Lepkowski (1978) found none to be clearly indicative of a threshold effect when compared with the single-hit and log-normal models. He concluded that unless a theoretical basis for their existence is found, the threshold concept does not provide a practical basis for carcinogenic risk estimation at low doses.

Another argument for the existence of a practical threshold for carcinogenesis depends upon the latent period, i.e., the length of time between initial exposure and the occurrence of a tumor. Druckrey (1967) has demonstrated that the median, or average, carcinogenic latency period in experimental animals is inversely related to the dose level. Thus, lowering the dose increases the median time to tumor occurrence. Jones and Grendon (1975) have concluded that if this

TABLE 10-3. Hypothetical Example: Dependence of No-Observed-Effect Level upon Size of Study Sample[a]

No-observed-effect level	Probability of no-effect level, by study sample size at each dose		
	$N = 5$	$N = 20$	$N = 80$
D_3	0.435	0.036	<0.001
D_2	0.301	0.257	0.007
D_1	0.215	0.525	0.440
$<D_1$	0.049	0.182	0.552

[a]Assumed true dose–response: $P(D_1) = 0.01$, $P(D_2) = 0.05$, $P(D_3) = 0.10$.

median latency occurs beyond the normal life span, a practical threshold for the carcinogenic effect will have been reached. According to Guess and Hoel (1977), however, although the median time to tumor occurrence is beyond the normal life span, there may be a significant chance of developing a tumor before the normal life span has elapsed. Using a Weibull model to describe the random time to tumor occurrence in the absence of competing causes of death, Guess and Hoel have demonstrated that even when the average time to tumor occurrence is ten times greater than the normal life span, the probability of developing a tumor before life is ended is still 7.1×10^{-4}—much greater than a practical threshold would imply. Schneiderman *et al.* (1979) have also examined this issue in the presence of competing causes of death. They arrived at the same conclusion as Guess and Hoel.

Falk (1978), among others, argues in favor of the existence of thresholds for chemical carcinogens that are essential components of the mammalian system. He argued that concentrations of such essential chemicals as nickel and estrogens, which are necessary for normal physiological activity, must be below their carcinogenic threshold concentrations, since cancer is not necessarily induced, despite their presence. However, many epidemiological studies of breast cancer in women have clearly shown that the length of time a woman is exposed to her own naturally produced estrogen is related to her risk of breast cancer; a long exposure duration, either early age at menarche or late age of menopause, increases her breast cancer risk (Pike *et al.,* 1983). Thus, even essential levels of certain compounds may increase the risk for cancer while preserving other normal physiological functions.

A still unsettled controversy surrounds the existence of thresholds for specific toxic agents. Mantel (1963) suggests that the answer to this question may be immaterial. A practical threshold can be expected to exist for a variety of reasons. The likelihood of such toxic effects may be affected by dose-dependent rates of absorption, distribution, metabolism, and excretion of the toxic agent. Activation and deactivation may require enzyme reactions that can be induced by the agent itself or by some other compound, and cellular repair mechanisms may affect the action of mutagens and carcinogens. However, many researchers advocate the use of the no-threshold assumption when extrapolating mutagenic or carcinogenic effects, unless knowledge of mechanisms warrants otherwise.

As previously mentioned, different persons exposed to the same level of a toxic agent may be at quite different levels of risk. Their risk expectation is affected to various degrees by numerous environmental and genetic factors. A number of epidemiological and experimental studies of carcinogenesis have identified many such modifying factors. Genetic studies have also indicated substantial heterogeneity in susceptibility to human cancer. Such environmental and genetic heterogeneity can have a profound effect on low-dose extrapolation. As an example, consider the situation in which each member of the population at risk has an individual Weibull dose–response relationship:

$$P(D; \lambda) = 1 - \exp(-\lambda D^k)$$

where the susceptibility parameter λ is assumed to vary among the population members and the shape parameter k is an assumed constant. If the variability of λ can be described by a gamma probability distribution having mean μ and variance σ^2, then the population dose–response relationship can be described as follows:

$$P(D) = 1 - (1 + \mu D^k/\gamma)^{-\gamma}$$

where $\gamma = (\mu/\sigma)^2$. As the population heterogeneity decreases, i.e., $\sigma^2 \rightarrow 0$, then γ becomes large and the population dose–response relationship becomes the individual Weibull dose response, with $\lambda = \mu$ as the average susceptibility. However, when the heterogeneity is large, i.e., γ is small, then the population and individual dose–response relationships can be quite different (Fig. 10-4). Figure 10-4 compares the individual dose response ($\gamma = \infty$) and two population curves ($\gamma = 2$ and $\gamma = 1/2$), assuming $k = 2$, and the three curves are equal at $D = 10$ units. This figure shows that as the heterogeneity increases, the population dose–response relation flattens out at high doses due to the increasing proportion of subjects with low susceptibilities.

If the extrapolation dose–response model does not take this heterogeneity into account, then biased estimates of low-dose risk may be produced. Table 10-4 presents the results when the Weibull dose–response model is fitted to a heterogeneous population dose–response relationship. This example assumes that $k = 2$, the average susceptibility in the population is $\mu = 0.007$, and the measure of heterogeneity γ ranges between 8 (little heterogeneity) and 1/8 (great heterogeneity).

This table shows how biased parameter estimates and low-dose risk estimates can result when a measure of population heterogeneity is not incorporated into an assumed dose–response relationship. When there is substantial heterogeneity

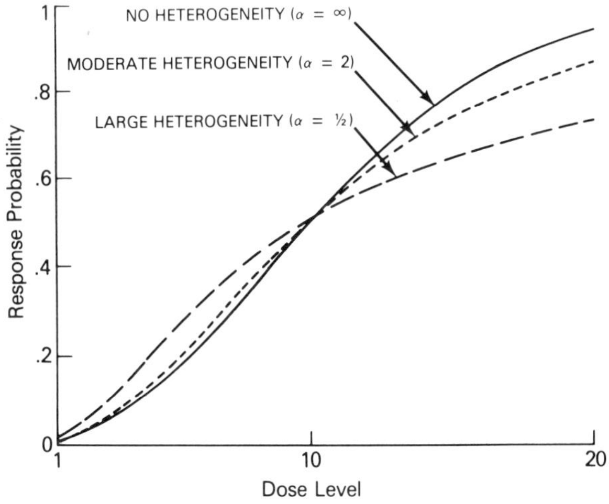

FIGURE 10-4. Dose–response relation modified by population heterogeneity.

TABLE 10-4. Results of Fitting Weibull Dose–Response Model to Actual Heterogeneous Dose–Response Relationship

Heterogeneity measure	Parameter estimates		Estimated risk at $D = 0.1$
	$\hat{\lambda}(0.007)^a$	$\hat{k}(2.0)^a$	$\hat{p}(0.1) = (7 \times 10^{-5})^a$
8	0.008	1.9	1.07×10^{-4}
4	0.010	1.8	1.51×10^{-4}
2	0.012	1.7	2.60×10^{-4}
1	0.016	1.5	5.22×10^{-4}
1/2	0.021	1.3	1.08×10^{-3}
1/4	0.025	1.1	1.99×10^{-3}
1/8	0.016	0.9	3.04×10^{-3}

[a]Actual values for population dose response.

($\gamma = 1/8$), the estimated risk at $D = 0.1$ is more than 40 times greater than the actual risk. This represents only one example of the potential for bias. Distributions of population susceptibility parameters other than the gamma model assumed in this example can produce biases in the opposite direction, i.e., underestimates of the true low-dose risk. This concern for the variability of susceptibility or sensitivity to toxic agents may have its greatest impact upon extrapolation of the dose–response relation observed in an experimental study conducted on genetically and environmentally homogeneous animals to the presumed dose–response relation in genetically and environmentally heterogeneous humans. The average experimental animal may be well representative of its "population," but the average human may not be.

COMPARISONS OF SPECIFIC MODELS

Discrimination among Dose–Response Models

Given a postulated functional form of the dose–response relationship, the frequency with which toxic effects occur may be used to estimate the unknown parameters. In additon, this estimated dose–response relationship can be extrapolated to provide either an estimate of risk probabilities at lower dose levels or an estimate of the dose level associated with any probability of risk. Implicitly, this approach presumes that the true dose–response relation can be realized within the postulated functional form used in the estimation and extrapolation procedure. Although this presumption is often not critical for interpolation within the range of observed response rates, it may be extremely critical for extrapolation outside this observable range.

It might be thought that the basis for selecting one particular model over the others would be provided by the observed dose response. However, this is often not the case, since many dose–response models appear to be similar over the range of observable response rates. Tables 10-5 and 10-6 compare the dose–

TABLE 10-5. Comparison of Dose–Response Relationships over Range of Observable Response Rates for Log-Normal, Log-Logistic, and Single-Hit Models[a]

	Percent responders, by model		
Dose level	Log-normal	Log-logistic	Single-hit
16	98	96	100
8	93	92	99
4	84	84	94
2	69	70	75
1	50	50	50
1/2	31	30	29
1/4	16	16	16
1/8	7	8	8
1/16	2	4	4

[a]From Food and Drug Administration Advisory Committee on Protocols for Safety Evaluation (1971).

response relationships of the more commonly used models: in Table 10-5, the log-normal, log-logistic, and single-hit models; in Table 10-6, the multihit, Weibull, and multistage models.

In Table 10-5, the parameters for these models were chosen to make the response rates equal at dose levels of 1 and 1/4; the values are $a = 0$ and $b = 1.657$ for the log-normal model in Eq. (6), $a = 0$ and $b = -2.754$ for the log-logistic model in Eq. (7), and $\lambda = 0.693$ for the single-hit model in Eq. (9). In Table 10-6, the parameters for the models were chosen to make the response rates equal at dose levels of 2 and 0.5. The values are $k = 2$ and $\lambda = 1.678$ for the multihit model in Eq. (10), $k = 1.5$ and $\lambda = 0.669$ for the Weibull model in Eq. (12), and $\lambda_1 = 0.298$, $\lambda_2 = 0.322$, and $k = 2$ for the multistage model in Eq. (11). These tables clearly demonstrate that an inordinately large set of experimental or

TABLE 10-6. Comparison of Dose–Response Relationships over Range of Observable Response Rates for Multihit, Weibull, and Multistage Models

	Percent responders, by model		
Dose level	Multihit	Weibull	Multistage
4	99	99	100
3	96	97	98
2	85	85	85
1	50	49	46
0.75	36	35	33
0.50	21	21	21
0.25	7	8	9

TABLE 10-7. Extrapolation of Dose–Response Relationships to Low Dose Levels for Log-Normal, Log-Logistic, and Single-Hit Models

	Percent responders, by model		
Dose level	Log-normal	Log-logistic	Single-hit
0.01	5.0×10^{-2}	4.0×10^{-1}	7.0×10^{-1}
0.001	3.5×10^{-4}	2.6×10^{-2}	7.0×10^{-2}
0.0001	1.0×10^{-7}	1.6×10^{-3}	7.0×10^{-3}

observational data would be required to indicate which of the models provides a significantly better fit to an observed dose response.

If the estimated dose–response relation is to be used to predict the response rate that would be expected from an exposure level within the range of observable rates, then the models within each of the two sets compared will give similar results. However, extrapolation to exposure levels expected to give very low response rates is highly dependent upon the selection of model, as shown in Tables 10-7 and 10-8. These tables extend the dose–response relationships of Tables 10-5 and 10-6 to much lower dose levels. The further one extrapolates from the observable response range, the more divergent the models become. At level 1/1000 of the dose giving a 50% response, the single-hit model gives an estimated response rate 200 times greater than that of the log-normal model, and the multistage model gives an estimated response rate more than 210 times that of the multihit model.

Krewski and Van Ryzin (1981) examined the extrapolation characteristics of six of the more commonly used dose–response models. They applied these models to 20 sets of toxic response data that were taken from the report of the Scientific Committee, Food Safety Council (1978, 1980). The toxic responses were both carcinogenic and noncarcinogenic. Of the 19 data sets with an observed convex (i.e., upward curvature) dose response, models provided estimates of virtually safe dose (VSD) at a response rate of $P = 10^{-5}$ or smaller in the following order: single-hit < multistage < Weibull, log-logistic, and multihit < log-normal. That is, the Weibull, log-logistic, and multihit models produce VSDs of approx-

TABLE 10-8. Extrapolation of Dose–Response Relationships to Low Dose Levels for Multihit, Weibull, and Multistage Models

	Percent responders, by model		
Dose level	Multihit	Weibull	Multistage
0.01	1.4×10^{-2}	6.7×10^{-2}	3.0×10^{-1}
0.001	1.4×10^{-4}	2.1×10^{-3}	3.0×10^{-2}
0.0001	1.5×10^{-6}	6.7×10^{-5}	3.0×10^{-3}

TABLE 10-9. Comparison of Virtually Safe Doses (VSD) Leading to an Excess Risk of 10^{-6} for Various Dose–Response Extrapolation Models Applied to Data from Tomatis *et al.* (1972)

Extrapolation model	VSD[a] (ppm DDT in daily diet)	Goodness-of-fit statistic of model to observed data χ^2 (d.f.) and p value		
Log-normal	6.8×10^{-1}	3.93	(2)	0.14
Weibull	5.0×10^{-2}	3.01	(2)	0.22
Multihit	1.3×10^{-2}	3.31	(2)	0.19
Log-logistic	6.6×10^{-3}	3.45	(2)	0.18
Multistage	2.5×10^{-4}	—[b]		
Single-hit	2.1×10^{-4}	5.10	(3)	0.16

[a]97.5% lower confidence limit on VSD.
[b]No goodness-of-fit statistic, since the number of parameters equals the number of data points.

imately the same order of magnitude, the single-hit model produces the smallest VSD, and the log-normal model produces the largest VSD. In addition, the difference between the extremes, i.e., the single-hit and log-normal models, is often several orders of magnitude.

Table 10-9 and Fig. 10-5 illustrate the behavior of these models when applied to the incidence of liver hepatomas in mice exposed to various levels of DDT (Tomatis *et al.* 1972). In this example, each of the six dose–response models fits the observed data nearly equally well. Therefore, the data in the observable response range, i.e., between DDT concentrations of 2 and 250 ppm (mg/kg) in the daily diet, cannot discriminate among these models. Based on the goodness-of-fit statistics, the Weibull model fits the best ($p = 0.22$), but not significantly better than any of the other models. However, there is a significant difference among the VSDs estimated from these models. The log-normal model estimates a VSD 3000 times greater than that estimated by the single-hit model. Therefore,

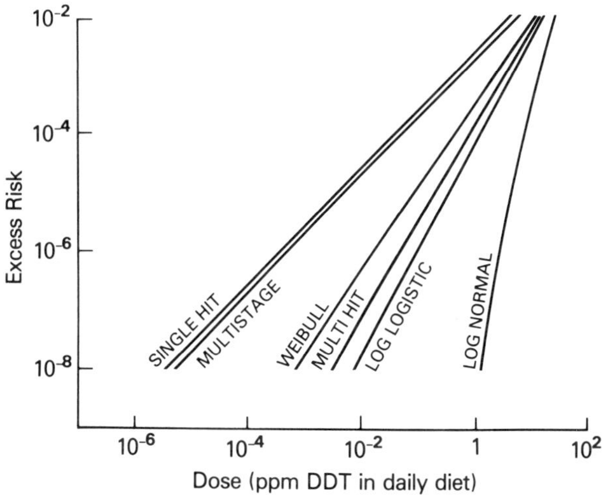

FIGURE 10-5. Extrapolation variation from use of different models.

these experimental data leave the true VSD open to wide speculation. Figure 10-5 provides a graphic display of these estimated dose–response models over a range of risk levels from 10^{-2} to 10^{-8}. The divergence of these models becomes more apparent as the dose and risk levels become smaller. The analyses of Krewski and Van Ryzin (1981) show that this result is a common occurrence.

A discussion of the statistical procedures used for fitting these mathematical dose–response models to experimental or observational data is beyond the scope of the chapter. However, it should be noted that upper statistical confidence limits on risk estimates, or lower confidence limits on VSD estimates, are commonly used to incorporate a measure of the statistical sampling error. Krewski and Van Ryzin (1981) have written a general discussion of these statistical procedures.

The fact that an experimental or observational study conducted at exposure levels high enough to give measurable response rates cannot clearly discriminate among these various models, along with the fact that those models show substantial divergence at low exposure levels, presents one of the major difficulties for the problem of low-dose extrapolation. Brown (1978b) has suggested that the general multistage model be used to provide estimates of both sampling and model variability for this problem, since the multistage model has the extrapolation characteristics of most other models, depending upon the number of stages used and the corresponding parameter values.

The relative likelihood analysis proposed by Brown (1978a,b) provides a graphic display of this wide variation in the VSD extrapolation estimates. Figure 10-6 shows the relative likelihood curve of the VSD (desired risk = 10^{-6}) for the multistage model applied to the data of Tomatis *et al.* (1972), which, for illustrative purposes, are limited to two stages. Since the generality of the multistage model leads to potentially wide variation in the VSD estimates, this model is

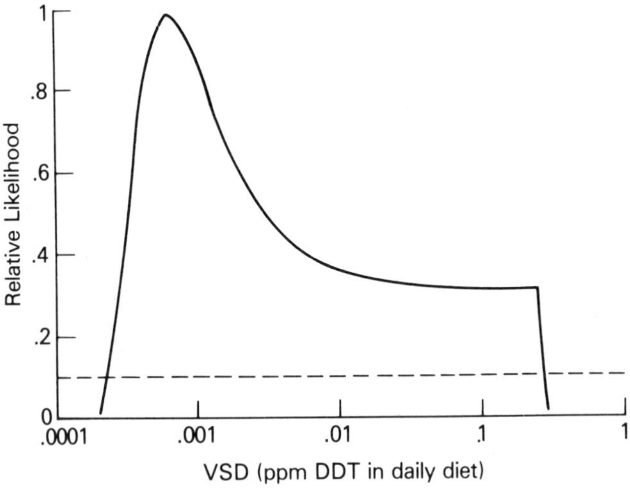

FIGURE 10-6. Relative likelihood curve for VSD based on multistage model.

particularly suitable for such a relative likelihood analysis. Figure 10-6 indicates that the most likely (i.e., the relative likelihood = 1) VSD is 0.0006, whereas VSD estimates of 0.00023 and 0.27 are only 10% as likely as the most likely VSD, based on these experimental dose–reponse data. The relative likelihood curve for the multistage model will often exhibit this unusual shape when the observed data are consistent with dose–response relationships at low dose levels that are proportional either to dose or to some higher power of dose.

Since many experimental dose–response studies are conducted with a limited number of animals (usually 100 or fewer) at each dose level over response rates ranging from 10 to 90%, it would seem likely that the wide variation in VSD might be reduced by testing more animals and using lower dose levels. However, this theory was weakened by the "megamouse" study of dietary exposure to 2-acetylaminofluorene (2-AAF) conducted at the National Center for Toxicological Research. One of the purposes of this massive study, involving more than 24,000 mice, was to describe the carcinogenic dose response for 2-AAF down to excess risks on the order of 1% (Cairns, 1980).

The results of this study provide an example of the additional information that one might expect to gain by testing large numbers of animals at lower than usual dose levels. The incidence of both bladder and liver neoplasms in animals that either died naturally or were sacrificed after being exposed for approximately 24 months is shown in Table 10-10 (Farmer *et al.,* 1980).

These data were examined to see if additional experimental results at dose levels giving low response rates would lead to a reduced variation of the VSD estimates. Two extrapolation models, the multistage and the log-normal models, were applied to these data in a series of calculations. In each case, both models fit the observed data very well. First, the VSDs leading to an excess risk of 10^{-6} were estimated using the controls and the four highest dose groups (60–150 ppm). Then, the VSDs were reestimated by adding the next lower dose, one at a time. These VSD estimates are shown in Table 10-11.

These results show that the inclusion of additional low-dose data has little effect on the VSD estimates. For bladder neoplasms, which show a highly convex

TABLE 10-10. Incidence of Bladder and Liver Neoplasms in Mice Fed 2-Acetylaminofluorene Continuously for 24 Months[a]

Dose level (ppm)	Bladder neoplasms[b]		Liver neoplasms[b]	
0	2/759	(0.3)	20/762	(2.6)
30	9/2105	(0.4)	164/2109	(7.8)
35	5/1357	(0.4)	128/1361	(9.4)
45	4/881	(0.5)	98/888	(11.0)
60	6/756	(0.8)	118/758	(15.6)
75	13/586	(2.2)	118/587	(20.1)
100	51/297	(17.2)	76/297	(25.6)
150	236/313	(75.4)	126/314	(40.1)

[a]From Farmer *et al.* (1980).
[b]Number of mice with neoplasms/number of mice examined (values in parentheses give percent of total).

TABLE 10-11. Virtually Safe Doses (VSD) for 2-AAF Based on Multistage and Log-Normal Models Applied to Different Dose Level Combinations

	VSD (ppm) by type of neoplasm and extrapolation model[a]			
	Bladder neoplasms		Liver neoplasms	
Dose levels used (ppm)	Multistage	Log-normal	Multistage	Log-normal
0, 60–150	3.07×10^{-2}	34.4	3.50×10^{-4}	4.84×10^{-1}
0, 45–150	4.12×10^{-2}	34.5	3.82×10^{-4}	5.27×10^{-1}
0, 35–150	4.48×10^{-2}	34.5	3.99×10^{-4}	4.03×10^{-1}
0, 30–150	3.63×10^{-2}	34.5	4.32×10^{-4}	3.84×10^{-1}

[a]97.5% lower confidence limit on VSD leading to an excess risk of 10^{-6}.

dose–response relation, the lower confidence limit on the VSD based on the multistage model is increased only 18% (from 3.07×10^{-2} to 3.63×10^{-2}), whereas that based on the log-normal model is hardly changed. For liver neoplasms, which show a nearly linear dose response, the lower confidence limit on the VSD is increased only 23% for the multistage model and is decreased for the log-normal model. The differences in the VSD estimates from these two extrapolation models is little affected by these additional low-dose data: for bladder neoplasms, the additional data reduce the difference from a log-normal/multistage ratio of 1120 ($34.4/3.07 \times 10^{-2}$) to 950 ($34.5/3.63 \times 10^{-2}$); for liver neoplasms, this ratio is reduced from 1380 to 890. Therefore, these additional low-dose data, based on substantial numbers of animals, have little effect on the VSD estimates for a particular extrapolation model. More importantly, they have little effect on reducing the variation in VSD estimates among different models.

EXTRAPOLATION OF HUMAN EPIDEMIOLOGICAL DATA

The dose–response extrapolation models discussed in the previous sections have been focused primarily on low-dose extrapolation of experimental data. However, they are also applicable to the extrapolation of human epidemiological data. To quantify the effects of factors related to specific human diseases, two epidemiological approaches are commonly used: (1) prospective cohort studies in which populations exposed to various levels of a toxic agent are followed and their future disease occurrences are compared and (2) retrospective case–control studies in which past exposure to a toxic agent is measured in samples of individuals with and without the disease in question. Mathematical models of dose–response relation can be fit to the results from either type of study to estimate the quantitative relationship between level of exposure and occurrence of disease.

One of the limitations of the case–control approach is that only relative risks (i.e., the disease risk of individuals exposed to various toxicant levels relative to the risk for unexposed individuals) can be measured. The estimation of absolute risks for exposed individuals requires the additional knowledge of the absolute risk for the unexposed individuals. The relevant dose–response methodology commonly assumes that the relative risk for a particular exposure level is constant over all other disease occurrence factors (e.g., sex, age, or other environmental exposures). That is, the risk for an exposed 20-year-old male relative to the risk for an unexposed 20-year-old male is equivalent to the risk for an exposed 30-year-old female relative to the risk of a 30-year-old unexposed female. The validity of this assumption will depend upon the mechanism of action for the particular toxic agent in question and how this mechanism might be modified by other intrinsic and extrinsic factors. For example, multistage theories of carcinogenesis lead to additive risk dose–response models that may be modified by other factors in an additive or multiplicative manner (Armitage and Doll, 1961; Day and Brown, 1980).

For long-term chronic exposure to a toxic agent, the relationship of risk to the rate and duration of exposure is often of importance when estimating risk for different exposure situations. One commonly used assumption is that risk is dependent upon total cumulative exposure. Thus, for example, a person exposed daily to 1 mg of the toxic agent for 20 years is assumed to have the same risk as another person exposed to 10 mg daily for 2 years. Again the mechanism of toxic action will determine the validity of this assumption. For example, if the toxic agent accumulates at the target site and the response becomes evident when the accumulated level attains some critical level, then this total dose assumption may be warranted. However, physiological processes, such as detoxification or elimination from the target site, are likely to be dependent upon the accumulated level, and thus may modify this simple total dose relationship.

The multistage theory of carcinogenesis predicts that cancer risk is dependent upon the dose rate and duration of exposure, but not necessarily leading to a relationship with total dose—the product of rate and duration. Whittemore and Keller (1978), Whittemore (1977), and Day and Brown (1980) have discussed these multistage theories and have indicated that the risk of cancer is likely to be the product of two different functions of dose rate and duration. In an analysis of cigarette smoking and lung cancer, Doll (1971) found that lung cancer incidence rates rise in proportion to approximately the fourth power of duration of smoking. In a later study, Doll and Peto (1978) observed that incidence rises with approximately the first or second power of the daily number of cigarettes smoked.

Besides being a function of both dose rate and duration, the multistage theory also predicts that cancer risk may be a function of the age at which exposure first begins and the length of time following cessation of exposure. Whittemore (1977) and Day and Brown (1980) have demonstrated how risk may be a function of these two factors, depending upon the stage at which the process of carcinogenesis is affected by the toxic agent. For example, exposure at a young age to a carcinogen affecting an early stage of the disease process (i.e., an initiator) is predicted by the theory to have a greater effect on future cancer risk than the same exposure at a later age. The converse is predicted for exposure to a carcinogen that affects a late stage. Table 10-12 shows the magnitude of differences in the excess cancer risk predicted by the multistage theory, depending upon the age at initial exposure to the carcinogen. This example assumes that the exposure duration is 10 years

TABLE 10-12. Relative Excess Age-Specific Cancer Risk at Age 60 Predicted by Multistage Model of Carcinogenesis[a]

Age at initial exposure (years)	Excess risk according to stage affected by carcinogen[b]	
	First	Penultimate
0	671	1.0
10	44.7	1.8
20	10.3	3.8
30	3.8	10.3
40	1.8	44.7
50	1.0	671

[a]Exposure duration 10 years. Five stages in carcinogenic process.
[b]Excess risks relative to risk of (1) initial exposure at age 50 for first stage affected or (2) initial exposure at age 0 for penultimate stage affected.

and that the process of carcinogenesis comprises five stages. This table shows that 10 years of exposure to a carcinogen that initiates the disease process (i.e., affects the first stage) leads to a cancer risk at age 60 that is more than 600 times greater for a person first exposed at birth than for a person first exposed at age 50. The reverse is shown for a carcinogen acting at a late (i.e., the penultimate) stage; a much greater risk is imparted by exposure at a late age than at an earlier age. Therefore, the multistage theory of carcinogenesis predicts another level of complexity in extrapolating cancer risk from one exposure situation to another, since a limited duration (e.g., 10 years) of exposure to the same dose rate will not necessarily produce the same excess cancer risk in two otherwise identical individuals whose exposures occur during different stages of their life.

SUMMARY AND CONCLUSIONS

High-dose to low-dose extrapolation within a single species—laboratory animal or human—is necessary to estimate the effects of low-level exposure to toxic agents known to be associated with undesirable effects at high dose levels. Mathematical dose–response models are a necessary evil in this process, since the response rates at low doses, expected to be approximately 10^{-6}, are too small to be measured accurately with limited study sample sizes. The mathematical dose–response models proposed for these extrapolations appear similar to one another in the range of observable response rates; yet, how different they become at lower, unobservable response rates—the region of primary interest! This is the single, most important limitation of this extrapolation methodology. An estimate of risk at a specific low dose, or an estimate of the dose leading to a prespecified level of risk, is highly dependent upon the mathematical form of the presumed dose–response relation. This chapter has demonstrated that differences of 3–4 orders of magnitude are not uncommon. "New" models, unless based upon strong mechanistic information, will not alleviate the difficulties. In a Bayesian approach along the lines suggested by Altshuler (1977), judgments about the plausibilities of different functional forms, based on toxicological and biological information, could be incorporated into the extrapolation process. The contribution from statisticians and model-builders has reached an impasse, and more accurate extrapolations are not possible without additional information on the mechanisms of action of the toxic agents.

Pharmacokinetic information on the fate of a toxic agent once it enters the body is beginning to be incorporated into the high-to-low-dose extrapolation process. Nonlinear kinetics may be an important determinant of the nonlinear dose–response relationships often observed in experimental studies of toxic agents. As noted above, Gehring *et al.* (1978) have shown that the metabolism of inhaled vinyl chloride is a saturable process that provides one explanation of the concave liver carcinogenesis dose–response relation observed in animal studies. In a study of urethane-induced pulmonary adenomas, White (1972) showed that the convex relationship between the amount of urethane injected into a mouse lung and the number of subsequent lung adenomas could be explained by nonlinear kinetics of excretion. Such pharmacokinetic models and dose–response studies of the kinetics of physiological processes might considerably strengthen the ability to extrapolate from high to low dose levels. This avenue of investigation holds potentially great promise for the future.

Other sources of uncertainty in high-to-low-dose extrapolation include the possible existence of thresholds, heterogeneous sensitivity to the toxic agent among members of the exposed population, and nature of mechanisms of action for carcinogens, i.e., whether the agent initiates the process or acts at a later stage. The existence of a single threshold for the entire exposed population should allow for estimation of a clearly safe level of exposure. However, its estimation could be associated with a high degree of uncertainty. Heterogeneity among individual thresholds and sensitivity to the toxic agent induce additional uncertainty in high-to-low-dose extrapolations. Population heterogeneity, if not accounted for, can bias extrapolated risk estimates by 1–2 orders of magnitude. The relationship of dose rate and duration of exposure indicates that identical exposure patterns (i.e., the same dose rate and duration) do not necessarily lead to identical levels of risk. Thus, uncertainty in the mechanism of toxic action induces another potentially large uncertainty into risk extrapolations.

Therefore, all these sources of uncertainty—dose–response model, pharmacokinetic behavior of the toxic agent, thresholds, heterogeneity, and mechanisms of action—may lead to enormous variation in estimates of risk from high-to-low-dose extrapolations.

REFERENCES

Abbott, W. S. 1925. A method of computing the effectiveness of an insecticide. *J. Econ. Entomol.* **18**:265–267.

Albert, R., and B. Altshuler. 1973. Considerations relating to the formulation of limits for unavoidable population exposures to environmental carcinogens. Pp. 233–253 in J. Ballou, R. Busch, D. Mahlum, and C. Sanders, eds. *Radionuclide Carcinogenesis.* AEC Symposium Series, Conference 720505, National Technical Information Service, Springfield, Virginia.

Altshuler, B. 1977. A Bayesian approach to assessing population risks from environmental carcinogens. Pp. 31–45 in A. Whittemore, ed. *Environmental Health, Quantitative Methods.* Society for Industrial and Applied Mathematics, Philadelphia.

Anderson, M. W., D. G. Hoel, and N. L. Kaplan. 1980. A general scheme for the incorporation of pharmacokinetics in low-dose risk estimation for chemical carcinogenesis; example—vinyl chloride. *Toxicol. Appl. Pharmacol.* **55:**154–161.

Armitage, P., and R. Doll. 1954. The age distribution of cancer and a multistage theory of carcinogenesis. *Br. J. Cancer* **8:**1–12.

Armitage, P., and R. Doll. 1961. Stochastic models for carcinogenesis. Pp. 19–38 in J. Newman, ed. *Proceedings of the Fourth Berkeley Symposium on Mathematical Statistics and Probability,* Volume 4. University of California Press, Berkeley.

Berkson, J. 1944. Application of the logistic function to bioassay. *J. Am. Stat. Assoc.* **39:**134–167.

Bliss, C. I. 1934. The method of probits. *Science* **79:**38–39.

Bliss, C. I. 1935. The calculation of the dosage–mortality curve. *Annu. Appl. Biol.* **22:**134–167.

Brown, C. 1976. Mathematical aspects of dose–response studies in carcinogenesis—The concept of thresholds. *Oncology* **33:**62–65.

Brown, C. 1978a. Statistical aspects of extrapolation of dichotomous dose response data. *J. Natl. Cancer Inst.* **60:**101–108.

Brown, C. C. 1978b. Variability of risk extrapolation in dose–response experiments [Abstract]. *Environ. Health Perspect.* **22:**183–184.

Brown, C., T. R. Fears, M. H. Gail, M. Schneiderman, R. E. Tarone, and N. Mantel. 1978. Letter to the editor. Reply by J. Cornfield. *Science* **202:**1105–1108.

Burch, P. R. J. 1960. Radiation carcinogenesis: A new hypothesis. *Nature* **185:**135–142.

Busvine, J. R. 1938. The toxicity of ethylene oxide to *Candra oryzae c. granaria, Tribolium castaneum,* and *Climex lectularis. Ann. Appl. Biol.* **25:**605–632.

Cairns, T. 1980. The ED_{01} study: Introduction, objectives, and experimental design. *J. Environ. Pathol. Toxicol.* **3:**1–8.

Chand, N., and D. Hoel. 1974. A comparison of models for determining safe levels of environmental agents. Pp. 681–700 in F. Proschan and R. J. Serfling, eds. *Reliability and Biometry.* Society for Industrial and Applied Mathematics, Philadelphia.

Cornfield, J. 1977. Carcinogenic risk assessment. *Science* **198:**693–699.

Crump, K. S. 1977. Low dose extrapolation utilizing the age distribution of cancer. Presented at Joint Statistical Meeting, Chicago, Illinois, August 15–18. *Biometrics* **34:**155.

Crump, K. S. 1979. Dose response problems in carcinogenesis. *Biometrics* **35:**157–167.

Crump, K. S., D. Hoel, C. Langley, and R. Peto. 1976. Fundamental carcinogenic processes and their implications for low dose risk assessment. *Cancer* **36:**2973–2979.

Day, N. E., and C. C. Brown. 1980. Multistage models and primary prevention of cancer. *J. Natl. Cancer Inst.* **64:**977–989.

Doll, R. 1971. The age distribution of cancer. Implications for models of carcinogenesis. *J. R. Stat. Soc.* **134:**133–155.

Doll, R., and R. Peto. 1978. Cigarette smoking and bronchial carcinoma: Dose and time relationships among regular smokers and lifelong non-smokers. *J. Epidemiol. Commun. Health* **32:**303–313.

Druckrey, H. 1967. Quantitative aspects of chemical carcinogens. Pp. 60–78 in *Potential Carcinogen Hazards from Drugs,* U.I.C.C. Monograph series, Volume 7. Springer-Verlag, New York.

Falk, H. L. 1978. Biologic evidence for the existence of thresholds in chemical carcinogenesis. *Environ. Health Perspect.* **22:**167–170.

Farmer, J. H., R. L. Kodell, D. L. Greenman, and G. W. Shaw. 1980. Dose and time response models for the incidence of bladder and liver neoplasms in mice fed 2-acetylaminofluorene continuously. *J. Environ. Pathol. Toxicol.* **3:**55–68.

Finney, D. J. 1949. The choice of a response metameter in bio-assay. *Biometrics* **5:**261–272.

Finney, D. J. 1952. *Statistical Methods in Biological Assay.* Griffin & Co., London. 661 Pp.

Finney, D. J. 1971. *Probit Analysis,* 3rd ed. Cambridge University Press, London. 333 Pp.

Fisher, J. C., and J. H. Holloman. 1951. A new hypothesis for the origin of cancer foci. *Cancer* **4:**916–918.

Food and Drug Administration Advisory Committee on Protocols for Safety Evaluation. 1971. Panel on carcinogenesis report on cancer testing in the safety evaluation of food additives and pesticides. *Toxicol. Appl. Pharmacol.* **20**:419–438.

Freese, E. 1973. Threshold in toxic, teratogenic, mutagenic and carcinogenic effects. *Environ. Health Perspect. Experimental Issue* **6**:171–178.

Gaddum, J. H. 1933. *Reports on Biological Standards* III. *Methods of Biological Assay Depending on a Quantal Response.* Special Report Series 183. Medical Research Council, London.

Gaylor, D. W., and R. L. Kodell. 1980. Linear interpolation algorithm for low dose risk assessment of toxic substances. *J. Environ. Pathol. Toxicol.* **4**:305–312.

Gahring, P. J., and G. E. Blau. 1977. Mechanisms of carcinogenesis: Dose response. *J. Environ. Pathol. Toxicol.* **1**:163–179.

Gehring, P. J., P. G. Watanabe, and J. D. Young. 1977. The relevance of dose dependent pharmacokinetics in the assessment of carcinogenic hazard of chemicals. Pp. 187–203 in H. H. Hiatt, J. D. Watson, and J. A. Winston, eds. *Origins of Human Cancer,* Book A: *Incidence of Cancer in Humans.* Cold Spring Harbor Laboratory, Cold Spring Harbor, New York.

Gehring, P. J., P. G. Watanabe, and C. N. Park. 1978. Resolution of dose response toxicity data for chemicals requiring metabolic activation: Example—vinyl chloride. *Toxicol. Appl. Pharmacol.* **44**:581–591.

Gillette, J. R. 1976. Application of pharmacokinetic principles in the extrapolation of animal data to humans. *Clin. Toxicol.* **9**:709–722.

Guess, H. A., and K. S. Crump. 1976. Low dose extrapolation of data from animal carcinogenicity experiments—Analysis of a new statistical technique. *Math. Biosci.* **32**:15–36.

Guess, H. A., and K. S. Crump. 1978. Best-estimate low-dose extrapolation of carcinogenicity data. *Environ. Health Perspect.* **22**:149–152.

Guess, H. A., and D. G. Hoel. 1977. The effect of dose on cancer latency period. *J. Environ. Pathol. Toxicol.* **1**:279–286.

Hartley, H. O., and R. L. Sielken. 1977. Estimation of "safe doses" in carcinogenic experiments. *Biometrics* **33**:1–30.

Hoel, D. G. 1980. Incorporation of background response in dose–response models. *Fed. Proc.* **39**:67–69.

Hoel, D. G., D. Gaylor, R. Kirschstein, U. Saffiotti, and M. Schneiderman. 1975. Estimation of risks of irreversible delayed toxicity. *J. Toxicol. Environ. Health* **1**:133–151.

Hoel, D. G., N. L. Kaplan, and M. W. Anderson. 1983. The implication of nonlinear kinetics on risk estimation in carcinogenesis. *Science* **219**:1032–1037.

Jones, H. B., and A. Grendon. 1975. Environmental factors in the origin of cancer and estimation of the possible hazard to man. *Food Cosmet. Toxicol.* **13**:251–268.

Krewski, D., and J. Van Ryzin. 1981. Dose response models for quantal response toxicity data. Pp. 201–231 in M. Csorgo, D. Dawson, J. N. K. Rao, and E. Saleh, eds. *Current Topics in Probability and Statistics.* Elsevier/North-Holland, New York.

Lepkowski, W. 1978. Extrapolation of carcinogenesis data. *Environ. Health Perspect.* **22**:173–181.

Mantel, N. 1963. The concept of threshold in carcinogenesis. *Clin. Pharmacol. Ther.* **4**:104–109.

Mantel, N. 1978. Letter to the editor and reply by K. S. Crump. *Cancer* **38**:1835–1838.

Mantel, N., and W. Bryan. 1961. "Safety" testing of carcinogenic agents. *J. Natl. Cancer Inst.* **27**:455–470.

Mantel, N., W. Heston, and J. Gurian. 1961. Thresholds in linear dose–response models for carcinogenesis. *J. Natl. Cancer Inst.* **27**:203–215.

Mantel, N., N. Bohidar, C. Brown, J. Ciminera, and J. Tukey. 1975. An improved Mantel–Bryan procedure for 'safety' testing of carcinogens. *Cancer Res.* **35**:865–872.

Marshall, J. H., and P. G. Groer. 1977. A theory of the induction of bone cancer by alpha radiation. *Radiat. Res.* **71**:149–192.

National Research Council. 1980a. *The Effects on Populations of Exposure to Low Levels of Ionizing Radiation.* A report of the Committee on the Biological Effects of Ionizing Radiation. National Academy of Sciences, Washington, D.C.

National Research Council. 1980b. *Drinking Water and Health,* Volume 3. A report of the Safe Drinking Water Committee. National Academy of Sciences, Washington, D.C.

Nordling, C. O. 1953. A new theory on the cancer industry mechanism. *Br. J. Cancer* **7**:68–72.

Peto, R., P. Lee, and W. Paige. 1972. Statistical analysis of the bioassay of continuous carcinogens. *Br. J. Cancer* **26**:258–261.

Peto, R., M. C. Pike, N. E. Day, R. G. Gray, P. N. Lee, S. Parish, J. Peto, S. Richards, and J. Wahrendorf. 1980. Guidelines for simple, sensitive significance tests for carcinogenic effects in long-term animal experiments. Pp. 311–425 in *Long-Term and Short-Term Screening Assays for Carcinogens: A Critical Appraisal.* IARC Monographs on the Evaluation of the Carcinogenic Risk of Chemicals to Humans, Annex to Supplement 2. International Agency for Research on Cancer, Lyon.

Pike, M. C., M. D. Krailo, B. E. Henderson, J. T. Casagrande, and D. G. Hoel. 1983. 'Hormonal' risk factors, 'breast tissue age,' and the age-incidence of breast cancer. *Nature* **303**:767–770.

Rai, K., and J. Van Ryzin. 1979. Risk assessment of toxic environmental substances using a generalized multi-hit dose response model. Pp. 99–117 in N. Breslow and A. Whittemore, eds. *Energy and Health.* Society for Industrial and Applied Mathematics, Philadelphia.

Rai, K., and J. Van Ryzin. 1981. A generalized multi-hit dose–response model for low-dose extrapolation. *Biometrics* **37**:341–352.

Rall, D. P. 1978. Thresholds. *Environ. Health Perspect.* **22**:163–165.

Riggs, D. S. 1963. *The Mathematical Approach to Physiological Problems*. MIT Press, Cambridge, Massachusetts.

Schneiderman, M. A. 1974. Safe dose? Problems of the statistician in the world of trans-science. *J. Wash. Acad. Sci.* **64**:68–78.

Schneiderman, M. A., N. Mantel, and C. C. Brown. 1975. From mouse to man—or how to get from the laboratory to Park Avenue and 59th Street. *Ann. N. Y. Acad. Sci.* **246**:237–246.

Schneiderman, M. A., P. Decoufle, and C. C. Brown. 1979. Thresholds for environmental cancer: Biologic and statistical considerations. *Ann. N. Y. Acad. Sci.* **329**:92–130.

Scientific Committee, Food Safety Council. 1978. Proposed system for food safety assessment. *Food Cosmet. Toxicol.* **16**(Suppl. 2):1–136.

Scientific Committee, Food Safety Council. 1980. Quantitative risk assessment. Pp. 137–160 in *Proposed System for Food Safety Assessment.* Food Safety Council, Washington, D.C.

Tomatis, L., V. Turusov, N. Day, and R. T. Charles. 1972. The effects of long term exposure to DDT on CF-1 mice. *Int. J. Cancer* **10**:489–506.

Turner, M. 1975. Some classes of hit theory models. *Math. Biosci.* **23**:219–235.

Van Ryzin, J. 1980. Quantitative risk assessment. *J. Occup. Med.* **22**:321–326.

White, J. 1972. Studies of the mechanism of induction of pulmonary adenomas in mice. Pp. 287–307 in L. E. LeCam, J. Neyman, and E. L. Scott, eds. *Proceedings of the Sixth Berkeley Symposium on Mathematical Statistics and Probability,* Volume 4. University of California Press, Berkeley.

Whittemore, A. S. 1977. The age distribution of human cancers for carcinogenic exposures of varying intensity. *Am. J. Epidemiol.* **106**:418–432.

Whittemore, A. S., and J. B. Keller. 1978. Quantitative theories of carcinogenesis. *Soc. Ind. Appl. Math. Rev.* **20**:1–30.

Worcester, J., and E. B. Wilson. 1943. The determination of LD_{50} and its sampling error in bioassay. *Proc. Natl. Acad. Sci. USA* **29**:79–85.

11

Extrapolation from Animal Data

Edward J. Calabrese

It is the premise of this chapter that there is a biological basis upon which toxicologists may be able to extrapolate from animals to humans. Its foundation lies in evolutionary theory, with the phylogenetic continuity of animal species.

EVOLUTIONARY THEORY AS THE PRACTICAL BASIS OF PREDICTIVE TOXICOLOGY

All life on earth is similarly organized, regulated, and controlled under the guidance of a strikingly uniform genetic code that ensures highly accurate reproductive continuity and structural development. Cellular structure and biochemistry are remarkably alike across the entire animal kingdom, starting with the lipoprotein cell membrane, which affects the absorption of xenobiotics into the cell, to metabolic processes such as glycolysis, the Krebs cycle, and numerous other aspects of intermediary metabolism.

The cellular similarity among animals serves as the basis upon which scientists have extrapolated or inferred functions from one species to another. In fact, much of what is now known about the basic principles of genetics, including DNA replication and repair processes, protein synthesis, and cellular aging, are derived from models such as bacteria, insects, and rodents. The Krebs cycle was extensively studied and elucidated in the pigeon because of the high metabolic rate of its breast muscle. Furthermore, the rationale behind the current genetic

Much of the material in this chapter was derived from the author's book (Calabrese, 1983), with permission of the publishers.

EDWARD J. CALABRESE • Division of Public Health, University of Massachusetts, Amherst, Massachusetts 01003.

toxicology movement assumes that the genome is fundamentally similar across the plant and animal kingdoms. In fact, the findings are so consistent as to be a validation of their own underlying assumptions.

Continuity is evident not only on the cellular and subcellular levels, but also organismically. In fact, the principles of allometry, as reflected in scaling phenomena, illustrate that aquatic and terrestrial animals display consistent and highly predictable relationships of biological parameters as mathematical functions of body weight. Investigators in the Soviet Union have found what they call a general biological regularity, whereby more than 100 highly diverse biological parameters are linearly related to body weight. In fact, on the basis of such an impressive and consistent series of predictive regularities, Krasovskii (1976) said it is logical that responses to toxicants also vary according to body weight. Subsequent tests by Krasovskii have strongly supported the hypothesis that susceptibility to toxicological agents may also be predicted by allometric relationships.

The basis of these relationships among terrestrial animals, including the common predictive laboratory animals, is in large part their need to regulate body temperature. Many biological characteristics, such as respiratory rate, lung volume, heart rate, blood flow rate, size of vessels and capillaries, and oxygen utilization, are functions of that need. As noted above, these characteristics are mathematical functions of body weight and are very accurately predicted across species. Since these biological parameters will also markedly affect the pharmacokinetics of xenobiotics, it follows that such pharmacokinetic relationships may also be capable of being predicted by allometric relationships. Once again, recent studies have provided strong validation for this theoretical foundation (Boxenbaum, 1980; Weiss *et al.*, 1977).

A second major need of all terrestrial species is the conservation of water. How they have adapted to this challenge has shaped to a considerable extent the biological basis of interspecies differences in metabolism of xenobiotics. Water conservation is best achieved metabolically if foreign compounds are converted to more water-soluble substances via phase 1 and 2 reactions. The net result is that more of the substance can be dissolved in the urine and therefore less water is needed to excrete the compounds. Despite the fact that all common models and humans face the same basic need for water conservation, there is considerable latitude in how it may be resolved. For instance, there are a number of potential options available to an organism to convert a lipophilic substance to a more water-soluble form. Each species may be conserving water in a somewhat different manner (i.e., through alternative metabolic routes and different rates of degradation), and these diverse strategies may create considerably different toxicological problems. The evolutionary paradigm therefore provides the theoretical basis upon which interspecies comparisons can be made. Unfortunately, it also is the nemesis of accurate extrapolation, because in addition to establishing the unity and continuity of life, it also establishes the biological basis of diversity. Therein lies the principal problem for accurate interspecies prediction.

A major goal of modern toxicology is to assess the causes of interspecies differences in susceptibility to toxic agents, with the ultimate hope of trying to improve the present capability of predicting human responses on the basis of animal studies. The key word is *improve.* There is abundant evidence that a wide range of factors contribute to differential species susceptibility to xenobiotics.

Differential metabolism is the most important factor affecting interspecies variation in susceptibility to toxic agents. Other factors contributing to the variability include absorption, plasma protein binding, biliary excretion, and intestinal flora.

GUIDING PRINCIPLES FOR ANIMAL EXTRAPOLATION

The most striking principle guiding interspecies predictions deals with the recognition of scaling factors, since numerous biological parameters are a mathematical function of body weight. These relationships are quite consistent over a broad range of species. The so-called biological regularity provides the theoretical foundation for comparing species on a dose per weight basis (mg/kg of body weight) or on a dose per surface area basis (mg/m^2).

Another generalization derived from the scaling phenomenon is that common laboratory animals metabolize drugs considerably faster than humans, but that their receptor sensitivity, at least for a number of drugs, appears to be quite similar across species (Calabrese, 1983).

These two generalizations provide a way for attempting to establish a linkage among species—a sort of common denominator. The first implies that barring any major specific cause of differential susceptibility, such as a defective enzyme system or difference in tissue distribution, everything should come out about the same across species when adjustments are made by scaling. These generalizations have proven exceptionally successful, as in the studies of Pinkel (1958), Freireich *et al.* (1966), and, more recently, Goldsmith *et al.* (1975), who reported strikingly excellent predictions of interspecies responses to toxic agents. These researchers used antimetabolites that were not subject to microsomal degradation and could be effectively excreted because of a sufficient number of polar groups. This technique may not be especially helpful, however, if there are important metabolic or other differences affecting susceptibility.

The second generalization is designed to complement the failings of the first one. Even if differential metabolism occurs, quantitative relationships between species can be established if the receptor sites are equally sensitive and plasma levels of the drug are carefully monitored. This approach, however, may work only if the drug is converted to pharmacologically inactive metabolites. It becomes appreciably more difficult to deal with if the parent compound becomes bioactivated, which can occur quite frequently.

GENERALIZATIONS OF PRACTICAL VALUE

These initial generalizations are attempts at finding a common denominator among the species to facilitate the extrapolation process. Is there so much diversity and individual idiosyncrasy that generalizations beyond those initial two are not really possible? In my opinion, no. Guidance can be provided by knowledge of the large number of examples of similarities and differences among species (Calabrese, 1983), for example:

1. The rat and dog are clearly the most efficient biliary excreters, whereas the guinea pig and monkey are quite inefficient. The toxicological implications of these observations are enormous when one considers that many xenobiotics display tissue-specific toxicity, including carcinogenicity.
2. The rodent placenta is considerably more porous than that of humans because of the prolonged development of the inverted yolk sac in the rodent. In general, this makes the rat much more susceptible to teratogens than are species in which the inverted yolk sac develops more quickly.
3. The metabolism of certain types of compounds is unique to specific types of animals. For example, cats and other felines are poor conjugators of glucuronic acid, dogs and their relatives are poor acetylators, while pigs are not proficient conjugators of sulfate.
4. Human skin is more resistant to the dermal absorption of foreign substances than is the skin of most animal models because of its thicker stratum corneum. Thus, the shaved skin of a rabbit or rat is much more easily penetrable by xenobiotics than is human skin.
5. Excision DNA repair is a function of life span. Although humans display a much more efficient repair capability than shorter lived species, the difference disappears after adjusting for age. This is another version of the scaling phenomenon.
6. The use of evolutionary relationships to predict human responses offers some potential for predictive toxicology. For example, monkeys are the best simulators of humans with regard to plasma protein binding, a number of metabolic patterns, especially certain amino acid conjugation patterns for arylacetic acid, and susceptibility to teratogenic agents.

These generalizations show that there are general patterns that can assist in the development of a rational and objective methodology for selecting or rejecting animal models and in the evaluation of published work.

ANIMAL MODELS FOR HIGH-RISK GROUPS

The preceding sections focused on extrapolation from animal models to normal humans. A second type of extrapolation is the prediction of response to toxicants in high-risk segments of the population on the basis of responses of normal humans or of animal models simulating such high-risk groups. This is a particularly important process, because high-risk groups are the first to experience adverse responses to stressor agents.

The use of animal models of potential high-risk groups is much more tenuous and qualitative than are comparisons with normal human populations. It is difficult enough to extrapolate from normal rats to normal humans. Superimposed on the usual uncertainties is the need for the animal to develop and express certain disease characteristics or genetic deficiencies just like those of the human. Furthermore, new demands are made on the disease model as a predictive entity as a greater understanding of the human disease emerges.

It is especially difficult to find animal models to simulate potential human groups at high risk for some common hereditary blood disorders, such as acatalasemia, sickle cell disease, G-6-PD deficiency, and thalassemia.

There are animals with red cell catalase deficiency, e.g., certain laboratory strains of mice (Feinstein *et al.*, 1968) and presumably all dogs (Paniker and Iyer, 1965); erythrocyte G-6-PD deficiency, e.g., several strains of sheep (Moranpot, 1972); sickle cell disease, deer (Amma *et al.*, 1974), genet (Ball *et al.*, 1976), and sheep (Rees-Evans, 1968); and thalassemia, the Bulgarian rat (Sladic-Simic *et al.*, 1969). Thus, it is possible that immediately useful data could be provided to decision-makers from toxicological studies with such models. However, each of these models is lacking in predictive potential. For example, all known cases of sickling in animal models occurs after oxygenation and increased pH of the blood, whereas human erythrocyte sickling takes place with the totally opposite stimulus (Amma *et al.*, 1974; Ball *et al.*, 1976; Rees-Evans, 1968).

The G-6-PD deficiency of sheep is apparently sufficiently similar to that of the human A variant—the most frequent genetic variant. Unfortunately, the predictive utility of this model is severely limited because the red cells of the sheep have major differences from those of humans in several other enzyme activities affecting oxidant stress. More specifically, sheep have 5.1 times more erythrocyte glutathione peroxidase activity than do humans, but only 20% of the normal human values for red cell catalase and glutathione reductase (Agar *et al.*, 1974; Maral *et al.*, 1977). In addition, the low G-6-PD activity of sheep is a biochemical characteristic of selective substrains, there being a lack of a high or normal G-6-PD activity in these sheep (Maronpot, 1972). Thus, there is a lack of adequate intrastrain comparison to evaluate the hypothesis that a G-6-PD deficiency alone enhances ozone toxicity, in contrast to the genetic polymorphism of the enzyme in the human population.

People often have very high expectations for the predictive capabilities of the animal model. It is expected to provide not only qualitative accuracy, but quantitative accuracy as well. This is an especially difficult request, since the red cell contains at least six different enzymes that affect susceptibility to oxidant stress. To find a model that adequately matches more than six enzymes may be expecting too much.

Thus, there are no animal models adequate to evaluate the hypotheses that environmental oxidants such as ozone and nitrogen dioxide may pose a greater risk to persons with any of the above-mentioned hereditary blood disorders. This lack does not suggest that the hypotheses are not valid, but rather that they require testing in another system, namely humans. If there is no appropriate model and concern about a specific high-risk group is sufficiently high, decision-makers must face the ethical issue of testing the substance in humans either in high-risk groups in clinical toxicological settings or, a considerably more expensive undertaking, in epidemiological studies. Ignoring the problem because of its controversial nature is not the answer. If the substance has generated sufficient social concern, it must be directly addressed in ways that are morally and medically responsible.

These inadequacies of present models illustrate the need to find or develop novel approaches to the use of animal models for predicting responses in high-risk humans. Although common laboratory models have often been proved inadequate for the type of predictions required, there is good potential in at least several general areas.

The use of ecological principles to find such models is one promising approach. For example, since humans heterozygous for thalassemia and G-6-PD deficiency have a selective advantage in areas of endemic malaria, similar types of adaptations may have occurred in other animals living in the same habitat. In fact, two kinds of malarial parasites—*Plasmodium berghei* and *P. vinckei*—are endemic in rodents of the Congo, Nigeria, and other parts of Africa. A survey of wild rodents in these areas may reveal animal models that have also adapted to malarial environments by evolving red cell alterations comparable to those of humans (Bannerman *et al.,* 1974).

Since there is no adequate predictive model for sickle cell disease, Castro *et al.* (1973) decided to create a new model type by placing human cells with hemoglobin S in the rat after its phagocytizing capacity had been eliminated. This model offers the potential of evaluating the responses of human sickle cells in experimental settings, since transfused sickle and normal cells maintained the same relative survival times in the rat model as in the human. However, considerably more development is required before this model can have any practical application for the regulatory agencies.

This transfusion model deserves considerable attention because of its possible utility in studies of the other common red cell hereditary disorders, such as thalassemia and G-6-PD deficiency. This would not necessarily diminish the problems inherent in the disruption of the normal function of the reticuloendothelial system and complement that occurs after chemical treatment to ensure that human red cells are not rapidly rejected upon transfusion.

When extrapolation of data from animal studies to humans is required, one question is invariably raised: Which human? Mankind is highly diverse—genetically, culturally, and developmentally—and has markedly differential susceptibilities to pollutant-induced adverse health effects. No one animal model can predict the responses of a group as diverse as humans. However, the diversity among the potential models suggests that different systems may be found to reasonably simulate selected high-risk groups of humans or normally functioning persons. Many biochemical indices, such as enzyme activity levels, dermal thickness, repair processes, and hemoglobin variants, display such a range and diversity within the animal kingdom that it may be possible to find one or more that respond similarly to some human subgroup.

QUANTITATIVE RISK ASSESSMENT

The third type of extrapolation is the prediction of effects that will occur outside the range of observed responses. Typically, this involves the prediction of effects at realistic low levels when there are data only on exceptionally high and seemingly unrealistic exposures. This particular extrapolation problem has emerged as a leading scientific policy issue in the regulation of carcinogens. Is there a safe level of exposure below which no risk is present? Is there no threshold, and thus is there some degree of risk regardless of the dose?

Numerous mathematical models have been developed to predict the risk at various levels. Despite the fact that different groups and agencies have selected one model or other for their risk assessment calculations, considerable uncertainty shrouds them all—some more than others. No model sufficiently incorporates an understanding of the process of carcinogenesis, as currently accepted, while recognizing population characteristics, including differential susceptibility. Of particular concern is that in quantitative risk assessments based exclusively on animal studies (typically the rat) investigators have assumed that both the animal model and human respond in an identical fashion.

Table 11-1 compares the biochemical and physiological differences between humans and rats that may result in differential susceptibility. A quick scanning of the table reveals the large number and wide range of differences between them, whether the concern is absorption, tissue distribution, biliary excretion, intestinal flora, enterohepatic circulation, mechanisms of conjugation, or other considerations. Despite such differences, the results obtained from downward extrapolation of the rat data are immediately transferred to the human. Although there may be no good alternative in light of our limited understanding, there seems to be a disproportionate amount of reliance on the risk evaluations that are based on this procedure. Whenever regulations are to be established, proof of the validity of one's assertions should be demanded. This is not typically the case with quantitative risk assessment, however, because it is difficult to prove either toxicologically or epidemiologically that exposure to X micrograms of a suspected human carcinogen results in a cancer risk of one in one million over a 70-year lifetime. Yet this is a typical way of expressing an acceptable risk derived from animal data using downward extrapolation techniques. Of several attempts that have been made to validate the predictions of quantitative risk assessments, e.g., for arsenic in drinking water and for aflatoxin, DDT, and dieldrin in food, the predictions and the actual incidence of human cancers have been orders of magnitude apart (Calabrese, 1983).

Regulators must see quantitative risk assessment for what it is: a new, rapidly developing science that shows promise of helping delineate the extent of human risks, especially with respect to the difficult problem of risks from chronic exposure to low levels of carcinogens. However, the science is still in its childhood or adolescence and should not be expected to provide mature guidance for some time. Yet, because of the extreme interest in this aspect of extrapolation, quantitative risk assessment is showing precocious development, and a number of attempts have been made to improve the consistency and reliability of its predictions. Nonetheless, it is still very questionable whether regulatory agencies should accept this methodology as the basis for setting exposure limits. Although it is convenient to have a number on which to base a risk of one in one million, this number should be seen for what it is with all its limitations.

TABLE 11-1. Some Biochemical, Physiological, and Morphological Differences between Rats and Humans That May Be Toxicologically Significant[a]

Factor	Rat	Human
Skin characteristics		
Stratum corneum	—	Much thicker than rat
Dermal vasculature	—	Much thicker than rat
Sweat glands	Missing from general body surface; eccrine sweat glands located in foot pads to moisten frictional surfaces	Numerous coiled, tubular sweat glands (100–600/m^2)
Hair follicles	Densely haired, up to 4000 hairs/cm^2	Many fewer hairs; 40–70 hairs/cm^2 on skin of the trunk and limbs
Dermal absorption based on the above characteristics	Considerably more efficient absorber than humans for a variety of organic compounds	—
Respiratory parameters		
Histamine content (μg/g)	15.8	27.7
Exogenous histamine catabolism (%)	44.2	29.2
Histamine release (μg/g)		
Compound 48/80	17.1	43.2
Cotton dust	0.0	16.1
Lung morphometry		
Branching angle	Decreases with increasing depth in the lung	Increases with increasing depth in the lung
Symmetry	Less than humans	—
Diameter ratio of daughter branches at bifurcation	Greater than humans	—
Number of diversions of tracheobronchial tree	More variable than humans	—
Mucus flow patterns (mm/min)	13.5	15
Bronchial glands	Absent	Numerous
Position of lung relative to ground	Horizontal	Vertical
Breathing	Obligate nose-breathers	—
Gut flora location	Numerous flora in stomach and proximal small intestine	Little or no flora in stomach and proximal small intestine
	Numerous flora in distal small intestine, large intestine, rectum, and feces	Similar to rat
Estimated β-glucuronidase activity (units)		
Proximal intestine	Very high (304.0)	0.02
Distal small intestine	Very high (1342.0)	0.9
Plasma protein binding	Generally not as extensive binding as in the human; a number of	

TABLE 11-1. (*continued*)

Factor	Rat	Human
	researchers believe that the extent of human binding cannot be predicted with the rat model	
Biliary excretion	The rat is perhaps the most efficient biliary excreter, whereas limited evidence suggests that the human is not an efficient excreter of intermediate weight compounds	—
Metabolism		
Conjugations		
Sulfate	Less active than human	—
Glucuronidation	More active than human	—
Acetylation	Effective	Both effective and slow acetylation phenotypes
Deacetylation	Displays a relatively low ability	Human data inadequate to assess
Amino acid with carboxylic acid substrates		
Benzoic acid	80–100% conjugation with glycine	100% conjugation with glycine
Phenylacetic acid	Strongly favors glycine conjugation	Strongly favors glutamine conjugation
Rhodanese activity (liver)	Considerably more active than in humans	—
Epoxide hydrase	Less active in humans	—
Red blood cell enzymes, which prevent oxidant stress (in values relative to humans, which are given as 1)		
Glutathione peroxidase	10.2	1
Glutathione reductase	0.2	1
Catalase	0.2	1
G-6-PD	2.4	1
Superoxide dismutase	1.7	1
Methemoglobin reductase	2.4	1
Comparison of rat versus human for 23 substances, according to qualitative and quantitative similarity of metabolic pathway	Good predictor, 4 of 23; invalid predictor, 8 of 23	—
Concentration of urine	Typical laboratory rat has the ability to concentrate its urine approximately 2 times as much as that of humans, as indicated by urine:plasma ratios; the desert rat will concentrate its	—

TABLE 11-1. (*continued*)

Factor	Rat	Human
	urine 4–5 times more than humans[b]	
Dermatotoxicity		
Eyes, skin	Practical reasons (e.g., lack of docility) preclude widespread use of the rat as a predictive model for dermatotoxicity; the rabbit is the model of choice for historical reasons and practical considerations, e.g., its docility, large skin surface area, and large, nonpigmented eyes	—
Allergic hypersensitivity	The rat is *not* a good model for allergic hypersensitivity, since, unlike humans, it does not produce anaphylactic antibodies in response to the diversity of allergens; the guinea pig is favored in studies for qualitative predictions	—
DNA repair	In absolute terms, less efficient in excision repair; when adjusted for the influence of life span, little difference between these species is found; more efficient than humans in postreplicative repair	—
Teratogenicity	Prolonged dependence of the rat (up to the 20–25th somite) on the inverted yolk sac placenta during organogenesis as compared to higher mammals, especially humans (5th somite), thereby making the rodent generally much more susceptible to teratogens than are humans	—
High-risk animal models		
Respiratory disease		
Asthma	Rats are not considered appropriate; with the exception of humans, dogs are the only animal that develop a defined hypersensitivity disease related to aeroallergens	—
Bronchitis	The rat is not considered an appropriate model, because its version of chronic respiratory	—

TABLE 11-1. (*continued*)

Factor	Rat	Human
	disease involves bronchitis displays, excessive inflammation, and involvement of the pulmonary parenchyma	
Cardiovascular disease		
Atherosclerosis	The rat is generally not considered a very effective model, since it is very resistant to developing this disease	—
Hypertension	Numerous predictive rat strains exist; the rat is the animal of choice	—

[a]From Calabrese (1983).
[b]Dickes (1970).

CONCLUSION

It is not possible to evaluate comprehensively the usefulness of animal models for predictive toxicology. Nonetheless, extrapolation from animal data to predict risk to humans is a science. It involves a rational process in which differences and similarities are dissected and evaluated. This analysis is only as good as the research on which the assessment is based.

The demands of society that predictions become more precise and quantitative are usually well beyond the current state of the art. In addition, U.S. legislation, such as the Toxic Substances Control Act and the Federal Insecticide, Fungicide, and Rodenticide Act, requires premarket testing before a product can be acceptable for interstate commerce, and thus emphasizes the need for accurate predictive animal models.

Predictive toxicology is the vehicle by which society may be protected from chemically induced tragedies. Without appropriate models to predict human responses to toxic agents, however, untold numbers of individuals may become the victims of the chemical products of our technological society. To build the knowledge necessary to meet society's demands for perfection in toxicological predictions, strong support is needed.

REFERENCES

Agar, N. S., M. Gruca, and J. D. Harley. 1974. Studies on glucose-6-phosphate dehydrogenase, glutathione reductase, and regeneration of reduced glutathione in the red blood cells of various mammalian species. *Aust. J. Exp. Biol. Med. Sci.* **52**:607–614.

Amma, E. L., G. D. Sproul, W. Wong, and T. H. J. Heisman. 1974. Mechanism of sickling in deer erythrocytes. *Ann. N. Y. Acad. Sci.* **241**:605–613.

Ball, S., C. M. Hawkey, J. M. Hime, I. F. Keymer, and M. R. Brambell. 1976. Red cell sickling in genets. *Comp. Biochem. Physiol.* **A 54**:49–54.

Bannerman, R. M., J. A. Edwards, and M. Kreimer-Birnbaum. 1974. Investigation of potential models of thalassemia. *Ann. N. Y. Acad. Sci.* **232**:306–322.

Boxenbaum, H. 1980. Interspecies variation in liver weight, hepatic blood flow, and antipyrine intrinsic clearance: Extrapolation of data to benzodiazepines and phenytoin. *J. Pharmacokin. Biopharm.* **8**:165–176.

Calabrese, E. J. 1983. *Principles of Animal Extrapolation.* Wiley, New York.

Castro, O., J. Orlin, M. W. Rosen, and S. C. Finch. 1973. Survival of human sickle-cell erythrocytes in heterologous species: Response variations in oxygen tension. *Proc. Natl. Acad. Sci. USA* **70**:2356–2359.

Dickes, S. E. 1970. Pp. 16–30 in *Mechanisms of Urine Concentration and Dilution in Mammals.* Edward Arnold, London.

Feinstein, R. N., J. Braun, and J. B. Howard. 1968. Nature of the heterozygote blood catalase in a hypocatalasemic mouse mutant. *Biochem. Genet.* **1**:277–285.

Freireich, E. J., E. A. Gehan, D. P. Rall, L. H. Schmidt, and H. E. Skipper. 1966. Quantitative comparison of toxicity of anticancer agents in mouse, rat, hamster, dog, monkey and man. *Cancer Chemother. Rep.* **50**:219–244.

Goldsmith, M. A., M. Slavik, and S. K. Carter. 1975. Quantitative prediction of drug toxicity in humans from toxicology in small and large animals. *Cancer Res.* **35**:1354–1364.

Krasovskii, G. N. 1976. Extrapolation of experimental data from animals to man. *Environ. Health Perspect.* **13**:51–58.

Maral, J., K. Puget, and A. M. Michelson. 1977. Comparative study of superoxide dismutase, catalase, and glutathione peroxidase levels in erythrocytes of different animals. *Biochem. Biophys. Res. Commun.* **77**:1525–1535.

Maronpot, R. R. 1972. Erythrocyte glucose-6-phosphate dehydrogenase and glutathione deficiency in sheep. *Can. J. Comp. Med.* **36**:55–60.

Paniker, N. V., and G. Y. Iyer. 1965. Erythrocyte catalase and detoxication of hydrogen peroxide. *Can. J. Biochem.* **43**:1029–1039.

Pinkel, D. 1958. The use of body surface area as a criterion of drug dosage in cancer chemotherapy. *Cancer Res.* **18**:853–856.

Rees Evans, E. T. 1968. Sickling phenomenon in sheep. *Nature* **217**:74–75.

Sladic-Simic, D., P. N. Martinovich, N. Zivkovic, M. Kahn, and H. Ranney. 1969. A thalassemia-like disorder in Belgrade laboratory rats. *Ann. N. Y. Acad. Sci.* **165**:93–99.

Weiss, M., W. Sziegoleit, and W. Förster. 1977. Dependence of pharmacokinetic parameters on the body weight. *Int. J. Clin. Pharmacol. Biopharm.* **15**:572–575.

12

Approaches to Route Extrapolation

James R. Withey

The route of exposure to a toxic agent may affect the site and the magnitude of the toxic response. The greatest potency and most rapid response for any agent will usually be observed when the route of administration is intravenous, followed, in descending order, by the inhalation, intraperitoneal, subcutaneous, intramuscular, intragastric, and topical routes (Klaassen and Doull, 1980). Interactions at or near the site of uptake can also affect both the rate and extent of uptake and thus modify the intensity of the toxic effect (Martin *et al.*, 1971). This chapter examines the reasons for such route specificity and explores suggested approaches in which data obtained from one route may be extrapolated to another.

When determining standards for industrial contaminants such as those found in workroom air, regulatory agencies usually use dose–response data from the most common routes of exposure—pulmonary and dermal. To regulate levels of xenobiotic compounds in drinking water, however, data on oral exposures are desirable. In some cases, data on the relevant route of administration have either been unavailable or unusable for one reason or another. For example, when searching the toxicology data to develop ambient water quality criteria for many compounds, such as ethyl benzene, hexachloropentadiene, or chlorinated naphthalenes (U.S. Environmental Protection Agency, 1979), for which the oral dosing route was used, one can find no report of studies. The low aqueous solubility and immeasurably small toxic response at low dose levels of these compounds preclude the design of ideal or appropriate studies for this purpose. On the other hand, toxicological investigations have been conducted on inhalation exposures to these compounds in animals and humans. Organizations such as the American

JAMES R. WITHEY • Environmental and Occupational Toxicology Division, Environmental Health Directorate, Ottawa, Ontario, Canada K1A OL2.

Conference of Governmental Industrial Hygienists (1962, 1971, 1977) have used data from such studies to develop threshold limit values (TLVs) for substances in workroom air.

THE STOKINGER–WOODWARD APPROACH

Nearly 30 years ago, Stokinger and Woodward (1958) proposed methodology to establish drinking water standards for substances for which there were no toxicological data on intragastric exposures. Their model assumed that, for any given inhaled dose, e.g., the TLV, an equivalent dose from ingested water could be calculated by using estimates of the daily air and water intake combined with appropriate absorption factors for uptake by the inhalation and gastrointestinal routes.

To illustrate the method, Stokinger and Woodward started with the TLV for barium and its compounds (0.5 mg/m^3 as barium) and, since TLVs are based on the assumption that a worker inhales 10 m^3 of air per 8-hr day, calculated the quantity of barium inhaled during a 1-day exposure to the TLV for this compound. They also assumed 0.75 as the factor expressing the amount of inhaled barium that passes into the bloodstream. Thus, the total *amount* of barium assumed to reach the systemic circulation would be $10 \times 0.5 \times 0.75 = 3.75$ mg/day. This amount was presumed to be noninjurious to health, regardless of exposure route, and might be considered equivalent to a no-observed-adverse-effect level (NOAEL).

To calculate the concentration, in parts per million (wt/wt) in water, that would yield an equivalent uptake by ingestion, they assumed that the maximum daily intake of water would be 2 liters and that a reasonable absorption factor for the gastrointestinal uptake of barium would be 90%. Thus, a maximum daily oral intake of $3.75 \times 100/90$ or 4.17 mg of barium daily in 2 liters of water, or 2 ppm (wt/wt), should be equivalent to the acceptable daily intake. More recently, the U.S. Environmental Protection Agency recommended that the calculation should include a weighting factor of 5/7 to reflect the 7-day water exposure as opposed to the 5-day weekly industrial exposure.

The TLV is derived for an exposure to a time-weighted average concentration for 8 hr/day, 5 days/week. It may be based not only on the systemic toxicity of a substance, but also on eye, mucosal, or skin irritation, narcosis, nuisance (e.g., odorous), or other organoleptic effects (American Conference of Governmental Industrial Hygienists, 1977). It should not, therefore, be used as a basis for comparing the systemic toxicity of chemicals without a thorough understanding of the information incorporated into the setting of individual guidelines (Cornish, 1980). TLVs are based on the best available information from industrial experience, experimental studies, or, when possible, a combination of the two.

Limitations of the Stokinger–Woodward Method

Absorption Factors

Among the many uncertainties that should be borne in mind when applying the Stokinger–Woodward approach, the greatest is perhaps the choice of absorp-

TABLE 12-1. Limits for Inorganic Substances in Water Calculated from Threshold Limits for Air[a]

Substance	Absorption factor: Inhalation	Absorption factor: Ingestion	Limit for water (ppm)	Threshold limit for air (mg/m^3)
Arsenic	0.2	0.8	0.6	0.5
Barium	0.75	1.0	2.0	0.5
Cadmium	0.25	0.03	4.0	0.1
Chromium	0.75	0.06	6.0	0.1
Cyanide	0.6	0.8	19.0	5.0
Fluoride	0.1	1.0	1.25	2.5
Lead	0.2	0.2	1.0	0.2
Manganese	0.1	0.2	15.0	6.0
Selenium	0.8	0.8	0.5	0.1
Vanadium	0.5	0.5	2.5	0.5
Zinc	0.25	0.25	75.0	15.0

[a]From Stokinger and Woodward (1958).

tion factors for both the inhalation and gastrointestinal uptake. Stokinger and Woodward stated that the exact data needed at low concentrations to derive precise figures for the absorption factor are either nonexistent or rough approximations. They suggested that "estimates of the magnitude of these factors must be made, in many instances, on toxicologic judgment" (Stokinger and Woodward, 1958). In their tabulated data, reproduced in Tables 12-1 and 12-2, the absorption factors vary from 0.03 to 1.0, and there is no rationale presented for either their selection or wide variation.

Stokinger and Woodward did not insist that their model, with its necessary

TABLE 12-2. Limits for Organic Substances in Water Calculated from Threshold Limits for Air[a]

Substance	Absorption factor: Inhalation	Absorption factor: Ingestion	Limit for water (ppm)	Threshold limit for air[b] (mg/m^3)
Acetone	0.5	1.0	6000	2400
Acrylonitrile	0.75	0.85	240	45
Allyl alcohol	0.75	0.8	55	5
Aniline	0.5	1.0	47	5
Benzene	0.35	0.4	340	80
Carbon tetrachloride	0.3	0.5	240	65
Ethylene diamine	0.75	0.8	140	25
Formaldehyde	0.8	0.8	30	6.25
Methyl bromide	0.3	0.5	240	40
Methylchloroform	0.1	0.5	2700	2714
Phenol	1.0	1.0	95	19
Pyridine	0.1	0.5	30	30

[a]From Stokinger and Woodward (1958).
[b]The data units have been changed from ppm to mg/m^3.

assumptions, could be applied without "additional tests of reasonability." In their worked example with barium, for example, they cited a report that established the association between the exposure to barium and its compounds and industrial disease. "Baritosis" has been frequently reported in the industrial hygiene literature (Leschke, 1943). The average lethal dose of barium for an adult is 550–600 mg (Sollman, 1953), and 30-mg doses of soluble barium chloride have been administered three or four times daily under medical supervision without adverse effect (Nelson, 1953). Thus, a maximum single dose of approximately 4 mg, calculated by the Stokinger–Woodward method, would seem to provide a fairly large margin of safety, not only for healthy adults, but for the population at large.

FIRST-PASS EFFECTS

If reliable absorbtion factors are available, the Stokinger-Woodward approach will provide a reasonable estimate of the doses required to result in the same *amount* of uptake to the systemic circulation, no matter what route of administration is used. Even in the ideal case, where absorption factors have been determined with some degree of precision, there can be no assurance that the time course of the toxicant concentration at target sites will be independent of the route of administration. Indeed, where extensive metabolism is known to occur as a direct consequence of the route of administration, the toxic insult may be amplified or attenuated, depending on whether active or inactive metabolites are formed. Such is the case when a toxicant is totally absorbed into the portal circulation, which occurs after it is administered intragastrically or intraperitoneally. If hepatic metabolism is considerable, then only a fraction of the dose will be available for distribution by the systemic circulation to the other major organs. This so-called "first-pass effect" would not only increase the apparent rate of elimination from the central pharmacokinetic compartment as reflected by blood concentration values, but also, in the extreme case, could allow the complete metabolism of the administered dose and reduce the bioavailability to zero (Levy and Gibaldi, 1975). Since there are other sites of uptake, where interactions with enzymes or binding sites reduce the effective bioavailability, a brief review of the important factors that can affect the uptake and pharmacokinetics after dosing by the common routes of administration would facilitate discussion of the equivalence of dose to give the same toxic potency by different routes.

Toxicokinetic Considerations

The most usual sites of uptake involved in the assessment of potential hazards to humans are the lungs and skin (from industrial and environmental exposure) and the gastrointestinal tract (from the presence of xenobiotics in the food chain). Other routes of administration used by toxicologists are the intravenous, intramuscular, and subcutaneous routes. No special mechanisms are usually involved in the uptake of xenobiotics, which enter the bloodstream by the same processes by which oxygen, foods, and other nutrients are absorbed (Klaassen, 1980).

For routes other than the intravenous route, substances will usually diffuse from the site of administration as neutral molecules through epithelial and other membranes to the bloodstream. The systemic circulation will then transport the absorbed molecules to organs and tissues of the body and, ultimately, to the sites of toxic action. Excellent summaries of the processes involved are available (Barr, 1968; Himmelstein and Lutz, 1979; National Research Council, 1980).

Intravenous and Route Uptake Processes

Most reviewers agree that the initial approach to evaluating the pharmacokinetics of a xenobiotic should be to examine the elimination of the substance after the administration of an intravenous (i.v.) dose (Gibaldi and Perrier, 1975; Gladtke and von Hattingberg, 1979; Levy and Gibaldi, 1975; Wagner, 1971; Withey, 1978). This is often the best approach for quantifying the effect and understanding the mechanism of action. Generally speaking, the avoidance of complicating factors that affect uptake by routes of administration other than i.v. will facilitate examination of the pharmacokinetics of elimination and its variation with dose. This approach can provide valuable information on the metabolism and distribution of the administered compound. As the dose is increased, changes in the apparent rate of elimination after i.v. administration will usually be indicative of capacity-limited metabolic processes or saturation of depots to which the compound is distributed. When such mechanisms are involved, an approach suggested by Loo and Riegelman (1968) is precluded. These investigators proposed that the pharmacokinetic model and values for the elimination rate coefficient be obtained from infusion data to obtain precise values for the apparent rate of elimination after any route of administration.

The Gastrointestinal Route

Some of the factors that can affect the rate of absorption from the gastrointestinal tract are summarized in Fig. 12-1. Any one of these factors could reduce the portion of the administered dose that reaches the systemic circulation. These factors have been reviewed by Schanker (1971) and Wagner (1971).

A number of pharmacokinetic relationships that are useful in the calculation of the amount of agent that reaches the systemic circulation have been derived (Barr, 1968; Gibaldi and Perrier, 1975; Wagner, 1971). For example, it can readily be shown that the total amount of the dose absorbed is directly proportional to the area under the blood concentration–time curve (*AUC*) and that this in turn is related to the apparent volume of distribution V_d, the elimination rate coefficient k_e (metabolism and excretion), and the amount that reaches the systemic circulation Q_b:

$$AUC = \int_0^\infty C_t \, dt = Q_b / V_d k_e \qquad (1)$$

The rate of absorption does not appear in this relationship, since the *AUC* depends only on the *amount* of uptake and not on the rate. The product *FD* of

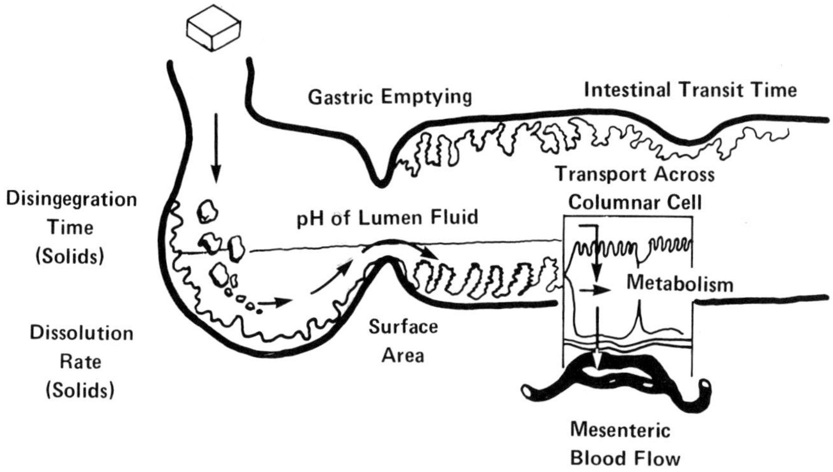

TOXICANT AND DOSAGE FORM

Physical Properties of Toxicant

- Lipid solubility, partition coefficient
- pKa
- Solubility
- Stability
- Molecular weight

Properties of Dosage Form

- Disintegration time
- Dissolution rate
- Surface area of particles
- Crystal size
- Polymorphic form
- Solvates
- Salt form

Pharmacologic Properties of Toxicant

- Effect on blood flow
- Effect on peristalsis

Effects of Other Substances

- Adsorbants
- Antacids
- Complexing agents
- Food

GASTROINTESTINAL FUNCTION

Factors Effecting Transit to Site

- Gastric emptying
- Intestinal transit
- Motility

Properties of Lumen Fluid

- pH
- Enzymes
- Complexing agents (mucin)
- Solubilizing agents (bile)
- Viscosity

Factors of the Site of Absorption

- Surface area
- Specialized transport
- Local blood flow
- Intestinal metabolism

FIGURE 12-1. Factors affecting gastrointestinal uptake. [Adapted from Barr (1968), with permission.]

the fraction F of the administered dose D also equals the amount of uptake Q_b. For an orally administered dose, the factors that effectively reduce the value of F may be considered as those that are a consequence of interactions within the extracorporeal system of the gastrointestinal tract and those involved after uptake. Among the former that should be seriously considered are degradation by the gastrointestinal fluids and enzymes, that part of the dose that is not absorbed due to the physicochemical properties of the agent, and that bound to food or other components within the gastrointestinal tract. Since absorption from the gastrointestinal tract usually takes place via the portal circulation, interactions that occur after absorption will involve metabolism within the portal circulation or the liver and binding to tissue sites in the intestine, portal circulation, or the liver. When these factors are predominantly involved, Barr (1968) has suggested that a term Φ should be used to modify Eq. (1):

$$AUC = (1 - \Phi)FD/V_d k_e \tag{2}$$

The factor Φ is important in assessing the toxicological insult of compounds that are subject to extensive hepatic metabolism and that are administered by the oral or intraperitoneal routes. This is due to the fact that an appreciable fraction of the administered dose will be metabolized during the first pass through the liver and only a small fraction of the absorbed dose will reach the systemic circulation. The term "systemic availability" has been suggested to indicate the extent of absorption of *unchanged* dose into the systemic circulation (Levy and Gibaldi, 1975).

There have been reports of large differences in systemic availability when the oral and intravenous routes are compared. Organic nitrates have been shown to be completely degraded to inactive metabolites after one pass through the liver (Needleman *et al.,* 1972), and similar results have been reported for lidocaine (Boyes *et al.,* 1970). Only 0.5–14% of the given dose of phenol was estimated to reach the systemic circulation after oral administration (Cassidy and Houston, 1980). Other drugs, such as aspirin (Rowland *et al.,* 1967) and salicylamide (Barr, 1969), are also well absorbed, but poorly available.

On the other hand, an excellent example of the equivalence of similar oral and intravenous doses is provided by the data on valproic acid (Perucca *et al.,* 1978), which are plotted in Fig. 12-2. This anticonvulsant drug conforms to a one-compartment pharmacokinetic analysis and has a relatively long elimination half-life of about 13 hr after administration by either the oral or i.v. route. Its systemic availability, as measured by the relative AUC after oral and i.v. dosing in six human subjects, was 1.00 ± 0.10. There is, clearly, no question here that there are any interactions with the molecules of valproic acid as a consequence of the factors given in Fig. 12-1.

The amount of an orally or intravenously administered dose that reaches the systemic circulation can also be reduced as a consequence of interactions between the xenobiotic and enzymes or components within the gastrointestinal tract. For example, there is a thousandfold reduction in systemic availability when isopro-

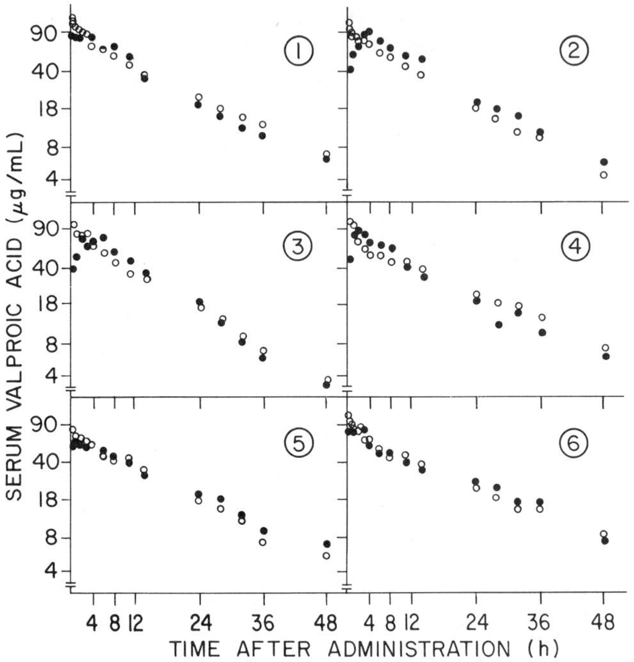

FIGURE 12-2. Serum valproic acid concentration curves following single (●) oral and (○) intravenous 800-mg doses in six subjects. [From Perucca *et al.* (1978) with permission.]

terenol is administered orally due to the formation of an inactive ethereal sulfate during transfer across the wall of the gastrointestinal tract (Dollery *et al.,* 1971).

Interactions that can affect the fraction of the dose reaching the systemic circulation can also occur at sites within the lung. An investigation of the protective role of pulmonary enzymes against environmental phenols (Cassidy and Houston, 1980) suggests that for phenol itself, only 40–76% of a dose administered to rats via the jugular vein reaches the systemic circulation. The authors point out that when the dose is administered by this route, it must traverse the lung before reaching the arterial system for distribution throughout the body.

The Pulmonary Route

The dynamics of uptake via the lungs is unique in that, for most cases where the atmosphere is inhaled at a constant concentration, the kinetics of the process is zeroth order. Consequently, for small molecules of a gaseous or volatile xenobiotic, "steady-state" or "plateau" concentrations within the central pharmacokinetic compartment are usually reached fairly rapidly after the initiation of the exposure. Such molecules diffuse rapidly across alveolar membranes, and an equilibration of the xenobiotic in alveolar air with the blood concentration enables the latter to reach a steady-state concentration that is related to the solubility of the vapor in blood and air according to Henry's law (Goldstein *et al.,* 1974). Such is the case, for example, with vinyl chloride (Withey, 1976). For gases with a low blood-to-air solubility, the rate of transfer will depend largely on the rate of blood flow through the pulmonary capillary beds (perfusion-limited); for gases with a high blood-to-air solubility, the ventilation rate and depth of respiration (tidal volume) will be the most important factors (ventilation-limited). Thus, the important parameters that affect the rate of uptake and the steady-state blood level are the solubility of the compound in blood, the respiration rate, the tidal volume, the cardiac output, and the blood flow per unit volume within the alveolar capillary beds. For xenobiotics of larger molecular size that also have a lower blood solubility, the elimination rate will usually be slower and the steady-state level will be achieved more slowly. The time to equilibrium is solely dependent on the rate of elimination (Withey and Collins, 1979).

Comparisons of Inhalation and Intragastric Uptake

Apart from continuous inhalation exposure, there are very few "real-life" situations in which constant steady-state blood levels will be generated. For a one-compartment model, such conditions might be engineered if the xenobiotic is delivered intravenously at a constant linear rate that equals the rate of elimination (McNamara *et al.,* 1979). A similar steady-state situation can be realized if multiple oral doses of a compound are administered at appropriate dose intervals relative to the elimination half-life (Wagner *et al.,* 1965). Thus, an equilibrium level C_{av}^{∞} will be achieved after the oral administration of a few doses when the dosing interval τ is equal to the half-life of the monoexponential terminal phase for a system that behaves as a one-compartment model (Withey, 1978). Of course, any oral dose administered chronically will yield "equilibrium" levels that result in an essentially "constant" insult to the target site. For a one-compartment toxicokinetic model, the equilibrium or average asymptotic blood concentration C_{av}^{∞} is related to the administered dose D, the fraction of the dose absorbed F, the dose interval τ, the volume of distribution V_d, and the first-order rate coefficient k_e for the overall loss of the compound from the blood, by

$$C_{\text{av}}^{\infty} = FD/V_d k_e \tau \tag{3}$$

FD/V_d may be estimated by the method suggested by Wagner and Nelson (1963), and k_e can be derived from the terminal linear phase of the semilogarithmic plot of the blood concentration–time curve obtained after the administration of a single dose. Thus, it would not seem unreasonable to presume that when a C_{av}^{∞} observed after multiple oral dosing closely approximates the plateau blood concentration arising from an inhalation exposure, similar areas under the curve for the temporal blood concentration–time relationship would represent a similar toxicological insult.

The examples shown in Fig. 12-3 illustrate a number of features that have been discussed above and allow further discussion of the important variables when considering a comparison of the toxicological insult arising from repeated inhalation and orally ingested doses. The toxicological equivalence of the same amount of uptake by these two routes is essential if the Stokinger–Woodward method is applied. The curves in this figure show the blood concentration–time

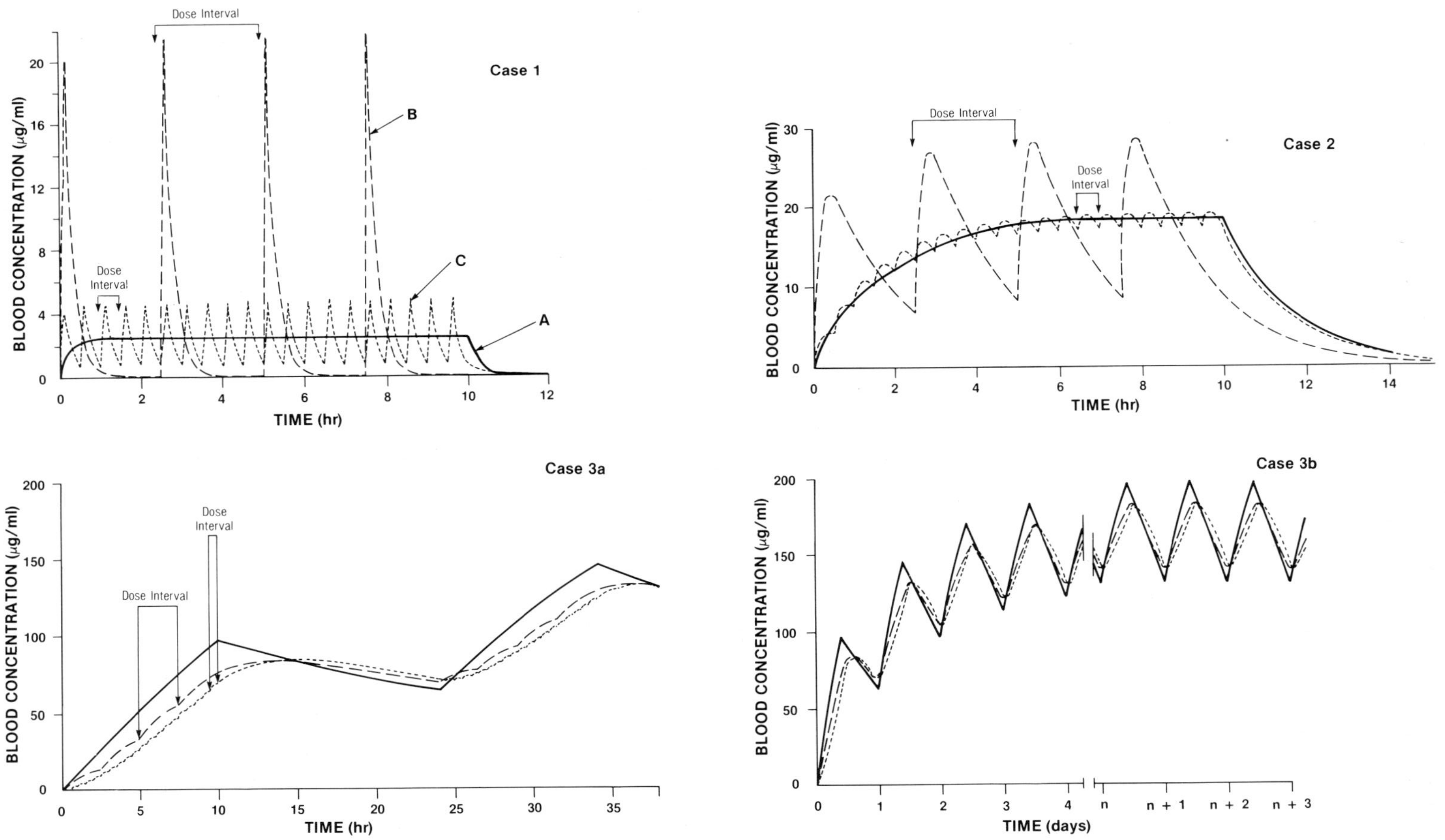

FIGURE 12-3. Temporal blood concentration relationships for uptake by the pulmonary and gastrointestinal routes: (—) 10-hr inhalation exposure; (— —) four equal divided doses, administered intragastrically over 10 hr; (· · ·) 20 equal doses, intragastrically administered over 10 hr. The different elimination rates used to calculate the data for Cases 1, 2, and 3 are given in the text.

profile during and after a 10-hr exposure involving multiple oral dosing or inhalation exposure.

In case 1 of Fig. 12-3, curve A was derived by assuming that the xenobiotic in the vapor phase was taken up by a zeroth-order rate process to give a steady-state blood concentration that was 95% complete within 30 min of initiating the exposure. This is a reasonable assumption for a low-molecular weight compound such as an anesthetic gas. If the xenobiotic is assumed to follow a one-compartment pharmacokinetic model, irrespective of the route of administration, then this would require that the first-order elimination rate coefficient k_e has a value of approximately 4.61 hr^{-1} or a half-life of 0.15 hr. A human with a blood volume of 71.4 ml/kg and a breathing rate of 10 liters/min will breathe 1.0 mg of the xenobiotic/min during exposure to an assumed vapor concentration of 100 mg/m^3 (1000 ppm). Thus, the rate of uptake will be 1.0 mg in 5.355 liters of blood/min or 0.1867 mg/liter per min = 11.20 mg/liter per hr = k_0, the zeroth-order rate of uptake. The total amount of uptake during a 10-hr exposure will therefore be 112 mg/liter, or a total of 599.76 mg. The equation that relates the blood concentration with time during the exposure will be

$$C_t = A(1 - e^{-k_e t}) \tag{4}$$

where $A = k_0/k_e$, and the corresponding postexposure equation is

$$C_t = Ae^{-k_e t} \tag{5}$$

The rate of uptake during the exposure will be

$$dC_t/dt = k_0 - k_e C_t \tag{6}$$

so that the steady-state, or plateau, level will be reached when $k_0 = k_e C_t$ and, in this example, it will have a value of 2.43 μg/ml.

Curve B was generated by assuming that the total dose of the xenobiotic taken up during the inhalation exposure, i.e., 599.76 mg, is administered as four equal oral doses, beginning at time zero, with a dose interval of 2.5 hr over a 10-hr dosing period. It was also assumed that the first-order rate coefficient for elimination in the one-compartment pharmacokinetic model was one-tenth of the value for the absorption coefficient. Thus, $k_a = 10k_e = 46.1\ hr^{-1}$ ($t_{1/2} = 0.015$ hr) and the equation for the blood levels arising from the first dose is

$$C_t = \frac{k_a D}{(k_a - k_e)V_d}(e^{-k_e t} - e^{k_a t}) \tag{7}$$

where D is the magnitude of the dose and V_d is the volume of distribution (assumed, in these examples, to be equal to the blood volume).

Curve C was generated in a similar manner by assuming that the same total dose was administered as 20 divided doses of 29.988 mg at a dosing interval of 0.5 hr over a 10-hr period, beginning at time zero.

For repeated oral dosing, the blood concentration at any time C_t for the 10-hr dosing period, the 14-hour postdosing period, and for successive days is given by the expression

$$C_t = \frac{k_a D}{(k_a - k_e)V_d}\left[\left\{\frac{(1 - e^{-N\tau_0 k_e})e^{-\tau_2 k_e}(1 - e^{-(m-1)\tau k_e})e^{-(n-1)\tau_0 k_e}}{(1 - e^{-\tau_0 k_e})(1 - e^{-\tau k_e})} + \frac{1 - e^{-n\tau_0 k_e}}{1 - e^{-\tau_0 k_e}}\right\} e^{-k_e S} - \left\{\frac{(1 - e^{-N\tau_0 k_a})e^{-\tau_2 k_a}(1 - e^{-(m-1)\tau k_a})e^{-(n-1)\tau_0 k_a}}{(1 - e^{-\tau_0 k_a})(1 - e^{-\tau k_a})} + \frac{1 - e^{-n\tau_0 k_a}}{1 - e^{-\tau_0 k_a}}\right\} e^{-k_a S}\right] \quad (8)$$

where k_a, k_e, D, and V_d have the same meaning as in Eq. (7); τ is the long dosing period, i.e., 24 hr; τ_0 is the short dosing period, i.e., 2.5 hr for the four-dose regimen and 0.5 for the 20-dose regimen; τ_2 is the time between the last dose on one day and the first dose on the next, i.e., 14 hr; S is the time after the last dose; N is the number of daily doses, i.e., 4 or 20; n is the dose under consideration; and m is the day under consideration.

These equations were also used to construct the curves shown in Fig. 12-3 for case 2 and 3 conditions. In case 2, the elimination rate coefficient, after oral or inhalation dosing, was decreased to 0.599 hr^{-1}; for case 3, it was decreased further to 0.0288 hr^{-1}. In all cases, the absorption rate coefficient for oral dosing was ten times the elimination rate coefficient. In every case, the administered dose over the 10-hr dosing period was the same for all routes and the uptake to the systemic circulation was assumed to be 100%.

A number of important observations are immediately apparent upon inspection of the three curves in Fig. 12-3, case 1, and the parameters in Table 12-3. First, dosing by the oral route, given as four divided doses, shows brief excursions to very high blood concentrations (to values of 21.68 μg/ml) relative to the limiting maximum levels of 4.87 and 2.43 μg/ml for the 20 divided oral doses and the vapor phase exposure, respectively. Second, it is apparent that the more the oral dose is divided, the narrower is the range between the limiting maximum and minimum values and the more closely does the blood level–time plot approximate the vapor phase curve. Third, for these regimens and rate coefficients there is no likelihood of bioaccumulation, i.e., each successive 24-hr dosing period would yield curves identical to those shown in Fig. 12-3, case 1. It is clear that, given a threshold of toxic effect, say at 10 μg/ml, only the four divided oral doses would produce levels of toxicological significance and, even in the absence of a threshold, the four divided oral doses would yield short duration effects at least 5–10 times greater than the other regimens. All three regimens yield identical areas under the blood level–time curve (Table 12-3), i.e., the *amounts* of uptake are the same.

In Fig. 12-3, case 2, in which the elimination rate coefficient is decreased by a factor of approximately ten, a steady-state level for the inhalation exposure is barely reached during the 10-hr exposure. Although the vapor exposure curve is

TABLE 12-3. Significant Parameters for the Dosing Routes and Regimens Illustrated in Fig. 12-3

Parameter	Inhalation			Oral					
				Four doses			Twenty doses		
	Case 1	Case 2	Case 3	Case 1	Case 2	Case 3	Case 1	Case 2	Case 3
Absorption rate coefficient (hr^{-1})	11.2[a]	11.2[a]	11.2[a]	46.1	5.99	0.288	46.1	5.99	0.288
$t_{1/2}$ (min)	—	—	—	0.9	6.9	144	0.9	6.9	144
Elimination rate coefficient (hr^{-1})	4.61	0.599	0.0288	4.61	0.599	0.0288	4.61	0.599	0.0288
$t_{1/2}$ (min)	9.0	69	1440	9.0	69.0	1440	9.0	69	1440
Total *AUC* (0–24 hr)[b]	24.3	186.9	3888	24.30	186.9	3888	24.30	187.0	3888
AUC (0–10 hr)[b]	23.77	55.8	1642	24.30	172.0	1665	24.30	157.4	1558
Limiting maximum concentration (μg/ml)	2.43	18.69	194.3	21.68	28.72	182.1	4.87	19.36	181.7
Limiting minimum concentration (μg/ml)	2.43	18.69	129.8	0.0003	8.96	139.1	0.69	17.49	143.1

[a]The units for the inhalation uptake rate are mg/liter per hr.
[b]The units for *AUC* are μg/hr per ml and the areas were determined for steady-state conditions.

complex over the 10-hr exposure period, the dosing regimen of 20 equal divided doses gives a blood level profile that follows its curvature over the dosing period. Limiting peak levels are also increased compared with those in case 1, but not by as much for the four equal divided doses. Again, for a xenobiotic having these toxicokinetic characteristics, there would appear to be no likelihood of bioaccumulation for these regimens and routes of exposure.

Case 3(a) in Fig. 12-3 illustrates the effects of further reducing the elimination rate coefficient by a factor of about 20. Clearly, the inhalation exposure does not yield a constant or plateau blood concentration either during the first 10-hr exposure or at any time during repeated exposure, even after steady-state conditions have been reached [case 3(b), Fig. 12-3]. Very little difference is observed in the relative blood levels either for the two oral dosing regimens or for the inhalation exposure. Over a 24-hr period, the areas under the blood level–time curve are the same, regardless of the route or regimen used.

It is apparent from the preceding examples and discussion that route extrapolation, even when pharmacokinetic parameters are known, should be considered on a case by case basis. The slower the elimination rate, the more closely will a multiple oral dosing regimen mimic an inhalation uptake; for very slow elimination rates, the number of doses administered per day will be irrelevant.

Although the respective equations are more complex, it is not too difficult to derive similar curves for substances that follow two-compartment or even more complex toxicokinetic models. An appropriate volume of distribution required for the evaluation of Eq. (8) can usually be obtained from single-dose studies. The dose D, in Eq. (8), will usually require modification by multiplying it by a suitable factor F, the fraction of the dose absorbed, or even $(1 - \phi)F$, where the function ϕ represents the fraction that does not reach the systemic circulation, as a consequence of interactions within the gastrointestinal tract, the portal circulation, or the liver.

It is important to consider that equivalent values of AUC for different routes of administration may not, in some cases, represent equivalent effects. Indeed, this question has been addressed in several publications that have considered the bioavailability and therapeutic efficacy of orally administered drug dosages (Ritschel, 1972; Withey, 1973). In the cases considered in these publications, the formulation of a drug dosage form may have affected the toxicokinetic rates of uptake, distribution, metabolism, and elimination so that the areas under the blood concentration–time curves for a single dose of several different formulations could be identical. However, the individual values of the peak blood concentration and the time to reach the peak concentration may be very different. In such cases, Ritschel (1972) proposed that the rate of bioavailability, which includes both these parameters, should be considered when assessing therapeutic efficacy. These same considerations should also be applied when assessing toxicological effects.

Percutaneous Uptake

Percutaneous uptake involves a special consideration of the skin and its permeability. Specific aspects that can affect the rate and extent of uptake of substances that come into contact with the skin appear to be the anatomical location of the area exposed to the toxicant, the physical nature of the toxicant (liquid, vapor, aerosol, or gas), lipid solubility characteristics, and molecular size of the xenobiotic. Other factors, such as solvents that radically affect skin permeability, have been well reviewed (Klaassen and Doull, 1980). Enzyme activity in the skin, particularly with respect to monooxygenase systems, has also been considered (Ullrich *et al.,* 1977).

Comparative Uptake Studies: Inhalation and Percutaneous

Studies comparing uptake via the skin with other routes are few. Because of its industrial importance, the percutaneous uptake of styrene has been compared with pulmonary uptake both in the rat (Shugaev, 1969) and in humans (Rihimaki and Pfaffli, 1978). In the former study, the tails of rats were immersed in pure liquid styrene for 1 hr and inhalation of the vapor by the test animals was carefully avoided. Concentrations of styrene in the liver and brain were between 50 and 70% of the concentrations found in these organs after a few hours of inhalation exposure to a vapor concentration of styrene of 11.8 mg/liter. In the second study, Rihimaki and Pfaffli (1978) exposed the total body area of human volunteers, equipped with full-face respirators to prevent uptake by the inhalation route, to 600 ppm of styrene vapor for 3.5 hr. They found that the percutaneous uptake was comparable to an inhalation exposure of approximately 20 ppm of styrene vapor for the same period.

Since comparisons of the uptake by the percutaneous and inhalation routes have attracted some attention as a consequence of their relevance in the assessment of total industrial exposure, it is instructive to review the rationale and calculations used in these comparisons. Rihimaki and Pfaffli multiplied the ventilation rate (10 liters/min during light work) by the length of exposure and the retained fraction (inhaled air concentration minus the exhaled air concentration). An average pulmonary retention of 60% (Astrand, 1975) was used in the case of styrene. These authors mention that alveolar air concentrations used in these calculations will be smaller by a factor of 1.5 due to the diluting effect of pulmonary dead space (Sherwood, 1972). The percutaneous uptake was assessed by adding the total amount of metabolites excreted in the urine (in micromoles) to the amount excreted in expired air. The percutaneous penetration rate (in micromoles per square centimeter per hour) was also calculated by dividing the total uptake by the product of the total exposed surface area of the subject (approximately 19,000 cm^2) and the duration of exposure.

Comparisons of the percutaneous and inhalation uptake for other solvents were also reported by Rihimaki and Pfaffli (1978). At an atmospheric concentration of either 300 or 600 ppm, the uptake of xylene, styrene, or tetrachloroethylene was approximately 2% of the amount estimated to be absorbed by the respiratory route. The percutaneous uptake of toluene and 1,1,1-trichloroethane was 0.9 and 0.08%, respectively, of that from the respiratory route at 600 ppm. They also commented that xylene vapor at 600 ppm displays ten times more efficient skin penetration than did pure liquid xylene when corrections for the differences in the total surface area exposed and concentration are made. In another study, the uptake of 1,1,1-trichloroethane was measured after both hands (where the skin is considered to be relatively more permeable than the skin on other parts of the body) were immersed in pure liquid 11 times for 10-min periods during a workday. The investigators reported that the uptake corresponded to a 2-hr inhalation exposure to an atmospheric concentration of 10–20 ppm (Fukabori *et al.*, 1976). After both hands were immersed in pure liquid styrene for 15 min, the percutaneous absorption was reported to be roughly the same as the pulmonary absorption observed after exposure to 100 ppm for the same period (Engstrom *et al.*, 1977).

In the foregoing examples, the comparisons of uptake by the percutaneous and pulmonary routes were made on the basis of calculated amounts. No consideration was given either to the magnitude of blood concentrations or to the rate processes involved. Gases and vapors penetrate the skin and appear in the systemic circulation with a shorter lag period than is observed with liquids or solutions (Scheuplein and Blank, 1971). Whereas the skin exposure to 1,1,1-trichloroethane gave blood concentrations that continued to rise during the entire exposure period, blood levels rose very rapidly to yield a steady-state concentration within minutes of initiating an inhalation exposure (Rihimaki and Pfaffli, 1978). A delay of several minutes was observed before detectable blood levels were observed after the percutaneous absorption of liquid toluene (Guillemin *et al.*, 1974) and xylene (Engstrom *et al.*, 1977).

Areas under Blood Level–Time Curves

Uptake, based on the area under the blood concentration–time curve following administration of the dose via the pulmonary route to rats, has been used to compare the bioavailability of vinyl chloride monomer from an aqueous solution administered intragastrically (Withey and Collins, 1976). With the knowledge that an inhalation exposure yields an equilibrium blood level that is proportional to the exposure concentration, the *AUC* for an inhalation exposure approximates the product of the equilibrium blood concentration and the exposure time. If the relationship derived for the *AUC* and magnitude of the dose holds for an intragastrically administered dose of vinyl chloride dissolved in water, equivalent *AUC* values would be generated by administering an intragastric dose of 0.9 mg of vinyl chloride as an aqueous solution or by exposing a rat to an atmospheric concentration of 1.97 ppm for 24 hr. It is apparent from the nature of the toxicokinetic behavior of vinyl chloride that the inhalation insult would arise from a prolonged low blood concentration, whereas the single bolus dose given as an aqueous solution in this example would yield relatively higher blood levels for only a matter of minutes. Clearly, this example involves an assumption, in terms of the toxicological insult, that may not hold in the case of vinyl chloride, since it has been demonstrated that alternative metabolic pathways exist for this compound, some of which are saturable (Hefner *et al.*, 1975). Since the concentration of vinyl chloride increases at organ sites where metabolizing enzymes interact with the substrate, the amounts of the metabolite produced will change and so will the toxicological effect.

Evidently, the American Conference of Governmental Industrial Hygienists has recognized that the methodology using the product of the time-weighted-average vapor concentration and the duration of exposure in order to calculate the threshold limit value (TLV) has its limitations. Special considerations are now given to the setting of limits for short-term exposures to high concentrations.

SUMMARY

The methods discussed in this chapter for the quantitation of toxicological insult for the same compound administered by different routes are relatively straightforward when equating amounts that reach the systemic circulation. Even when extensive metabolism of a toxicant is involved, it is possible to calculate equivalent doses that will give rise to similar toxic effects for compounds administered by different routes, provided that sufficient information is available to allow the calculation of the fraction of the administered dose that is metabolized prior to the toxicant reaching the systemic circulation. However, as a consequence of the pharmacokinetics involved, it is emphasized that, in many cases, due consideration must be given to the temporal relationship of the concentration of the toxicant in the systemic circulation.

REFERENCES

American Conference of Governmental Industrial Hygienists. 1962. Threshold limit values for 1962. *Am. Ind. Hyg. Assoc. J.* **23**:419–423.

American Conference of Governmental Industrial Hygienists. 1971. *Documentation of the Threshold Limit Values for Substances in Workroom Air,* 3rd ed. American Conference of Governmental Industrial Hygienists, Cincinnati, Ohio.

American Conference of Governmental Industrial Hygienists. 1977. *TLV (Threshold Limit) Values for Chemical Substances and Physical Agents in the Workroom Environment with Intended Changes for 1977.* American Conference of Governmental Industrial Hygienists, Cincinnati, Ohio.

Astrand, I. 1975. Uptake of solvents in the blood and tissues of man: A review. *Scand. J. Work Environ. Health* **1**:199–218.

Barr, W. H. 1968. Principles of biopharmaceutics. *Am. J. Pharmacol. Ed.* **32**:958–981.

Barr, W. H. 1969. Factors involved in the assessment of systemic or biologic availability of drug products. *Drug Inf. Bull.* **3**:27–45.

Boyes, R. N., H. J. Adams, and B. R. Duce, 1970. Oral absorption and disposition kinetics of lidocaine hydrochloride in dogs. *J. Pharmacol. Exp. Ther.* **174**:1–8.

Cassidy, M. K., and J. B. Houston. 1980. Protective role of intestinal and pulmonary enzymes against environmental phenols. *Br. J. Pharmacol.* **69**:316.

Cornish, H. H. 1980. Solvents and vapors. Pp. 468–496 in J. Doull, C. D. Klaassen, and M. O. Amdur, eds. *Casarett and Doull's Toxicology,* 2nd ed. Macmillan, New York.

Dollery, C. T., D. S. Davies, and M. E. Conolly. 1971. Differences in the metabolism of drugs depending upon their route of administration. *Ann. N. Y. Acad. Sci.* **179**:108–114.

Engstrom, K., K. Husman, and V. Riihimaki. 1977. Percutaneous absorption of *m*-xylene in man. *Int. Arch. Occup. Environ. Health* **39**:181–189.

Fukabori, S., K. Nakaaki, J. Yonemoto, and O. Tada. 1976. On the cutaneous absorption of methyl chloroform. *J. Sci. Labour* **52**:67–80.

Gibaldi, M., and D. Perrier. 1975. *Pharmacokinetics.* Marcel Dekker, New York.

Gladtke, E., and H. M. von Hattingberg. 1979. *Pharmacokinetics: An Introduction.* Springer-Verlag, Berlin.

Goldstein, A., L. Aronow, and S. M. Kalman. 1974. *Principles of Drug Action,* 2nd ed. Wiley, New York.

Guillemin, M., J. C. Murset, M. Lob, and J. Riquez. 1974. Simple method to determine the efficiency of a cream used for skin protection against solvents. *Br. J. Ind. Med.* **31**:310–316.

Hefner, R. E., P. G. Watanabe, and P. Gehring. 1975. Preliminary studies of the fate of vinylchloride monomer in rats. *Ann. N. Y. Acad. Sci.* **246**:135–138.

Himmelstein, K. J., and R. J. Lutz. 1979. A review of the applications of physiologically based pharmacokinetic modeling. *J. Pharmacokin. Biopharm.* **7**:127–145.

Klaassen, C. D. 1980. Absorption, distribution and excretion of toxicants. Pp. 28–55 in J. Doull, C. D. Klaassen, and M. O. Amdur, eds. *Casarett and Doull's Toxicology,* 2nd ed. Macmillan, New York.

Klaassen, C. D., and J. Doull. 1980. Evaluation of safety: Toxicologic evaluation. Pp. 11–27 in J. Doull, C. D. Klaassen, and M. O. Amdur, eds. *Casarett and Doull's Toxicology,* 2nd ed. Macmillan, New York.

Leschke, E. 1943. *Clinical Toxicology.* W. Wood, Baltimore, Maryland.

Levy, G., and M. Gibaldi. 1975. Pharmacokinetics. Pp. 1–34 in *Handbook of Experimental Pharmacology,* Volume 28, Part 3. Springer-Verlag, New York.

Loo, J. K. C., and S. Riegelman. 1968. New methods for calculating the intrinsic rate of drugs. *J. Pharmacol. Sci.* **57**:918–928.

Martin, E. W., S. F. Alexander, D. J. Farage, and W. E. Hassan. 1971. *Hazards of Medication.* Lippincott, Philadelphia.

McNamara, P. J., J. T. Slattery, M. Gibaldi, and G. Levy. 1979. Accumulation kinetics of drugs with nonlinear plasma protein and tissue binding characteristics. *J. Pharmacokin. Biopharm.* **7**:397–405.

National Research Council. 1980. *Principles of Toxicological Interactions Associated with Multiple Chemical Exposures.* National Academy Press, Washington, D.C.

Needleman, P., S. Lang, and E. M. Johnson, Jr. 1972. Organic nitrates: Relationship between biotransformation and rational angina pectoris therapy. *J. Pharmacol. Exp. Ther.* **181**:489–497.

Nelson, W. E. 1953. *Textbook of Pediatrics,* 5th ed. Saunders, Philadelphia.

Perucca, E., G. Gatti, G. M. Frigo, and A. Crema. 1978. Pharmacokinetics of valproic acid after oral and intravenous administration. *Br. J. Pharmacol.* **5**:313–318.

Rihimaki, V., and P. Pfaffli. 1978. Percutaneous absorption of solvent vapours in man. *Scand. J. Work Environ. Health* **4**:73–85.

Ritschel, W. A. 1972. Bioavailability in the clinical evaluation of drugs. *Drug Intell. Clin. Pharm.* **6**:246–256.

Rowland, J., S. Riegelman, P. A. Harris, S. D. Sholkoff, and E. J. Eyring. 1967. Kinetics of acetyl salicyclic acid disposition in man. *Nature* **215**:413–414.

Schanker, L. S. 1971. Drug absorption. Pp. 22–43 in B. N. La Du, H. G. Mandel, and E. L. Way, eds. *Fundamentals of Drug Metabolism and Drug Disposition.* Williams & Wilkins, Baltimore, Maryland.

Scheuplein, R. J., and I. H. Blank. 1971. Permeability of the skin. *Physiol. Rev.* **51**:702–747.

Sherwood, R. J. 1972. Comparative methods of biological monitoring of benzene exposures. Pp. 29–52 in *Proceedings of the Third Conference on Environmental Toxicology.* AMRL-TR 72-130. Aerospace Medical Research Laboratory, Wright-Patterson Air Force Base, Dayton, Ohio.

Shugaev, B. B. 1969. Concentration of hydrocarbons in tissues as a measure of toxicity. *Arch. Environ. Health* **18**:878–882.

Sollman, T. 1953. *A Manual of Pharmacology,* 7th ed. Saunders, Philadelphia.

Stokinger, H. E., and R. L. Woodward. 1958. Toxicologic methods for establishing drinking water standards. *J. Am. Water Works Assoc.* **50**:515–529.

Ullrich, V., A. Hildebrandt, I. Roots, R. W. Estabrook, and A. H. Conney, eds. 1977. *Microsomes and Drug Oxidants.* Pergamon Press, New York.

U.S. Environmental Protection Agency. 1979. EPA Water Quality Criteria: Request for Comments, Part V. March 15. *Fed. Reg.* **43:**15926–15981.

Wagner, J. G. 1971. *Biopharmaceutics and Relevant Pharmacokinetics.* Drug Intelligence, Hamilton, Illinois.

Wagner, J. G., and E. Nelson. 1963. Percent absorbed time plots derived from blood level and/or urinary excretion data. *J. Pharm. Sci.* **52:**610–611.

Wagner, J. G., J. I. Northam, C. D. Alway, and O. S. Carpenter. 1965. Blood levels of drug at the equilibrium state after multiple dosing. *Nature* **207:**1301–1302.

Withey, J. R. 1973. Bioavailability and therapeutic efficacy. *Rev. Can. Biol.* **32:**21–30.

Withey, J. R. 1976. Pharmacodynamics and uptake of vinyl chloride monomer administered by various routes to rats. *J. Toxicol. Environ. Health* **1:**381–394.

Withey, J. R. 1978. Pharmacokinetic principles. Pp. 97–118 in *Proceedings of the First International Congress on Toxicology.* Academic Press, New York.

Withey, J. R., and B. T. Collins. 1976. A statistical assessment of the quantitative uptake of vinyl chloride monomer from aqueous solution. *J. Toxicol. Environ. Health* **2:**311–321.

Withey, J. R., and P. G. Collins. 1979. The distribution and pharmacokinetics of styrene monomer in rats by the pulmonary route. *J. Environ. Pathol. Toxicol.* **2:**1329–1342.

13

Application of *in Vivo* Data on Chemical–Biological Interactions

Larry S. Andrews, Stephen L. Longacre, and Robert Snyder

Chemical–biological (CB) interactions are governed by both the structure of the chemical and the properties of the biological system. The ability of a drug or chemical to be absorbed from the lung, skin, or gastrointestinal tract and to enter the body is a CB interaction. Unless the chemical structure is fat-soluble enough to dissolve in the lipoprotein membrane, or small enough to pass through pores, or it mimics an endogenous substance that is actively transported, the chemical cannot move across membranes or skin. Ultimately, for a chemical to exert an effect it must be absorbed and enter the blood. Once in the blood, a chemical's further disposition is governed by its pharmacokinetic characteristics, i.e., its distribution and elimination from body compartments.

One of the first CB interactions encountered in the body is reversible binding to plasma proteins—an interaction that can influence transport to other tissues. The partitioning of the chemical into body tissues will involve not only the solubility of the chemical, but also blood supply to and the clearance of the chemical from the tissue. These processes are generic to animals, although differences between species certainly exist. Until a comprehensive data base for these general CB interactions is developed, they will be of little use in predicting the toxicity of the chemical or in estimating risk. Few conclusions about the toxicological poten-

LARRY S. ANDREWS • ARCO Chemical Co., Newtown Square, Pennsylvania 19073. **STEPHEN L. LONGACRE** • Toxicology Department, Rohm & Haas Co., Spring House, Pennsylvania 19477. **ROBERT SNYDER** • Rutgers University College of Pharmacy, Busch Campus, Piscataway, New Jersey 08854. *Present address of L. S. A.:* Agricultural Research Division, American Cyanamid Co., Princeton, New Jersey 08540.

tial of a chemical may be drawn with certainty simply on the basis of observations that a chemical is absorbed, binds to plasma proteins, and is distributed to a variety of tissues.

A second class of CB interactions includes drug–receptor interactions. For example, a chemical with cholinomimetic activity will interact with the cholinergic receptor by a specific drug–receptor binding. Both chemical structure and the spatial configuration of the receptor confer ability to bind to the receptor and are quite specific. Moreover, animals respond predictably when exposed to a substance with cholinergic activity. Knowledge that a chemical binds to cholinergic receptors enables one to predict reliably toxic signs of overexposure.

Many other specific toxicological CB interactions have been identified in recent years. Interactions that have been most closely associated with chemically induced toxicity and risk assessment involve the metabolism of the chemical and the concept of detoxification. The cytochrome P450-dependent microsomal mixed-function oxidase is an enzyme system responsible for the metabolism of many chemicals (Schenkman and Kupfer, 1982). Detoxification often results from the introduction of a polar moiety into a chemical structure followed by a more rapid excretion from the body relative to the chemical itself. Not infrequently, mixed-function oxidase-catalyzed reactions produce chemically reactive electrophiles termed "biological reactive intermediates" (BRIs), which react with cellular materials such as deoxyribonucleic acid (DNA), protein, and lipid. The metabolism of chemicals to electrophiles has been termed biological activation, and the subsequent reaction of the electrophiles with nucleophilic groups on cellular molecules may represent an increase in toxicity rather than detoxification.

Specific CB interactions have been widely discussed as indices for interpreting the toxicological potential and risk of a chemical. In the following sections, we consider a number of specific CB interactions and discuss their value in assessing the toxicity and risk associated with exposure to a chemical.

METABOLISM, DETOXIFICATION, AND BIOACTIVATION

Most chemicals that are absorbed and cause toxicity are hydrophobic and would remain in the body indefinitely were it not for phase I and phase II metabolic enzyme systems. Phase I metabolic enzymes catalyze the introduction of polar moieties into hydrophobic molecules, thereby facilitating their ultimate excretion from the body (Williams, 1959; Williams and Parke, 1964). Phase II metabolic reactions are synthetic processes that introduce further polarity into the chemical so that it may be more readily excreted (Dutton, 1966; Parke, 1968). The products of phase I and phase II metabolism are termed metabolites.

Many phase I reactions are mediated by cytochrome P450-containing mixed-function oxidase enzyme systems, which have been extensively reviewed by Sato and Omura (1978) and Schenkman and Kupfer (1982). Mixed-function oxidases are a family of enzymes, consisting primarily of the hemoprotein cytochrome P450. They are located in the smooth endoplasmic reticulum of liver as well as in most other tissues. These enzymes catalyze the oxidation or reduction of a variety of chemical structures. A postulate mechanism for mixed-function oxidative reactions can be shown as follows:

$$\text{Chemical}_{\text{red}} + \text{Cyt P450}_{\text{ox}} \quad \text{Chemical–Cyt P450}_{\text{ox}}$$
$$\text{NADPH} + \text{H}^+ \rightarrow \text{Chemical–Cyt P450}_{\text{red}}$$
$$\text{P450 reductase}$$
$$\text{Chemical–Cyt P450}_{\text{red}} + \text{O}_2 \rightarrow \text{Chemical}_{\text{ox}} + \text{Cyt P450}_{\text{ox}} + \text{H}_2\text{O}$$

The chemical is reversibly bound by the oxidized form of cytochrome P450. The resultant complex is next reduced by an electron supplied by NADPH–cytochrome P450 reductase. The reduced chemical–P450 complex binds a molecule of oxygen, and upon addition of further reducing equivalents, one atom of oxygen is introduced into the chemical and the other is reduced to water. Upon release of the oxidized product, cytochrome P450 returns to its oxidized state and is again capable of binding another drug molecule. As a result of this process, oxygen is introduced into any chemical that contains a favorably positioned C–H, N–H, S–H, or C–X (X = halogen) bond.

Mixed-function oxidases are inducible enzyme systems that can be stimulated by prior exposure of animals to chemicals (Conney, 1967; Snyder and Remmer, 1979). Many chemicals, such as benzene, are capable of increasing their own metabolism (Gonasun *et al.,* 1973; Norpoth *et al.,* 1974; Snyder *et al.,* 1967). Drugs like phenobarbital and other barbituates or environmental chemicals such as polychlorinated biphenyls can increase not only their own metabolism, but also the metabolism of a variety of other chemical structures. It has become increasingly evident that there is an entire family of cytochrome P450 (Guengerich, 1977; Johnson, 1979). Not only do the different forms of cytochrome P450 catalyze the production of varying amounts of metabolities, but inducing agents can act to change the spectrum of metabolites produced (Frommer *et al.,* 1974; Holder *et al.,* 1974; Zampaglione *et al.,* 1973). Each of these metabolites will have its own toxicological properties and potency.

Even as the activity of mixed-function oxidases can be elevated by inducing agents, the presence of one chemical can inhibit the metabolism of another chemical, i.e., competitive inhibition. For example, toluene competitively inhibits the oxidative metabolism of benzene by competing with available catalytic sites on cytochrome P450 (Andrews *et al.,* 1977; Sato and Nakajima, 1979). Heavy metals, such as lead, in the environment can decrease levels of cytochrome P450 by reducing levels of heme available for synthesis of this hemoprotein (Alvares, 1978). The net effect of heavy metals on xenobiotic metabolism is a noncompetitive inhibition.

Phase II biotransformations are synthetic reactions in which sulfate, glucuronate, acetate, glutathione, or other moieties are added onto the chemical itself or onto phase I metabolites. These reactions are catalyzed by families of enzymes termed sulfo-, glucuronyl-, acetyl-, and glutathione-transferases. Isozymes of these enzymes react preferentially with different families of substrates. Chemically mediated induction or inhibition may modify the yield of metabolites from phase II reactions.

Any theoretical prediction of the toxicity of a chemical would have to utilize a large volume of metabolic and pharmacokinetic information on the chemical and each of its metabolites. Although metabolism is frequently associated with elimination of a toxic chemical from the body, i.e., detoxification, there are many instances for which the introduction of a polar moiety into the molecule increases the toxicity observed (Snyder *et al.,* 1982). The conversion of a chemical into a more toxic form is termed biological activation and can be mediated by both phase I and phase II enzyme systems.

Figure 13-1 illustrates important phase I reactions in the biological activation of benzo(*a*)pyrene to an electrophilic metabolite. The metabolism of benze(*a*)pyrene has been reviewed in detail by Yang *et al.* (1978) and Estabrook *et al.* (1978). In the first step, oxygen is inserted across the 7,8 position to form the 7,8 epoxide. Epoxide hydrolase may then catalyze the formation of the 7,8 dihydrodiol. The electrophilic metabolite (7,8-diol-9,10-epoxide), which reacts with and is covalently bound to cellular nucleophiles, is produced by an epoxidation reaction across the 9,10 position.

Figure 13-1 also shows that phase II reactions can produce BRIs. The carcinogen 2-acetylaminofluorene (2-AAF) undergoes *N*-hydroxylation to *N*-hydroxy-2-AAF. In the activation step, sulfotransferases, glucuronyltransferases, and acetyltransferases can catalyze production of a reactive *N-O*-ester. The *N-O*-ester spontaneously breaks down to yield a highly electrophilic nitrene species that can react with nucleophilic groups on DNA, ribonucleic acid (RNA), and protein (Barry *et al.,* 1969; Lin *et al.,* 1975; J. A. Miller and Miller, 1977; E. C. Miller and Miller, 1982; Poirier *et al.,* 1967).

The examples of biological activation given in Figure 13-1 are but two of many that have been identified. For benzo(*a*)pyrene and 2-AAF, the toxic endpoints associated with the biological activation are mutation and cancer. Biological activation has also been associated with a variety of other toxicological endpoints, including damage to the liver, lung, kidney, peripheral and central nervous system, and bone marrow (Gillette, 1974a,b; Ingelman-Sundberg, 1980; Jollow *et al.,* 1977; J. A. Miller, 1970; Mitchell and Jollow, 1975; Nebert and Jensen, 1979; Sims, 1980; Synder *et al.,* 1980). Figure 13-2 illustrates a number of postulated reactive intermediates and the toxicological endpoint with which they are associated.

Metabolism is a specific CB interaction, which must be considered to have the dual nature of detoxification and activation. The rate of metabolism of a chemical is not indicative of its toxicological potential. If a chemical is activated to an electrophilic species that reacts with nucleophilic sites on cells, it may only be concluded that the potential for toxicity exists.

USE OF COVALENT BINDING DATA IN INTERPRETING RESULTS OF TOXICITY STUDIES

A considerable amount of data on BRIs and their implications for toxicity have been amassed since the pioneering studies of Boyland (1950) and E. C.

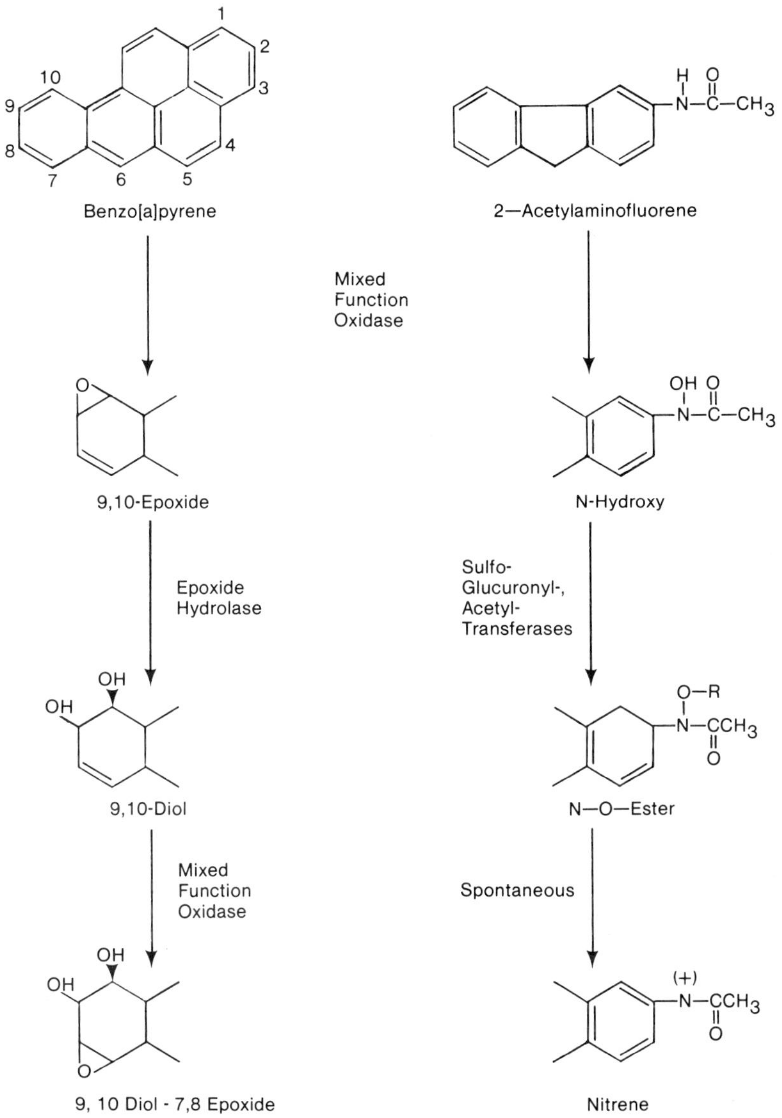

FIGURE 13-1. Phase I reactions in the biological activation of benzo(*a*)pyrene to an electrophilic metabolite.

Miller and J. A. Miller (1947). Notable is the work of Gillette, Mitchell, and Jollow, who studied the relationship between covlent binding and the hepatoxicity of bromobenzene and acetaminophen. Today there is evidence relating covalent binding of BRIs to such diverse forms of toxicity as cellular necrosis, mutagenesis, carcinogenesis, teratogenesis, hypersensitivity, and blood dyscrasias. These studies have been extensively reviewed and were the subject of a symposium (Snyder *et al.*, 1982).

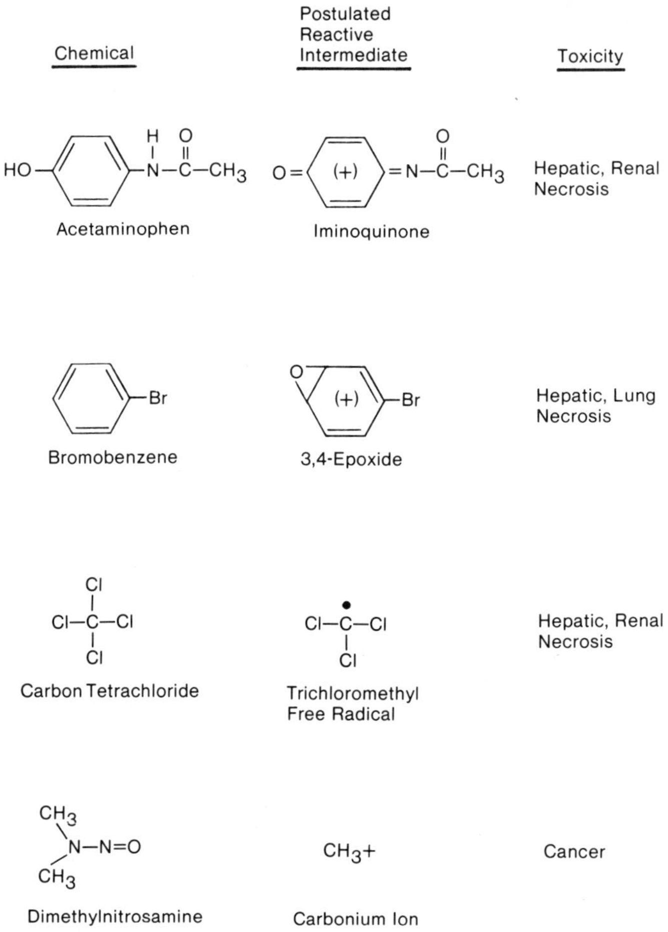

FIGURE 13-2. Postulated reactive intermediates and associated toxicological endpoints. [From Jollow *et al.* (1977).]

For purpose of this discussion, covalent binding and toxicity other than carcinogenicity are described first. Pioneering studies of hepatoxicity caused by bromobenzene (Mitchell *et al.*, 1971; Zampaglione *et al.*, 1973), acetaminophen (Jollow *et al.*, 1973; Mitchell *et al.*, 1973), and furosemide (Thorgeirsson *et al.*, 1976) showed that necrosis follows, and may stem from, excessive covalent binding to protein.

Acetaminophen (4-hydroxyacetanilide) is a good example of a xenobiotic compound for which there is a good correlation between the degree of cytotoxicity and the amount of covalently bound metabolites observed in the target tissue. It is a safe and effective analgesic and antipyretic when used at therapeutic doses. In overdose quantities, however, acetaminophen causes fulminant heptic necrosis in both laboratory animals and in humans (Mitchell *et al.*, 1973; Prescott *et al.*, 1971; Proudfoot and Wright, 1970). Acetaminophen-induced hepatotoxicity probably results from the covalent binding of a reactive intermediate of acetaminophen to hepatic macromolecules (Jollow *et al.*, 1973; Mitchell and Jollow, 1974).

The pathways of acetaminophen metabolism are shown in Fig. 13-3. The major metabolic pathways are those leading to the nontoxic glucuronide and sulfate conjugates. A minor pathway, mediated by mixed-function oxidase, converts acetaminophen to a chemically reactive intermediate that is preferentially and quantitatively detoxified by conjugation with hepatic glutathione. The glutathione conjugate undergoes further metabolism and appears in the urine as the mercapturic acid conjugate. Thus, as long as adequate amounts of hepatic glutathione are present, the reactive intermediate will be detoxified and will not covalently bind to tissue macromolecules. In the overdose situation, however, operation of the mixed-function oxidase-mediated pathway acts to deplete stores of hepatic glutathione. With this protective mechanism diminished, a reactive intermediate,

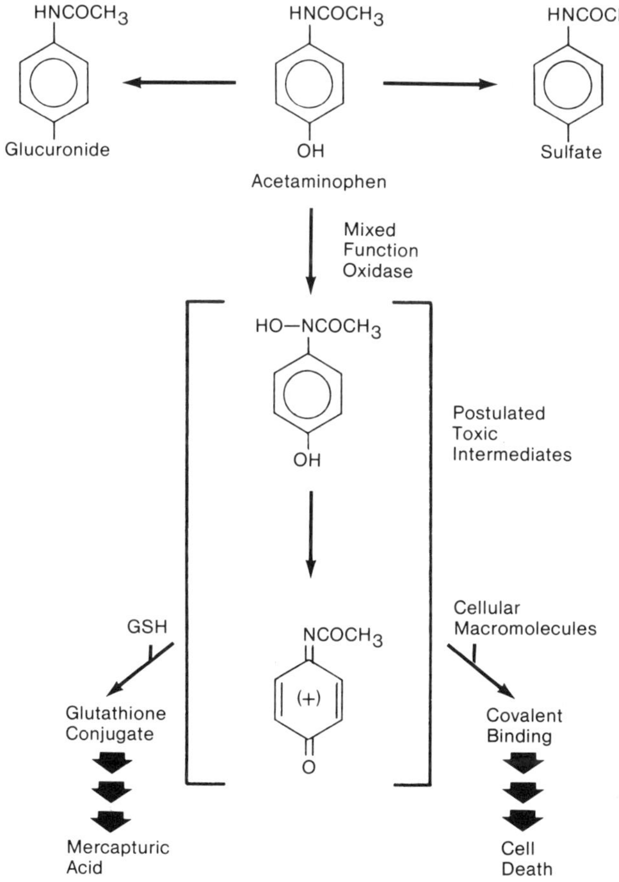

FIGURE 13-3. Pathways of acetaminophen metabolism.

presumably benzoiminoquinone (Fig. 13-2), covalently binds to hepatic macromolecules—an action believed to initiate events leading to cell death.

Several lines of evidence support the contention that acetaminophen-induced hepatotoxicity is mediated by the covalent binding of acetaminophen metabolites to hepatic macromolecules. First, there is a very good correlation between the amount of covalently bound acetaminophen metabolites and the extent of acetaminophen-induced hepatic necrosis (Potter *et al.,* 1974). Apparently there is a dose threshold for both toxicity and covalent binding. This is probably due to the activity of the glutathione detoxification pathway. Second, following hepatotoxic doses of acetaminophen, the highest concentrations of covalently bound metabolites were observed in liver, the target tissue, whereas negligible amounts were observed in heart and muscle—tissues showing no adverse effects toward acetaminophen (Jollow *et al.,* 1973). Significant covalent binding was found in the kidney—a tissue that also exhibits toxicity from acetaminophen. Third, pretreatment regimens that alter the rate of acetaminophen metabolism also alter both the extent of hepatic necrosis and the amount of covalent binding in a similar fashion. For example, pretreatment of mice with either phenobarbital or 3-methylcholanthrene potentiates acetaminophen-induced hepatic necrosis and markedly increases the amount of covalent binding observed in liver. Conversely, pretreatment of mice with either piperonyl butoxide or cobaltous chloride inhibited acetaminophen-induced hepatic necrosis and decreased the amount of covalent binding in liver (Mitchell *et al.,* 1973). More recent studies by Jollow and co-workers have shown that pretreatment regimens that modify the rates of acetaminophen glucuronide and sulfate conjugate formation alter both the extent of acetaminophen-induced hepatic necrosis and the amount of covalent binding in liver (Jollow *et al.,* 1982). For example, pretreatment of hamsters with galactosamine, which inhibits acetaminophen glucuronide formation, was shown to both potentiate acetaminophen-induced hepatic necrosis and increase the amount of covalent binding observed (Smith and Jollow, 1976). Apparently, inhibition of acetaminophen glucuronide formation by galactosamine resulted in a greater fraction of the dose of acetaminophen being metabolized through the cytochrome P450-mediated toxic pathway, which led to a potentiation of toxicity.

In contrast, 2-hydroxyacetanilide (2-HAA) and 3-hydroxyacetanilide (3-HAA) covalently bind to hamster liver protein in amounts similar to those produced by toxic doses of acetaminophen, but do not produce hepatic necrosis (Roberts and Jollow, 1978, 1979a,b, 1980). Covalent binding resulting from either 2-HAA or 3-HAA was apparently mediated by the mixed-function oxidase in an analogous fashion to acetaminophen, since pretreatment of hamsters with cobaltous chloride greatly reduced the amount of covalent binding in liver and *in vitro* studies indicated that activation of 2-HAA and 3-HAA to their respective reactive intermediates by hepatic microsomes was dependent on NADPH and O_2 (Roberts and Jollow, 1979a, 1980). Thus, studies on acetaminophen have shown a good correlation between total covalent binding and toxicity, but such a relationship is not evident for 2-HAA or 3-HAA. Collectively, these studies indicate that determination of total macromolecular covalent binding, without regard to characterization of the nature of the binding, is not adequate for assessing the degree of chemically induced cytotoxicity.

Recent studies have explored the relationship between covalent binding and chemically induced hematotoxicity. It has been known for many years that exposure to benzene can cause serious blood dyscrasis (Snyder and Kocsis, 1975). Although the exact mechanism of benzene-induced hematotoxicity is not known, it is widely believed that a metabolite of benzene, perhaps a BRI, is responsible for the observed toxicity. A metabolite of benzene has been shown to bind covalently to mouse liver and bone marrow tissue residues in a dose-dependent manner (Snyder *et al.*, 1978a) as well as to rat liver DNA (Lutz and Schlatter, 1977), an event consistent with benzene's suspected leukemogenic actions (Snyder *et al.*, 1978b). In addition, Longacre *et al.* (1981) have shown that in inbred mice, decreased susceptibility to benzene toxicity was correlated with a decreased amount of covalently bound benzene metabolites in bone marrow and several other tissues. Furthermore, Sammett *et al.* (1979) have shown that partially hepatectomized rats were protected from benzene toxicity, and smaller amounts of covalently bound metabolites were found in bone marrow of these animals compared to sham-operated controls.

Nevertheless, in all the studies cited above, the liver always contained an approximately tenfold greater concentration of covalently bound benzene metabolites than did bone marrow and did not shown any adverse effects from benzene. The fact that bone marrow, the organ expressing benzene toxicity, has been shown to contain less covalently bound metabolites than liver might be explained by data suggesting that benzene is most harmful to rapidly proliferating cells (D'Souza *et al.*, 1979; Sammett *et al.*, 1979; Snyder *et al.*, 1978a). Since the majority of the cells in the liver are in the G_0 phase and are not normally undergoing cell division, it is reasonable that benzene does not produce liver toxicity. Bone marrow, which is a rapidly proliferating tissue, is sensitive to benzene, but displays minimal covalent binding. If a BRI is responsible for the hematotoxicity of benzene, then the BRI may react with a limited number of highly sensitive target molecules.

Covalent binding of chemicals to DNA appears to be of significant value for predicting genotoxic responses. The following studies explore some of the issues regarding covalent binding to DNA. However, these issues must be better defined before we can begin to use these methods in lieu of extensive testing programs currently in vogue.

Bolt and co-workers (1982) assessed the data on a series of halogenated ethylenes with respect to covalent binding to protein and nucleic acids, bacterial mutagenicity after metabolic activation, and carcinogenicity. Among the compounds for which covalent binding data were available, vinyl chloride was shown to bind covalently to protein and nucleic acids, both *in vitro* and *in vivo,* and was both mutagenic and carcinogenic. Vinylidene chloride, which was found to bind to protein *in vivo,* was mutagenic to bacteria after activation, but did not demonstrate a high degree of alkylation of nucleic acids. Reports of vinylidene chloride-induced carcinogenicity in animal bioassays have been conflicting. Similarly, the data for trichloroethylene show that it binds to protein both *in vitro* and *in vivo,* but the covalent binding to nucleic acids is extremely low *in vivo,* and the carcinogenicity and mutagenicity data are both equivocal, since the compound was active in mice at high doses, but not in rats. Perchloroethylene, which was demonstrated to bind to protein both *in vivo* and *in vitro,* did not bind to nucleic acids *in vivo* and was not mutagenic. As was true for trichloroethylene, data for carcinogenicity are equivocal. Finally, vinyl bromide, which was bound to protein *in vitro* and to nucleic acids both *in vitro* and *in vivo,* was mutagenic and also carcinogenic.

All these compounds are metabolized via the formation of epoxides, i.e., halooxiranes. The extent to which adducts are formed and either mutagenesis or carcinogenesis is observed is initially a function of the stability of the epoxide and its ability to reach DNA prior to inactivation rather than the rate of metabolic activation. Thus, among vinyl chloride (vc), vinyl fluoride (vf), and vinyl bromide (vb), the relative rates of metabolism were vc > vb > vf, whereas the relative rates of formation of preneoplastic foci produced over a 10-week exposure period in rats were vc > vf > vb. By the same token, halogenated ethylenes containing more than one halogen are less oncogenic and their oxirane derivatives are much less stable than those of the monohalogenated compounds. Nevertheless, a high degree of chemical reactivity is required for nonenzymatic covalent binding to DNA and, although the monohalooxiranes are rather highly reactive, they appear to retain sufficient stability to reach the target site for carcinogenesis.

It is well established that BRIs can covalently bind to DNA and that these interactions may cause miscoding, which can cause mutagenic effects (Lawley, 1980). Among sites at which covalent binding has been observed are O-6, N-7, C-8, and N-3 of guanine; N-1, N-3, N-7, and the 6-amino position of adenine; N-1 and N-3 of cytosine; and O-4 and possibly C-6 of thymine (Farber, 1978; Lawley, 1980). It has been strongly argued that the N-7 position of guanine is a major binding site for metabolically activated aflatoxin B_1 (J. A. Miller and Miller, 1977) and dialkyl nitrosamines (Magee, 1980). Lawley (1980) has reported that the greatest perturbation of DNA results from alkylation at the O-6 position of guanine and is the most likely site for development of carcinogenesis. A comparison of several alkylating agents studied in mice demonstrated a relationship between the degree of carcinogenicity and binding to O-6 of guanine. Furthermore, once binding occurred, there was resistance to DNA repair.

Blackburn *et al.* (1981) described the relationship between the levels of carcinogen–protein complexes and histological changes that occur during carcinogenesis in rat liver. In this study, after rats were fed ^{14}C-labeled 2-AAF for up to 30 weeks, premalignant liver lesions were observed. Periodically, rats were killed and liver cytosolic proteins were separated into molecular weight ranges by gel filtration. Initially, the carcinogen was bound to a relatively small protein, whereas with continued treatment, the carcinogen was bound to a protein about ten times larger. Similar effects were seen with the hepatocarcinogens 3-methyl-4-dimethylaminoazobenzene and ethionine, but not with the noncarcinogenic analogs fluorene and aminoazobenzene. The relationship between the two proteins does not appear to be direct, since they do not immunologically cross-react. The loss of both the small bound and free proteins during liver carcinogenesis induced by the three carcinogens studied could reflect the inactivation and loss of a growth regulator.

Poirier *et al.* (1982) measured both acetylated and deacetylated guanine C-8 adducts in liver and kidney DNA of rats fed radiolabeled 2-AAF for 3, 7, or 28 days. For each group, C-8 adducts were measured in liver and kidney periodically after 2-AAF treatment was terminated. After feeding 2-AAF for only 3 or 7 days, there was a rapid loss of C-8 adducts from liver. In contrast, feeding 2-AAF for 28 days resulted in a slow and less complete depletion of radiolabel from DNA. Thus, long-term feeding of 2-AAF results in an apparent loss in the capacity to remove the adduct. These results probably reflect both altered metabolic activity and DNA repair activity caused by 2-AAF treatment.

Beland *et al.* (1982) have investigated the nature and persistence of DNA adducts in rat liver and kidney after multiple doses of *N*-hydroxy-2-acetylaminofluorene (N-OH-2-AAF). In this study, radiolabeled N-OH-2-AAF was administered to groups of rats at 14-day intervals. Rats were killed 1 or 14 days following 1–7 treatments with radiolabeled compounds. Specific DNA adducts were determined following hydrolysis of DNA. The major adduct found in male rat liver, and the only adduct found in female rat liver and in kidney of either sex, was *N*-(deoxyguanosin-8-yl)-2-aminofluorene. As was found in the study of Poirier *et al.* (1982), the adduct was slowly removed from liver and kidney of both male and female rats. The finding that DNA adducts persist in rat kidney and in female rat liver was unexpected because these organs are resistant to the carcinogenic action of N-OH-2AAF; yet persistence of adducts may be important in tumorigensis (Pound and McGuire, 1978a,b).

INFORMATION NECESSARY FOR MORE PREDICTIVE USE OF COVALENT BINDING DATA IN RISK ASSESSMENT

If covalent binding information is to be used reliably as a predictive tool in safety evaluation, it will be necessary to address critical issues such as those proposed in the following questions.

Which Molecules Serve As Targets?

Many electrophiles will react with a variety of cellular molecules, but only a limited spectrum of macromolecules will be receptors for toxicity. Thus, much of the covalent binding to tissue may not be associated with the toxic response. Furthermore, demonstration of covalent binding at any of several sites is indicative of chemical reactivity, but is not synonymous with the identification of the target site.

How Much Covalent Binding Must Occur before a Response Is Assumed to Be Mediated by a Reactive Metabolite?

The answer to this question will vary with the toxicity produced and the target molecule. Consider the hypothetical case where 50% of the target molecules must be bound to cause clinically observable toxicity and the reactive metabolite binds to the target molecule in a one-to-one stoichiometry. Furthermore, assume that covalent binding to the target molecule always results in inactivation of the target and that binding to the target represents a large percentage of the total covalent binding to protein in the cell. If the cell contains 2 nmole of the target molecule, then a covalent binding level of 1 nmole/mg protein would be of toxicological significance, whereas, if the cell contains 10 nmole of the target molecule, then this covalent binding level may not be toxicologically significant. Such a hypothetical example is obviously an oversimplification, since covalent binding will vary with time as a result of repair or turnover of the target molecule. The toxicological significance of the covalent binding will depend upon the number of target molecules, the rate of target inactivation resulting from covalent binding, and rate of target molecule repair.

How Persistent Must the Covalent Binding Be to Result in a Response?

This is clearly a major problem when dealing with DNA alkylation, since replication of the cell is likely to result in a change in the nucleic acid sequence of DNA, i.e., a mutation, unless DNA repair occurs prior to cell division. Unless the DNA is altered in a position where repair may be hindered, the persistence of covalent binding is critical to the production of a toxic response. The persistence of protein alkylation is more closely related to the half-life of the protein. Thus, enzymic proteins critical to cellular function may be rendered inactive by covalent binding, leading to decreased capacity of critical metabolic pathways. The resulting metabolic lesions will persist until the normal levels of enzyme are restored.

Is Covalent Binding a Necessary Prerequisite for an Adverse Response?

It is not necessary that a covalent bond be formed to modify cellular structure and produce toxicity. Recknagel (1967) postulated that the reactive metabolite of carbon tetrachloride is the trichloromethyl free radical (Fig. 13-2), which abstracts a methylene hydrogen from unsaturated fatty acids in membranes. The free radical that is produced leads to a free radical chain reaction, lipid peroxide formation, and, ultimately, hepatic centrilobular necrosis (Recknagel, 1967). Free radical destabilization of a cellular structure without covalent binding has also been proposed for the neurotoxicity of 6-hydroxydopamine (Kostrzewa and Jacobowitz, 1974), the mutagenicity of benzo(*a*)pyrene (Lesko *et al.,* 1978), the mutagenicity of 2-AAF (Andrews *et al.,* 1978), and many other toxicities (Mason, 1979). Nevertheless, covalent binding of many chemical carcinogens appears to be associated with their oncogenic response (Jollow *et al.,* 1977; Snyder *et al.,* 1982).

CONCLUSIONS

The concept of metabolic activation and toxicity induced by biologically reactive intermediates has arisen as a result of the last two decades of toxicological research. The challenge today, and likely for the foreseeable future, is: How do we apply this information for risk assessment purposes? Given our present state of knowledge, there is no clear answer to this question.

It would be a mistake, however, for the scientific and regulatory communities to conclude that the problem is too complex to approach. Such a conclusion could lead to inaccurate risk assessments based solely on the observation of covalent binding to macromolecules. Covalent binding indicates the ability of a xenobiotic metabolite to interact with macromolecules, but it does not mandate the production of a toxic response.

This brief discussion of issues relating to covalent binding and toxicity has identified a number of research areas that, with further development, could lead to important information for risk assessment. Essential to the practical use of covalent binding information in safety evaluation is the better identification of the target for toxicity. Recent studies are beginning to relate both the quantitative and qualitative aspects of covalent binding to carcinogenic potential. These studies not only offer the prospect of application for risk assessment, but also the potential to address fundamental mechanisms of toxicity and carcinogenesis.

REFERENCES

Alvares, A. P. 1978. Interactions between environmental chemicals and drug biotransformation in man. *Clin. Pharmacokin.* **3**:462–477.

Andrews, L. S., E. W. Lee, C. M. Witmer, J. J. Kocsis, and R. Snyder. 1977. Effects of toluene on the metabolism, disposition and hematopoietic toxicity of ^{3}H-benzene. *Biochem. Pharmacol.* **26**:293–300.

Andrews, L. S., J. A. Hinson, and J. R. Gillette. 1978. Studies on the mutagenicity of *N*-hydroxy-2-acetylaminofluorene in the Ames *Salmonella* mutagenesis test system. *Biochem. Pharmacol.* **27**:2399–2408.

Barry, E. J., D. Malejka-Giganti, and H. R. Gutmann. 1969. Interaction of aromatic amines with rat liver proteins *in vivo.* III. On the mechanism of binding of the carcinogens, *N*-2-fluorenylacetamide and *N*-hydroxy-2-fluorenylacetamide, to the soluble proteins. *Chem. Biol. Interact.* **1**:139–155.

Beland, F. A., K. L. Dooley, and C. D. Jackson. 1982. Persistence of DNA adducts in rat liver and kidney after multiple doses of the carcinogen *N*-hydroxy-2-acetylaminofluorene. *Cancer Res.* **42**:1348–1354.

Blackburn, G. R., J. P. Andrews, R. P. Custer, and S. Soraf. 1981. Early events during liver carcinogenesis involving two carcinogen:protein complexes. *Cancer Res.* **41**:4039–4049.

Bolt, H. M., R. J. Laib, and J. G. Filser. 1982. Reactive metabolites and carcinogenicity of halogenated ethylenes. *Biochem. Pharmacol.* **31**:1–4.

Boyland, E. 1950. The biological significance of metabolism of polycyclic compounds. *Biochem. Soc. Symp.* **5**:40–54.

Conney, A. H. 1967. Pharmacological implications of microsomal enzyme induction. *Pharmacol. Rev.* **9**:317–366.

D'Souza, M., R. Snyder, and J. J. Kocsis. 1979. Benzene inhibits ovarian hypertrophy in the hemi-spayed rat. *Toxicol. Appl. Pharmacol.* **40**:A40.

Dutton, G. J., ed. 1966. *Glucuronic Acid, Free and Combined: Chemistry, Biochemistry, Pharamacology and Medicine.* Academic Press, New York.

Estabrook, R. W., J. Werringloer, J. Capdevila, and R. A. Prough. 1978. The role of cytochrome P450 and the microsomal electron transport system: The oxidative metabolism of benzo(*a*)pyrene. Pp. 285–316 in H. A. Gelboin and P. O. P. Ts'o, eds. *Polycyclic Hydrocarbons and Cancer,* Volume 1. *Environment, Chemistry, and Metabolism.* Academic Press, New York.

Farber, E. 1978. Experimental liver carcinogens: A perspective. Pp. 357–375 in H. Remmer, P. Bannasch, H. M. Bolt, and H. Popper, eds. *Primary Liver Tumors.* MTP Press, New York.

Frommer, U., V. Ullrich, and S. Orrenius. 1974. Influence of inducers and inhibitors on the hydroxylation pattern of *n*-hexane in rat liver microsomes. *FEBS Lett.* **41**:14–16.

Gillette, J. R. 1974a. A perspective on the role of chemically reactive metabolites of foreign compounds in toxicity. I. Correlation of changes in covalent binding of reactive metabolites with changes in the incidence and severity of toxicity. *Biochem. Pharmacol.* **23**:2785–2794.

Gillette, J. R. 1974b. A perspective on the role of chemically reactive metabolites of foreign compounds in toxicity. II. Alterations in the kinetics of covalent binding. *Biochem. Pharmacol.* **23**:2927–2938.

Gonasun, L. M., C. Witmer, J. J. Kocsis, and R. Snyder. 1973. Benzene metabolism in mouse liver microsomes. *Toxicol. Appl. Pharmacol.* **26**:398–406.

Guengerich, F. P. 1977. Separation and purification of multiple forms of microsomal cytochrome P-450. Activities of different forms of cytochrome P-450 towards several compounds of environmental interest. *J. Biol. Chem.* **252**:3970–3979.

Holder, G., H. Yagi, P. Dansette, D. M. Jerina, W. Levein, A. Y. H. Lu, and A. H. Conney. 1974. Effects of inducers of epoxide hydrase on the metabolism of benzo(*a*)pyrene by liver microsomes and a reconstituted system: Analysis by high pressure liquid chromatography. *Proc. Natl. Acad. Sci. USA* **71**:4356–4360.

Ingleman-Sundberg, M. 1980. Bioactivation or inactivation of toxic compounds? *Trends Pharmacol. Sci.* **1980**:176–179.

Johnson, E. F. 1979. Multiple forms of cytochrome P-450: Criteria and significance. Pp. 1–26 in E. Hodgson, J. R. Bend, and R. M. Philpot, eds. *Reviews in Biochemical Toxicology,* Volume 2. Elsevier/North-Holland, New York.

Jollow, D. J., J. R. Mitchell, W. Z. Potter, D. C. Davis, J. R. Gillette, and B. B. Brodie. 1973. Acetaminophen-induced hepatic necrosis. II. Role of covalent binding *in vivo. J. Pharmacol. Exp. Ther.* **187**:195–202.

Jollow, D. J., J. J. Kocsis, R. Snyder, and H. Vainio, eds. 1977. *Biological Reactive Intermediates: Formation, Toxicity, and Inactivation.* Plenum Press, New York. 514 pp.

Jollow, D. J., S. Roberts, V. Price, and C. Smith. 1982. Biochemical basis for dose response relationships in reactive metabolite toxicity. Pp. 99–113 in R. Snyder, D. V. Parke, J. J. Kocsis, D. J. Jollow, G. G. Gibson, and C. M. Witmer, eds. *Biological Reactive Intermediates* II: *Chemical Mechanisms and Biological Effects.* Plenum Press, New York.

Kostrzewa, R. M., and D. M. Jacobowitz. 1974. Pharmacological actions of 6-hydroxy-dopamine. *Pharmacol. Rev.* **26**:199–288.

Lawley, P. D. 1980. DNA as a target of alkylating carcinogens. *Br. Med. Bull.* **36**:19–24.

Lesko, S. A., R. J. Lorentzen, and P. O. P. Ts'o. 1978. Benzo(*a*)pyrene metabolism: One-electron pathways and the role of nuclear enzymes. Pp. 261–269 in H. V. Gelboin and P. O. P. Ts'o, eds. *Polycyclic Hydrocarbons and Cancer,* Volume 1. Academic Press, New York.

Lin, J. K., J. A. Miller, and E. C. Miller. 1975. On the structures of hepatic nucleic acid-bound dyes in rats given the carcinogen *N*-methyl-4-aminoazobenzene. *Cancer Res.* **35**:844–850.

Longacre, S. L., J. J. Kocsis, and R. Snyder. 1981. Influence of strain differences in mice on the metabolism and toxicity of benzene. *Toxicol. Appl. Pharmacol.* **60**:398–409.

Lutz, W. K., and C. M. Schlatter. 1977. Mechanism of carcinogenic action of benzene. Irreversible binding to rat liver DNA. *Chem. Biol. Interact.* **18**:241–245.

Magee, P. N. 1980. Metabolism of nitrosamines: An overview. Pp. 1081–1090 in M. J. Coon, A. H. Conney, R. W. Estabrook, H. V. Gelboin, J. R. Gillette, and P. J. O'Brien, eds. *Microsomes, Drug Oxidations, and Chemical Carcinogenesis.* Academic Press, New York.

Mason, R. P. 1979. Free radical metabolites of foreign compounds and their toxicological significance.

Pp. 151–200 in E. Hodgson, J. R. Bend, and R. M. Philpot, eds. *Reviews in Biochemical Toxicology,* Volume 1. Elsevier/North-Holland, New York.

Miller, E. C., and J. A. Miller. 1947. Presence and significance of bound aminoazo dyes in livers of rats fed *p*-dimethylaminoazobenzene. *Cancer Res.* **7**:468–480.

Miller, E. C., and J. A. Miller. 1982. Reactive metabolites as key intermediates in pharmacologic and toxicologic responses. Examples from chemical carcinogenesis. Pp. 1–21 in R. Snyder, D. V. Parke, J. J. Kocsis, D. J. Jollow, G. G. Gibson, and C. M. Witmer, eds. *Biological Reactive Intermediates* II: *Chemical Mechanisms and Biological Effects.* Plenum Press, New York.

Miller, J. A. 1970. Carcinogenesis by chemicals: An overview—G. H. A. Clowes memorial lecture. *Cancer Res.* **30**:559–576.

Miller, J. A., and E. C. Miller. 1977. The concept of reactive electrophilic metabolites in chemical carcinogenesis: Recent results with aromatic amines, safrole, and aflatoxin B_1. Pp. 6–24 in D. J. Jollow, J. J. Kocsis, R. Snyder, and H. Vanio, eds. *Biological Reactive Intermediates: Formation, Toxicity, and Inactivation.* Plenum Press, New York.

Mitchell, J. R., and D. J. Jollow. 1974. Biochemical basis for drug-induced heptatoxicity. *Isr. J. Med. Sci.* **10**:312–318.

Mitchell, J. R., and D. J. Jollow. 1975. Role of metabolic activation in chemical carcinogenesis in drug-induced hepatic injury. Pp. 395–416 in W. Gerok and K. Sickinger, eds. *Drugs and the Liver.* Schattaner, Stuttgart.

Mitchell, J. R., W. D. Reid, B. Christine, J. Moskowitz, G. Krishna, and B. B. Brodie. 1971. Bromobenzene-induced hepatic necrosis: Species differences and protection by SKF 525-A. *Res. Commun. Chem. Pathol. Pharmacol.* **2**:877–888.

Mitchell, J. R., D. J. Jollow, W. Z. Potter, D. C. Davis, J. R. Gillette, and B. B. Brodie. 1973. Acetaminophen-induced hepatic necrosis. I. Role of drug metabolism. *J. Pharmacol. Exp. Ther.* **187**:185–194.

Nebert, D. W., and N. M. Jensen. 1979. The Ah locus: Genetic regulation of the metabolism of carcinogens, drugs, and other environmental chemicals by cytochrome P-450-mediated monooxygenases. *CRC Crit. Rev. Biochem.* **6**:401–438.

Norpoth, J., U. Witting, and M. Springorum. 1974. Induction of microsomal enzymes in the rat liver by inhalation of hydrocarbon solvents. *Int. Arch. Arbeitsmed.* **33**:315–321.

Parke, D. V. 1968. *The Biochemistry of Foreign Compounds.* Pergamon Press, New York.

Poirier, L. A., J. A. Miller, E. C. Miller, and K. Sato. 1967. *N*-Benzoyloxy-*N*-methyl-4-aminoazobenzene: Its carcinogenic activity in the rat and its reactions with proteins and nucleic acids and their constituents *in vitro. Cancer Res.* **27**:1600–1613.

Poirier, M. C., B. True, and B. A. Laishes. 1982. Formation and removal of (guan-8-yl) DNA-2-acetylaminofluorene adducts in liver and kidney of male rats given dietary 2-acetylaminofluorene. *Cancer Res.* **42**:1317–1321.

Potter, W. Z., S. S. Thorgeirsson, D. J. Jollow, and J. R. Mitchell. 1974. Acetaminophen-induced hepatic necrosis. V. Correlation of hepatic necrosis, covalent binding and glutathione depletion in hamsters. *Pharmacology* **12**:129–143.

Pound, A. W., and L. J. McGuire. 1978a. Repeated partial hepatectomy as a promoting stimulus for carcinogenic response of liver to nitrosamines in rats. *Br. J. Cancer* **37**:585–594.

Pound, A. W., and L. J. McGuire. 1978b. Influence of repeated liver regeneration on hepatic carcinogenesis by diethylnitrosamine in mice. *Br. J. Cancer* **37**:595–602.

Prescott, L. F., N. Wright, P. Roscoe, and S. S. Bronn. 1971. Plasma-paracentamol half-life and hepatic necrosis in patients with paracetamol overdosage. *Lancet* **i**:519–522.

Proudfoot, A. T., and N. Wright. 1970. Acute paracetamol poisoning. *Br. Med. J.* **3**:557–558.

Recknagel, R. O. 1967. Carbon tetrachloride hepatotoxicity. *Pharmacol. Rev.* **19**:145–208.

Roberts, S. A., and D. J. Jollow. 1978. Acetaminophen structure–toxicity relationships—Why is 3-hydroxyacetanilide not hepatotoxic? *Pharmacology* **20**:259.

Roberts, S. A., and D. J. Jollow. 1979a. Acetaminophen structure–toxicity studies: Lack of liver necrosis after 2-hydroxyacetanilide. *Pharmacology* **21**:220.

Roberts, S. A., and D. J. Jollow. 1979b. Acetaminophen structure–toxicity studies—*In vivo* covalent binding of a non-hepatotoxic analog 3-hydroxyacetanilide. *Fed. Proc.* **38**:426.

Roberts, S. A., and D. J. Jollow. 1980. Acetaminophen structure–toxicity studies: *In vivo* covalent binding of a non-hepatotoxic analog, 2-hydroxyacetanilide. *Fed. Proc.* **39**:748.

Sammett, D., E. W. Lee, J. J. Kocsis, and R. Snyder. 1979. Partial hepatectomy reduces both the metabolism and toxicity of benzene. *J. Toxicol. Environ. Health* **5**:785–792.

Sato, A., and T. Nakajima. 1979. Dose-dependent metabolic interaction between benzene and toluene *in vivo* and *in vitro*. *Toxicol. Appl. Pharmacol.* **48**:249–256.

Sato, R., and T. Omura. 1978. *Cytochrome P-450*. Academic Press, New York. 233 pp.

Schenkman, J. B., and D. Kupfer, eds. 1982. *Hepatic Cytochrome P-450 Monooxygenase System*. Pergamon Press, New York.

Sims, P. 1980. The metabolic activation of chemical carcinogens. *Br. Med. Bull.* **36**:11–18.

Smith, C. L., and D. J. Jollow. 1976. Potentiation of acetaminophen-induced liver necrosis in hamsters by galactosamine. *Pharmacology* **18**:156.

Snyder, R., and J. J. Kocsis. 1975. Current concepts of chronic benzene toxicity. *CRC Crit. Rev. Toxicol.* **3**:265–288.

Snyder, R., and H. Remmer. 1979. Classes of hepatic microsomal mixed function oxidase inducers. *Pharmacol. Ther.* **7**:213–244.

Snyder, R. F., F. Uzuki, L. Gonasun, E. Bromfeld, and A. Wells. 1967. The metabolism of benzene *in vitro*. *Toxicol. Appl. Pharmacol.* **11**:346–360.

Snyder, R., E. W. Lee, and J. J. Kocsis. 1978a. Binding of labeled benzene metabolites to mouse liver and bone marrow. *Res. Commun. Chem. Pathol. Pharmacol.* **20**:191–194.

Snyder, R., E. W. Lee, and J. J. Kocsis. 1978b. Minireview: Bone marrow depressant and leukemogenic actions of benzene. *Life Sci.* **21**:1709–1722.

Snyder, R., S. L. Longacre, C. M. Witmer, J. J. Kocsis, L. S. Andrews, and E. W. Lee. 1980. Biochemical toxicology of benzene. Pp. 123–153 in E. Hodgson, J. R. Bend, and R. M. Philpot, eds. *Reviews in Biochemical Toxicology*, Volume 3. Elsevier/North-Holland, New York.

Snyder, R., D. V. Parke, J. J. Kocsis, D. J. Jollow, G. G. Gibson, and C. W. Witmer, eds. 1982. *Biological Reactive Intermediates*. II. *Clinical Mechanisms and Biological Effects*. Plenum Press, New York. 1476 pp.

Thorgeirsson, S. S., H. A. Sasame, J. R. Mitchell, D. J. Jollow, and W. Z. Potter. 1976. Biochemical changes after hepatic injury from toxic doses of acetaminophen or furosemide. *Pharmacology* **14**:205–217.

Williams, R. T. 1959. *Detoxification Mechanisms: The Mechanism and Detoxication of Drugs, Toxic Substances and Other Organic Compounds*. Wiley, New York.

Williams, R. T., and D. V. Parke. 1964. The metabolic fate of drugs. *Annu. Rev. Pharmacol.* **4**:85–114.

Yang, S. K., J. Duetsch, and H. V. Gelboin. 1978. Benzo(*a*)pyrene metabolism, activation and detoxification. Pp. 205–231 in H. V. Gelboin and P. O. P. Ts'o, eds. *Polycyclic Hydrocarbons and Cancer*. Volume 1. *Environment, Chemistry and Metabolism*. Academic Press, New York.

Zampaglione, N., D. J. Jollow, J. R. Mitchell, B. Stripp, M. Hamrick, and J. R. Gillette. 1973. Role of detoxifying enzymes in bromobenzene-induced liver necrosis. *J. Pharmacol. Exp. Ther.* **187**:218–227.

14

Computer-Assisted Prediction of Toxicity

Kurt Enslein

By using statistical structure–activity relationship (SAR) models of toxicological endpoints, it is possible to estimate the expected toxicity from structural and physical characteristics of untested chemicals. Thus, these models can be used for setting testing and developmental priorities, obtaining additional information on given compounds, and, to a lesser degree, determining what substructural factors are most involved in a particular toxicological effect.

The rationale for the development of these models rests in a body of work that has been ongoing for some years, especially in the area of medicinal chemistry. In that field, Hansch (1969), among others, demonstrated that drug design could be materially aided by means of structure–activity equations. Later, Free and Wilson (1964) applied similar principles to this same problem, but introduced indicator parameters (variables given the value 0 if a certain characteristic is absent or 1 if it is present) to portray substructural fragments (Craig, 1973). In addition to Free and Wilson, structure–activity workers in classical medicinal chemistry used in their analyses closely related compounds and almost always limited their field of interest to therapeutic effects, with the notable exception of Wishnok *et al.* (1978), who developed SAR models of carcinogenesis.

Apart from these early efforts, it was not until approximately 1975, partly as a result of a Gordon Research Conference organized by Dr. Paul N. Craig, that it became evident to a few investigators that toxic effects, i.e., the undesirable actions of a chemical, could be considered as the other side of the coin from therapeutic effects, and hence that the same SAR principles could possibly be applied to the modeling of toxicological endpoints. A number of approaches have in fact been used. Except for the work by Enslein (e.g., Enslein, 1980; Enslein and Craig,

KURT ENSLEIN • Health Designs Inc., Rochester, New York 14604.

1978, 1982; Enslein *et al.*, 1983a,b; Enslein and Craig, 1982) and Jurs and co-workers (e.g., Chou and Jurs, 1979; Jurs *et al.*, 1979; Yuan and Jurs, 1980; Yuta and Jurs, 1981), most studies have been limited to research in relatively closely related series. Of particular interest is the work of Smith *et al.* (1978), Purcell *et al.* (1973), Kaufman (1979), and Loew *et al.* (1979), who applied SAR principles to polycyclic aromatic hydrocarbons, for example, or developed *ab initio* methodology. The studies by Enslein are an extension of these same ideas, but applied to diverse structures as contrasted to closely related series. One could consider this methodology to be a cross between the Hansch and Free–Wilson approaches. The work by Jurs and his co-workers typifies the results obtainable from heterogeneous data sets using learning-machine methodology as applied to mutagenicity and carcinogenicity. Throughout our developments there has been an insistence on the use of well-documented and well-developed statistical techniques. When possible, an attempt has been made to stay within the bounds of recognized technology.

To construct SAR models, one must decide on the parameters that will be used for describing the chemicals. Although some early work (e.g., Enslein and Craig, 1978) was based on substructural keys developed by the CIDS (U.S. Army Chemical Information Data System) in Edgewood, Maryland, it was discovered that the availability of the programs for the generation of these keys was quite limited. Investigations then progressed to the CROSSBOW system (Eakin *et al.*, 1974), which is based on WLN (Wiswesser line notation)—a linear description of chemical structure. A computer program that has been available for some time was first produced by ICI (Imperial Chemical Industries, Ltd.) and later by Fraser-Williams, a firm in Cheshire, England. The original program contained a set of 149 descriptors. That number was subsequently increased to approximately 309 and later, by adding structures considered potential carcinogenic moieties by the Food and Drug Administration (FDA) Bureau of Foods, to approximately 350. Other parameters that have been investigated include molecular weight, octanol–water partition coefficient, molor refractivity, and molecular connectivity indices. These are all described in this chapter.

Since the initial preparation of this manuscript, there has been considerable progress in the development of the SAR art as applied to toxicological endpoints. For example, two versions of a new substructural key scheme (MOLSTAC©) have been devised to circumvent the many limitations of the CROSSBOW system. Further sophistication has been introduced for the identification of troublesome compounds and independent parameters, etc. It is to be expected that the SAR field will continue to evolve rapidly with the application of three-dimensional modeling methods and the ability to calculate important parameters taking the lead in making it possible to develop better and more meaningful SAR models.

CAUTIONS REGARDING APPLICABILITY OF MODELS

The estimates produced by these SAR models are by their nature the results of statistical computations. As such, they are known to incorporate a certain degree of unavoidable error. The same kind of error is of course inherent in the measurements on which these models are based. There is nevertheless a great temptation to assume that these estimates, because they are produced by a formalized computational technique, represent truth. But they are only estimates, just as the results from a single measurement are also only estimates of reality. Moreover, any given estimate must not be viewed by itself, but only in the context of other estimates for the same endpoint. In other words, these models can be used for deriving priorities for a set of compounds, but not as a way of establishing absolute values for the probabilities of these toxicological endpoints.

A convenient way of "calibrating" the relative probabilities is to incorporate into the set of compounds whose toxicity is to be estimated several related compounds for which the relevant toxicity has been measured. These compounds then provide references, or benchmarks, that can serve as the basis from which the meaning of the other estimates can be evaluated. In a sense, the calibration method provides a way of establishing relative scales for the test compounds. An example of this technique is shown in Table 14-1. To fully understand this table, the reader is referred to the section on teratogenesis later in this chapter.

STATISTICAL METHODOLOGY

In the development of any statistical model, be it econometric, diagnostic, or toxicological, the starting point is a data base of objects, in this case chemicals, for which the endpoint in question has been measured or otherwise obtained. For each of these chemicals, one then needs descriptors, which become the candidate parameters for the statistical equation, i.e., the independent parameters. The development of each model then consists in identifying from among the candidates those parameters that, in combination, "best" model the desired endpoint

TABLE 14-1. Ranked Estimates of Teratogenicity[a]

Rank	CAS number[b]	Compound	Comment	Terato-genesis	Marker[c]
1	57-22-7	Vincristine	In model, score = 0.85	0.90	X
2	118-52-5	1,3-Dichloro-5,5-dimethylhydantoin	—	0.88	—
3	50-00-0	Formaldehyde	In model, score = 0.85	0.876	X
4	58-08-02	Caffeine	In model, score = 0.85	0.839	X
5	54-71-7	Pilocarpine hydrochloride	In model, score = 0.40	0.343	X
6	58-93-5	Hydrochlorothiazide	—	0.24	—
7	96-69-5	4,4-Thiobis(6-*tert*-butyl)-*m*-cresol	—	0.174	—
8	110-82-7	Cyclohexane	—	0.174	—
9	56-81-5	Glycerine	In model, score = 0.0	0.03	—

[a] Based on the teratogenicity model.
[b] Chemical Abstracts Service.
[c] X refers to compounds used in the model data base.

(see explanation below). With few exceptions, essentially the same methodology has been used in the development of all the different models.

Regression and Discriminant Analysis

The main statistical techniques used in the development of these models are stepwise regression and stepwise discriminant analysis (Enslein *et al.,* 1977), which are quite similar. In fact, two-group discriminant analysis can really be formulated in terms of a regression problem. Thus, these two techniques have been used interchangeably during this development.

There has been much discussion in the literature concerning the biased equations (i.e., equations with coefficients higher or lower than they would have been without correlation between parameters) that result from the stepwise aggregation of a set of candidate parameters (Hocking, 1976). To avoid this problem, some investigators begin with a relatively large set of parameters, using as a criterion *F*-to-enter—a variance ratio (Enslein *et al.,* 1977) set at a low value to select all those variables that could possibly make any contribution to the explanation of the endpoint. One subsequently removes from the equation those variables, one by one, that contribute least to the equation. This removal process is stopped when there are no variables left in the equation with an *F*-to-remove (another variance ratio) smaller than 1.7 (Costanza and Afifi, 1979). The equation resulting from this process is then subjected to ridge-regression analysis (Marquardt and Snee, 1975) as well as several other regression diagnostic procedures (Belsley *et al.,* 1980), whereby the ridge trace for each of the parameters is examined by an automated algorithm. Parameters that contribute to the near singularity of the equation are then removed, since they really do not add information, but could contribute to the instability of the equation. As a final step, the stepwise forward and stepdown regression process is repeated, this time using only the variables that survived the ridge-regression procedure.

Frequently outlier compounds are also removed as they are identified at various stages of this process; but since one is dealing with relatively large data bases, the effect of a few outlier compounds will mostly be minimal.

The decision whether to use regression or discriminant analysis is based initially on the endpoint to be modeled. If the endpoint is continuous, regression is chosen. If it is categorical, then one uses discriminant analysis.

Resulting Estimates

As a parallel to the endpoint to be modeled in regression models, the resultant estimates are also continuous parameters, for example mg/kg. In discriminant analysis models, the calculation results in a statement of probability that a compound has a certain endpoint, e.g., the probability that a compound may be a carcinogen.

The evaluation of the equations proceeds as follows. One obtains (through a computer program) the fragment keys and physical constants that relate to the test compound. These keys and constants are then evaluated by the estimating equation, and a subsequent transformation is performed to produce the final estimate.

To obtain the continuous endpoint (e.g., oral LD_{50} in the rat), one proceeds as follows:

Calculate antilog of derived scores $(1/C) = X$.

Then calculate LD_{50} (in mg/kg) $= 1000 \times MW/X$, where MW is molecular weight.

To obtain the probability of a discrete endpoint such as carcinogenicity:

Calculate negative score (indefinite) $= -7.42 +$ (presence of terminal oxygen, not carbonyl, $\times$ 3.04) $-$ (presence of three-branch nitrogen atom $\times$ 0.63) $+ \cdots$.

Calculate positive score (definite) $= -6.55 +$ (presence of terminal oxygen, not carbonyl, $\times$ 0.16) $+$ (presence of three-branch nitrogen atom $\times$ 4.19) $+ \cdots$.

The presence of a substructure is assigned the value 1; the absence, the value 0. The coefficients are derived by the statistical process. The coefficients are obtained from the discriminant equation.

Then calculate the following exponential equation:

$$\text{Probability of carcinogenicity} = \frac{\exp\,(\text{definite score})}{\exp\,(\text{definite score}) + \exp\,(\text{indefinite score})}$$

Model Verification

Ideally, one would be able to make clear-cut statements about the predictive quality of these SAR models. Unfortunately, the only test methods available to evaluate these models are relatively weak. Several of these methods are described.

Subset Verification

In this method one reserves a randomly selected subset of compounds from the total data set. The model is designed (trained) with one subset and tested with the other. One then makes judgments about the quality of the model by comparing the estimated values with the measured values for the particular endpoint. This method has several limitations. For example, it assumes that the measured values are in fact exact, which of course they never are. Inasmuch as a random number generator is used to select the test subset, that subset and the training set are homogeneous. Therefore, any major predictive discrepancies would have to be the result of a sampling problem. Nevertheless, this method is frequently used for the evaluation of statistical models. Although it is not believed to carry very much weight, results based on this method are presented for several of the models.

KS Method

For the discriminant analysis models consisting of two distinct groups, one can test whether the misclassifications fit a binominal distribution by comparing the actual cumulative error distribution with the expected distribution by means of a two-sample Kolomogorov–Smirnov (KS) test (Siegel, 1956). If in fact the two distributions cannot be statistically distinguished from one another, the estimated probabilities can be regarded as "accurate." One still cannot make any strong statements about predictability, however.

Error Matrices

In these matrices, probabilities of classification of the compounds obtained in the model can be classified and compared to the given categorical probabilities. This is one way to measure the fit of the model to the training set. It is probably as good an estimate of the expected performance of the model as can presently be obtained. In view of the fact that some probabilities will inevitably be near 0.5, i.e., chance, the question does arise as to how wide an uncertainty band should be postulated around 0.5. Limits of 0.5 ± 0.1 and ± 0.2 have arbitrarily been chosen.

Clustering

The clustering technique is used from time to time to determine whether a set of compounds is homogeneous so that a single model can be constructed. During the use of regression or discriminant analysis, one sometimes finds substantial groups of outliers. One must then determine whether there is a single more or less homogeneous group of compounds or multiple groups. In those cases, the clustering technique to identify the subgroups is clearly applicable. Since homogeneity was not an issue in any of the models discussed in this chapter, clustering techniques were not used.

ESTIMATOR FOR RAT ORAL LD_{50}

The following model is based on a study conducted by Enslein and Craig (1978).

Description

Rat oral LD_{50} in mg/kg is estimated by evaluation of a regression in which substructural fragments and molecular weight are used as independent parameters (see Fig. 14-1).

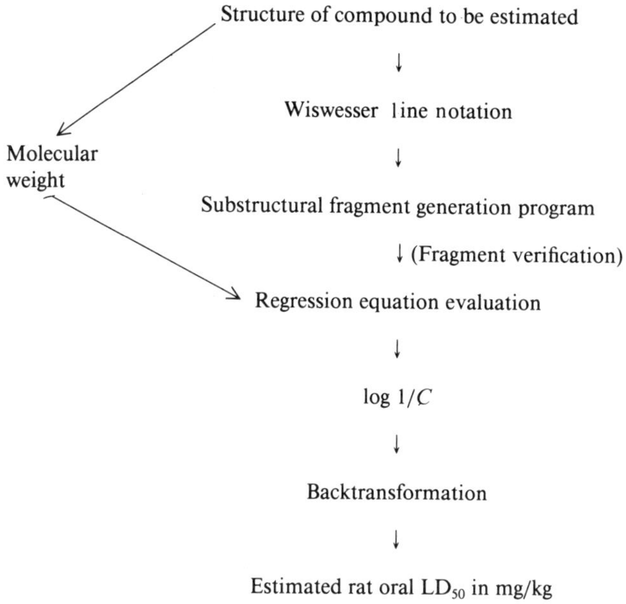

FIGURE 14-1. Computer-based process for estimating LD_{50}.

Data Used in Development of Estimating Equations

The chemicals used in the development of the equations were drawn from the April 1979 tape version of the Registry of Toxic Effects of Chemical Substances (RTECS), published by the National Institute for Occupational Safety and Health (NIOSH) (Lewis, 1978). This compendium lists the most toxic effects reported in the literature. The tape is reasonably free of errors and represents the most comprehensive compilation of LD_{50} data available.

The tape included approximately 3600 compounds with rat oral LD_{50} values. A randomly selected subset of 2000 of these formed the basis of the SAR equation. Removal of duplicates, some outliers, and restriction of range reduced the actual number of compounds to 1851.

The terms used in estimating the equation included 93 substructural fragments, one log molecular weight, and one constant.

Estimation Range

The log $1/C$ of the compounds used for model development ranged from 1.25 to 4.75, where $1/C = 1000 \times MW/LD_{50}$ and the LD_{50} is expressed in mg/kg. The molar form (i.e., moles/kg) is preferred and used in all computations.

Performance Characteristics

From the group of 3600 compounds, 567 were randomly selected. These compounds were *not* used in the development of the estimating equation. The

TABLE 14-2. Quantities and Simple Statistics of Residuals for LD_{50} Estimator

Predicted log 1/*C*	*N*	Quantile 1	5	10	25	75	90	95	99
1.0–1.5	3	−0.12	−0.12	−0.12	−0.12	1.74	1.74	1.74	1.74
1.5–2.0	78	−1.34	−1.05	−0.71	−0.42	0.26	0.54	0.79	1.00
2.0–2.5	224	−1.34	−0.87	−0.67	−0.45	0.34	0.76	1.03	1.74
2.5–3.0	143	−1.99	−1.25	−0.78	−0.45	0.36	0.86	1.45	2.71
3.0–3.5	97	−1.80	−1.20	−0.81	−0.46	0.52	1.17	1.56	2.51
3.5–4.0	18	−1.63	−1.63	−1.60	−1.07	0.50	1.48	1.64	1.64
4.0–4.5	4	−1.02	−1.02	−1.02	−1.02	0.85	0.85	0.85	0.85

Predicted log 1/*C*	Average	Median	SD	Minimum	Maximum
1.0–1.5	0.72	0.55	0.94	−0.12	1.74
1.5–2.0	−0.059	−0.12	0.66	−1.37	1.00
2.0–2.5	−0.015	−0.059	0.58	−1.49	1.94
2.5–3.0	−0.017	−0.07	0.77	−2.33	2.76
3.0–3.5	0.081	0.067	0.81	−1.80	2.51
3.5–4.0	−0.21	−0.35	0.98	−1.63	1.64
4.0–4.5	−0.082	−0.077	1.07	−1.02	0.85

log 1/C values were then calculated for the compounds by evaluating their rat oral LD_{50} with the estimating equation. These predicted values were then subtracted from the corresponding values in RTECS, and residuals obtained. Quantiles and simple statistics were then calculated for various ranges of predicted log 1/C. The results are shown in Table 14-2.

The following example shows how one estimates the confidence bounds for a predicted value. Suppose the predicted log 1/C value is 2.7. Then enter Table 14-2 at line 2.5–3.0, where the 10th and 90th quantiles have the values of −0.78 and +0.86, respectively. Thus, one can state that not more than 10% of the true values will be less than 1.92 (i.e., 2.70 − 0.78) and not more than 10% will be higher than 3.56 (i.e., 2.70 + 0.86). Limits for other bounds can be determined in a similar fashion. Tighter confidence bounds can be developed with a larger independent data base, especially at the extremes.

The Estimating Equation

The first 30 most important terms of the equation are shown in Table 14-3. The *F* value is a variance ratio indicating the relative importance of each of the terms of the equation.

Following are the regression statistics for the rat oral LD_{50} model described in Table 14-3:

Regression statistics

log 1/*C* mean	2.533
Standard deviation	0.755

TABLE 14-3. Rat Oral LD_{50} "Equation." Terms Listed in Order of Decreasing Importance[a]

WLN key no.	Frequency in 1851 compounds	Description	Regression coefficient	F
11	114	More than one sulfur atom	0.821	150.7
293	22	Tin	1.175	78.6
5	158	Terminal oxygen (not carbonyl)	0.458	62.5
—	1851	Log molecular weight	0.681	53.5
10	185	One sulfur atom	0.362	52.2
309	61	Substituent carbamate	0.548	38.5
20	223	Alkyl chain $(CH_2)_n$ or $CH_3(CH_2)_{n-1}$, where $n = 3$–9	−0.250	31.6
341	79	Aromatic nitro	0.399	31.3
256	9	Mercury	1.114	29.3
28	153	One −NH-group chain fragment	0.334	29.1
315	161	Haloalkane	0.323	21.2
104	233	One heteroatom in one ring	0.202	18.6
166	106	Chain dialkylamino	0.255	16.4
250	2	Lead	−1.164	16.0
111	26	More than one single heterocyclic ring	−0.574	15.6
107	56	One heteroatom in more than one ring	−0.477	15.4
30	102	One $-NH_2$-group chain fragment	0.236	14.5
36	205	More than one −O-group chain fragment	0.211	14.4
130	99	Bilinkage	0.271	13.9
50	280	Substituent generic halogen	0.212	12.6
37	195	One −OH-group chain fragment	−0.163	12.2
193	14	Substituent sulfonamide	−0.561	12.0
34	54	Unusual carbon-atom chain fragment	0.278	11.5
26	30	Fluoride	0.435	11.4
322	4	Aziridine	0.934	10.1
330	16	Fused aromatic-unsaturated lactone	0.539	10.1
344	83	Germinal-dihaloalkane	−0.315	10.0
201	4	Substituent *N*-nitro	−0.934	9.0
112	65	One single carbocyclic ring	0.321	9.0
282	13	Tin	−0.484	8.6
⋮	⋮	⋮	⋮	⋮
Constant			0.522	

[a] F is the variance ratio indicating the relative importance of each of the terms of the equation.

Range	1.25–4.73
R^2 (multiple correlation coefficient)	0.449
Residual mean square	0.33
Standard error of estimate	0.57

ESTIMATOR FOR PROBABILITY OF MUTAGENICITY

Most models to estimate the probability of mutagenicity deal only with congeners. However, Tinker (1981) has recently published a model based mostly on the Ames *Salmonella*/microsome assay using chemical structures of different classes. He uses a different method of describing chemical structure, based on a method of Hodes *et al.* (1977). The quality of this model's performance is on the same general order as the models described here.

Description

The probability of mutagenicity as derived from the Ames test (conducted in *Salmonella typhimurium* strains) is produced by a discriminant equation. The estimating equation uses substructural fragments as independent parameters. The computer-based estimation process, which is used for the remaining predictors in this chapter, is shown in Fig. 14-2.

Data Used in the Development of the Estimating Equation

The files of the Environmental Mutagenesis Information Center (EMIC) at Oak Ridge, Tennessee, were examined. These files contain most of the papers ever published in the field of mutagenesis. Data of sufficient quality, meeting the EPA Gene-Tox criteria, were collected from those files. Data were separately coded for each of the five Ames *Salmonella* strains TA98, TA100, TA1535, TA1537, and TA1538, with and without S9 liver homogenate activation.

Structure of compound to be estimated

↓

Wiswesser Line Notation

↓

Substructural fragment generation program

↓ (Fragment verification)

Discriminant equation evaluation

↓

Estimated probability of mutagenicity

FIGURE 14-2. Computer-based process for estimating the probability of mutagenicity.

A compound was called a nonmutagen if it had been tested in at least three strains and was judged negative in all strains tested. For a compound to be called a mutagen, it had to be positive in at least two strains. In addition, if a compound was tested by the National Toxicology Program (NTP), those results overrode any derived from EMIC, since the NTP criteria were more stringent.

The number of chemicals used in development of the estimating equation included 264 mutagens and 208 nonmutagens for a total of 472 compounds. The terms used in the equation included 59 substructural fragments and are constant.

Performance Characteristics

The discriminant equation was used to calculate the probability of mutagenicity for the 472 compounds used for the development of that equation. The resulting probabilities were then classified into one of three regions: positive, negative, and indeterminate. The indeterminate region spans the probability near chance, i.e., probability of 0.5 ± 0.2. The results are shown in Table 14-4.

The false positives are compounds that are actually nonmutagens, but are classified as mutagens by the equation. The false negatives are chemicals that are in fact mutagens, but are called nonmutagens by the equation. The 25 + 27 = 52 compounds in the middle region represent 11% of the compounds; these cannot be classified by the discriminant equation.

From the data base, 55 compounds were randomly selected, but were not used in the development of the discriminant equation. Their probability of mutagenicity was also calculated with the discriminant equation and the results classified (Table 14-5).

Although the number of compounds was small, these results parallel those of Table 14-4 well.

The Estimating Equation

The first 30 most important terms of the discriminant equation are shown in Table 14-6. The F value is a variance ratio indicating the relative importance of each of the terms of the equation.

TABLE 14-4. Classification Matrix for Mutagenicity Model: Design Compounds

	Classification by discriminant equation					
	Positive		Indeterminate[a]		Negative	
"Actual" classification	N	%	N	%	N	%
Positive	218	82.6	25	9.5	21	8.0[b]
Negative	13	6.3[c]	27	13.0	168	80.8

[a] Probability 0.300–0.699.
[b] False negatives.
[c] False positives.

TABLE 14.5. Classification Matrix for Mutagenicity Model: Test Compounds

	Classification by discriminant equation					
	Positive		Indeterminate[a]		Negative	
"Actual" classification	*N*	%	*N*	%	*N*	%
Positive	25	78.1	5	15.6	2	6.3[b]
Negative	2	8.7[c]	6	26.1	15	65.2

[a] Probability 0.300–0.699.
[b] False negatives.
[c] False positives.

ESTIMATOR FOR PROBABILITY OF CARCINOGENICITY

Estimators for the probability of carcinogenicity have been developed by Jurs and co-workers (1979), whose work has been mostly concentrated on nitrosoamines, by Smith *et al.* (1978), whose efforts have been devoted mostly to polycyclic aromatic hydrocarbons, but whose methods could be applied to other series, and by Loew *et al.* (1979) and Kaufman (1979), who have attempted to use quantum mechanical and *ab initio* calculations to derive estimates of carcinogenicity.

Description

The probability of carcinogenicity, as designated by the committees that write the monographs published by the International Agency for Research on Cancer (IARC, 1972–1978), is estimated by a discriminant equation. The estimating equation includes substructural fragments and molecular weight as independent parameters. The computer-based estimation process is similar to that shown for the mutagenesis model.

Data Used in Development of Estimating Equations

Volumes 1–17 of the IARC monographs were used as sources of data. Certain compounds, such as polymers and inorganics, have not been included in the data base. The IARC classifications "noncarcinogen" and "indefinite carcinogen" have been combined and labeled "indefinite." The IARC categories "definite animal" and "definite human" carcinogen have been combined and labeled "definite." The remaining two categories, i.e., animal and human suspected carcinogens, have not been used for the development of the predictive equation.

The number of chemicals used in developing the estimating equation included 120 indefinites and 223 definites for a total of 343. The terms in the estimating equation included 78 substructural fragments, molecular weight, and

TABLE 14-6. Mutagenesis Model "Equation." Terms Listed in Order of Decreasing Importance[a]

WLN key no.	Occurrences in data base: Mutagen	Occurrences in data base: Nonmutagen	Description	Discriminant coefficients (Mutagen / Nonmutagen)	F
3	60	12	Branching terminal nitro group $-NO_2$	4.158	65.5
315	23	8	Haloalkane	6.136	47.9
99	20	26	Carbocyclic six-membered ring	−3.947	35.1
331	38	9	Fused polynuclear aromatic hydrocarbon	4.009	35.0
59	13	2	More than one $-NH_2$ substituent group	6.000	26.0
41	1	18	One O (CID) chain fragment COOH	−4.379	25.4
103	32	11	Heterocyclic rings other than five- and six-membered	3.396	24.7
116	0	2	One carbo/carbo fusion in more than one ring system	−12.295	20.8
182	4	5	Substituent secondary amide	−7.293	18.6
138	29	24	Two benzene rings	−2.839	17.7
19	33	26	Ethyl/ethylene chain fragment group	−2.361	15.3
168	3	0	Chain hydroxylamine	12.092	15.0
5	48	3	Terminal oxygen (not carbonyl)	3.126	14.3
26	0	5	Chain fluorine	7.795	14.0
180	18	1	Substituent biphenyl	5.136	13.8
190	9	3	Substituent azo and diazo	4.543	13.5
209	2	5	Substituent ureas	−6.130	13.5
73	51	24	Single occurrence of oxygen in a ring heteroatom	2.070	13.1
137	30	68	One benzene ring	−1.637	12.8
66	23	14	More than one $-OH$ substituent group	2.569	12.4
20	8	14	Alkyl chain fragment (CH_2) or $CH_3(CH_2)$	−2.286	12.0
14	14	13	One double bond, excluding $-C{=}S$, $-N{=}$, or $-C{=}U$	2.537	11.1
161	22	0	Chain *N*-nitroso	3.542	10.2
163	2	5	Chain guanidine	−4.852	10.2
35	9	6	One −O-group chain fragment	3.247	10.1
31	3	0	More than one $-NH_2$-group chain fragment	6.669	9.9
54	9	0	Substituent fluorine	3.779	9.2
	2	3	More than two heteroatoms in a ring	−5.452	9.2

TABLE 14-6. (*continued*)

WLN key no.	Occurrences in data base: Mutagen	Occurrences in data base: Nonmutagen	Description	Discriminant coefficients: Mutagen	Discriminant coefficients: Nonmutagen	*F*
8	55	14	Three-branch nitrogen atom		2.156	8.8
58	23	8	One $-NH_2$-group substituent		2.668	8.0
.	.	.	.		.	.
.	.	.	.		.	.
.	.	.	.		.	.
Constant					−1.894	

[a] *F* is the variance ratio indicating the relative importance of each of the terms of the equation.

the constant. One then evaluated the classification of the compounds by the discriminant equation by using that equation. The resulting probabilities were then classified into definite, indefinite, and indeterminate regions. The indeterminate region spans the probability near chance, i.e., 0.5 ± 0.2. The results are shown in Table 14-7.

The false positives are chemicals that, by IARC committee decision, are indefinite carcinogens, but are classified as definite carcinogens by the equation. The false negatives are chemicals labeled definite by IARC, but called indefinite by the equation. The $14 + 21 = 35$ compounds in the central region represent approximately 10% of the compounds; these cannot be classified by the discriminant equation.

The Estimating Equation

The first 30 most important terms of the estimating equation are shown in Table 14-8. The *F* value is a variance ratio indicating the relative importance of each of the terms in the equation.

TABLE 14-7. Classification Matrix for Carcinogenicity Model

	Classification by discriminant equation					
	Non- or indefinite		Indeterminate[a]		Definite	
IARC classification	*N*	%	*N*	%	*N*	%
Noncarcinogenic or indefinite	93	77.5	14	11.7	13	10.8 [b]
Definite	8	3.6 [c]	21	9.4	194	87.0

[a] Probability 0.300–0.699.
[b] False negatives.
[c] False positives.

TABLE 14-8. Carcinogenis Model "Equation." Terms Listed in Order of Decreasing Importance[a]

WLN key no.	Occurrences in data base: Indefinite	Occurrences in data base: Definite	Description	Discriminant coefficients: Definite-indefinite	F
114	10	12	One carbo/carbo fusion	11.70	41.0
137	41	21	One benzene ring	−4.51	38.1
122	0	5	One hetero/hetero fusion	15.64	31.7
103	10	25	Heterocyclic rings other than five- and six-membered	11.62	30.3
86	9	10	Single occurrence of carbonyl	−7.33	21.4
13	2	0	More than one C=S group	−17.88	20.8
8	18	58	Three-branch nitrogen atom	4.82	20.5
246	2	0	Iron	−13.12	19.8
204	5	4	Phenylazo or phenylhydrazono substituent fragment	−11.46	18.0
113	1	1	More than one single carbocyclic ring	−16.24	17.7
321	5	6	Epoxide	−8.01	17.5
79	6	13	One occurrence of nitrogen in one ring	−11.06	17.4
227	4	4	Chromium	−8.15	17.1
335	0	8	Combination key	9.10	16.9
60	2	8	One −N= or HN= group	9.79	16.7
115	10	28	More than one carbo/carbo fusion	6.52	16.4
190	13	9	Azo or diazo substituent fragment	−10.01	16.1
216	5	0	Arsenic	−8.32	16.0
119	10	10	More than one carbo/hetero fusion	−6.64	14.6
MW	—	—	Molecular weight	0.00984	14.6
127	1	1	True bridge indicator	14.10	13.8
75	3	10	Single occurrence of oxygen in one ring	−8.32	13.1
102	20	33	Heterocyclic six-membered ring	6.51	12.7
61	16	14	More than one −N= or HN= group	10.41	12.3
101	13	30	Heterocyclic five-membered ring	5.55	12.1
104	21	35	One heteroatom in one ring	−4.52	11.8
90	3	7	More than one exocyclic bond	−8.81	11.7
18	18	42	More than one methyl/methylene group (chain fragment)	−2.66	11.3
172	3	2	Phenoxy chain fragment	−8.41	11.2
134	52	96	One ring system	−3.63	10.9
⋮	⋮	⋮	⋮	⋮	⋮
Constant				0.87	

[a] F is the variance ratio indicating the relative importance of each of the terms of the equation.

ESTIMATOR FOR PROBABILITY OF TERATOGENICITY

To the best of our knowledge, the equation shown in Table 14-10 currently represents the only attempt to estimate teratogenicity from structural equations.

Description

The probability of teratogenicity is estimated by a discriminant equation that includes substructural fragments as independent parameters. The computer-based estimation process is similar to that shown for the mutagenesis model. Three sources of data were used in development of the estimating equation: Schardein (1976), Shepard (1980), and the Environmental Teratogenesis Information Center (ETIC) in Oak Ridge, Tennessee. Just as for EMIC, ETIC has collected most of the papers in the literature relevant to teratogenicity.

Each chemical, in order to be considered for entry into the data base, had to have been evaluated after 1969 in at least two mammalian species in two separate laboratories. Data on 670 compounds met these criteria and were thus acquired. Each compound was then scored from 0 to 1, reflecting teratogenic potential. These scores were reviewed by four teratologists, and any disagreements were resolved.

Chemicals with scores between 0 and 0.25 were classified as "nonteratogens;" chemicals with scores between 0.75 and 1.00 were classified as "teratogens." Compounds with scores between 0.26 and 0.74 were not used in the development of the discriminant equation. The number of chemicals used in development of the estimating equation included 195 nonteratogens and 235 teratogens, for a total of 430 compounds. The terms in the estimating equation included 68 substructural fragments and one constant.

Performance Characteristics

The discriminant equation was used to calculate the probability of teratogenicity for the 430 chemicals used to develop that equation. The resulting proba-

TABLE 14-9. Classification Matrix for Teratogenicity Model

"Actual" classification	Classification by discriminant equation					
	Positive		Indeterminate[a]		Negative	
	N	%	*N*	%	*N*	%
Teratogen	161	69.4	49	21.1	22	9.5 [b]
Nonteratogen	15	7.7 [c]	46	23.7	133	68.6

[a] Probability 0.300–0.699.
[b] False negatives.
[c] False positives.

TABLE 14-10. Teratogenesis "Equation." Terms Listed in Order of Decreasing Importance[a]

WLN key no.	Frequency: Nonteratogen	Frequency: Teratogen	Description	Difference coefficient	F
198	9	15	Substituent methoxy	4.693	29.3
118	25	20	One carbo/hetero fusion	−3.747	29.1
112	14	4	One single carbocyclic ring	−4.465	25.1
99	29	57	Carbocyclic six-membered ring	2.157	21.4
185	3	0	Substituent *N*-substituted acylhydrazides	−9.869	18.4
18	62	51	More than one methyl/methylene group chain fragment	−1.710	17.1
193	9	5	Substituent sulfonamide	−4.517	16.3
15	1	4	More than one double-bond excluding −C=S, −N=, or −C=O	6.001	16.0
41	18	13	One COOH (acid) group chain	−2.425	14.3
312	0	4	Organohalogen mustards	7.994	13.8
133	5	5	Inorganic	−4.787	12.7
2	7	0	Positive charge	−4.329	12.4
10	20	22	One sulfur atom	2.284	11.7
7	15	3	Four-branch carbon atom	−2.877	11.1
143	53	56	One heterocyclic ring	2.657	11.0
141	18	13	Two carbocyclic rings	−2.567	10.8
5	3	12	Terminal oxygen	3.088	10.7
167	2	1	Chain methoxy	−6.523	10.2
188	5	13	Substituent barbiturate	4.001	10.2
105	40	41	Two heteroatoms in one ring	−1.972	9.9
27	0	2	Iodine	7.730	9.5
251	0	2	Lithium	7.730	9.5
38	13	4	More than one −OH-group chain fragment	−2.455	9.3
111	16	19	More than one single heterocyclic ring	−2.557	8.8
110	40	46	One single heterocyclic ring	−2.050	8.7
43	6	10	One COO (ester) chain fragment	2.454	8.6
50	19	40	Substituent generic halogen	1.690	8.6
88	1	3	Multiple occurrence of carbonyl in more than one ring	5.109	8.5
⋮	⋮	⋮	⋮	⋮	⋮
Constant				0.739	

[a] F is the variance ratio indicating the relative importance of each of the terms of the equation.

bilities were then classified into positive, negative, and indeterminate regions. The indeterminate region spans the probability near chance, i.e., 0.5 ± 0.2. The results are shown in Table 14-9.

The false positives are chemicals that are actually nonteratogens, but are classified as teratogens by the equation. The false negatives are chemicals called teratogens by the described decision process, but nonteratogens by the equation. The 49 + 46 = 95 compounds in the central region represent the ~22% of the chemicals that cannot be classified by the discriminant equation.

The Estimating Equation

The first 30 most important terms of the estimating equation are shown in Table 14-10. The *F* value is a variance ratio indicating the relative importance of each of the terms in the equation.

INTERSPECIES AND OTHER MODELS

The same principles used to develop the models explained above have also been used to estimate the rat oral LD_{50} from the mouse oral LD_{50} and chemical structure. To date this preliminary work has been limited to 160 compounds. The results indicate that 75% of the variability in the rat oral LD_{50} could be explained by the resulting SAR model equation. This experiment needs to be extended to a larger number of compounds as well as to species with greater differences than the mouse and rat.

It is also possible to develop other models for use with different routes of administration and different dosage forms—indeed, for any pair of different end points as long as data for a set of compounds are available for both of them. Also, models for many other end points, such as aquatic toxicity, subchronic toxicity, and other measures of carcinogenicity and mutagenicity, could be developed, given sufficient amounts of data.

Since the models described in this chapter were developed, a number of additional descriptive parameters have appeared in the literature and should be incorporated. These include estimates of a large variety of physical parameters, better estimators of octanol–water partition coefficients, estimators for molecular surface area and volume, as well as more efficient algorithms for the calculation of molecular connectivity indices.

SUMMARY

Through the use of SAR equations, it is possible to estimate the probable level of toxicity, and teratogenicity. The accuracy of the predictions ranges from 80 to 90%.

In the models, substructural keys and physical characteristics are used as descriptors. The most significant subset of the candidate parameters is selected by means of stepdown regression and discriminant analysis techniques, together with ridge-regression and other diagnostic techniques. Limitations of the models lie mostly in the restricted sets of compounds for which data exist and from which SAR equations are developed. But the equations are currently useful for setting priorities and as substitutes for rat oral LD_{50} assays.

REFERENCES

Belsley, D. A., Kuh, E., and R. E., Welsch. 1980. *Regression Diagnostics.* Wiley, New York.

Chou, J. T., and P. C. Jurs. 1979. Computer-assisted structure–activity studies of chemical carcinogens. *N*-Nitroso compounds. *J. Med. Chem.* **22**:792–797.

Costanza, M. C., and A. A. Afifi. 1979. Comparison of stopping rules in forward stepwise discriminant analysis. *J. Am. Stat. Assoc.* **74**:777–785.

Craig, P. N. 1973. Comparison of Hansch and Free–Wilson approaches to structure–activity correlations. *Adv. Chem.* **114**:115–129.

Eakin, D. L., E. Hyde, and G. Parker. 1974. The CROSSBOW System. *Pestic. Sci.* **5**:319–326.

Enslein, K. 1980. Statistical estimates of toxicological and other biological endpoints using structure–activity relationships. Poster Paper at 1st SETAC Conference, Washington, D.C., 25 November 1980.

Enslein, K., and P. N. Craig. 1978. A toxicity estimation model. *J. Environ. Pathol. Toxicol.* **2**:115–132.

Enslein, K., and P. N. Craig. 1982. Carcinogenesis: A predictive structure activity model. *J. Toxicol. Environ. Health* **10**:521–530.

Enslein, K., H. S. Wilf, and A. Ralston, eds. 1977. Chapters 4 and 5 in *Statistical Methods for Digital Computers.* Wiley, New York.

Enslein, K., T. R. Lander, and J. R. Strange. 1983a. Teratogenesis: A statistical structure–activity model. *Teratogen. Carcinogen. Mutagen.* **3**:289–309.

Enslein, K., T. R. Lander, M. E. Tomb, and W. G. Landis. 1983b. Mutagenicity (Ames): A structure–activity model. *Teratogen. Carcinogen. Mutagen.* **3**:503–513.

Free, S. M., and J. W. Wilson. 1964. A mathematical contribution to structure and activity studies. *J. Med. Chem.* **1**:295–399.

Hansch, C. 1969. Biological correlations—Hansch method. *Acc. Chem. Res.* **2**:232–239.

Hocking, R. R. 1976. The analysis and selection of variables in linear regression. *Biometrics* **32**:1–49.

Hodes, L., G. F. Hazard, R. I. Geran, and S. Richmann. 1977. A statistical-heuristic method for automated selection of drugs for screening. *J. Med. Chem.* **20**:469–475.

International Agency for Research on Cancer. 1972–1978. *Monographs on the Evaluation of the Carcinogenic Risk of Chemicals to Humans,* Volumes 1–17. International Agency for Research on Cancer, Lyon.

Jurs, P. C., J. T. Chous, and M. Yuan. 1979. Computer-assisted studies of chemical carcinogens—A heterogenous data set. *J. Med. Chem.* **22**:476–483.

Kaufman, J. J. 1979. Quantum chemical and physiochemical influences on structure–activity and drug design. *Int. J. Quantum Chem.* **16**:221–241.

Lewis, R. J., Sr., ed. 1978. *Registry of Toxic Effects of Chemical Substances.* National Institute for Occupational Safety and Health, Cincinnati, Ohio.

Loew, G., B. S. Sudhinda, and J. E. J., Ferrel. 1979. Quantum chemical studies of polycyclic aromatic hydrocarbons and their metabolites: Correlations to their carcinogenicity. *Chem. Biol. Interact.* **26**:75–89.

Marquardt, D. W., and R. D. Snee. 1975. Ridge regression in practice. *Am. Statistician* **29**:3–20.

Purcell, W. P., G. E. Bass, and J. M. Clayton. 1973. *Strategy of Drug Design—A Guide to Biological Activity.* Wiley, New York.

Schardein, J. L. 1976. *Drugs as Teratogens.* CRC Press, Cleveland, Ohio.

Shepard, I. H. 1980. *Catalog of Teratogenic Agents,* 3rd ed. Johns Hopkins University Press, Baltimore, Maryland.

Siegel, S. 1956. Pp. 127–136 in *Nonparametric Statistics for the Behavioral Sciences.* McGraw-Hill, New York.

Smith, I. A., G. D. Berger, P. G. Seybold, and M. P. Serve. 1978. Relationships between carcinogenicity and theoretical reactivity indices in polycyclic aromatic hydrocarbons. *Cancer Res.* **38**:2968–2977.

Tinker, J. 1981. Relating mutagenicity to chemical structure. *J. Chem. Inf. Comput. Sci.* **21**:3–7.

Wishnok, J. S., M. G. Archer, E. S. Edelman, and W. M. Rand. 1978. Nitrosamine carcinogenicity: A quantitative Hansch–Taft structure–activity relationship. *Chem. Biol. Interact.* **20**:43–54.

Yuan, M., and P. C. Jurs. 1980. Computer-assisted structure–activity studies of chemical carcinogens. Polycyclic aromatic hydrocarbons. *Toxicol. Appl. Pharmacol.* **52**:294–312.

Yuta, K., and P. C. Jurs. 1981. Computer-assisted studies of chemical carcinogens. Aromatic amines. *J. Med. Chem.* **24**:241–251.

IV

Interpretation of *in Vitro* Experimental Data for Evaluation of Hazards to Humans

15

Assessment of the Hazard of Genetic Toxicity

Ronald W. Hart and David Brusick

A variety of toxic substances have structural features that permit them to reach the genetic material, where they produce damage. If such damage is not satisfactorily repaired, serious consequences for the organism can ensue. The theoretical and empirical bases for this conclusion are summarized in this chapter.

Methods for ascertaining genotoxic activity involve both *in vivo* and *in vitro* systems. Inferences for human health may be based on observations of genetic toxicity in those systems as well as on correlations between this toxicity and other forms of activity, e.g., carcinogenicity and mutagenicity.

Results of chromosomal and clinical studies suggest that 10% of all human diseases are related to one of three forms of genetic damage (National Research Council, 1975):

1. Human birth defects (e.g., Down syndrome, Klinefelter syndrome, and Turner syndrome) arising from either chromosomal nondisjunction occurring in meiosis or translocations between nonhomologous chromosomes during gametogenesis, both of which can result from DNA damage in either the egg or the sperm.
2. Recessive gene disorders (e.g., Tay–Sachs disease, cystic fibrosis, xeroderma pigmentosum, ataxia telangiectasia, and sickle cell anemia) arising from recessive gene mutation.
3. Dominant-gene disorders (e.g., bilateral retinoblastoma and physical malformations) arising from either direct damage to DNA or mutation in a

RONALD W. HART • National Center for Toxicological Research (HFT-1), Jefferson, Arkansas 72079. **DAVID BRUSICK** • Division of Molecular Toxicology, Hazelton Laboratories, Kensington, Maryland 20895.

single dominant gene of either parent (see also Hook, 1983; Matsunaga, 1982). At present, there are no data to indicate what fraction of these germ-cell-related disease categories are due to new mutations.

In addition to possible effects on germ cells, genetic damage in the non-germ (somatic) cells of the body has been associated with such diverse diseases as arteriosclerosis, arthritis, autoimmune disorders, and cancer, as well as aging (Brash and Hart, 1978). It is a reasonable expectation that if cancer is related to alterations in somatic cell genes, then the rate at which at least this somatic cell disease occurs can serve as a barometer for germ cell-related diseases that may not be expressed for many generations in the future.

The present economic impact of damage to the human genome is immense. It is generally estimated that from 1 to 2.5% of the U.S. population is affected by more than 2000 known genetic disorders that fall within the three categories listed above for germ cell damage (National Research Council, 1975). Genetic disorders associated with somatic cells have been estimated to account for approximately 25% of our present health care budget. The health and economic impact of a new or widely used chemical agent should therefore include an analysis of the effects of that agent on the DNA of both the germ and somatic cells. Results of these analyses could be used to prevent further increases in genetic disease among humans and in the associated health care costs; however, this philosophy of gene protection is not widely held among those individuals and institutions responsible for chemical safety evaluation programs.

The assessment of genetic hazard from environmental exposures to man-made chemicals appears to be rather straightforward, since only three pieces of quantitative information are required: (1) the intrinsic genetic toxicity of the chemical; (2) the ability of the biologically active form of the chemical to reach the critical target site per unit of applied dose; and (3) the number of critical target site "hits" required to produce the endpoint measured *in vivo.* However, upon closer inspection of what is involved in quantifying these three phenomena, we find that there is a fundamental lack of the information required to perform a risk analysis for most compounds. Thus, an analysis using current techniques will be at best an estimate based upon limited data and numerous reasonable presumptions. Except for the most widely studied mutagens, very few quantitative risk analyses have been attempted.

Historically, there has been little interest outside the radiation field in performing risk analyses for mutation induction in humans. The reasons for this are not clear, since some of the chemicals that can be considered likely candidates (e.g., ethylene oxide, ethylene dibromide, and epichlorhydrin) have substantial data bases, including *in vivo* mammalian studies for heritable effects.

TERMINOLOGY

The fundamental problem of genetic risk rests with the ability to identify a chemical as a somatic or germ cell mutagen in humans. Since the mutagenicity of chemicals cannot be demonstrated in humans, one must rely on animal models. A "mammalian germ cell mutagen" can be operationally defined as an agent that induces a positive response in one of the *in vivo* assays detecting heritable effects (e.g., the mouse specific-locus mutation assay, the mouse heritable-translocation assay, or one of the dominant mutation tests in rodents). These tests have been shown to measure heritable genetic damage in mammalian germ cells, and they also detect the effects on DNA associated with genetic diseases studied in humans (Ehling, 1980; Generoso *et al.,* 1980; Russell *et al.,* 1981a).

Genetic correlate diseases (GCDs) are disease states that appear to require a genetic lesion for their initiation and that might be amenable to a risk analysis with genetic techniques. There is, however, no proven cause-and-effect relationship between the lesion in the DNA and the onset of the GCDs. The presumed events responsible for GCDs are somatic cell mutations, which are strongly associated with such dysfunctions as cancer, heart disease, and developmental toxicity. Thus, the induction of somatic cell genetic alterations implies corresponding risk for one of the GCDs. Better documentation of the requirement for genetic lesions in these dysfunctions is needed before they can be placed in a category equivalent to human diseases already proven to result from dominant or recessive mutations or chromosome anomalies.

Terms used in subsequent sections describing the development of genetic hazard assessment will be defined as they are introduced. These operational definitions are essential to the basic philosophy proposed for interpretation of results from short-term *in vivo* and *in vitro* assays.

HUMAN HEALTH EFFECTS ASSOCIATED WITH EXPOSURE TO GENOTOXIC AGENTS

Over the past 10 years an increasing number of health effects in humans have been attributed to the effects of genetic toxicity. However, the strength of evidence associating such effects with this toxicity varies greatly.

Documented Health Effects

Humans carry an intrinsic disease burden directly attributable to inheritance (McKusick, 1978). It has been estimated that at least 0.5% of all newborns are directly affected by this human genetic load (National Research Council, 1975). The inherited genetic burden can be divided into two general classes of mutations, genetic and chromosomal, each with several subclasses (e.g., recessive, dominant, autosomal, and X-linked gene mutations, as well as numerical and structural chromosome mutations). Such subclasses predict the inheritance pattern, expressivity, and to some extent severity of the associated disease. Hence, these subclassifications must be considered in risk analysis if it can be shown that a toxic agent preferentially induces one class over another.

Because of the long time required for expression of new genetic alterations in the germ cells of a heterogeneous population, the fate of induced mutations introduced into the human gene pool is extremely difficult to estimate epidemiologically. Some new recessive alleles, for example, may not be expressed for many generations. Thus, there is only a limited recognition of the extent to which genetic diseases are presently being induced by man-made environmental chemicals. Recognition of this limitation is especially important if a high proportion of new mutations induce subtle alterations in gene activity rather than the catastrophic symptoms typically associated with most genetic diseases in humans.

Presumed Human Health Effects

This group of health effects consists primarily of the GCD, i.e., toxic effects in humans that correlate with genetic toxicity, but for which cause-and-effect relationships have not been established. The major GCD is cancer. In fact, most validation efforts undertaken to date have been designed to quantify the correlation of mutagenesis with mammalian carcinogenesis in order to facilitate more rapid analyses of chemicals for their carcinogenic activity (OSTP, 1985a).

There are a number of theories about tumor origin—such as the theory regarding oncogenes—which relate genetic changes with carcinogenesis (OSTP, 1985b). Generally, the results from a wide range of test batteries of genetic toxicity support a correlation between genetic toxicity and carcinogenesis (OSTP, 1985a).

Health Effects Suggesting a Cause Related to Genetic Toxicity

Although many teratogens act through mechanisms that appear not to involve genetic toxicity (e.g., vitamin deficiencies, hormonal regulation of organ growth, and interference with sequenced enzyme activity during organogenesis), a significant body of evidence suggests that there may be an involvement of altered DNA integrity (i.e., mutagens) or DNA replicaton in teratogenesis (Manson, 1981). Examples of the latter include hydroxyurea (blocks S-phase replication), cytosine arabinoside (blocks DNA synthesis), and thalidomide (alters DNA and RNA synthesis). These agents apper to produce cytotoxic effects in the embryo at rapidly differentiating sites, resulting in localized necrosis followed by regenerative processes lacking the fidelity of normal differentiation. In the former case, many chemical mutagens have been shown to be teratogens at levels not producing lethality in the fetus or the mother. Gene mutation or chromosome aberrations might alter the normal differentiation capacity of a critical stem cell, ultimately resulting in the absence or dysfunction of an embryo component.

Other toxic phenomena have also been linked with DNA damage; however, the association is rather speculative at this time. Some theories explaining aging, the etiology of certain types of heart disease, and some neurological diseases have been built on notions of genetic damage (Benditt, 1977; Burnet, 1974; Hart and Turturro, 1981).

METHODS TO IDENTIFY THE HAZARD OF GENETIC TOXICITY

The three approaches currently used to evaluate the probable hazard of these agents for humans are discussed in this section.

Epidemiology

This is the oldest and most directly applicable approach. Because of constraints imposed by the lack of suitable exposed and control groups, historical exposure data, and financial resources, however, this approach is essentially not feasible. Consequently, there are only limited data regarding induced genetic effects in humans (Brusick *et al.,* 1981; National Research Council, 1975; Oftedal and Brøgger, 1986). We do not have sufficient details concerning the frequencies of mutant alleles (of all classes) in the human gene pool to evaluate with certainty any small changes in these frequencies that might be generated from human exposure to genotoxic chemicals.

Sufficient data may become available in the future through major epidemiological studies with serum protein maps, as proposed by Neel *et al.* (1980) and Anderson (1977), as well as through other techniques. But such studies will prove to be major financial undertakings, and their success will depend upon the suitability of the population selected for the analysis.

Animal Models

In vivo animal studies in which exposed gametes, embryos, or the F_1 generation are analyzed directly for induced genetic alterations provide the most scientifically rigorous data base upon which risk analyses for human populations can be developed. There has been a series of recent reviews of information using different models (Green *et al.,* 1985; Lee *et al.,* 1983; Preston *et al.,* 1981; Russell *et al.,* 1981a,b; Valencia *et al.,* 1984); however, the data base from *in vivo* models for germ cell effects is still not extensive enough to make generalizations about chemical classes, and the data base is expanding slowly because of the relatively high cost of the assays.

In addition to germ cell (heritable) results *in vivo,* effects in somatic cells can also be detected *in vivo.* It has not been unequivocally established that somatic cell damage can be directly extrapolated to the initiation of somatic cell diseases (e.g., cancer, terata, and heart disease) or as an indicator of germ cell alterations. There have been some attempts to establish the extent of extrapolation from somatic effects to germ cell effects *in vivo* (Sobels, 1982; Streisinger, 1983), but there are still problems to be overcome before a workable solution is devised.

In Vitro and Submammalian Models

The use of *in vitro* tests and submammalian models offers a flexible and cost-effective means to evaluate significant samples of environmental chemicals for genetic toxicity; however, the resultant data are not very reliable for extrapolations in making risk analyses.

Williams (1979) reviewed a large group of batteries of *in vitro*/submammalian tests to identify chemical mutagens and chemical carcinogens. Although each battery may vary from the others, numerous features appear to be common to all of them.

1. Most test batteries attempt to use target organisms from numerous phylogenetic levels in order to minimize any effects of species specificity.
2. Most test batteries use tests that, in combination, cover the basic types of genetic damage.
3. Most batteries rely heavily on tests for which the most data have been published to demonstrate that the technique is capable of responding to a broad range of chemical classes and genotoxic mechanisms.

There are several important limitations associated with the *in vitro*/submammalian test battery approach to chemical evaluation. Although these limitations do not prevent the use of *in vitro*/submammalian batteries for risk analysis, they clearly point out the need to use these models cautiously. The major limitations are as follows:

1. Mixed responses are to be expected, even when a battery of reliable assays is used; however, few of the test battery approaches offer a method for interpreting and categorizing the chemical when mixed responses are obtained. In fact, most of the batteries have been designed to detect positive responses, not to interpret a test-response profile. In one system that considers mixed responses (Brusick, 1981), a numerical value is attached to the tests and test responses. The final interpretation involves chemical categorization, which is calculated by considering both the positive and negative data derived from the application of a test battery (OSTP, 1985a).
2. Batteries of *in vitro*/submammalian tests are not amenable to an evaluation of specialized metabolism or repair in the *in vivo* target tissue. Integrated metabolic effects occurring in the intact organism cannot be mimicked in these tests.
3. Most *in vitro*/submammalian systems cannot characterize activity modulators such as synergistic effects, promotion phenomena, or immune suppression. Many of these modulators are presumed to influence the potency of a given genotoxic agent.
4. *In vitro*/submammalian systems are constructed to be sensitive indicators for a specific toxic endpoint, but the methods by which the data from these tests can (if at all) be factored into a quantitative assessment of risk have not been established.

In summary, it appears that persons conducting studies and evaluating data in genetic toxicology will be required to make far-reaching decisions on a heterogeneous data base ranging from *in vitro* test results to results from epidemiological studies in humans. The approach to hazard assessment presented in the remainder of this chapter addresses these limitations.

ASSESSMENT OF GENETIC RISK

When assessing hazard for the majority of new chemicals, the best one can hope to achieve is a "degree of concern" for the agent with regard to human health. In order to reach the appropriate degree of concern, or estimate of risk, for a given chemical it is necessary to use all the available data, including *in vitro*/submammalian test results as well as data from animal models or exposures of humans. The following three subsections describe the theoretical considerations and formulation strategy for an approach to determine the degree of concern and the categorization of human risk for genotoxic agents.

DNA As the Critical Target

To develop cause-and-effect relationships between toxic hazards and genetic effects, one must identify which types of genetic damage are responsible for the induction of specific toxicological effects. For example, genetic diseases identified in the human population all arise from dominant or recessive single-gene mutations and structural or numerical chromosome aberrations. Therefore, an analysis directed toward identifying genetic hazards in humans must involve an evaluation of gene mutation, chromosome aberrations, or specific DNA lesions directly leading to these events. The macromolecular events involved in the production of chromosome aberrations or gene mutations include DNA–adduct formation, errors in DNA repair, replication, and synthesis, as well as possibly other phenomena such as sister chromatid exchange (SCE) and mitotic recombination. Significant progress has been made in recent years in both measuring these endpoints and understanding the role of these processes in the induction of somatic and germ cell mutagenesis.

Although DNA, RNA, protein, and lipid are each possible targets for significant molecular alterations induced by toxic substances, there are three reasons to believe that DNA is the primary target macromolecule: (1) the size of cellular DNA, (2) the presence of DNA in unique copy, and (3) the role of DNA as the initial template for all macromolecules, including itself. These intuitive indications that DNA may serve as a principal target for toxic substances are supported by numerous published reports (OSTP, 1985b). When attempting to understand the role of DNA damage in any number of toxicological endpoints, one may interpret most of these studies more broadly than simply the relationship of DNA damage to cancer. The key findings are as follows (OSTP, 1985b):

1. Virtually all major classes of chemical carcinogens are mutagens, either directly or in the presence of metabolic activation.
2. The extent of binding of a chemical agent or its subsequent rate of

removal correlates well with the carcinogenic potential of the agent when the rate of DNA replication is considered.

3. Individuals defective in any of the various forms or steps of DNA repair are predisposed to cancer and other degenerative disease at a frequency inversely related to the degree of deficiency.
4. The direct enzymatic reversal of ultraviolet (UV) light-induced DNA damage results in a corresponding decrease in the resultant number of tumors induced by a given influence of UV.
5. Both mutation frequency and transformation frequency decrease with increasing time for repair prior to DNA replication.

From the results of these and other supportive studies, it is reasonable to infer that DNA damage plays a role in carcinogenesis and other physiological dysfunctions.

DNA damage may be induced by the interaction of cellular DNA with either endogenously or exogenously generated nucleophilic agents or free radicals. The forms of such damage vary as a function of the agent that induces them, the target macromolecule, and specific site of interaction. Thus, the interaction of an electrophilic agent with DNA may lead to many types of damage, including adducts, phosphotriesters, cross-links, strand breaks, hydrations, and dimers (Lawley, 1974; Singer, 1977). Table 15-1 summarizes the general classes of genetic damage. Studies of the mechanisms with which specific agents interact with cellular DNA to produce distortions will lead to a better understanding of how alterations in cellular DNA eventually lead to the three major types of somatic and germ cell mutations.

In the assessment of chemically induced genetic toxicity, it must be remembered that DNA damage occurs spontaneously and accumulates as a function of

TABLE 15-1. Major DNA Lesions Produced by Chemical Interaction and Their Genotoxic Consequences

Primary lesion	Description	Consequence
Alkylation	Covalent adduct formed, involving the genotoxic agent and a DNA base or phosphodiester bridge	Alteration of base pairing, loss of the base, stimulation of error-prone repair
Intercalation	Noncovalent stacking of the genotoxic agent between adjacent base pairs in the DNA helix	Alteration of DNA transcription replication or repair
Cross-linkage	Formation of two covalent bonds between bases within (intrastrand) or between (interstrand) DNA strands	Dimer formation, alteration of replication
Breakage	Scission of either a single or both strands of the DNA helix	DNA rearrangements forming chromosome aberrations after mitotic cell division

time in cellular DNA *in vivo* (Turturro and Hart, 1985). Three examples of naturally occurring DNA damage are briefly reviewed below.

The first mechanism involves loss of a base or base pair due to depurination or depyrimidination. Depurination of cellular DNA is estimated to be tenfold more frequent than depyrimidination. Once the base is lost from the helix, the apurinic/apyrimidinic sites are highly labile, resulting in strand scission (Lindahl and Anderson, 1972; Lindahl and Hyberg, 1972). Repair endonucleases appears to accelerate the rate of chain cleavage (Grecz and Bhatarakamo, 1977).

A second class of spontaneously occurring DNA damage arises from the interaction of free radicals with cellular DNA (Myers, 1973). There are several mechanisms for producing intracellular free radicals that alter the integrity of the DNA structure by covalent binding DNA cross-linking. DNA cross-linking by aldehydes resulting from lipid peroxidation is well documented and may be a dominant mechanism *in vivo* (Tappel, 1973).

The third type of exogenous DNA damage *in vivo* is a decrease in the 5-methylcytosine content of cellular DNA. The biological consequence of this form of damage is potentially significant, but this has yet to be proven (Brash and Hart, 1978).

The significance of spontaneous mutation mechanisms in mutagenesis risk assessment is not completely known, although the frequency of the endogenous events must be considered in constructing model systems to measure induced changes.

The Genetic Toxicity Pathway

To evaluate the hazards to humans resulting from genetic lesions, one must identify the types of genetic damage that are responsible for genetic risk to humans. For example, genetic diseases identified in the human population arise from gene mutations and from chromosome aberrations (Hook, 1983). The somatic mutation theories of cancer and some forms of teratogenesis also implicate these endpoints. Therefore, a reliable analysis for genetic hazards must involve at least an evaluation of the induction of gene mutation or chromosome aberrations. The macromolecular events involved in the production of chromosome aberrations or gene mutations may, however, involve associated processes of DNA binding, repair, DNA replication and synthesis, and possibly other phenomena, such as sister chromatid exchange (SCE) and mitotic recombination.

This concept can be illustrated in a simplistic fashion using a hypothetical pathway from exposure of the test organisms to the production of gene mutations or chromosome alterations. This pathway, shown in Fig. 15-1, demonstrates the interrelationship of the various steps involved in the production of gene mutation and chromosome aberration, the two toxicologically critical endpoints.

Designation of a material as a hazardous genotoxic agent must be based on a demonstration that the material induces gene mutation or chromosome aberrations—not simply that it initiates one of the intervening steps. This is a technically narrow approach because it essentially limits classification of a chemical

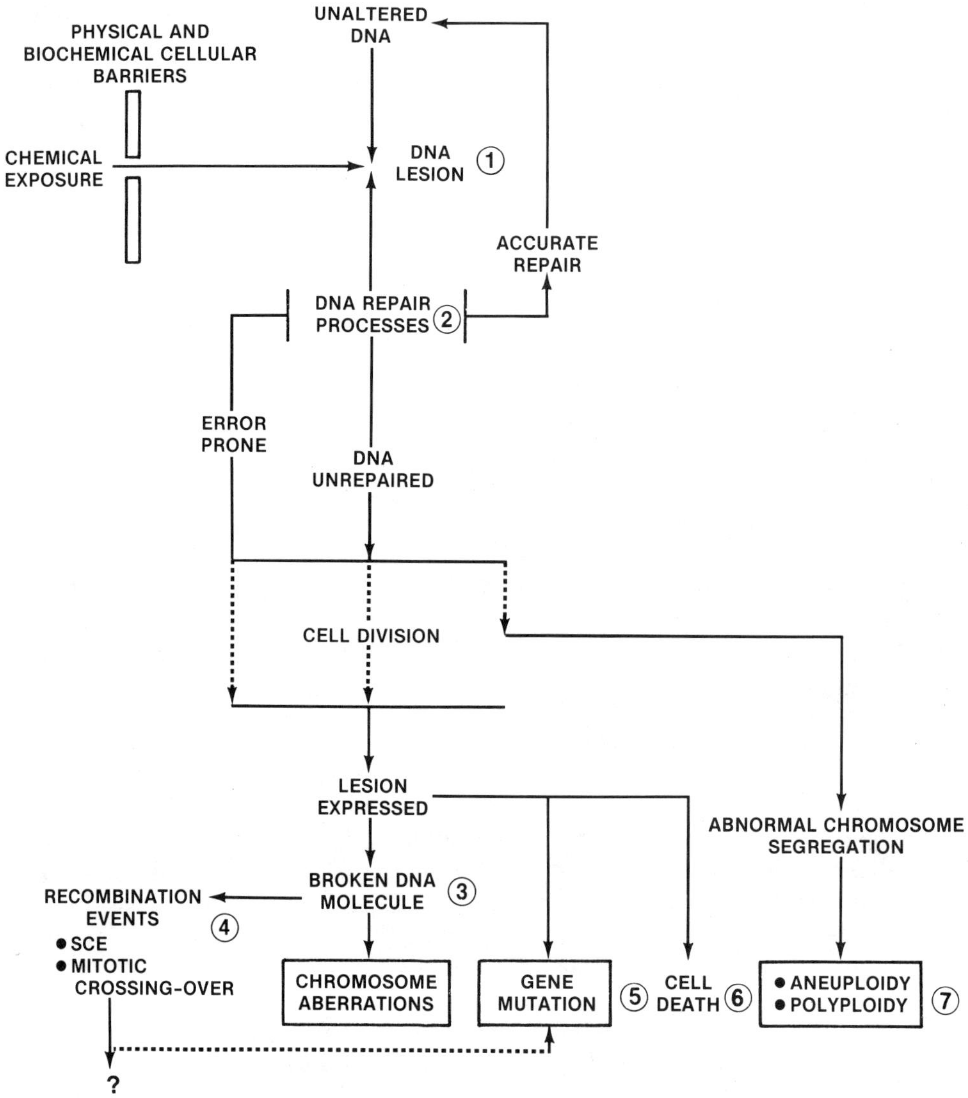

FIGURE 15-1. A hypothetical pathway leading to stable heritable genetic endpoints. The final steps in the process are indicated by the closed boxes. Contributions of possible protein lesions (e.g., in the spindle, for chromosome segregation) are not shown. Genetic testing involves assay systems that measure the primary genetic endpoints directly, or measure phenomena along the pathway that are indicative of DNA damage capable of leading to one of the primary endpoints. These indirect measures of DNA damage are numbered as follows: 1,2, alteration of DNA integrity; 3, induction of DNA exchange or rearrangement; 4,5, alteration of chromosome integrity; 6, alteration of DNA base sequence; 7, alteration of chromosome segregation.

as a hazard only if it can be shown to induce gene or chromosome mutations. However, it is justified by the cause-and-effect relationship of these endpoints to initiation of human disease. Chemicals that only induce DNA repair synthesis or SCEs, for example, cannot be accorded the same health hazard status as chemicals that induce gene mutation or chromosome aberrations, because no known human diseases can be directly linked to these nonmutational endpoints. Consequently, under the proposed approach to hazard assessment, test systems become classified as indicators of genetic hazard in accordance with their association with mutation or chromosome aberration induction. In the following scheme, class 1 is the least indicative, and class 3 the most indicative for human hazard:

Class 3. Assays that directly measure gene mutation or structural and numerical chromosome aberrations.
Class 2. Assays that measure SCE, recombination, DNA repair synthesis, DNA strand breakage, or DNA cross-linking.
Class 1. Assays that measure inhibition or interference with DNA synthesis or replication.

This assay stratification is not necessarily the consensus of genetic toxicologists, but it does rest upon the presumption that a chemical cannot be classified as a potential human hazard unless it can be shown to induce lesions relevant to human disease. Such a philosophy provides the symmetry and logic trail that is amenable to legal interpretation of safety assessment data.

Demonstration that a candidate chemical induces intermediate lesions in the genotoxic pathway, although sufficient to categorize the agents as genetically toxic, is insufficient to establish a defensible cause-and-effect relationship for presumed health effects in humans.

Recommendations for the Collection and Interpretation of Test Data

A direct approach for evaluating different levels of evidence is described in Fig. 15-2. Consideration is given both to the rationale underlying the genotoxic pathway shown in Fig. 15-1 and to specific test system phylogenetic factors believed to be relevant to the extrapolation of data across species lines.

The information in Fig. 15-2 can be consolidated and quantified by placing it in a two-way evaluation table that stratifies the predictive ability of test data from each individual test according to class 1, 2, or 3. Class 3 is the most predictive of genetic risk, since gene mutation and chromosome alterations are assessed directly. The second stratification involves the phylogenetic complexity of the test organism and hence the relevance of the target cell to human cells and to its repair pathways. There are five strata ranging from the least relevant to the most relevant (human) response. This method of data evaluation permits the classification of a chemical using the maximum limits of the test data that might be available for review (i.e., all test results can be factored into this assessment).

For tested materials, there may be data from as few as one test and as many as ten or more. A classification scheme flexible enough to accommodate such vari-

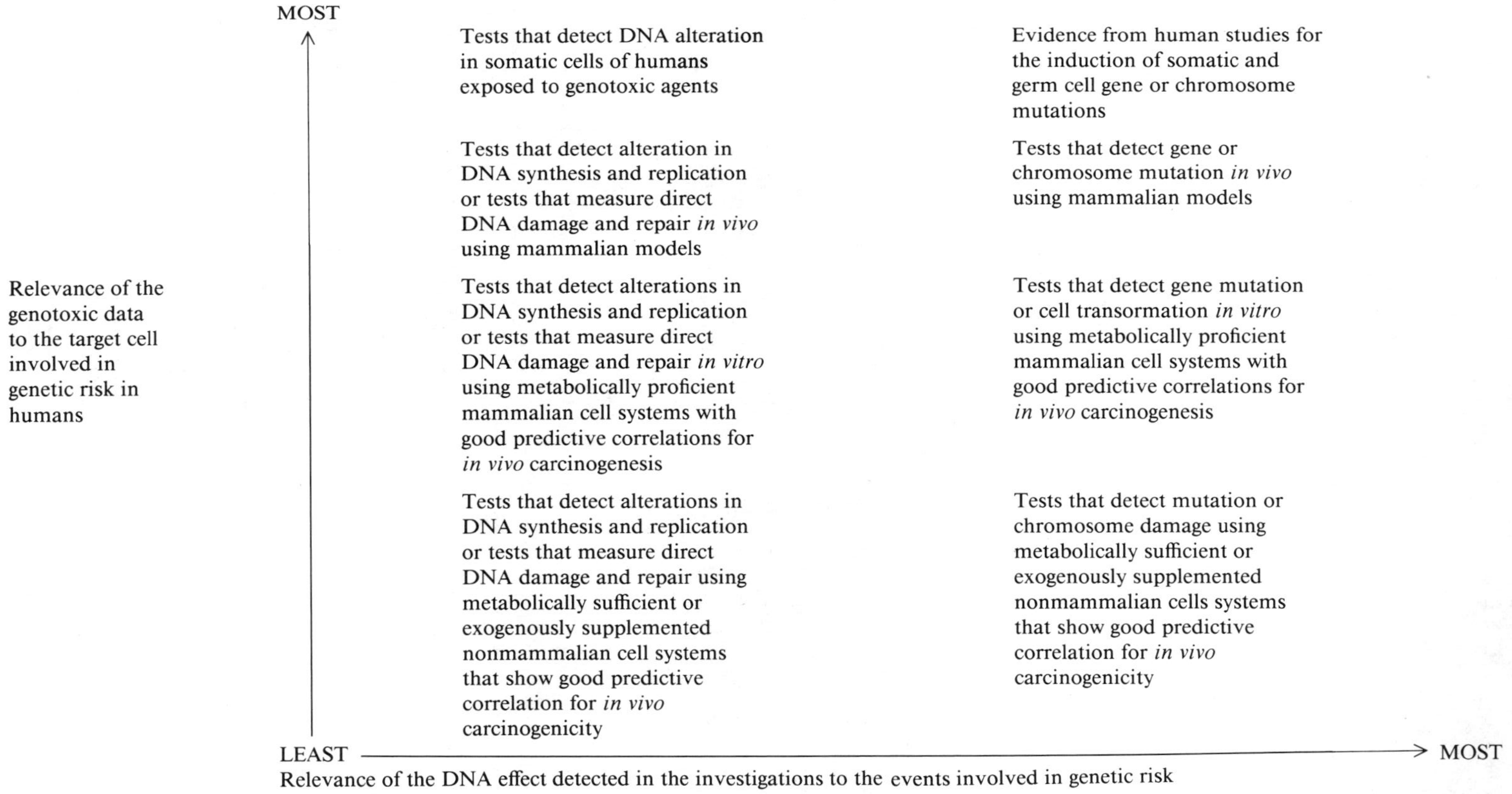

FIGURE 15-2. Two-dimensional analysis of the relevance of test results.

ation is outlined in Table 15-2 and Fig.15-3. This scheme uses the two-dimensional stratification concept to rank test data by the seriousness of the lesions detected in bioassays and the relative phylogenetic similarity of the assay technique to humans. Figure 15-3 permits one to obtain a consensus of the data by producing combinations of numbers and letters representing the extent of the test data. Table 15-2 uses these combinations to classify the chemicals according to their potential hazard. In many cases, the data on a chemical will not be sufficient to extend the classification of a chemical beyond genetically toxic or mutagenic. In a few cases, a chemical may be defined as a potential mutagen in humans. As previously noted, however, the value of this approach lies in its capacity for assessing an entire data base for a chemical, including *in vitro* and *in vivo* data from rodents and humans.

Negative short-term results will not provide reliable assurance of nonmutagenicity and noncarcinogenicity because of the uncertain etiology of disease. The most one might conclude from negative test results from an array of tests that cover all or a great proportion of the combinations in Table 15-1 is that the test chemical is not likely to be an initiating carcinogen.

TABLE 15-2. A Proposed Scheme for Classifying Chemicals on the Basis of Test Results

Chemical classification	Responses required for classification of hazard potential[a]	Implications for GCD (carcinogenic effects)
Genetically toxic agent	Positive response in at least one combination	None, unless other conditions defined below are met
Mutagenic agent	Positive response in at least one class 3 combination	Carcinogenic potential is possible, and the level of probability increases from 3A to 3D tests
Animal cell somatic mutagen[b]	Positive responses in at least one test each from 1C and 2C and at least two tests from 3C	Relevance of test systems, correlation coefficients, and nature of the DNA lesions all point toward carcinogenic activity
Possible animal (human) mutagen[c]	Positive responses in at least two tests each from 2C and 3C plus a positive response in 2D, 3D, or 2E	Relevance of test systems, correlation coefficients, nature of the DNA lesions, and *in vivo* chemical pharmacodynamics all point toward *in vivo* tumorigenicity
Potential human mutagen	Positive results or evidence in 3E	Probable increased risk of tumor induction if exposure persists

[a] See Fig. 15-3 for hazard test categories.
[b] The test material has biological properties in common with materials known to produce cancer in laboratory animals, but the pharmacodynamics in *in vivo* exposure is not known.
[c] The test material both has relevant genotoxic activity in mammalian cells and has been shown to express that activity under *in vivo* test conditions.

Stratification of assays by complexity of bioassay system			Least relevant indicator of somatic mutation →		Most relevant indicator of somatic mutation
Least relevant assay			Class 1	Class 2	Class 3
↓	Plant and prokaryotic systems	A	1A	2A	3A
	Lower euckaryotic microorganisms and insect systems	B	1B	2B	3B
	Mammalian *in vitro* systems	C	1C	2C	3C
	Mammalian *in vivo* systems	D	1D	2D	3D
Most relevant assay	Human *in vivo* data	E	1E	2E	3E

FIGURE 15-3. Data analysis table for categorizing hazard from positive test results.

LIMITATIONS OF QUANTITATIVE AND QUALITATIVE EXTRAPOLATION

Quantitative and qualitative extrapolation is an essential part of risk assessment. It can be dealt with properly only with generalization and, to some extent, speculation.

The limitations of extrapolation are certainly not unique to genetic toxicology. They are a subject of great controversy among toxicologists for their role in studies of experimental carcinogenesis and teratology. For instance, data from animal studies suggest that agents such as aspirin (which is teratogenic in rodents), phenobarbital and saccharin (which are carcinogenic in rodents), and radiation (which is mutagenic in rodents) should be hazardous for humans; yet, extensive human exposure and epidemiological analyses for these materials have failed to confirm the predictions.

Genetic toxicity data ought to be readily extrapolatable under specified conditions. For example, a direct-acting alkylating agent that requires no metabolic activation and is not known to be specifically deactivated *in vivo* will have relatively constant DNA-binding kinetics, regardless of the source of the DNA (e.g., bacteria, insect, rodent, or human). Thus, if one can define mutation induction based on the frequency of alkylations per base pair, an estimate of somatic induction should be possible across phylogenetic levels simply by measuring the binding products in the DNA of presumptive target cells (Lee, 1978).

This approach, which is currently under investigation, may eventually provide a systematic method for extrapolation. However, unforeseen problems have arisen in this otherwise apparently straightforward method. For example:

1. The number of transcribed genes per unit length of the genome changes substantially as one proceeds up the phylogenetic scale. Almost 100% of

the genome of bacteria such as *Escherichia coli* codes for structural or associated potential gene target. As species move up the phylogenetic scale toward humans, the number of functional (transcribed) genes per unit length of the genome is strikingly reduced. In humans it has been estimated that only 1% of the entire genome codes for structural genes. Thus, the critical target size versus total mass of DNA appears to be very disproportionate among different species (Malling and Valcovic, 1978).

2. The mutation-processing steps appear to be different in different species. In bacteria, gene mutation induction is directly involved with specific inducible repair enzymes and mutation-processing genes (Mount *et al.*, 1982). Similar repair steps and mutation-processing genes appear absent or are significantly more complex in eukaryotic cells. This raises questions concerning how the DNA lesions are processed in phylogenetically different organisms.

Consequently, evidence for quantitative extrapolation of data from *in vitro* submammalian and *in vivo* rodent studies directly to humans has not been adequately documented. Considerably more research is needed in this area of genetic toxicology, especially with *in vivo* and *in vitro* tests in which small groups of animals are exposed to the test material under anticipated environmental conditions and the resultant genetic effects are measured in primary or low-passage cells removed from the treated animal. In many cases, this can be accomplished with noninvasive techniques permitting direct comparisons between animal model systems and exposed human populations (Brusick *et al.*, 1981).

CONCLUSIONS

A relative degree of hazard can be derived from genetic testing using data derived from fundamental concepts about relevance of the test organisms and the type of induced genetic lesions to those same phenomena known to produce genetic effects in humans.

The present state of the art in genetic toxicology does not permit accurate quantitative risk analysis to be performed. Because research in this area is making signficant progress, however, such hazard evaluations may be possible in the not-too-distant future.

REFERENCES

Anderson, L., and N. G. Anderson. 1977. High resolution two-dimensional electrophoresis of human plasma proteins. *Proc. Natl. Acad. Sci. USA* **74**(12):5421–5425.

Benditt, E. P. 1977. The origins of atherosclerosis. *Sci. Am.* **235**:74–85.

Brash, D. E., and R. W. Hart. 1978. DNA damage and repair *in vivo. J. Environ. Pathol. Toxicol.* **2**:79–114.

Brusick, D. 1981. Unified scoring system and activity definitions for results from *in vitro* and submammalian mutagenesis test batteries. Pp. 273–286 in E. D. Copanhaven, C. R. Richmond, and P. J. Walsh, eds. *Health Risk Analysis.* Franklin Institute Press, Philadelphia.

Brusick, D., F. J. deSerres, R. B. Everson, M. L. Mendelsohn, J. V. Neel, M. D. Shelby, and M. D. Waters. 1981. Monitoring the human population for mutations and DNA damage. Pp. 111–140 in A. D. Bloom, ed. *Guidelines for Studies of Human Populations Exposed to Mutagenic and Reproductive Hazards.* March of Dimes Foundation, New York.

Burnet, F. M. 1974. *Intrinsic Mutagenesis: A Genetic Approach to Aging.* Medical and Technical Publishing, Lancaster, England.

Ehling, U. H. 1980. Induction of gene mutations in germ cells of the mouse. *Arch. Toxicol.* **46:**123–138.

Generoso, W. M., J. B. Bishop, D. G. Gosslee, G. W. Newell, C. J. Sheu, and E. von Halle. 1980. Heritable translocation test in mice. *Mutat. Res.* **76:**191–215.

Grecz, N., and S. Bhatarakamo. 1977. Apurinic acid endonuclease implicated in DNA breakage in *Escherichia coli* subjected to mild heat. *Biochem. Biophys. Res. Commun.* **77:**1183–1188.

Green, S., A. Auletta, J. Fabrikant, R. Kapp, M. Manandhar, C. Shen, J. Springer, and B. Whitfield. 1985. Current status of bioassays in genetic toxicology—The dominant lethal assay. A report of the United States Environmental Protection Agency Gene-Tox Program. *Mutat. Res.* **154:**49–67.

Hart, R. W., and A. Turturro. 1981. Evaluation and longevity-assurance processes. *Naturwissenschaften* **68:**552–557.

Hook, E. B. 1983. Perspectives in mutation epidemiology. 3. Contribution of chromosome abnormalities to human morbidity and mortality and some comments upon surveillance of chromosome mutation rates. *Mutat. Res.* **114:**389–423.

Lawley, P. D. 1974. Alkylation of nucleic acids and mutagens. Pp. 17–33 in L. Prakash, F. Sherman, M. Miller, C. Lawrence, and H. Taber, eds. *Molecular and Environmental Aspects of Mutagens.* Thomas, Springfield, Illinois.

Lee, W. R. 1978. Dosimetry of chemical mutagens in eukaryote germ cells. Pp. 177–202 in A. Hollaender and F. J. de Serres, eds. *Chemical Mutagens: Principles and Methods for Their Detection,* Volume 5. Plenum Press, New York.

Lee, W. R., S. Abrahamson, R. Valencia, E. S. von Halle, F. E. Wurgler, and S. Zimmering. 1983. The sex-linked recessive lethal test for mutagenesis in *Drosophila melanogaster.* A report of the United States Environmental Protection Agency Gene-Tox Program. *Mutat. Res.* **123:**183–279.

Lindahl, T., and A. Anderson. 1972. Rate of chain breakage at apurinic sites in double-stranded deoxyribonucleic acid. *Biochemistry* **11:**3618–3623.

Lindahl, T., and B. Hyberg. 1972. Rate of depurination of native deoxyribonucleic acid. *Biochemistry* **11:**3610–3618.

Malling, H. V., and L. R. Valcovic. 1978. New approaches to detection of gene mutations in mammals. Pp. 149–171 in G. Flamm and M. Mehlman, eds. *Advances in Modern Toxicology,* Volume 4. Hemisphere Press, New York.

Manson, J. M. 1981. Developmental toxicity of alkylating agents: Mechanism of action. Pp. 95–129 in M. R. Juchau, ed. *The Biochemical Basis of Chemical Teratogenesis.* Elsevier/North-Holland, New York.

Matsunaga, E. 1982. Perspectives in mutation epidemiology. 1. Incidence and prevalence of genetic disease (excluding chromosome aberrations) in human populations. *Mutat. Res.* **99:**95–128.

McKusick, V. A. 1978. *Mendelian Inheritance in Man: Catalogs of Autosomal Recessive and X-Linked Phenotypes, 5th ed.* John Hopkins University Press, Baltimore, Maryland.

Mount, D. W., J. W. Little, B. Markham, H. Ginsburg, C. Yanish, and S. Edmonston. 1982. Mechanisms of mutagenesis in bacteria. Pp. 105–111 in T. Sugimura, S. Kondo, and H. Takebe, eds. *Environmental Mutagens and Carcinogens.* Alan R. Liss, New York.

Myers, L. S. 1973. Free radical damage of nucleic acids and their components by ionizing radiation. *Fed. Proc.* **32:**1832–1894.

National Research Council. 1975. *Genetic Screening, Programs, Principles, and Research.* National Academy of Sciences, Washington, D.C.

Neel, J. V., C. Satoh, H. B. Hamilton, M. Otake, K. Goriki, T. Kageoka, M. Fugita, S. Neriishi, and A. Asakawa. 1980. Search for mutations affecting protein structure in children of atomic bomb survivors: Preliminary report. *Proc. Natl. Acad Sci. USA* **77**(7)**:**4221–4225.

Oftendal, P., and Brøgger, A., eds. 1986. *Risk and Reason: Risk Assessment in Relation to Environ-*

mental Mutagens and Carcinogens. Progress in Clinical and Biological Research, Volume 208. Alan R. Liss, New York. 189 pp.

OSTP (Office of Science and Technology Policy). 1985a. Chapter 2: Short-term tests for potential carcinogens. *Fed. Reg.* **50:**10372; 10403–10410.

OSTP (Office of Science and Technology Policy). 1985b. Chapter 1: Current views on the mechanisms of carcinogenisis. *Fed. Reg.* **50:**10372; 10379–10403.

Preston, R. J., W. Au, M. A. Bender, J. G. Breaven, A. V. Carrano, J. V. Heddle, A. F. McFee, S. Wolff, and J. S. Wassom. 1981. Mammalian *in vivo* and *in vitro* cytogenetic assays. A report of the United States Environmental Protection Agency Gene-Tox Program. *Mutat. Res.* **87:**143–188.

Russell, L. B., P. B. Selby, E. S. von Halle, W. Sheridan, and L. Vakovic. 1981a. The mouse specific-locus test with agents other than radiations. Interpretation of data and recommendations for future work. *Mutat. Res.* **86:**329–354.

Russell, L. B., P. B. Selby, E. S. von Halle, W. Sheridan, and L. Vakovic. 1981b. Use of the mouse spot test in chemical mutagenesis: Interpretation of past data and recommendations for future work. *Mutat. Res.* **86:**355–379.

Singer B. 1977. Sites in nucleic acids reacting with alkylating agents of differing carcinogenicity or mutagenicity. *J. Toxicol. Environ. Health* **2:**1279–1295.

Sobels, F. H. 1982. The parallelogram. An indirect approach for the assessment of genetic risks from chemical mutagens. Pp. 323–327 in K. C. Bora, G. R. Douglas, and E. R. Nestmann, eds. *Progress in Mutation Research,* Volume 3. Elsevier, Amsterdam.

Streisinger, G. 1983. Extrapolations from species to species and from various cell types in assessing risks from chemical mutagens. *Mutat. Res.* **114:**93–105.

Tappel, A. L. 1973. Lipid peroxidation damage to cell components. *Fed. Proc.* **32:**1870–1874.

Turturro, A., and R. W. Hart. 1984. DNA repair mechanisms in aging. Pp. 19–46 in D. G. Scarpelli and G. Migaki, eds. *Comparative Pathobiology of Major Age-Related Diseases.* Alan R. Liss, New York.

Valencia, R., S. Abrahamson, W. R. Lee, E. S. von Halle, R. C. Woodruff, F. E. Wurgler, and S. Zimmering. 1984. Chromosome mutation test for mutagenesis in *Drosophila melanogaster.* A report of the United States Environmental Protection Agency Gene-Tox Program. *Mutat. Res.* **134:**61–88.

Williams, G. M. 1979. The status of *in vitro* test systems utilizing DNA damage and repair for the screening of chemical carcinogens. *J. Assoc. Off. Anal. Chem.* **62:**857–863.

16

Evaluation of Xenobiotic Metabolism

James R. Gillette and Ronald W. Estabrook

In the broadest sense of the word, a toxicant is any substance that causes changes in the function, structure, or replication of cells or the maintenance of homeostasis in a tissue of any living organism. Ordinarily, a toxicant is considered to be a substance in the environment that gains entrance into the body by inhalation, ingestion, injection, or absorption. But a toxicant may also be formed within the organism. Indeed, it may be a neurotransmitter, a hormone, or a metabolic acid that at low concentrations plays a vitally important role in maintaining the well-being of the organism, but at higher concentrations triggers an abnormal response.

To identify such biologically active substances from either an endogenous or exogenous source, studies in animals are performed according to elaborate protocols in which an exogenous substance is administered as a concentrated solution or suspension in a single dose or in multiple doses for periods ranging from a day to several years. Even the most subtle changes in the concentrations and activities of endogenous substances, cellular morphology, or animal behavior are then assessed by sophisticated methods and instruments. Such studies obviously can reveal whether an exogenous substance, either directly or indirectly, can cause changes in test animals under the conditions of the experiment. However, such experiments may not provide conclusive scientific estimates of incidence rates of the effects of such chemicals when one considers another animal strain or species or a given group of humans.

The dose that causes a 50% incidence rate ID_{50} in a given strain of animals seldom will be the same as the ID_{50} in another strain of the same animal species or for a given strain of another animal species, including humans. Moreover,

JAMES R. GILLETTE • Laboratory of Chemical Pharmacology, National Heart, Lung, and Blood Institute, Bethesda, Maryland 20892. RONALD W. ESTABROOK • Department of Biochemistry, Southwestern Medical School, University of Texas Health Science Center, Dallas, Texas 75235.

exposure to toxic substances in the environment frequently occurs at levels significantly lower than those used for testing. This requires an extrapolation of the log dose–response data, often by many orders of magnitude. Thus, one must remain skeptical of the belief described by Mantel and Bryan (1961) and by Whittemore (1978) that the incidence rate obtained at a low dose in one population of animals may be similar to that obtained in another population of animals, even if it were possible to demonstrate that the ID_{50} values were identical in the two groups of animals tested.

The slope of a log dose–incidence rate curve is a measure of the variance of the response within the population of experimental animals tested. Because toxicological tests are much more sensitive and easily interpreted when the slopes of the log dose–incidence rate curves are large, the strains of animals selected for such tests must be as genetically homogeneous as possible. The human species, however, is very heterogeneous and comprises many subpopulations. Thus, a given test animal strain may be reflecting the response of only a fraction of the human population, whereas results obtained with another test animal strain or species of animals would be representative of another human subpopulation.

Because heterogeneous human subpopulations are widely distributed on earth, it seems unlikely that the variance in incidence of the response to any toxicant by the total human population will ever be precisely known. Since there are approximately 5 billion of us on earth, however, it seems probable that the range of differences could and will be very large. Indeed, it seems reasonable that the range of incidence rates among the subpopulations may be nearly as large as the range of incidence rates that occur among the various species and strains of laboratory animals used to test toxicants. The endpoint used by investigators as a measure of toxicity may be remarkably different when comparing results obtained from studies in humans relative to those from experiments in laboratory animals. It may well be that virtually any incidence of toxicity obtained with virtually any species or strain of animal may mimic that obtained with one of the human subpopulations. Even contradictory results may be relevant to different subpopulations. It is prudent, therefore, to assume that qualitative results obtained in toxicity studies with any animal species are possibly relevant to some fraction of the total human population. Since it is a breach of ethics to expose humans to known or presumed toxic chemicals in prospective studies, the proportion of the total human population represented by the sensitive subpopulation cannot be known unless or until extensive epidemiological studies are evaluated from the exposure history of a significant number of humans.

Epidemiological studies, however, also have their limitations. They are inherently rather insensitive, especially when similar symptoms may arise from several different causes. They are also difficult to evaluate when the exposures to a given toxicant are difficult to assess and relatively uniform throughout the human population.

It is not always possible, nor even desirable, to exclude entirely a chemical from the environment, even though the chemical has been shown to cause toxic responses in laboratory animals and other test systems. Many of the toxic chemicals may be produced naturally by microorganisms, plants, and animals; some are products of combustion; some have been developed as pesticides or medicines; and some are formed during the synthesis of economically important chemicals. Therefore, it is important to be able to identify the human subpopulations and the animal species that are likely to be unusually sensitive to the toxic effects of given chemicals.

CATEGORIES OF TOXIC CHEMICALS

One can subdivide toxic chemicals into three general groups.

1. The first group is made up of chemicals that initiate an adverse response by direct interaction with an essential cellular constituent required for the maintenance of the metabolic function or genomic expression of the cell. Included in this group are many heavy metal ions, such as those of mercury, which react with sulfhydryl groups of proteins, and chemicals that are powerful alkylating agents, such as the nitrogen mustards. A subgroup of this type contains agents, such as asbestos, that can cause a physical modification of the cell (Doull *et al.,* 1980.)

2. The second group comprises chemicals that by themselves are not toxic, but require metabolic activation, leading to the formation of metabolites that are highly reactive to cellular constituents. The large number of examples in this category include polycyclic hydrocarbons, which are converted to epoxides that can then react with nucleophiles such as DNA, and halogenated alkanes, which are metabolically transformed to very reactive free-radical species (Snyder *et al.,* 1982).

3. The third group is also represented by chemicals that are not toxic *per se.* However, these chemicals participate in enzymatic reactions that result in the generation of other products of high toxicity. In this group are quinone-like substances that undergo reduction and oxidation, resulting in the formulation of oxygen reduction products, such as the superoxide anion radical or hydrogen peroxide, or result in a change in redox systems of various cells, such as the NADH/NAD^+, NADPH/NAD^+, or GSH/GSSR ratios (Kappus and Sies, 1981). A variant of this group is a chemical, such as carbon monoxide, that interferes with the transport of oxygen from the lungs to tissues, thereby leading to high levels of cellular lactic acid as a consequence of hypoxia.

Many problems arise when attempting to determine why some animal strains and species are more sensitive than others to different groups of toxicants. A given chemical may be responsible for manifesting several different types of toxicities through different mechanisms. Indeed, a given chemical rarely causes only one kind of response. Thus, it becomes important to elucidate the sequence of events by which the chemical causes a given toxic effect in a given animal species.

Despite the plethora of mechanisms by which a chemical can cause toxic response, there are three key steps to any mechanism:

1. The time course (i.e., the increase and decrease) of the concentration of the active form(s) of the toxicant at putative action (target) sites.

2. The kinetics of the interaction between the active form(s) of the toxicant and action sites.
3. The kinetics of the events that result from the interaction and lead to the observed biological response.

There can be species or strain differences in animals or differences among human subpopulations in any or all of the three parts of the mechanism as described. Moreover, any or all of the three parts may be modified by disease or by various treatments or environments. Thus, it is possible that steps 2 and 3 may be nearly identical in two groups of animals, but that the incidence of the toxic response may be markedly different because step 1 does not share a common limiting property. It is also possible, however, that all three parts of the mechanism are different in two groups of animals, but that the overall responses in the groups may be similar because the variables in the three parts cancel each other.

Each of the three parts may be complex. Indeed, many factors and pharmacokinetic relationships can affect the time course of the concentration of a parent compound in various organs of the body, including those that contain target cells. Some of these factors are described in Table 16-1.

Each of these factors can play a dominant role in governing the concentration of a compound within target cells, but their relative importance varies with the compound and the animal species tested. One of the objectives of disposition studies of a given foreign compound is to determine which factors play dominant roles, and therefore need to be studied, and which factors play trivial roles, and therefore may be ignored. For example, if the compound under study is actively transported into target cells, attempts to relate the concentration of the compound in blood to the incidence of toxicity in different animal species are likely to be disappointing. A species difference in the effectiveness of the transport system may be a dominant reason for species differences in the incidence of toxicity. Such differences are not usually revealed by pharmacokinetic studies of concentrations in blood, however, because only the gross disposition of the substance is assessed in such studies (Gibaldi and Perrier, 1975).

By contrast, when a substance enters and leaves cells by passive diffusion, pharmacokinetic studies are very useful in predicting the relationship between the intracellular and plasma concentrations of the unbound substance. With such substances, species differences in other factors, such as the rates at which the substance is absorbed or reversibility of its binding to tissue components, may become the dominant reason for species differences in the incidence of toxicities after single doses of the toxicant. In considering the effects of absorption rate and the reversible binding to tissues, however, the investigator should recognize that differences in these factors may not be important during multidose administration. After a single dose, the maximum concentration and the duration of effective concentrations of an unbound substance in cells can be affected markedly by the rate of absorption or the reversible binding of the chemical. After multidoses, however, the average concentration of unbound substance in cells will usually be independent of the chemical's reversible binding to cellular components and will

TABLE 16-1. Factors Affecting the Concentration of a Compound in Body Organs

- Single doses versus multiple doses
- Routes of administration
  - Rates of absorption
  - Bioavailability from site of administration
  - First-pass effects on the way to the systemic circulation
- Rates of distribution to organs
  - Blood flow through nonelimination organs
  - Diffusional clearance into and out of tissue cells
  - Active transport into target cells
  - Reversible binding to components in blood and organs and pH differences between blood and cells
- Elimination by excretion and metabolism:
  - Major organs of elimination (kidney versus liver)
  - Flow-limited versus diffusion-limited versus enzyme- (or carrier-) limited elimination in each organ
  - Concentration of unbound toxicant relative to K_m value of complexes between the compound and enzymes or carriers
  - Intrinsic clearance of enzymes (V_{max}/K_m) in each major organ:
    - Genetic factors that govern the amounts of constitutive enzymes and the induction of various forms of enzymes that catalyze the metabolism of foreign compounds
    - Presence of inducers or activators
    - Presence of reversible and irreversible inhibitors
    - Product inhibition
    - Cofactor concentrations
- Cycles
  - Flow cycles (blood–brain barrier and cerebrospinal flow)
  - Ion trapping followed by reabsorption (gastrointestinal cycle of weak bases)
  - Excretion followed by absorption (enterohepatic cycle; renal–bladder cycle)
  - Metabolic cycles:
    - Biliary metabolite followed by hydrolysis and reabsorption
    - Metabolic interconversion (*N*-demethylation followed by *N*-methylation)

usually be dependent on the absorption rate only to the extent that changes in that rate affect the amount of the dose that is absorbed (i.e., the bioavailability). Indeed, reversible binding to tissue components would be expected to affect the steady-state concentration of the unbound chemical only when the concentration approaches or exceeds the K_m values of the enzymes (or carriers) that participate in the elimination of the substance from the body. In addition, reversible binding to plasma proteins will affect the average steady-state concentration of unbound chemical when the blood flow through organs of elimination limits the clearance of the substance from the body (i.e., the first-pass effect).

METABOLIC ACTIVATION

If toxicities were caused only by parent substances, attempts to relate species differences in the disposition of the substances to differences in toxicity incidence would be rather simple. It would be important to study the total body clearance and the elimination half-lives of the chemicals from the body, but it would not be especially important to identify the products formed by the enzymes that metabolize the substances. Many years ago, however, it became evident that metabolites of foreign compounds also cause toxicities. Other than drugs that act reversibly with receptor sites in target tissues, the majority of chemicals now known or thought to initiate a toxic response require metabolic transformation by enzymes in the cell to form reactive metabolites. It is generally accepted that the responsible toxic metabolites have strong electrophilic characteristics providing a needed functional group for attack on a nucleophilic receptor. This electrophilic characteristic is frequently gained when the parent compound undergoes oxidation by enzymes that react with molecular oxygen to generate "active oxygen." Such enzymes are called oxygenases (i.e., oxygen-fixing enzymes). In special circumstances a second type of oxidative reaction, a peroxidatic reaction, can also result in a modification of the substrate. In this case hydrogen peroxide or organic peroxides, such as lipid peroxides, can serve as necessary cofactors for the reaction. The metabolic oxidation of many chemicals is known to result in the formation of epoxides, *N*-hydroxy metabolites, free radicals, and other reactive metabolites, as shown in Fig. 16-1.

There are at least two types of enzymes in mammalian tissues that are now known to oxidatively metabolize many of these chemicals. One type is represented by the family of heme proteins, the cytochromes P450, which serve as the oxygen- and substrate-interacting enzymes for many different types of chemicals (Coon *et al.,* 1980; Kato and Sato, 1982). This type of heme protein is widely distributed in tissues of the body, where it not only participates in the oxidation of xenobiotics, but also plays a central role in the metabolism of a number of endogenous compounds, such as steroids and fatty acids. A second type of oxygen-reacting enzyme is a flavoprotein—the microsomal amine *N*-oxidase—which catalyzes the metabolism of many secondary and tertiary amines (Zeigler, 1978, 1980). Both types of enzymes are membrane-bound and require the donation of reducing equivalents (electrons) from reduced pyridine nucleotides as a necessary prerequisite for initiation of oxygen activation for substrate metabolism. In addi-

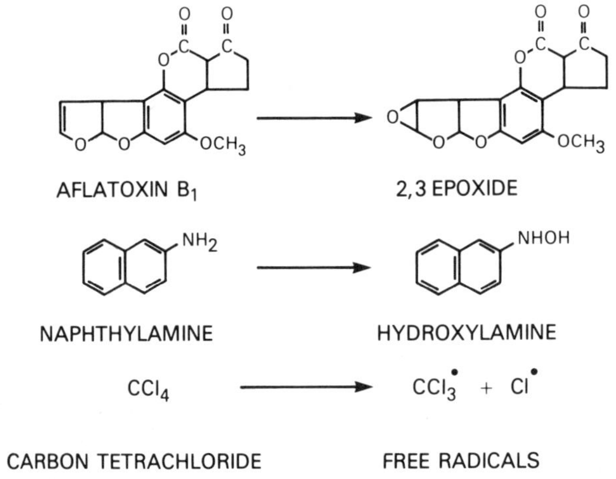

FIGURE 16-1. Three examples of metabolic activation of chemicals leading to reactive metabolites capable of initiating a toxic response.

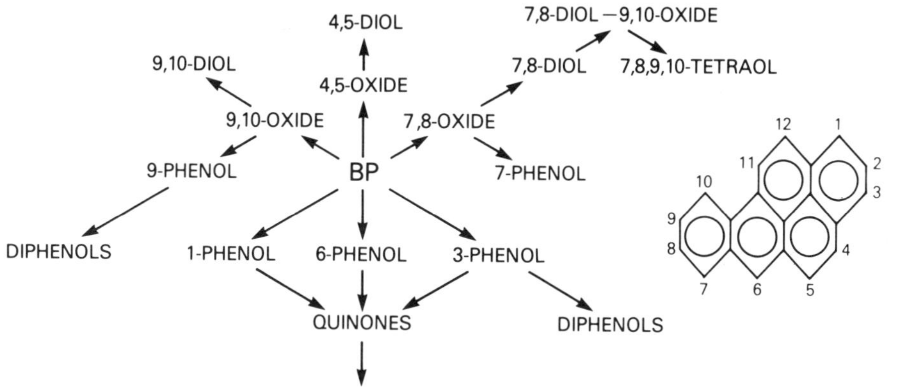

FIGURE 16-2. Multiple metabolites formed during the oxidative biotransformation of the polycyclic hydrocarbon benzo(*a*)pyrene.

tion, the remarkably high concentrations of P450s in many tissues and the ability to change the isozyme inventory of types of P450s following exposure of animals to different inducing agents provide an opportunity to evaluate the role of these enzyme systems in the metabolism of a number of different chemicals.

Greatest interest has focused on the function of cytochromes P450, largely because their heme protein properties make them amenable to study by a number of different biochemical and biophysical techniques. Many of the details of the enzymology associated with substrate interaction, kinetics of reduction, and reaction with oxygen have been described in reviews by Gustafsson *et al.* (1980) and White and Coon (1980). The key reactions of "oxygen activation" and product formation remain to be defined more accurately, although a number of promising hypotheses have been proposed and are currently under investigation. In addition to their function as oxygenases, cytochromes P450 have been shown *in vitro* to serve as both oxidases, leading to the formation of hydrogen peroxide, and as catalysts for the molecular oxygen-independent peroxidatic metabolism of some substrates. Likewise, the flavoprotein microsomal amine *N*-oxidase has been isolated and purified and many of the properties associated with its enzymatic function have been described (Ziegler, 1978, 1980).

The study of toxic metabolites generated during the *in vivo* action of enzymes such as cytochrome P450 or the amine *N*-oxidase is complicated, since both the amount and properties of these enzymes will change as a consequence of a large number of variables. Differences occurring in the profile of metabolites formed during the oxidative metabolism of a compound such as benzo(*a*)pyrene (Fig. 16-2) depend on many factors, including the species of animal selected for study, the organ used as the source of enzymes, the sex of the animal, age, diet, and degree of exposure to various chemicals that can preferentially induce one or more forms of the enzyme. Cytochrome P450 is known to represent a family of isozymes with markedly different physical properties and enzymatic activities. Likewise, the flavoprotein amine *N*-oxidase activity of a tissue can vary with age, sex, and alterations in the steroid hormone balance of the animal. Because of this plethora of perturbing factors, the inability to observe the formation of a toxic metabolite under one set of specified conditions does not exclude the possibility that metabolic activation of the same chemical will form a toxic product under a different set of conditions. In addition, differences in the type of reaction catalyzed, i.e., the oxygenases or peroxidase activity of P450, can lead to significantly different product profiles.

In some instances, the oxidative metabolism of a chemical will not lead directly to a toxic metabolite *per se,* but instead will result in the formation of a product that can participate in other enzymatic reactions to form a different type of toxic product. For example, the cytochrome P450-directed oxidation of some aromatic compounds can result in the formation of quinones. These quinone products can undergo enzymatic reduction to hydroquinones, which are readily oxidized by molecular oxygen to form the superoxide anion radical or hydrogen peroxide. The latter are recognized toxic agents that can react with sulfhydryl groups of proteins or initiate the scission of DNA. In this case, the quinone product of metabolism is not itself the toxic chemical, but is merely the facilitating agent leading to a different type of toxic chemical.

The means to cope with this diversity of enzymatic reactions when attempting to evaluate the potential toxicity of a chemical remain a central problem of toxicology. Differences between *in vitro* results and the reactions occurring *in vivo* require extensive study before a decision can be made that any given chemical is responsible for a toxic reaction. Clearly, metabolic activation is an important part of any evaluation of the toxicity of a chemical. However, detailed knowledge of the types of products formed and the factors regulating the enzymes responsible for their formation or inactivation remain an important objective of studies to determine the safety or hazard associated with chemical exposure.

CHARACTERISTICS OF REACTIVE METABOLITES

Because the half-lives of chemically reactive metabolites may vary from milliseconds to many hours, the strategies used to detect their formation differ markedly. Even the kinds of toxicity that might be expected to be caused by a given chemically reactive metabolite depend on the half-life of the metabolites within cells in the body. It is thus useful to group chemically reactive metabolites into five functional categories: ultrashort-lived, short-lived, transitional, long-lived, and ultralong-lived.

Ultrashort-lived, chemically reactive metabolites never leave the enzyme that catalyzes their formation, but frequently combine irreversibly with the enzyme and thereby inhibit it. The precursors of such metabolites have been called "suicide substrates." This unique class of enzyme inhibitors includes allyl compounds (De Matteis and Cantoni, 1979; Ortiz de Montellano *et al.,* 1978) and 5-fluorouracil (Singer and Orndarza, 1980). Since these reactive metabolites do not combine with any other cellular component, the toxic effects are limited to those resulting from a decrease in the activity of the specific enzyme. Because of the presumed specificity of these metabolites, the development of suicide enzyme inhibitors may thus have therapeutic significance.

Short-lived, chemically reactive metabolites never leave the cells in which they are formed. Any toxicity caused by such metabolites would thus be restricted to cells that contain the enzymes catalyzing their formation. Because these metabolites presumably could react in a nondiscriminating way with several different intracellular components, the resulting responses would rarely be therapeutically useful, but may be responsible for cellular toxicities. One example of this type of chemical would be carbon tetrachloride. Chemicals of this type also may undergo further metabolic modification, leading to secondary metabolites that may evoke either similar or different toxic effects or be innocuous.

Transitional chemical reactive metabolites may leave the cells in which they are formed and enter neighboring cells, but they are inactivated in blood and in other cells within the organ so rapidly that they never leave the organ in which they are formed. The half-lives of the metabolites in this category approach the residence time of the blood within the organ, which in liver is approximately 1–10 sec. As yet, no chemicals have been conclusively placed in this class, although their existence is theoretically possible.

Long-lived, chemically reactive metabolites may leave the cells in which they are formed and become distributed throughout the body, but are not rapidly excreted into bile or urine. Included in this category are bromobenzene-3,4-oxide (Lau *et al.,* 1984) and perhaps *N*-hydroxyacetominophen (Gillette *et al.,* 1982). Such metabolites could contribute to the toxicities in virtually any organ in the body. The organs that manifest the toxicity may either lack mechanisms to inactivate the reactive metabolite or may contain unique target substances that readily react with the metabolite.

Ultralong-lived, chemically reactive metabolites are sufficiently stable that they will not only be distributed throughout the body, but also will be excreted into urine or bile. Although such metabolites may cause toxicities in virtually any organ of the body, their high concentration in urine or bile may lead preferentially to toxicities in the urinary tract or in the intestines. Some bladder and intestinal cancers, for example, might be caused by such long-lived metabolites.

In addition to these general categories, metabolic cycles and sequential generation of short-lived metabolites may mimic the long-lived and ultralong-lived metabolic categories. For example, a short-lived metabolite may leave cells as a glucuronide, which is excreted into either bile or urine, where it undergoes hydrolysis to the short-lived metabolite. This pathway also offers the opportunity for action by bacteria in the intestine, leading to additional chemical derivatives that may have increased toxicity. Alternatively, short-lived epoxides may be converted to dihydrodiols and phenols, which are transported to other organs, where they are converted to other short-lived chemically reactive metabolites. Moreover, a given reactive metabolite may belong to different categories in different tissues, animals, or situations, because the half-life depends on the activities of the enzymes that convert it to stable metabolites. Furthermore, a given substance may be converted to different primary metabolites belonging to different categories. For example, it may be converted to several different epoxides that are converted to stable metabolites at different rates.

STRATEGIES FOR STUDYING THE REACTIVITY OF A TOXIC CHEMICAL

The strategies used to study the formation and toxic effects of chemically reactive metabolites will differ with the category of toxic chemical. It may be possible, for example, to isolate, identify, and synthesize ultralong-lived and long-lived metabolites and to evaluate their toxic potential after their administration to a test system. Moreover, the formation of transitional, long-lived, and ultra-long-lived metabolites might be assessed by measuring metabolites covalently bound to components of blood in living animals or humans (Ehrenberg *et al.*, 1977; Golkar, 1983).

These strategies, however, are clearly not feasible for studying the formation of chemically reactive metabolites in the short-lived and ultrashort-lived categories. Metabolites in these categories can be detected only indirectly by measuring adducts to proteins or nucleic acids in organs containing the enzymes that catalyze their formation. Even if it were possible to synthesize such metabolites, possible toxicities cannot be studied by administering the synthetic metabolite to animals, because there could be no way of determining how much, if any, of the exogenous toxic chemical was delivered to the target site of a target organ.

Several strategies have been developed to determine whether a given toxicity is caused by a chemically reactive metabolite or parent substance. In one strategy, animals are pretreated with inducers or inhibitors, and a radiolabeled compound of suspected toxicity is then administered. The amounts of radiolabel covalently bound to protein or nucleic acid in target organs, such as liver, in pretreated and untreated animals are then compared with the incidence or severity of the toxicity (Gillette, 1974a,b). If the toxicity is caused by a chemically reactive metabolite, differences in the amounts of covalently bound radiolabel between the groups of animals may correlate with differences in the severity of the toxicity, even when the covalent binding of the metabolite to protein is not the cause of toxicity.

Investigators should be well aware of the theoretical basis and the limitations of this strategy. First, the amount of radiolabel covalently bound to protein should be viewed only as a measure of the area under the curve AUC of the amounts of the chemically reactive metabolite in the organ versus time, multiplied by the rate constant k_p for the combination of the metabolite with the protein or nucleic acid. Differences in covalent binding, therefore, may occur either when there are differences in the fraction F of the dose that is converted to the chemically reactive metabolite or differences in the elimination constant of the metabolite k_e, because $AUC = F \times \text{dose}/k_e$. Since the relationships between the rate constants for the reactions of the reactive metabolite with protein or nucleic acid k_p and with putative target substances k_t will differ with the reactive metabolite and the target substance, there can be a considerable amount of covalently bound material in organs, even when the chemically reactive metabolite is nontoxic. *Thus, little importance should be placed on measurements of covalently bound material without corroborative toxicity studies.* Second, if, for example, ^{14}C in the parent compound is located in a group that is cleaved during metabolism, ^{14}C may enter the 1- and 2-carbon pools in the body and thus may be redistributed to label pools of various amino acids, such as methionine, glycine, or serine. Under these conditions, there would be covalently bound radiolabel in cellular constituents even when the parent substance is not converted to a chemically reactive metabolite. Third, many compounds are converted to several chemically reactive metabolites, only some of which may be toxic. Since all the chemically reactive metabolites may be covalently bound, it is frequently difficult to assess whether a change in the amount of covalently bound material in an organ is due mainly to changes in the *AUC* value of the toxic or the nontoxic metabolites.

LIMITATIONS AND PERTURBATIONS OF THE TOXIC RESPONSE

It is frequently possible to differentiate short-lived from long-lived chemically reactive metabolites by comparing the effects of inducers, activators, or inhibitors of hepatic enzymes on the covalent binding of the radiolabel in the liver and in other target organs. If an inducer increases the amount of dose that is converted to a chemically reactive metabolite in liver, the covalent binding of radiolabel in liver would be increased. But in extrahepatic organs the covalent binding of radiolabel would also be increased with a long-lived metabolite and either not changed or decreased with a short-lived metabolite. In this way, inducers can change the relative severities of the toxicity caused by short-lived metabolites in different organs. For example, 3-methylcholanthrene decreases pulmonary necrosis and increases hepatic necrosis caused by ipomeanol in rats (Boyd and Burke, 1978). The finding that an inducer increases covalent binding in extrahepatic organs, however, is not always definitive. A stable decomposition product of a short-lived, chemically reactive metabolite formed in the liver may be carried to other organs to be converted to another short-lived, chemically reactive metabolite. For example, it is possible that *o*-bromophenol, formed from bromobenzene in the liver, is carried to the kidney and lung, where it is converted to chemically reactive metabolites that become covalently bound.

THE ROLE OF METABOLIC STUDIES IN THE EVALUATION OF TOXICITY

When a toxic event is manifested shortly after the administration of a compound, the relationships between the metabolism of the compound and the intensity or duration of the event are rather easily established. But many important kinds of toxicity are not manifested until the compound has been administered repeatedly for various periods ranging from a few days to several years. Because the pattern and rates of metabolism in animals can change markedly during the course of treatment, the relationships between the metabolism of the compound and the incidence of the disease can be obscure.

Long-term tests are an inefficient use of time, money, and highly trained personnel. Therefore, there has been considerable effort to identify the sequence of events that occur during the development of delayed toxicities caused by foreign compounds. Clearly, the incidence or the severity of a toxic response can be affected by altering any of the events that lead to the alteration of the target cell. But by identifying the initial events, it is hoped that suitable short-term tests based on these events may be developed for the identification of potential toxicants.

For example, several short-term tests for mutagenesis have been developed in the hope that they may serve as suitable surrogates for long-term *in vivo* studies of carcinogenesis in animals. In most of these tests, substances are incubated *in vitro* with cell cultures of mammalian tissues or specifically constructed strains of microorganisms for various times ranging from a few minutes to several weeks. Because the test cells may lack the enzymes required for the formation of some mutagenic metabolites, the systems are frequently supplemented with cells from mammalian organs, mixtures of enzymes (such as the $9000 \times g$ supernatant of rat liver), or purified enzymes.

By comparing the effects of a compound in supplemented or unsupplemented systems, it is frequently possible to determine not only whether mutagens are present in the test systems, but also whether they are formed by metabolic activation of the parent compound leading to reactive products. But without studies on the metabolism of the substance, it is virtually impossible to identify the mutagenic substance, to determine the factors that govern the number of transformations in the system, and to relate this number to the number of transformed cells that would be expected to occur in organs of a given animal strain or species.

In the absence of a DNA repair system, the incidence of mutagenesis depends on the amount of metabolite that reacts with the relevant loci on the DNA in the test organism. Therefore, the number of transformations that occur should be proportional to the *AUC* of the concentration of mutagenic metabolite during the course of the experiment and thus to the activities of the enzymes that catalyze the formation and inactivation of the metabolite.

Through the use of various purified enzymes or mixtures of enzymes, with and without various cofactors, the *AUC* of the metabolite concentrations may be markedly changed. A battery of such experiments frequently reveals either the identity of mutagenic metabolites or the pathways along which they are formed. Once the mutagenic metabolites and their inactive secondary metabolites are known, various kinetic models may be constructed to demonstrate the many ways in which the amounts of mutagenic metabolite formed in the systems might be altered.

Many investigators believe that studies of the metabolism of substances by intact cells would be inherently superior to studies with purified enzymes or with mixtures of enzymes, because the cells would contain all the enzymes and cofactors required for the formation and inactivation of the mutagenic metabolites. Certainly, such studies are useful in confirming the relevance of the cell-free systems, but they do have their limitations and frequently provide misleading information. For example:

1. The relative formation rates of the various metabolites are more difficult to change with intact cells than they are in cell-free preparations. With isolated cells it is thus more difficult to elucidate which metabolites are mutagenic and which are not.
2. Unless the mutagenic metabolites are generated within the target cells, mutagenesis will depend on the relative rates of formation and inactivation of the mutagenic metabolites within the generating cells and of the diffusion of the metabolites through cellular barriers into the medium and thence into the target cells. Such studies of mutagenesis will therefore detect only long-lived, chemically reactive metabolites.
3. The relative abundance of the various enzymes within cells is markedly different in various cell types, for example, in the liver and in other complex organs such as the lung. Thus, the relative rates of formation and inactivation of the mutagenic metabolite can vary markedly within the heterogeneous population of cells in an organ *in vivo*. Because the recovery of cells during isolation from tissues is incomplete, it is impossible to tell whether the isolated cell population truly represents the entire cell population within the organ.
4. Cofactors required by the enzymes frequently change during the isolation of the cells, either because they diffuse out of the cells or because their rates of synthesis change *in vitro*.
5. The relative activities of the enzymes may change as the cells are grown or maintained on different kinds of media.

For these reasons the study of a chemical's toxicity in intact isolated cells may not truly reflect the contribution of these types of cells toward the metabolism of the substance in living animals.

EXTRAPOLATION FROM *IN VITRO* EXPERIMENTS TO LIVING ANIMALS

There are many reasons why studies in cell-free systems or in intact isolated cells may not truly represent the metabolism of substances in living animals.

In both types of studies, experiments are frequently terminated before significant amounts of the parent substance added to the incubation system are metabolized. Under these conditions, the addition of specific inhibitors of side reactions that lead to innocuous metabolites would not be expected either to change the formation rate for the first metabolite of the toxic pathway or to change the incidence of toxicity *in vitro*. By contrast, the single doses of almost all foreign compounds administered to living animals are completely eliminated from the body. Moreover, multidoses result in *quasi*-steady-state concentrations of the chemical. Thus, inhibitors of the side reactions may profoundly affect the amount of the single dose that is converted to the reactive metabolite or the steady-state concentration of the reactive metabolite during multiple-dose administration (in the body) and, therefore, the incidence of toxicity. Estimation of the importance of the side reactions in limiting the incidence of mutagenesis *in vivo* thus requires knowledge of the *total* metabolism of the parent compound. Failure to measure the disappearance of the substance during *in vitro* mutagenesis studies can also lead to misleading estimates of relative potency. A substance that is slowly metabolized to a mutagenic metabolite may appear to be less potent than one that is rapidly metabolized.

In vitro studies of mutagenesis and metabolism are frequently conducted at concentrations far exceeding those that would be reached *in vivo*. The *in vitro* concentrations also frequently exceed the K_m values of the various enzymes that catalyze the metabolism of the substance and its toxic metabolites. Thus, the relative rates at which the various metabolites are formed in the *in vitro* system may be governed mainly by the relative V_{max} values of the enzymes, whereas the relative formation rates of the metabolites *in vitro* may be governed mainly by the relative intrinsic clearance values, i.e., V_{max}/K_m. Thus, it is necessary to measure not only the formation rates for the primary metabolites, but also the K_m values of the reactions in order to simulate expected patterns of metabolites at different concentrations of the substance.

Because estimates of the volumes of distribution of a compound and its metabolites *in vivo* cannot be obtained from *in vitro* experiments, it is not possible to predict either the maximum concentrations or the half-lives of the parent compound and its metabolites *in vivo* solely from *in vitro* studies with cell-free preparations, intact cells, or organ perfusion. It is therefore impossible to predict from *in vitro* studies whether a given single dose of the parent substance will result in concentrations that approach or exceed the K_m values of the enzymes that catalyze the formation of the primary metabolites. Only after administration of the parent substance and measurement of the concentration of unbound substance in blood plasma can it be determined whether the formation rates for the various products approach the V_{max} values of the enzymes. Nevertheless, *in vitro* studies with isolated cells may be useful in obtaining gross estimates of the contribution of various tissues to the total body clearance of the substance and its metabolites. They therefore may be useful in estimating the average steady-state concentrations of unbound substance achieved during multidose administration of the substance, but such estimates would only be valid when the substance is eliminated from the body mainly by metabolism in a single organ, such as the liver.

In some instances the metabolism of an intravenously administered foreign compound is governed mainly by the blood flow through the liver rather than by the intrinsic activity of the various enzymes in the liver. This will occur when the concentration of the substance in the blood leaving the liver is much lower than that in blood entering the liver under steady-state conditions (Gibaldi and Perrier, 1975; Gillette, 1980). Induction of one or more of the enzymes may alter the relative amounts of the various metabolites, but would not appreciably change either the maximum concentration or the half-life of the parent substance in the systemic circulation. Obviously, changes in blood flow rate would change both the maximum concentration and the half-life of an intravenously administered substance. But less obviously, such changes in blood flow may also alter the relative amounts of the metabolites formed from the parent substance, because of the intercellular heterogeneity in the relative amounts of the enzymes present within the cells. When the parent substance is administered perorally, however, the amount of the dose that reaches the systemic circulation depends on the sum of the intrinsic clearances of the enzymes. Thus, induction may decrease the maximum concentration of the perorally administered substance in the systemic circulation even though it may not markedly change the half-life of the parent substance.

The maximum concentration of a primary metabolite in systemic blood after peroral administration of the parent compound may exceed the maximum concentration of the metabolite after intravenous administration of the parent chemical. But as long as the concentration of the parent compound in liver does not approach or exceed the K_m values of any of the hepatic enzymes, the average concentration of the metabolite in the liver and systemic blood will be independent of the route of administration of the parent chemical when the parent substance is eliminated from the body solely by the liver (i.e, the parent chemical is completely absorbed and is not metabolized or excreted by any other organ in the body) (Gillette, 1980). Much may be inferred from studies on the effects of the route of administration on the severity of the toxicity and measurements of the concentrations of the primary metabolites in blood. For example, if the *AUC* values of the primary metabolites are markedly different after intravenous or peroral administration, either the parent drug is incompletely absorbed or it is eliminated by other organs as well as by the liver. If the severity of the toxicity is lower after oral administration than after intravenous administration, but the *AUC* values of the metabolites are nearly identical, then the toxicity is probably not caused by a metabolite formed in the liver (however, the same toxic metabolite may be formed in another tissue directly from the parent compound).

Studies of the metabolism of the parent compound by isolated hepatocytes may provide gross estimates that can be used to predict whether an appreciable first-pass effect might occur *in vivo*. The blood flow through the liver of various laboratory animals usually ranges from ~1 to ~3 ml/g per min. If the sum of the V_{max}/K_m values of the enzymes per gram of hepatocytes (expressed as ml/g per min times the unbound fraction of substance in blood) exceeds 3 ml/g per min, the investigator may suspect that an appreciable first-pass effect might occur in the living animal.

Sometimes a toxic response is caused by a short-lived, chemically reactive metabolite formed in a target organ that is not the major organ of elimination of the parent substance. Since the toxic effect depends on the amount of the dose of parent substance that is converted to the toxic metabolite in the target tissue, the severity (or incidence) of the toxicity will be governed not only by the intrinsic clearance of the enzyme that catalyzes the formation of the chemically reactive metabolite, but also by the intrinsic clearances of the enzymes in the major organs of elimination. Studies with isolated target cells alone thus cannot provide all the information needed to predict the severity (or incidence) of the toxicity in the animal strains from which the target cells were obtained, even if it is possible to show that the activities of the enzymes in the isolated target cells are the same as those in the target cells of living animals.

For these reasons *in vitro* tests, as usually performed, will rarely provide accurate estimates of the relative potencies of carcinogens in humans. Indeed, the rank order of the potencies of carcinogens may vary with the animal species or the human subpopulation. Nevertheless, these tests, combined with studies of metabolism in the test systems and animals, should aid in identifying the general characteristics of groups (such as subpopulations of humans) that would be the most sensitive to the toxic effects of different carcinogens.

SUMMARY

The nature of the toxicity of any given chemical is complex. Because of imperfections in the present study methods, the importance of *any* information may be exaggerated. Thus, there is a need to develop more reliable and relevant approaches to developing methods that are predictive of effects in various human subpopulations. It is apparent that laboratory studies, comprising either *in vivo* or *in vitro* experiments, have serious limitations, but so do studies with a limited number of human subjects. No single approach to the study of toxicity is likely to provide all the information required for understanding the relative contributions of the three parts of the mechanism of toxicity when accounting for differences in the toxic response between strains and species of animals and between human subpopulations. Classical toxicity studies alone may reveal whether some form of a substance combines with action sites in sufficient amounts to evoke a toxic response in a test animal strain. But such studies cannot reveal whether the toxic form is the parent substance or one or more of its metabolites, nor can they reveal the effective concentration of the toxic form at the action sites, even when the toxic form is known. On the other hand, studies of drug metabolism and disposition may reveal the plethora of metabolites that may be formed from the parent substance and the kinetic constants that describe their concentrations in various organs of the body at various times during different dosage regimens. But such studies by themselves do not reveal which, if any, of the forms of the substance interact with action sites, thereby evoking a toxic response.

Clearly, there is a requirement for well-integrated programs of study designed to identify toxic forms of a substance and to evaluate the factors that govern their concentration at action sites within target organs in different groups of animals manifesting the toxicities. Once these are known, it will be possible to develop dosage schedules that will result in similar concentrations of the toxic form in target tissues of different groups of animals. Differences in the action sites and the subsequent events leading to the toxic response may then be evaluated in different strains and species of animals and in different human subpopulations. However, since it is unethical to administer irreversibly acting toxicants, such as suspected carcinogens, to humans, much of our knowledge about possible toxic chemicals must remain dependent on retrospective epidemiological studies. Direct knowledge about the pattern of metabolism and the *in vivo* kinetic parameters for the formation and elimination of the metabolites of such compounds in all human subpopulations may never be available, but clues to possible toxic metabolites sometimes may be obtained by comparing the patterns of metabolism in affected and unaffected individuals in similar cohorts. The suspicions resulting from such studies, however, should be confirmed by prospective animal experiments. At the present state of development in chemical toxicology, there is no valid way of acquiring all the necessary knowledge required to predict the patterns and parameters that reflect the formation of toxic metabolites.

Nevertheless, it seems plausible that future studies with pure enzymes isolated from human tissues may provide a stronger scientific base for the development of indirect approaches. For example, studies with pure enzymes should reveal which of them catalyzes the metabolism of a given procarcinogen and/or its carcinogenic metabolites. Studies of the enzyme-specific nontoxic substrates of the enzymes then could be used to assess the "effective" abundance of the enzymes in human subpopulations *in vivo.* With such data, it might be possible to calculate the approximate metabolic clearances that govern the average concentrations of the suspected carcinogen and its stable metabolites in target organs. This approach, however, still has several obvious limitations. It cannot provide estimates of the rate constants that describe the distribution and elimination of the parent compound or its metabolites, because it does not provide estimates of the volumes of distribution of the substance and its metabolites in various compartments of the body. The approach will also be virtually useless in predicting the areas under the curve *AUC* of short-lived, chemically reactive metabolites within the various cells of the body, because *in vivo* studies with surrogate substrates will not reveal the relative abundance of the enzymes catalyzing the formation and inactivation of the metabolite in any given group of cells, especially those in the smaller organs of the body. Studies designed to increase our knowledge of the factors that control the relative abundance of the enzymes in various tissues of humans representing different subpopulations will still be needed.

In this chapter we have briefly reviewed some of the reasons why studies to predict the potential carcinogenicity or toxicity of a chemical have been disappointing in predicting the incidence rates of toxicity caused by foreign compounds in humans. It is evident, however, that additional studies, in particular those on the metabolic activation of many different types of chemicals, will have to be performed and properly interpreted if valid risk assessments of toxicity are ever going to be anything more than qualitative.

REFERENCES

Boyd, M. R., and L. T. Buka. 1978. *In vivo* studies on the relationship between target organ alkylation and the pulmonary toxicity of a chemically reactive metabolite of 4-ipomeanol. *J. Pharmacol. Exp. Ther.* **207**:687–697.

Coon, M. J., A. H. Conney, R. W. Estabrook, H. V. Gelboin, J. R. Gillette, and P. J. O'Brien, eds. 1980. *Microsomes, Drug Oxidations, and Chemical Carcinogenesis,* Volumes I and II. Academic Press, New York. 1258 pp.

De Matteis, F., and L. Cantoni. 1970. Alteration of the porphyrin nucleus of cytochrome P-450 caused in the liver by treatment with allyl-containing drugs. Is the modified porphyrin N-substituted? *Biochem. J.* **183**:99–103.

Doull, J., C. D. Klaassen, and M. O. Amdur, eds. 1980. *Toxicology, The Basic Science of Poisons,* 2nd ed. Macmillan, New York.

Ehrenberg, L., S. Osterman-Golkar, D. Segerback, K. Svensson, and C. J. Calleman. 1977. Evaluation of genetic risks of alkylating agents. III. Alkylation of haemoglobin after metabolic conversion of ethylene oxide *in vivo. Mutat. Res.* **45**:175–184.

Gibaldi, M., and D. Perrier. 1975. *Pharmacokinetics.* Marcel Dekker, New York. 329 pp.

Gillette, J. R. 1974a. A perspective on the role of chemically reactive metabolites of foreign compounds in toxicity. I. Correlation of changes in the covalent binding of reactive metabolite with changes in the incidence of severity of toxicity. *Biochem. Pharmacol.* **23**:2785–2794.

Gillette, J. R. 1974b. A perspective on the role of chemically reactive metabolites of foreign compounds in toxicity. II. Alterations in the kinetics of covalent binding. *Biochem. Pharmacol.* **23**:2927–2938.

Gillette, J. R. 1980. Pharmacokinetic factors governing the steady-state concentrations of foreign chemicals and their metabolites. Pp. 191–217 in *Environmental Chemicals, Enzyme Function and Human Disease.* Ciba Foundation Symposium 76, Excerpta Medica, Elsevier/North-Holland, Amsterdam.

Gillette, J. R., S. D. Nelson, G. J. Mulder, D. J. Jollow, J. R. Mitchell, L. R. Pohl, and J. A. Hinson. 1982. Formation of chemically reactive metabolites of phenacetin and acetominophen. Pp. 931–950 in R. Snyder, D. V. Parke, J. J. Kocsis, D. J. Jollow, C. G. Gibson, and C. M. Witmer, eds. *Biologically Reactive Intermediates.* II. *Chemical Mechanisms and Biological Effects,* Part A. Plenum Press, New York.

Golkar, S. O. 1983. Tissue doses in man; implications in risk assessment. *Dev. Toxicol. Environ. Sci.* **11**:289–298.

Gustafsson, J. A., J. Carlstedt-Duke, A. Mode, and J. Rafter, eds. 1980. *Biochemistry, Biophysics and Regulation of Cytochrome P-450.* Elsevier/North-Holland, Amsterdam.

Kappus, H., and H. Sies. 1981. Toxic drug effects associated with oxygen metabolism: Redox cycling and lipid peroxidation. *Experientia* **37**:1233–1241.

Kato, R., and R. Sato, eds. 1982. *Microsomes, Drug Oxidations and Drug Toxicity.* Japan Scientific Societies Press, Tokyo.

Lau, S. S., T. J. Monks, K. E. Greene, and J. R. Gillete. 1984. Detection and half-life of bromobenzene-3,4-oxide in blood. *Xenobiotica* **14**:539–543.

Mantel, N., and W. R. Bryan, 1961. "Safety" testing of carcinogenic agents. *J. Natl. Cancer Inst.* **27**:455–470.

Ortiz de Montellano, P. R., B. A. Mico, and G. S. Yost. 1978. Suicidal activation of cytochrome P-450. Formation of a heme-substrate covalent adduct. *Biochem. Biophys. Res. Commun.* **83**:1.

Singer, T., and R. Orndarza, eds. 1980. *The Biochemical Basis of Drug Action.* Elsevier/North-Holland, New York.

Snyder, R., D. V. Parke, J. J. Kocsis, D. J. Jollow, C. G. Gibson, and C. M. Witmer, eds. 1982. *Biologically Reactive Intermediates.* II. *Chemical Mechanisms and Biological Effects.* Plenum Press, New York.

White, R. E., and M. J. Coon, 1980. Oxygen activation by cytochrome P-450. *Annu. Rev. Biochem.* **49**:315–356.

Whittemore, A. S. 1978. Qualitative theories of oncogenesis. *Adv. Cancer Res.* **27**:55–88.

Ziegler, D. M. 1978. Microsomal oxidases. Pp. 193–204 in S. Fleischer, Y. Hatafi, D. H. MacLennan, and A. Tzagoloff, eds. *The Molecular Biology of Membranes.* Plenum Press, New York.

Ziegler, D. M. 1980. Microsomal flavin-containing monooxygenase: Oxygenation of nucleophilic nitrogen and sulfur compounds. Pp. 201–227 in W. B. Jakoby, ed. *Enzymatic Basis of Detoxication,* Volume 1. Academic Press, New York.

V

Risk Analysis

17

Exposure Assessment

Stephen L. Brown

Assessing the significance of toxicological tests and observations for risks to both humans and other organisms is impossible without defining the exposures to the hazardous agent (chemical or otherwise). The risk assessor must use a collection of techniques to express the exposures of target organisms in terms that are compatible with the measurements of exposure reported in the literature on toxicological tests.

EXPRESSIONS OF EXPOSURE

There is no uniform, well-established procedure for expressing exposure. The possibilities include: concentrations in media, quantities available for absorption, rates of intake, concentrations in body tissues, body burden, and organ dose. Several of these possibilities fit the usual definitions of "dose" (or at least dose rate). A list of examples for each category appears as Table 17-1.

Clearly, the risk assessor has a variety of exposure measures from which to choose. However, the selection must be made to ensure that the units in which exposure is expressed are compatible with those that toxicologists have used to report the results of their experiments. If the exposure measures are not compatible, the risk assessment will at best be flawed and at worst be impossible.

Even with the variety of exposure measures listed in Table 17-1, further definitions are necessary for ensuring that the exposure assessment will lead to a meaningful risk assessment. Critical among these is the time pattern of exposure.

STEPHEN L. BROWN • Environ Corp., Washington, D.C. 20007.

TABLE 17-1. Different Ways of Expressing Exposures

Type of expression	Units	Type of agent
Concentration in media	mg/kg food	Chemical
	mg/liter water	Chemical
	mg/m^3 air	Chemical
	Fraction by weight in consumer products	Chemical
	$Fibers/cm^3$ air	Particulate
Quantity available for absorption	mg inhaled, total	Chemical
	mg/kg body weight inhaled	Chemical
	mg ingested, total	Chemical
	mg/kg body weight ingested	Chemical
	mg on skin, total	Chemical
	mg/cm^2 skin area	Chemical
	mg injected or implanted	Chemical
Rate of intake or exposure	mg/kg body weight	Chemical
	ingested per day	Chemical
	mg inhaled per year	Chemical
	R/hr	Radiation
Concentration in body tissues	mg/ml blood	Chemical
	$Fibers/cm^3$ lung tissue	Chemical
Body burden	g in bone	Chemical
	curies (Ci) in total body	Radioactive substances
Organ dose	mg to liver	Chemical
	rad to thyroid	Radiation
Other measures	decibels or dBA	Noise
	$cal/(cm^2\text{-}sec)$	Heat
	W/cm^2	UV or microwave radiation
	v/m	Electric fields
	G	Magnetic field
	psi (peak)	Explosive shock
	lamberts	Visible light

To interpret the toxicological significance of an exposure, one must know the following:

1. Duration of each exposure, if not instantaneous.
2. Frequency with which the exposure is repeated.
3. Variation of exposure level within each period of exposure.
4. Elapsed time to the observation of effects from the last (or sometimes first) exposure period.

Figures 17-1 and 17-2, adapted from Brown and Bomberger (1982), show several possible time patterns of exposure of an individual organism and the sig-

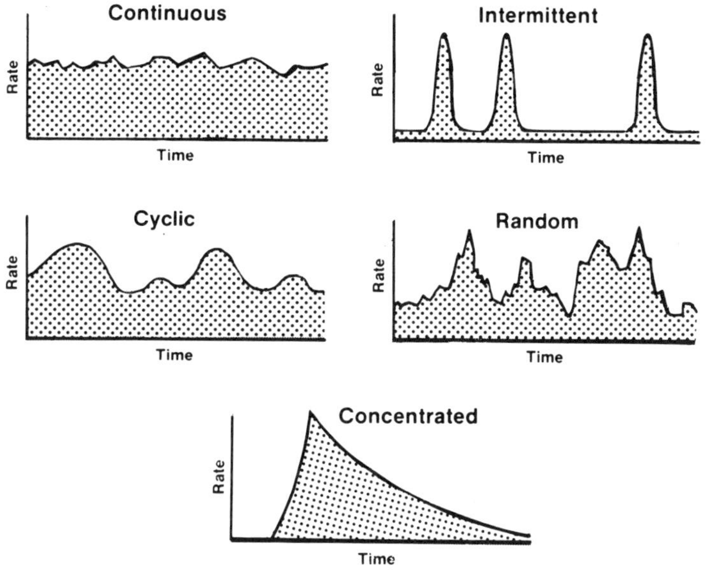

FIGURE 17-1. Many exposure assessments assume a rather constant level of exposure (top left), but real patterns can be quite different. [From Brown and Bomberger (1982).]

nificance of various interpretations of a given pattern, respectively. The ordinate, labeled "rate," could be concentration units (e.g., mg/m^3), dose rate units (mg/day inhaled), or any other time-varying rate of exposure. In Fig. 17-2, the different averaging times (1, 3, or 8 hr, or 1 year) show how the exposure assessor might express a time-varying exposure. The short-term averages might be appropriate for assessing acute toxic responses, whereas the long-term annual average might

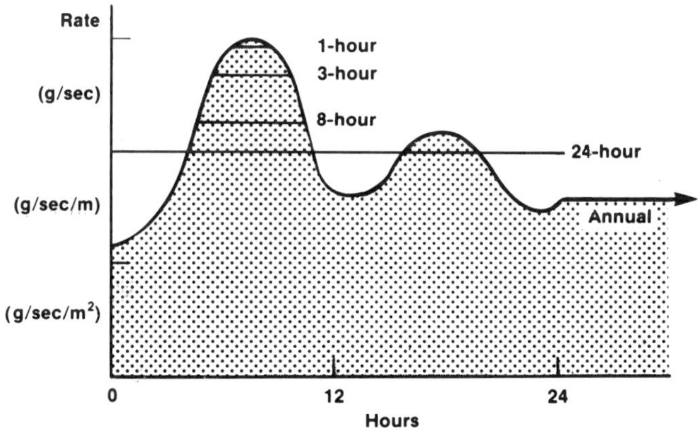

FIGURE 17-2. The heavy curve shows the instantaneous exposure rate (arbitrary units) over a specific 24-hr period, changing to an annual average rate beyond 24 hr. The short-term highest averages for 24, 8, 3, and 1 hr are also schematically indicated. These are often used to compare with standards. [From Brown and Bomberger (1982).]

be appropriate for assessing chronic toxicity (e.g., carcinogenicity), especially when a linear dose–response relationship is postulated.

To complete a risk assessment and characterize the number or extent of effects potentially related to an agent, the exposure assessor must also describe the geographic distribution of the population(s) at risk, so that, for example, the number of people exposed at given concentrations can be determined from the geographic distribution of the concentrations. Moreover, if the toxicological assessment suggests significantly different responses to the same doses for different ages, sex, race, sizes, and other characteristics of those population(s), then the distribution of exposures over such categories of the population must also be characterized.

Finally, if the risk assessment is to be used to decide what sources of exposure (e.g., automobile exhausts, food additives, or factory discharges) need to be controlled, then the exposure assessor may also be asked to estimate how much exposure comes from each source. Therefore, complete characterization of exposure may involve the following factors:

1. Specification of levels of exposure (Table 17-1).
2. Specification of route(s) of exposure.
3. Distribution of exposures over time.
4. Distribution of exposures over space (geography).
5. Distribution of exposure over segments of the population at risk.
6. Distribution of exposures over sources.

CONVENTIONAL APPROACHES

Most exposure assessment activities develop from concerns about a single chemical or other type of agent, often from laboratory, clinical, or epidemiological observations of adverse effects, but sometimes simply from observations of growing production levels or occurrence in the environment.

There are two principal ways of dealing with an exposure assessment under such circumstances: measurement-based approaches and calculation-based approaches. Although the two types share some characteristics, the former is principally inductive, whereas the latter is principally deductive.

In measurement-based approaches, assessors attempt to use data gathered directly (or as directly as possible) on exposure. Ideally, one might measure the exposure quantity of interest directly. For example, to estimate dermal exposures of farm workers to pesticides, sticky patch collectors are sometimes affixed to their clothing during either real or simulated applications of the pesticide. Cyclone-type samples can also be worn by miners and their pumps calibrated at normal breathing rates to simulate the respiratory exposures to mine dust. Most measurements are less direct, however, and the assessor must relate measured concentrations in air, water, or food to absorbed dose through some model of the exposure and absorption process. The advantage of the measurement approach is its foundation of real data rather than a framework of assumptions; its principal

TABLE 17-2. Mathematical Model for Chemical Exposures

Assume that we want to assess human exposures to a hypothetical chemical, ethyl methyl chicken-wire (EMC).

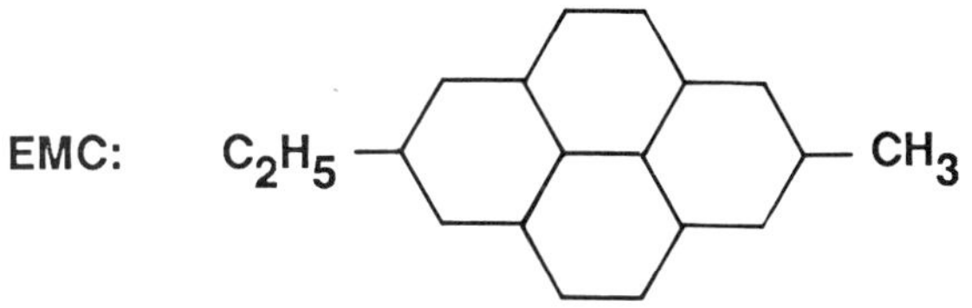

EMC occurs as a by-product in the manufacture of α-glupate, which is used as an intermediate in the production of moustache wax. The wax production process destroys all residual EMC as well as the α-glupate, so the only exposures to EMC occur in the α-glupate factory and in the surrounding communities (only one plant, located in West Overshoe, produces α-glupate).

EMC is relatively volatile, and between production and bottling of α-glupate, 63% of the EMC will reach the atmosphere. EMC contaminates α-glupate 1.7% by weight, and 200 lb of α-glupate are made per day (24-hr, 7-day/week operations). Therefore, $200 \times 0.017 \times 0.63 = 2.1$ lb of EMC is released to the atmosphere daily.

If the ventilation of the workplace is such that the air changes once every hour, then the amount of EMC in the plant environment is $2.1/24 = 0.088$ lb at any one time. If the manufacturing area is 40 ft $\times$ 50 ft $\times$ 10 ft, then $0.088/(40)(50)(10) = 0.088/20{,}000 = 4.4 \times 10^{-6}$ lb/ft^3 is the concentration. In metric units, this concentration is 70 mg/m^3. Therefore, the 20 workers on each 8-hr shift are exposed to a concentration of 70 mg/m^3, or have a daily exposure (assuming an inhalation rate of about 10 m^3 over the 8 hr) of 0.7 g. At 220 working days/year, that exposure translates to an annual dose of about 150 g. About four shifts—or 80 people—are required to run the plant year-round.

Outside the plant, the air is very still and the EMC diffuses in a hemisphere around the plant. If 0.088 lb/hr is leaving the plant, this amount must be passing through a surface of area $4\pi r^2$ at distances r from the plant each hour when a "steady-state" is reached. If the velocity of diffusion averages 0.5 mile/hr, then a shell 0.5 mile thick must contain 0.088 lb at any distance. Thus, the concentration is $0.088/(4\pi r^2)(0.5)$, with r in miles and concentration in lb/mile^3. Two miles from the plant, then, the concentration is $0.088/(4\pi)(4)(0.5) = 0.0035$ lb/mile^3. In metric units, it is 0.38 ng/m^3. If 5000 people live in a town 2 miles from the plant, then they will be exposed to a concentration of 0.38 ng/m^3. But they breathe air at a rate of 20 m^3/day for 365 days/year, so their annual dose is about $20 \times 365 \times 0.38 = 2.8$ μg/year. Total exposures for 80 workers would be $80 \times 150 =$ 12,000 g, whereas for 5000 community residents it would be only $5000 \times 2.8 \times 10^{-6} = 14$ mg. Such is the power of dilution in the atmosphere.

Of course, to keep the EMC from building up in the atmosphere, we need a "sink" for the EMC. If sunlight degraded (photolyzed) the EMC with a half-life of 10 days, the EMC would be essentially gone after ten half-lives, or 100 days, by which time it would have diffused to $100 \times 24 \times 0.5 =$ 1200 miles—not even halfway across the country.

Finally, suppose a big city (10 million people) were 100 miles from West Overshoe. Also, suppose that the cloud stopped diffusing upward after it had spread 2 miles, and expanded only radially. The concentration would fall off as $1/r$ instead of $1/r^2$, and the concentration in the city should be 1/50 that at 2 miles, or 7.6×10^{-12} g/m^3; the corresponding total exposure is $7.6 \times 10^{-12} \times 20 \times 365 \times 10^7 = 560$ mg.

So, of the $2.1 \times 365 = 770$ lb of EMC released each year, only 12,000 g or about 26 lb is inhaled by people, most of it by workers.

Problem: Suppose that everybody else in the world (4 billion people) lived 1000 miles from West Overshoe. Using the same logic ($1/r$ spread, 10-day half-life), what would be the total exposure of all 4 billion?

(*Answer:* About 3 mg)

disadvantage is that measurements are expensive to make, and unless many are made, extrapolation and interpolation to other situations are tenuous.

By contrast, calculation-based approaches begin with much less direct measurements of exposure—for example, chemical production volumes, chemical and physical properties, and so on. Then, ultimate distributions of exposures are estimated through a series of calculations or mathematical models that attempt to represent the behavior of the chemical in the real world. The models might include descriptions of fractions released to the environment, dispersion in the environment, chemical degradation, and bioaccumulation. The models predict rather than measure the needed variable of exposure. An example of exposure assessment based on a simple model is given in Table 17-2. The calculational approach eliminates the need for multiple measurements of final concentrations for doses and allows (usually) calculation of exposure at any geographic location or time. Conversely, this approach must depend on a series of assumptions and mathematical representations that may be exceedingly poor descriptions of real phenomena and that must be kept relatively simplistic to avoid excessive computational costs. The validity of the input data—whether measured or simply estimated—may also be questionable.

TABLE 17-3. An Environmental Exposure Assessment Report[a]

1. Introduction and executive summary
   - Purpose and scope
   - Summary
2. General information
   - Identity
   - Chemical and physical properties
3. Sources (materials balance)
   - Characterization of production and distribution
   - Uses and application rates
   - Disposal
   - Summary of environmental release
4. Exposure pathways and environmental fate
   - Transport and transformation
   - Identification of principal pathways of exposure
5. Estimated environmental concentrations
   - Summary of monitoring data
   - Estimates of environmental concentrations using models
   - Comparisons of monitored and modeled concentrations
6. Population-at-risk
   - Human populations
   - Nonhuman populations
7. Integrated exposure analysis
   - Development of exposure profiles/scenarios
   - Human dosimetry/monitoring
   - Calculation of exposures
   - Evaluation of uncertainty
   - Bibliography
   - Appendices

[a] From Brown and Suta (1982).

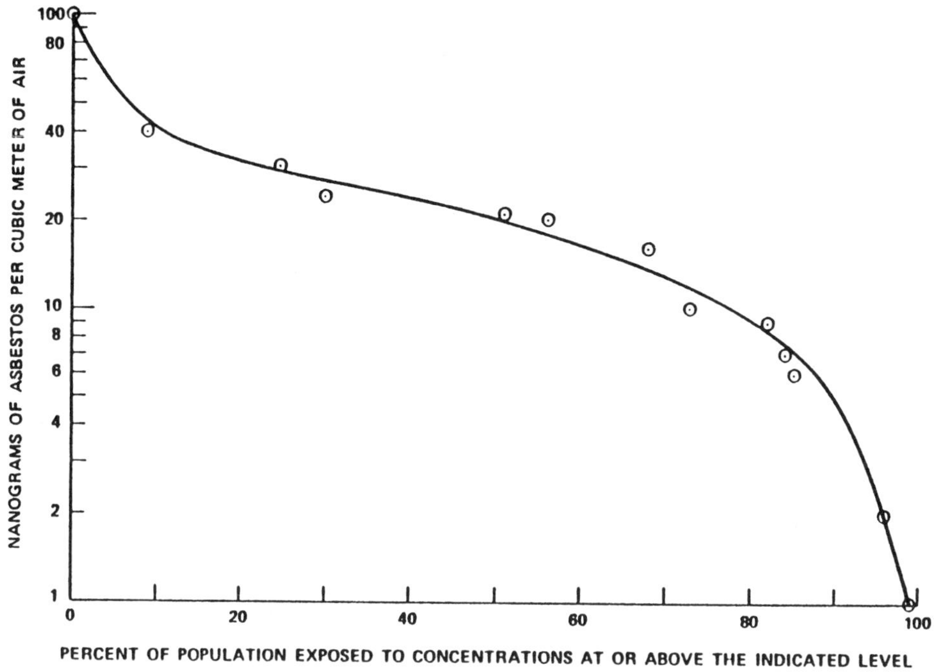

FIGURE 17-3. The data points represent estimates of the percent of people exposed at or above the indicated level to asbestos in ambient air, each corresponding to a group of people exposed at approximately identical levels in different situations (see text). [From Brown and Suta (1982).]

Of course, the best approach combines the positive features of both types of calibrating the calculations against exposure measurements in known situations and then using the model to extrapolate or interpolate to unknown situations. Brown and Suta (1982) have presented a general overview of conventional exposure assessment emphasizing the calculational approach, but including environmental monitoring as well. Their descriptions are oriented toward exposures that occur through the general environment rather than in the workplace, through the use of consumer products, or through consumption of foods, drugs, or cosmetics. However, many of the principles apply to any exposure assessment. Table 17-3 presents their outline of an environmental exposure assessment report.

Notice that concentration data (item 5) must be combined with population data (item 6) for an integrated exposure analysis (item 8) to be made. Brown and Suta considered exposures to environmental carcinogens, for which the key information may be the number of people (or percent of the population) exposed to various concentrations of the carcinogen. They constructed a graph of population exposed versus level of exposure for asbestos (Fig. 17-3). The "data points" in the figure represent different exposure situations, ranging from general background exposures at the right of the figure through community and consumer product exposures in the middle to high occupational exposures at the upper left. If a dose–response relationship can be obtained from the toxicological assessment, this diagram facilitates the estimation of asbestos-related disease incidence nationwide.

AN ALTERNATIVE APPROACH: "BACKWARD TRACING"

All the foregoing has implied a fundamental assumption about exposure assessment, namely, that the agent of concern has been well defined and that the assessor's job is to identify and quantify all human or ecological exposures to the agent (or at least some subset of those exposures). This section focuses on a different objective and approach: to characterize exposures to a range of agents from given exposure situations and then to combine the exposures for each agent from all the situations studied. We call this approach "backward tracing" for reasons that will soon become apparent.

The conventional approach is "forward tracing" in the following sense. All the sources of an agent—say a chemical substance—are identified, and the "life history" of the substance is traced forward in time as it moves from natural sources or raw materials through manufacturing, distribution, use, and disposal directly to humans or indirectly through the environment to humans and other organisms (Fig. 17-4).

A molecule of the substance is "followed" from its initial production or liberation by humans to its interaction with the exposed organism. Along the way there are losses resulting from conversion to other substances or from other processes that make the substance unavailable for further exposures.

By contrast, backward tracing starts with the exposed organisms, e.g., humans, and looks back in time to the origins of the substances to which they are exposed. In principle, the technique is applicable to single substances, but in practice it is only efficient when there is interest in assessing exposures to many substances. A case in point is the setting of priorities for toxicity testing: which substances, by virtue of their high potential for human exposures, should be tested for unknown but suspected toxicity? There may be little reason to select one chemical over another on the basis of toxic potential, but the potential for exposure over a wide range of substances may be very important.

The technique discussed below is applicable to human exposures. It is necessary to focus on the *activities* of humans that could provide the opportunity for exposures to chemicals. Humans breathe. Humans drink. Humans eat. Humans treat themselves for disease. Humans decorate their bodies. Humans take care of their households. Humans work. Humans recreate. And so on. Each of these activities can provide opportunities for exposures, which can be divided into four broad categories:

1. Environmental
2. Household (consumer product)
3. Occupational
4. Avocational

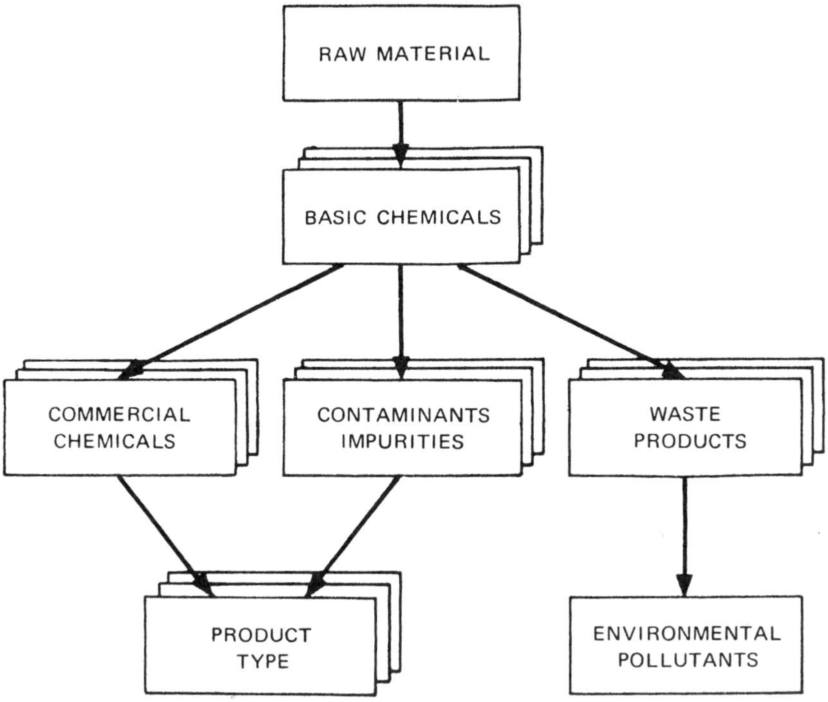

FIGURE 17-4. Forward tracing follows a chemical from raw materials through product uses and disposal to eventual human exposure. [From Stanford Research Institute (1971).]

Each of these categories can be further divided into major classes of exposure. In the environmental category, the classes are the media of exposure: air, water, or land. The exposures are calculated directly on the basis of occurrence (concentrations) in ambient air, drinking water, bathing/swimming water, or soils. Exposures through soil are usually counted via particulates suspended in air or dissolved or suspended solids in water, rather than direct exposures—e.g., dermal ones—to soil contaminants. However, one can also count environmental contaminants of food, e.g., pesticide residues, as "soil."

For household exposures, the classes are major product types or uses: foods (food additives, in particular), drugs, cosmetics, paints, soaps and detergents, adhesives, and so on. In the occupational area, the classes are ordinarily the type of employing industry (e.g., chemical plants, petroleum refineries, mines, auto repair shops, beauty parlors, and funeral homes). For avocational exposures, the classes are broad recreational or similar activity classes, e.g., crafts, gardening, and sports.

The major "use" classes can often be subdivided into minor use classes and product types. For example, in the major use class of cosmetics, the minor use class might be hair-care products and the product type, shampoos or even antidandruff shampoos. Figure 17-5 shows how the exposure category is traced backward to the product type.

The figure also shows that each product can be made up of a combination of

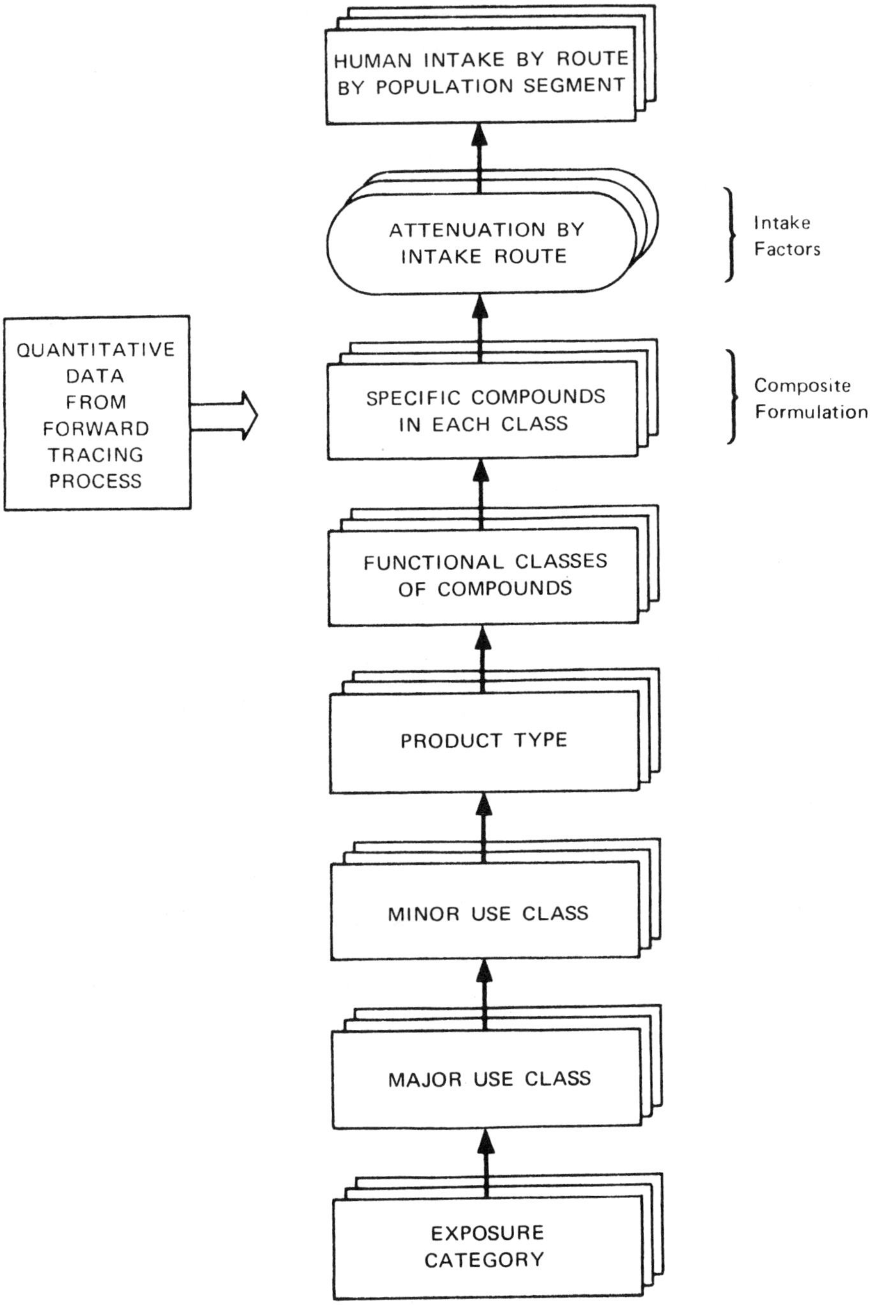

FIGURE 17-5. Backward tracing starts with the situations that cause exposures and works back to the chemicals through a hierarchy of use classes. [From Stanford Research Institute (1971).]

functional classes of substances—some of them well-defined compounds—that each "do" something in the product. For example, in the shampoo there might be detergents, foaming agents, perfumes, colors, antiredeposition agents, and bactericides or preservatives.

By identifying the specific compounds that can perform these functions and estimating the relative amounts of each used in the product type, one can then develop a "composite formulation" that shows the percentage (by weight) distribution of chemicals in that product type. For example, if shampoos generally contain about 10% detergent, and of the three possible detergent chemicals one is used twice as often as either of the others, then the composite formulation would show a content of 5% for the first detergent and 2.5% for the others.

The next step is to identify how much shampoo is used (probably by assuming that every ounce sold is eventually used). To make this estimate, census statistics can be used to determine the dollar sales volume for shampoo. Then a unit cost estimate can be applied to obtain the weight sold. The products of that number and the composite formulation percentages give the amounts of chemicals available for exposure.

Not all of the shampoo really results in exposures. By measurement and calculation, it is possible to estimate how much of the chemical might be absorbed dermally through the scalp, how much might be ingested accidentally, and how much might be volatilized and inhaled, especially for fragrances. These "exposure factors" attenuate the potential exposure levels to give intake by route.

Finally, one can obtain information on the distribution of shampoo use over different population segments and determine the distribution of exposures by route in a form similar to that shown in Figure 17-3. It is then a straightforward, although possibly time-consuming, task to add the exposures for a given chemical from different product types, minor and major use classes, and exposure categories.

EXAMPLE OF BACKWARD TRACING: SOAPS AND DETERGENTS

General Description and Methodology

Soaps and detergents are representative of a variety of chemical products used in the home in a form that allows chemicals to be available for intake (in contrast with the metals in, say, a refrigerator), yet that are not intended to enter the body. This is in contrast to chemicals used in drugs and food. Moreover, almost 50 lb per capita reaches the home each year, and there is ample opportunity for relatively high intakes of chemicals from this exposure mode.

Although some population groups, e.g., homemakers, are clearly exposed more frequently to cleaning products than are others, the principal hazard is probably presented by ingested soap and detergent residues on tableware, which applies to all persons.

The first step in estimating exposure to soap and detergent chemicals is to determine how much of the chemicals are used in what ways in the home. All the products can be classified into the following eight product types, each corresponding to a likely set of conditions of use:

1. Dry machine laundry detergent
2. Liquid machine laundry detergent
3. Bar laundry soap
4. Chip and flake laundry soap
5. Machine dishwashing detergent
6. Hand dishwashing detergent, dry
7. Hand dishwashing detergent, liquid
8. Toilet bar soap

Data from a census of the manufacturers can indicate quantities of these products that are sold.

On the basis of their chemical function, the chemicals used in soaps and detergents can be placed in the following classes: surfactants, builders, fillers, bleaches, germicides, antiredeposition agents, anticorrosion agents, hydrotropic agents, solvents, fluorescent brighteners, dyes, and perfumes. To the extent possible, the major chemical species within each functional class should be identified and the total usage by class or by compound in cleaning products should be estimated. By balancing chemical usage against product volume, also using data on typical product formulations, one can establish the composite formulation of each product type by specific compound, where possible, or by functional class. This permits allocation of chemical usage per capita by product type.

To determine the fractions of the chemicals used that might enter the body under the conditions of use for each product, a matrix of intake attenuation factors by product type and route of intake can be established, although these factors are so qualitatively determined that they can be quantified only to the closest order of magnitude. When the composite formulations are multiplied by the appropriate intake attenuation factors and the results summed over route and product, a list of intakes by chemical compound or group can be obtained.

Results

Estimates of average per capita intake have ben obtained for 50 compounds or chemical groups (Stanford Research Institute, 1971). Surfactants, both soaps and synthetic detergents, are high on the list, as are certain builders and fillers. Selected intake values are shown in Table 17-4.

In most cases, the principal contributions to the intakes result from the ingestion of dishwashing residues from plates and direct absorption of chemical from the toilet soap bar through the skin. The respiratory route does not appear to be of particular significance for soaps and detergents.

In addition to the chemicals for which estimates have been made, there may also be exposure to several hundred other known surfactants, to approximately 30 known perfumes, about 80 dyes, about 30 fluorescent brighteners, and about 15 other specified chemicals.

TABLE 17-4. Intakes of Selected Soap and Detergent Chemicals[a]

Product	Intake (mg/year)
Alkylbenzene sulfonates	5000
Alcohol ethoxylate sulfates	4000
Sodium sulfate	2000
Sodium tripolyphosphate	1000
Perfumes	200
Dyes	100
Fluorescent brighteners	10
Titanium dioxide	10
Germicides (e.g., triclocarban)	9
Polyvinyl alcohols	0.05

[a] From Stanford Research Institute (1971).

SOME OBSERVATIONS ON BACKWARD TRACING

Backward tracing has two major advantages: it naturally focuses attention on human activities likely to lead to significant exposures and forces systematic consideration of all chemicals reaching humans through those activities. On the other hand, backward tracing must of necessity be done exposure class by exposure class, and the total exposure for a given chemical will be underestimated until all classes have been investigated. For example, exposure to titanium dioxide might first be identified while studying pigments used in processed foods, but the true extent of exposure not discovered until the massive uses of titanium dioxide in paper, paint, and plastics have all been studied.

Several lessons can be learned from the use of backward tracing techniques:

1. Most exposure classes involve scores or hundreds of substances. Few product types are made with only a few ingredients, especially when one takes into account competing formulations.
2. There are great differences between classes comprising intentional exposure (especially food additives, drugs, and cosmetics) and classes with no intentional exposure, in terms of the portion of material used that results in actual exposure.
3. With the exception of the active ingredients of drugs, pesticides, and a few other product types, most high-volume chemicals have a variety of uses with potential for exposure. It may be a big mistake to assume that the most obvious use is also the highest contributor to exposure.
4. Exposure levels differ by tremendous factors. A study of approximately 4000 chemicals for the National Cancer Institute (Stanford Research Institute, 1971) indicated that exposures, based on total grams of exposure by any route for the entire U.S. population, ranged over about 15 orders of magnitude (10^{15})! Even when the tails of the distribution—water, dex-

trose, salt on one end, and catnip oil on the other—were excluded, reasonably interesting exposure levels ranged over seven or eight powers of 10.

5. Typical exposures are quite small when all chemicals are considered. The model exposures are on the order of 10^5–10^6 g/year for all U.S. citizens, implying average per capita exposures of 1 mg or so per year.
6. Per capita exposures range widely as well. Certain drugs may be used in 100 g/year quantities by a few hundred or a few thousand people with a rare disease, whereas everyone else experiences virtually no exposure, except for low levels encountered by pharmaceutical industry workers.
7. Although uncertainties in exposure estimates are large (a factor of 100 or more is not unusual), the relative exposures to different chemicals can often be confidently predicted for setting priorities.
8. Exposure assessment remains a largely ad hoc and creative discipline; prescribed procedures are rare and "detective work" is the rule.

ACKNOWLEDGMENT. This chapter is based in part on research performed pursuant to contract NIH-71-2045 with the National Cancer Institute, National Institutes of Health.

REFERENCES

Brown, S. L., and D. C. Bomberger. 1982. Releases of chemicals into the environment. Pp. 3–21 in *Proceedings of the Symposium on Models for Predicting the Fate of Chemicals in the Environment.* American Chemical Society, Washington, D.C.

Brown, S. L., and B. E. Suta. 1982. A general overview of approaches toward comprehensive exposure assessments. *Toxic Substances* **4**(1):23–37.

Stanford Research Institute. 1971. Cancer Hazards Ranking and Information System. Progress Report, November 1970 to October 1971. Contract NIH-NCI-71-2045. Stanford Research Institute, Menlo Park, California. 80 pp.

18

Comprehensive Risk Assessment

Robert G. Tardiff and Joseph V. Rodricks

The assessment of risks to public health is a highly complex undertaking. A single risk assessment comprises many discrete decisions—assumptions, interpretations, weighing of evidence—that are necessary if useful conclusions are to be reached concerning the existence and magnitude of risks to human health presented by a substance. This chapter describes those principles (some of which have been extracted from previous chapters) applicable to various facets of health risk assessment and integrates them into a functional and scientifically defensible approach to the interpretation of data bases comprised of *in vivo* and *in vitro* toxicological information and results from human studies.

For more than half a century, evaluation of the safe use of chemicals has been focused mainly on the development of toxicity data and on the application of professional judgment to the ad hoc interpretation of such data to derive acceptable levels of exposure for humans. Generally, this practice has taken the form of identifying from studies in laboratory animals the no-observed-effect level (NOEL) and dividing it by a safety factor (usually 100 for NOELs derived from chronic studies) reflecting the uncertainties of relating the data to humans under their conditions of exposure and the quality and appropriateness of the data base.

In recent years, alternative approaches to traditional safety evaluation schemes have emerged, in large measure because of more advanced understanding of toxicological manifestations and of critical reexamination of fundamental concepts. The growing recognition that toxicological methods measure injury and possess substantial limitations in identifying the absence of injury (i.e., safety) has redirected the focus of toxicological assessments to hazard assessment and risk estimation. Likewise, advances in chemical carcinogenesis—considered a specialized form of toxicity—have indicated that the pathogenesis for at least some

ROBERT G. TARDIFF and JOSEPH V. RODRICKS • Environ Corp., Washington, D.C. 20007.

carcinogens was unlikely to display a biological threshold. This realization prompted the adoption of the concept of risk per unit dose at very low levels of exposure and catalyzed considerable activity, at least for carcinogens, in the mathematical modeling of dose–response curves beyond the range of experimental observation (Rodricks and Tardiff, 1983).

More recently, the distinction between risk assessment and the establishment of "safe" exposure levels has been more sharply drawn: the former is clearly a scientific endeavor, whereas the latter is an exercise in societal decision-making to be performed by legally constituted representatives of the affected constituencies.

The following sections of this chapter concern the conduct of risk assessment. They do not analyze the multifaceted problem of deciding what level of risk is tolerable in various circumstances. They rely on all the information contained in the earlier chapters.

DEFINITIONS OF TERMS

Risk assessment is a systematic, multistep process of data evaluation designed to characterize the nature and magnitude of health damage posed by an environmental agent under various conditions of exposure. If scientific understanding and data were complete, risk assessment might be reducible to a simple set of quantitative estimates of the probability that a given agent will cause an adverse health effect in an individual or population group. Given the state of current understanding, however, any attempt to reduce risk assessment to a set of quantitative estimates would result in serious distortion. Since much of the information bearing importantly on the risks of a substance cannot be reduced to quantitative terms, other means must be found to incorporate such data into a risk assessment. Unless this is done, those who must base decisions on the result of risk assessments will not be fully informed. Thus, one objective of a comprehensive risk assessment is to ensure that all pertinent knowledge, both qualitative and quantitative, is recorded, evaluated, and presented in a fashion that is understandable to decision-makers. An overview of the strategy and planning for risk assessment has been described by Rodricks (1982).

A comprehensive risk assessment will have the following four major components (National Research Council, 1983a):

(1) *Hazard Evaluation** is the process of gathering, organizing, and evaluating all data that may reveal the types of adverse biological effects that may be produced by the sub-

*There appears to be some confusion in the literature about the meaning of several terms associated with risk assessment. In particular, many analysts do not distinguish between "hazard" and "toxicity" or between "hazard" and "risk." The common dictionary definition of "hazard" is a "potential source of danger." If this definition is accepted, then all substances are hazardous, and the goal of risk assessment is to identify the conditions of exposure under which the hazardous properties of a substance are likely to be realized and the range of conditions in which there is very little likelihood that the hazardous properties of a substance will become manifest. The definitions in the cited report, which we adopt, recognize this distinction between hazard and risk. Toxicity is one of the hazardous properties of chemicals, and it is the hazard with which this book is concerned.

stance under evaluation and the conditions under which they are produced. The evaluation should be critical, and include a characterization of the strength of evidence that the substance causes specific effects, the underlying mechanisms that contribute to those effects, and the extent to which observations obtained in one biological system (e.g., animals, microorganisms, etc.) can be said to apply to humans. All uncertainties need to be specified.

(2) *Dose–Response Evaluation* is the process of describing the observed dose–response relation for each of the documented biological effects of the substance and specifying the biological and statistical uncertainties in those relations. It also involves specifying how dose schedule and duration of exposure influence the dose–response relationship. The influence of the mechanism of biological action on the dose–response relationships should also be incorporated in the evaluation.

(3) *Human Exposure Evaluation* is the process of describing each of the sources of human exposure to the substance and the likely human intake of the substance by each route as a function of time. Discrete population groups are also defined according to magnitude and duration of exposure and by any appropriate demographic and biological characteristics. If risk depends upon exposure route, population groups may also be defined by route of exposure. For each group, a dose is defined (or, more likely, a range of doses).

(4) *Risk Characterization* involves combining the analyses of steps 1, 2, and 3 to yield measures of risk for each discrete population group along with descriptions of the strengths and weaknesses of the data supporting these measures.

(i) For some effects, the measure is the experimentally determined threshold dose divided by the human dose (hereinafter called the Margin-of-Safety, MOS). As the MOS increases in magnitude, the degree of health protection increases and the health risk decreases.

(ii) For other effects, the measure is the probability of effect at the human dose level (hereinafter called the risk).

As previously noted, the conduct of risk assessment is distinct from the process of determining the societal significance of the assessed risk and the appropriate means for its control. The latter activities fall in the domain of risk management, and it is important that risk assessment avoid consideration of risk management options, although perhaps cognizant of their existence (National Research Council, 1983a).

ANALYTIC STAGES

The assessment of health risks from chemical substances is a multistage process. It begins with a clear statement of the risk question that is to be examined. It then requires a collection and organization of all pertinent *information. Individual studies* pertaining to hazard, dose–response relationship, and exposure are then each *critically evaluated.* The total data base is then critically evaluated so that broad conclusions about hazard, dose–response relationship, and exposure can be drawn and attendant uncertainties specified. Finally, all these analyses are drawn together, and critical studies and pathological manifestations are identified in order to characterize risk.

These analytic stages are described in the next three sections with emphasis on hazard and dose–response evaluation, which typically create the greatest problems. Subsequent sections of this chapter concern human exposure assessment and the integration of all information and analysis to characterize risk.

DEFINING THE RISK QUESTION

At the outset, the dimensions of a risk assessment must be carefully structured to provide guidance in the acquisition of data and in the formulation of concepts by which to interpret those data. A risk question may be relatively narrow (e.g., the cancer risk to humans from using chlorine in the drinking water of Philadelphia) or quite broad (e.g., all risks to all U.S. population groups exposed to vinyl chloride through ingestion, inhalation, and skin penetration for all possible exposure durations). The scope of the risk assessment question may often be predicated upon risk management options (e.g., ability to control only route of discharge), by legislative mandates, or by use characteristics (e.g., a chemical intermediate used in a process that is virtually completely closed).

The risk question encompasses several components, the most important of which are the following:

1. What is the substance of interest (i.e., chemical definition)?
2. What exposure conditions, e.g., route and duration, are to be examined?
3. What hazard characteristics are to be examined? Cancer only? All adverse effects?
4. What population groups are to be considered? The general population? Workers only?

Some questions may be too narrow; others, too broad. A narrow focus (e.g., one route of exposure) may overlook significant sources of risk for the same substance and may not identify the source of exposure where substantial public health improvements might be made most cost-effectively. By contrast, a risk assessment of global perspective may be cumbersome, consume substantial resources, and lead to an array of choices so complex that decision-making becomes unduly impeded. Careful planning and analysis before initiating the risk assessment is vital to increasing the likelihood that the product of the undertaking has true utility in the decision-making process.

DATA COLLECTION

The search for and collection of data that bear on adverse biological effects should be compatible with the objectives of the risk assessment. Optimal minimum data requirements for risk assessments of varying types have been addressed by a committee of the National Research Council (1983b), and organizations such as the Environmental Protection Agency (Albert *et al.,* 1977), the U.S. Food and Drug Administration (1982), and the Organization for Economic

Cooperation and Development (1983) have prescribed data needs within the confines of their legal mandates.

In assessing human health risks, the information collected may derive from human studies and, more commonly, from studies in laboratory animals that are referable to humans (historically, these have been the rat, mouse, guinea pig, hamster, dog, and nonhuman primate). Some *in vitro* systems are also useful in generating data of value in the qualitative assessment of hazards to humans (Tardiff, 1978). Ideally, experimental conditions in whole-animal studies should be comparable in route of administration and duration of exposure to those anticipated in humans.

At times, information will be available only from an exposure route other than the one of interest for estimating risk. Such data should be included in the review, for they may be helpful in identifying some target organs and in determining relative potencies. In some cases, they may be the only data available.

The value of other forms of data (e.g., metabolism and mechanisms of action) for risk assessment may be substantial. Mechanistic data have an intrinsic value for assessing virtually all risks; they provide a fundamental understanding that permits extension to other exposure circumstances with substantial confidence because of the fundamental similarities of biological processes at the molecular level. Likewise, data describing toxicokinetics, toxicodynamics, and biotransformation characteristics are of generic utility in deriving insights about the likely hazardous properties of substances. Several approaches have been used with varying degrees of success to extrapolate toxicity findings from one route of exposure to another (Chapter 12); most depend in large measure on comparative metabolic information.

EVALUATION OF INDIVIDUAL STUDIES OF TOXICITY AND DOSE RESPONSE

Individual investigations of the adverse effects and biological behavior of substances vary in design, execution, and applicability to the specified risk assessment. Their specific natures will contribute significantly to the level of confidence that can be attributed to the conclusions about health risks as well as to the characterization of hazards. To facilitate the analysis of primary data, we recommend that findings from individual studies be classified into four parts: unequivocal, limited, equivocal, and uninterpretable. The characteristics of each and their utility are described in detail later in this chapter.

Evaluation of individual studies requires consideration of several factors associated with a study's hypothesis, design, execution, and interpretation. A valid study must address an answerable and clearly delineated hypothesis, and its basic design must be reasonably capable of testing the hypothesis.

Many substances in the human environment interact with biological systems to produce deleterious health effects after various types of exposures (e.g., acute and prolonged or repeated). The experimental study of toxicity must be designed to reveal exposure levels and conditions that are harmful and those that produce no adverse effects. Cause–effect relationships are essential to delineate, but are not as apparent with chronic as with acute exposures. The observable toxic response to a substance may be the result of its storage, the action of its metabolites in body tissue with subsequent mobilization and redistribution, or simply the repeated and added insult by the parent substance on target organ and molecules. In evaluating toxicological and epidemiological studies, consideration must be given to many factors, including characterization of the compound(s) under study, the test species, and their relationships to humans; the numbers of individuals in study groups; the number of study groups; the types of observations and methods of analysis; the nature of pathological changes; alterations in metabolic responses; sex and age of test species; and route of administration (see Chapter 4).

In general, such studies should attempt to achieve three objectives: comprehensive measurement of endpoints, including delayed effects; use of species most predictive of human responses; and identification of dose–response relationships to permit quantification of adverse effects. Although there are highly structured studies of toxicity in laboratory animals, there is no set of animal investigations that will be universally applicable for defining the toxic properties of substances.

Crucial Factors in Data Interpretation

The interpretation of toxicity data and their value in predicting risks to human health is associated with several factors, including the nature of the measured effects, the sensitivity of the biological methodology, the statistical sensitivity of the study, the reproducibility of the study results, and the quantitative nature of the study, i.e., dose–response relationship.

Observed effects must be judged as to whether or not they are adverse. Adverse effects from exposure to chemical and physical substances are defined as changes that result in impairment of functional capacity or in decrement of the ability to compensate for additional stress that are irreversible during exposure or following cessation of exposure, assuming such changes cause detectable decrement in the ability of the organism to maintain homeostasis, and that enhance the susceptibility of organisms to the deleterious effects of other influences (National Research Council, 1975b). A range of biological changes produced by a substance may not meet these definitions; those effects must be distinguished from the adverse effects that do.

The sensitivity of biological measurements to detect functional and anatomic harm varies greatly. The detection of adverse effects begins with gross observations of the intact animal in terms of growth, appearance, and activity.* The subsequent level of discrimination is that of the organ system in which biochemical and physiological changes are assessed. Sensitivity is further increased by examination of morphological changes at the cellular and subcellular levels. Further

*Although the detection of adverse effects requires some direct experience in intact organisms, there are occasions when adverse effects and the toxic potency of substances are predicted by analyses of the biological activity of structural analogs [see Chapter 14 and National Research Council (1982a)].

refinements are made at the molecular level, where interactions between a test substance and physiological molecules (e.g., DNA) can be observed. Several analytic methods can be used at each level of organization. Each of them has inherent sensitivity for identifying adverse effects—a factor to be considered when determining the value of each study.

The sensitivity of biological measurements is also influenced greatly by the susceptibility of the exposed organisms—a characteristic that is generally species-dependent. The underlying causes of differences in susceptibility are partly genetic (e.g., the occurrence or deletion of a target molecule or anatomic differences) and partly environmental (e.g., changes in rates of activation and detoxification brought about by environmental agents). The uncertainties associated with species differences can at times be attenuated by knowledge of comparative toxicokinetics, toxicodynamics, and metabolism. With such data, it may be possible to estimate the degree of similarity in response between humans and test organisms. Without such information, empirical associations across species are imprecise (i.e., no one species has responses parallel to the human across a wide range of the effects of chemicals). Therefore, there are substantial uncertainties in extrapolating adverse effects from nonhuman species to humans.

Statistical considerations are an important part of the interpretation of toxicological and epidemiological investigations. Sources of experimental and sampling error must be identified and dealt with so that meaningful conclusions can be reached. Studies of toxicity are rarely absolute in their identification of adverse effects and may incorrectly find either that an adverse effect is present when biologically it is not (a false positive or a type I error) or that an adverse effect is absent when one is actually present (a false negative or a type II error). False positives may occur as the result of chance and random variation (particularly if there are many replicates of point determinations) or of some uncontrolled factors in the conduct of studies. The basis for false negatives can be either a lack of biological responsiveness or inadequate statistical power. For a biologically responsive test, the sum of the rates of false positives and false negatives is generally constant, often between 15 and 20%. Individual study design can be modified to minimize either the false positives or the false negatives, but invariably at the expense of the corresponding characteristic. A detailed discussion of false-positive and false-negative rates and their impacts on conclusions about risks and safety has been provided by International Commission for the Protection Against Environmental Mutagens and Carcinogens, Committee No. 3 (1983). More extensive treatments focusing on chemical carcinogenesis assays have also been described (Fears *et al.,* 1977; Haseman, 1983; Salsburg, 1983). For the cancer bioassay, the optimum design for subsequent risk assessment should contain four dose groups and utilize a total of 150–300 animals (50–60 for control groups and 40–60 for treated groups) (Portier and Hoel, 1982, personal communication). Other statistical considerations in the interpretation of chronic assays for carcinogenicity have been described by Gart *et al.* (1979) and the International Agency for Research on Cancer (1980).

Identifiable factors that influence the outcome of laboratory animal studies are variation in composition of test material, interanimal and intergroup variations, quality of animal care and monitoring of test animals, and precision of laboratory techniques and instruments (National Research Council, 1977c). Their contribution to experimental error needs to be determined through independent replications. If major sources of variations are incorporated in the analysis, the width of the confidence intervals is likely to be quite narrow; if not, the confidence intervals will be misleading and add substantially to the uncertainty.

When evaluating intergroup variances, it is important to use historical controls to determine either gradual or sporadic fluctuations in control conditions that would influence the interpretation of results in the treated groups (e.g., raising the baseline on controls is likely to produce an artificially high NOEL, whereas lowering it may favor false-positive findings). The National Library of Medicine's Toxicology Document and Data Depository (TD3) provides a current source of information on many types of historical control data. Several statistical procedures exist for the incorporation of such data.

Dose–response relationships are quantifiable means of expressing toxicity and, hence, are essential components of the interpretation of toxicity data for eventual estimation of risks. Depending on the degree of control in an investigation, the dose–response relationship will convey different degrees of certainty about causal relationships. For example, in some epidemiological surveys, only associations between exposures and disease incidences can be concluded, because there is more than one uncontrolled variable. By contrast, laboratory animal studies offer greater opportunity to establish causal relationships unequivocally.

Two forms of dose–response relationships are often distinguished: (1) the dose–effect, or quantal, relationship, in which the number of responding individuals in a population varies with dose, and (2) the graded response, in which the severity of a lesion within an individual is modified with dose. Each relationship is based on the receptor theory of biological response. That theory specifies that molecular receptors mediate the initiation and continuation of adverse responses and that the nature and severity of the response are a function of the time of interaction with the receptor. An important concomitant assumption is that the dose and duration at the receptor are related to the administered dose.

Dose–response relationships are generally displayed graphically using either arithmetic, logarithmic, or power functions. Depending upon the nature of the substance and its toxicity and on the mathematical description of the data, the curve will be linear, sigmoid, rectilinear, or of other shape. The slope of the curve provides valuable information to determine in part the potency of the substance and will be of use in subsequent estimations of risk as well as for computing indices of comparative potency (Klassen and Doull, 1980).

Some data sets demonstrate no dose–response relationship. Such situations do not provide much assurance of causality or lack thereof. For example, a single positive response could be the result of spurious effect, and the determination of its biological significance will depend upon the results of confirmatory or other studies of varying design.

To obtain additional perspective on this subject, the reader is referred to World Health Organization (1978).

CRITICAL DATA POINTS

For toxicological and epidemiological studies that yield dose–response curves, emphasis is focused on selected data points. The NOEL is the maximum experimental dose at which no statistically significant effects are observed; it is used frequently as a baseline in performing safety evaluation of chemicals. For a particular compound, substantial uncertainty may be associated with the identification of a NOEL, which is likely to vary by target organ, route and duration of exposure, species, sex, study design, and investigator. Thus, its use must be properly qualified to provide meaningful interpretations. A modification of the NOEL is the no-observed-adverse-effect level (NOAEL), which is designed to separate adverse from nonadverse effects.

A forerunner of the NOEL is the no-effect level (NEL), which cannot be identified experimentally because every experiment has limited detection power. The term is sometimes erroneously confused with the NOEL.

The lowest-observed-effect level (LOEL) refers to the lowest dose that produced statistically significant effects. It is used in the absence of a NOEL as a baseline in performing safety evaluations of chemicals and is occasionally preferred to the NOEL, because it represents a definable point on the dose–response curve rather than the absence of effects. A modification of the LOEL distinguishes between adverse and nonadverse effects and is referred to as the lowest-observed-adverse-effect level (LOAEL).

In dose–response studies, the intensity of the effect and the frequency of response generally decrease with reductions in dose, and the biological reaction often reaches zero before the dose becomes equal to zero. That point has generally been referred to as an experimental threshold below which no adverse effects are anticipated and implies a discontinuity in the slope of a dose–response curve.

The experimental threshold is a function of the study design, especially its statistical power, the spacing of doses, and the sensitivity of the test group and the experimental methods [e.g., observing the appropriate group at risk, as noted by Tomatis *et al.* (1975)]. This type of threshold is not to be confused with a *population threshold,* which usually refers to the threshold that applies to a very large number of organisms, e.g., threshold dose for the mouse *(Mus musculus).* A population threshold is dependent upon the breadth of the distribution of sensitivities among all members of the population, including the highly sensitive and the highly resistant and also upon competing factors and substances that induce or facilitate biological damage.

There are valid biological reasons for believing that population thresholds are likely to exist, although their precise measurement for virtually all substances is beyond our technical capability. There are two biological bases for predicting their existence: (1) the presence of repair processes in most tissues of all organisms and the possibility that these processes may be more efficient with mild damage and (2) the presence of chemical detoxification processes in many tissues and, for some substances, the increased efficiency of these process at low levels of exposure. Some examples of approaches to identifying thresholds in laboratory animals for one toxic manifestation (carcinogenesis) by mechanisms of action include determining the critical quantity of molecules to activate receptors (Claus *et al.*, 1974), identifying proximate reactants (Newmann, 1974), studying the binding of toxicant to critical target molecules (Lutz and Schlatter, 1978), and examining toxicodynamic and toxicokinetic functionalities that govern the formation of proximate toxicants and their detoxification (Gehring *et al.*, 1978). Although promising, such approaches require further experimentation before their generic applicability can be determined. Nevertheless, population thresholds are not to be perceived as immutable, because of the likely competition by environmental factors for inducing similar adverse effects or for altering susceptibility to toxic effects.

Evaluation of Individual Studies

As individual studies are evaluated for possible inclusion in the comprehensive assessment of risks to human health, objective criteria are needed to place them in one of the following categories: *uninterpretable, equivocal, limited,* and *unequivocal.* These categorizations have value only for risk assessment and are not intended for other possible uses of the data.

Uninterpretable results are the progeny of studies in which the basic scientific method has failed. There may have been a nontestable hypothesis, an approach incompatible with the hypothesis, inappropriate measurements, absence of controls (except for observations from accidental poisonings and suicide attempts), extremely low statistical power, poor survival, intercurrent disease, or uncontrolled variables. Absence of information about study design and execution may also render a study uninterpretable.

Studies with completely negative findings are likely to fall into this category, at least when it is impossible to determine if the absence of observed effects was a function of inappropriate study design and conduct rather than a true lack of compound effect. Unless there are data to support the validity of a negative study, it would ordinarily be considered uninterpretable. An example of a properly designed negative study that produces results that can be utilized for risk assessment is a study that replicates a positive study in every detail except at lower doses and with the appropriate statistical power.

Equivocal findings characteristically derive from studies whose design and implementation are believed to be scientifically sound, but whose results are subject to multiple interpretations. Examples include marginal statistical significance at one point only (and no statistical significance for others), biological response not universally accepted as adverse or referable to humans (e.g., increase in endoplasmic reticulum, singular pathological manifestation in an organ peculiar to surrogate species, e.g., zymbal gland of the rat), or biologically implausible dose–response data, e.g., positive response at low dose, but no statistically significant effects at higher doses. Equivocal findings should not be confused with biologically significant changes that do not meet statistical conventions, such as vinyl chloride-induced hepatic angiocarcinoma, which occurs so rarely as to be highly indicative of causal relationships even when statistical significance is not achieved.

Limited evidence refers to unmistakable findings in well-designed and rigorously conducted studies that have been based on clearly articulated and experimentally feasible hypotheses, but have not been confirmed or extended by a broad range of methods. Studies that yield limited data might have clearly pathological manifestations at the high dose, but no dose–response relationship, or they may have dose–response information on target organ toxicity based on biochemical indexes, but with no histopathological examination. Such findings, especially without replication by alternative approaches, are of limited value and taken alone provide substantial uncertainty for risk assessment. Likewise, even singular studies that have been conducted with state-of-the-art methodology and have yielded dose–response toxicity are of limited value (although of greater certainty than the previous examples), for there remains the possibility that the effects might be species-specific (especially in laboratory animal studies) or that they were related to some unknown environmental factor (especially for epidemiological investigations).

Unequivocal evidence demonstrates clear, undisputed cause–effect relationships. It requires rigorous design and execution, validated methods, no uncertainty as to the detrimental nature of the observed effects, and internally corroborative findings, e.g., positive dose–response relationships and measurements by alternative methods.

For additional perspectives on the analysis of the value of results from individual toxicity studies, the reader is referred to publications by the Organization for Economic Cooperation and Development (1983) and by the U.S. Food and Drug Administration (1982).

COMPREHENSIVE EVALUATION OF TOXICITY DATA

At this point in the process, each of the studies providing data pertinent to identifying the various toxic properties of the agent will have been critically reviewed. It is essential that the risk assessor now attempt to characterize the entire body of scientific evidence upon which the toxic potential of a substance is to be judged. The quality and quantity of data available for different substances vary widely, and, for a given substance, some of its toxic properties and their dose–response characteristics are generally better characterized than others. Unless an attempt is made to evaluate the total set of available data, differences of these types will likely go unnoticed, and a false sense of confidence in the data may result.

For each major toxic endpoint, risk assessors must evaluate the *strength of the scientific evidence* upon which the characterization of toxicity is based. They must avoid simple labels (e.g., compound X is carcinogenic); rather, they should describe the total evidence, point out uncertainties, and characterize the strength of the scientific evidence. If this approach is taken, the assessors are relieved of the responsibility of deciding how much evidence is needed before a substance is labeled a "carcinogen" or "teratogen" for policy purposes, whether public or private. Of course, this approach places upon them the burden of characterizing the total evidence—a task not easily accomplished.

General Approach

The general approach described in the following paragraphs is not intended to provide a recipe for evaluation, but rather to suggest a mode of thinking about the evidence.

For some substances, data on toxicity and dose–response relationship may be available from numerous epidemiological and experimental studies. At the other extreme, evidence may be limited to a single experimental study. For most substances, the data base falls between these extremes. In some cases, all the data available may point in the same direction; in other cases, inconsistencies may be apparent. The goal of this stage of the evaluation process is to consider all such evidence and to judge its quality and strength. It is also necessary to describe the confidence with which results obtained in one biological system (e.g., laboratory animals) are likely to pertain to another system (in our case, humans). The following questions require examination:

1. What is the strength of the entire body of scientific evidence indicating that the substance under evaluation has the capacity to produce a given form of toxicity?
2. What is the likelihood that the substance has the capacity to produce a given form of toxicity in humans?

It is critically important not to reach conclusions about human risk at this stage. The second question concerns only the likelihood that an agent will display its toxic properties in exposed humans. It does not concern the ultimate question: What is the likelihood that humans will be affected under actual conditions of exposure? Thus, for example, evaluation of the total data base concerning the teratogenic properties of a substance—which may include data from epidemiological and whole-animal studies as well as studies on underlying mechanisms of action—should enable one to judge the strength of evidence on its teratogenicity and to estimate its teratogenic potential for humans.

Hazard evaluation stops at this point. After the dose–response evaluation and human exposure assessment phases are completed, a statement can be made about the probability that the agent will produce birth defects in humans under their actual exposure conditions. (This is the risk characterization stage of the assessment.)

At this point, the answer to the two critical hazard evaluation questions can be given in qualitative terms only. There is, moreover, no standardized terminology to characterize such scientific evidence. Rather than providing a standardized language, we have used terms meant only as examples.

After evaluating the total data base, a risk assessor might reach one of the seven conclusions listed below. Carcinogenicity is used as the endpoint in these examples, but the same types of characterization of evidence could be applied to other forms of toxicity.

1. The evidence for an agent's carcinogenicity in humans is as certain as it can be within the limits of current scientific achievement.
2. There is substantial scientific evidence to support the hypothesis that the agent is carcinogenic in humans.
3. There is a moderate amount of scientific evidence to support the hypothesis that the agent is carcinogenic in humans.
4. There is limited scientific evidence to support the hypothesis that the agent is carcinogenic in humans.
5. The scientific evidence is not adequate to determine whether the agent is carcinogenic in humans.
6. There is limited scientific evidence to support the hypothesis that the agent is not carcinogenic in humans.
7. There is substantial scientific evidence to support the hypothesis that the agent is not carcinogenic in humans.

Factors Used to Evaluate Total Data Base

Several factors should always be considered when selecting one of the many analytic approaches available for evaluating the total data base. The critical ones are *replication* of results, *reproducibility* of results, and *concordance* of results. It is also necessary to consider the *general body of scientific knowledge* concerning the particular toxic endpoint under evaluation.

These concepts are defined here as follows:

- An experimental result is said to be *replicated* if it is found in experiments of identical design.
- An experimental or epidemiological finding is said to be *reproducible* if it is produced in the same species under different conditions, such as different sexes, strains, dose groups, and routes of exposure.
- Experimental and epidemiological findings are said to be *concordant* if they are consistent across species.

For a given set of data pertinent to a given toxic endpoint (assuming each study has been critically evaluated), it is advisable to describe the *degree* of replication, reproducibility, and concordance. In general, as the degree of data replication, reproducibility, and concordance increases, it becomes more certain that a substance possesses the capacity to cause the specific toxic effect under review.

Moreover, it is necessary to consider the degree of correspondence between observations in experimental animals and expected responses in humans for a given form of toxicity, particularly when the only data available derive from animal studies and a judgment must be made about expected effects in humans. If certain types of animal carcinogens are likely to be similarly active in humans, inferences from animal data can be drawn in specific cases. Consequently, risk assessors must maintain their awareness of emerging scientific knowledge so they can make well-informed inferences from data on animals and other types of experimental information.

It is always important to separate true from apparent lack of data on replicability, reproducibility, and concordance. For example, several data sets may, upon superficial examination, appear to lack reproducibility, but more careful examination may reveal that this lack is not substantive. An attempt should be made to learn whether the apparent lack is due to differences in study design and conduct or to true differences in response. It is generally inappropriate to characterize toxicity tests as "positive" or "negative" and then to conclude that the degree of reproducibility of a given effect is low or absent simply because some tests yielded "positive" results and others yielded "negative" results. This oversimplification of test results can be avoided if each study has been critically evaluated so that the differences in various tests (e.g., in extent of examination of test animals, sample size, duration, and magnitude of exposure) are fully known. Once this examination has been completed, it becomes possible to assess the true degree of reproducibility (or lack thereof) for a given effect of a compound. This type of careful examination should also be applied when assessing degree of replication and concordance. Judging the degree of a response's concordance among various species is probably the most difficult aspect of data evaluation, so that the critical evaluation becomes especially important in order to avoid the oversimplification described above.

A final factor to be remembered when judging total data sets is the difference between absence of data and absence of replication, reproducibility, and concordance. A distinction should be made between the presence of positive evidence showing that an effect is not reproducible and the absence of negative evidence because no attempt has been made to reproduce an effect. Clearly, the failure to reproduce an effect (assuming that it is a true failure) may raise serious doubts about the toxic capacity of a substance, whereas the absence of data concerning the reproducibility of an effect should not raise such suspicions. Again, this same principle of evaluation applies not only to reproducibility, but also to judgments about replication and concordance.

Steps in the Process

Ordinarily, individual sets of data—from epidemiological, whole-animal, and *in vitro* studies—are evaluated separately. An attempt is then made to characterize the entire set.

After all pertinent epidemiological studies have been critically reviewed, the entire set of data is examined, and conclusions such as the following may be reached (again, using cancer as the example of a toxic manifestation):

A. A causal relationship between exposure to the suspect agent and human cancer (of a specific type) is established.
B. There is strong evidence for an association between exposure to the sus-

pect agent and human cancer, but a firm causal relationship is not established.

C. There is limited evidence for an association between exposure to the suspect agent and human cancer, and a causal relationship is not established.
D. There are no human studies adequate to permit a conclusion to be reached about an association between the suspect agent and human cancer.
E. There is no evidence of an association between exposure to the suspect agent and human cancer in adequately designed and conducted studies. (This conclusion would ordinarily require a full description of the extent to which an effect may have been missed.)

These conclusions are each based on expert judgment regarding the degree of reproducibility and concordance in the entire epidemiological data base.

A similar set of conclusions can be drawn after evaluating all relevant animal bioassay data, again based on the degree of replication (reproduction of the effect in experiments of identical design), reproducibility (reproduction of the effects in different strains, sexes, dose groups, etc., of the same species), and concordance (reproduction of the effect in different animal species). Conclusions of the following type might be appropriate:

A. There is strong evidence that the agent is carcinogenic in experimental animals.
B. There is adequate evidence that the agent is carcinogenic in experimental animals.
C. There is limited evidence that the agent is carcinogenic in experimental animals.
D. There are no studies adequate to permit any conclusion to be drawn about the carcinogenicity of the agent in experimental animals.
E. There is no evidence, based on adequate studies, that the substance is carcinogenic in experimental animals. (Again, the inherent limitations of negative data need to be specified.)

Judgments about categorizing evidence of animal carcinogenicity probably cannot be readily systematized. Generally, however, the first conclusion (A), i.e., strong evidence of carcinogenicity, would probably require not only a very high degree of reproducibility and concordance, but also tumor production at several sites and a high degree of malignancy produced in a relatively short period. The second conclusion (B), i.e., adequate evidence, might be reached if there were some degree of reproducibility and concordance, or, in the absence of these factors, if an especially serious form of cancer were produced in a single species, strain, sex, and tumor site. The third conclusion (C), i.e., limited evidence, might be reached if there were a low degree of reproducibility and concordance, or if tumor formation were limited to a single site in a single species, strain, and sex, and the tumors had a low degree of malignancy or none at all. Absence of sufficient test data or highly ambiguous test results might result in the fourth conclusion (D). The final conclusion on the list would be appropriate only if there were a relatively high degree of reproducibility and concordance in the negative data.

These general statements about degrees of experimental evidence are suggestions and are certainly not all-inclusive. The more formal published schemes for ranking animal evidence of carcinogenicity described below may be more suitable than the more general scheme suggested above.

Data from *in vitro* studies and data on the metabolism and toxicokinetic behavior of a suspect agent might also be evaluated and classified in the same way, although at times this may not be appropriate or necessary. Such data should, of course, be used to assist the full evaluation, but may best be used to qualify conclusions based on epidemiological and animal bioassay data. Consistent findings of mutagenic potential might, for example, raise the degree of suspicion about an agent when the data from animal studies alone were not adequate to justify a conclusion about carcinogenicity. Strong evidence of differences in metabolic patterns between test animals and humans might weaken inferences about the carcinogenic hazard for humans based on animal data when it becomes time to combine the conclusions from epidemiological and animal studies. Such information may be used in so many ways to qualify conclusions based on animal and epidemiological studies that we can only mention them briefly in this chapter.

COMBINING THE EVIDENCE

In the previous section we have provided sample conclusions about carcinogenicity based on different data bases, seven based on the total evidence for carcinogenicity and five each based on epidemiological or animal data only. Figure 18-1 shows how conclusions based on specific types of data might be used to

Conclusions regarding epidemiological evidence	Conclusions regarding animal evidence: A	B	C	D	E
A	1	1	1	1 or 2	1 or 2
B	1	2	2 or 3	2 or 3	2 or 3
C	2	2 or 3	4	4	4 or 7
D	2 or 3	3 or 4	4	5	6 or 7
E	4 or 5	4 or 5	6	7	7

FIGURE 18-1. Evidence to support various types of conclusions regarding the strength of evidence of carcinogenicity. The numbers 1–7 represent conclusions regarding total evidence. The letters A–E represent the total body of epidemiological and animal data. Evidence regarding genotoxicity and other biological measures of carcinogenic activity might modify the conclusions.

support the seven sample conclusions. The judgments expressed in this chapter are those of the authors; others might array evidence differently. Our purpose is only to illustrate the type of thinking that might underlie the process of weighting different types of scientific evidence.

OTHER SCHEMES

Several authors have proposed more formal schemes for categorizing evidence, mostly of carcinogenicity. For example, Squire (1981) proposed a method for ranking animal carcinogens "according to the strength of the experimental evidence and according to biological factors considered in assessing carcinogenic potency or potential human risk." His intent was to arrive at a regulatory classification for a carcinogen using the information provided by a standard two-species bioassay and short-term tests. In his proposed scheme, numerical scores are assigned to the evidence (Table 18-1), and the total score determines the regulatory classification (Table 18-2).

The Squire scheme includes consideration of potency (see dose–response relationships in Table 18-1). The inclusion of this factor might inappropriately confuse the "strength of the evidence of carcinogenicity" with the "strength of the carcinogen." This scheme is not designed to include epidemiological information.

Griesemer and Cueto (1980) developed a scheme for classifying the evidence for chemical carcinogenicity in animals using the NCI bioassays as their data source. They proposed nine evidence categories, ranging from "very strong evidence for carcinogenicity in two species" through "sufficient" or "equivocal evidence" in two or more species to "no evidence for carcinogenicity in two animal species." Their scheme includes a number of factors, such as the adequacy of the design and conduct of the test and the type of carcinogenic response observed. The strength of the evidence is based on the following three considerations, but the authors did not assign numerical weightings to them:

1. Number of species or strains with increased incidence.
2. Number and types of positive results in different experiments, considering differences in administration routes, doses, and/or sexes.
3. Degree of response, considering incidence, histological type, multiplicity of sites, and/or early onset.

The Griesemer and Cueto scheme presents a useful description of a process for considering evidence of carcinogenicity provided by the National Cancer Institute (NCI) and the National Toxicology Program (NTP) bioassays, but by design it is not a procedure for determining the degree to which the evidence predicts that a substance is likely to be a carcinogen in humans. Thus, their scheme is limited by its restriction to data from animal studies.

The International Agency for Research on Cancer (1979) has published a description of its approach to the evaluation of potential carcinogenic chemical hazards to humans. The assessment is based primarily on human and animal

TABLE 18-1. Scoring of Evidence of Carcinogenicity[a]

Factor	Score
Number of different species affected	
Two or more	15
One	5
Number of histogenetically different types of neoplasms in one or more species	
Three or more	15
Two	10
One	5
Spontaneous incidence in appropriate control groups of neoplasms induced in treated groups	
Less than 1%	15
1–10%	10
10–20%	5
More than 20%	1
Dose–response relationships (cumulative oral dose equivalent per kilogram of body weight per day for 2 years)[b]	
Less than 1 μg	15
1 μg to 1 mg	10
1 mg to 1 g	5
More than 1 g	1
Malignancy of induced neoplasms	
More than 50%	15
25–50%	10
Less than 25%	5
No malignancy	1
Genotoxicity, measured in a battery of tests	
Positive	25
Incompletely positive	10
Negative	0

[a] From Squire (1981).
[b] Based on estimated consumption of 100 g of diet/kg body weight. Scoring could also be developed for routes such as inhalation.

evidence, although mutagenicity and chemical structure can also be considered. The combined evidence is used to arrive at a definitive statement of the carcinogenicity of a compound in animals and in humans. Differences of opinion are decided by a majority vote of the IARC working group assembled to review the evidence on specific chemicals. The relative weighting of the evidence from animal studies is based on four categories:

- *Sufficient evidence:* [This classification] is indicated by an increased incidence of malignant tumor . . . in [several] species or strains, . . . in multiple experiments (preferably with different routes of administration or different dose levels), or . . . to an unusual degree with regard to incidence, site or type of tumor, or age at onset.
- *Limited evidence:* The data suggest a carcinogenic effect but are limited because . . . the studies involve a single species, strain, or experiment; . . . the experiments are

TABLE 18-2. Ranking of Carcinogens Based on Scores from Table 18-1[a]

Total factor score	Class	Regulatory options
86–100	1	Restrict or ban
71–85	2	—
56–70	3	—
41–55	4	—
<41	5	Several options (no action, limited use, labeling, public education)

[a] From Squire (1981).

restricted by inadequate dosage levels, inadequate duration of exposure to the agent, inadequate period of follow-up, poor survival, too few animals, or inadequate reporting; or . . . the neoplasms produced often occur spontaneously or are difficult to classify as malignant by histological criteria alone (e.g., lung and liver tumors in mice).

- *Inadequate evidence:* Because of major qualitative or quantitative limitations, the studies cannot be interpreted as showing either the presence or absence of a carcinogenic effect.
- *Negative evidence:* Within the limits of the tests used, the chemical is not carcinogenic. (International Agency for Research on Cancer, 1979)

Evidence from humans can fall into one of three categories:

- *Sufficient evidence:* [This category applies when] carcinogenicity is indicated by a causal association between exposure and . . . cancer [in humans].
- *Limited evidence:* A possible carcinogenic effect in humans is indicated, although the data are not sufficient to demonstrate a causal association.
- *Inadequate evidence:* The data are qualitatively or quantitatively insufficient to allow any conclusion regarding carcinogenicity for humans. (International Agency for Research on Cancer, 1979)

On the basis of the evidence, the suspected carcinogen is placed into one of four groups:

1. Sufficient evidence is available from human studies.

2A. Almost sufficient or suggestive evidence is available [from studies on] humans or animals.

2B. As in group 2A, but the evidence is less strong.

3. Classification is not possible. (International Agency for Research on Cancer, 1979).

It is evident that the IARC approach depends heavily on scientific judgment and that considerations of individual factors are not formalized. For example, the IARC does not discuss the critical determination of the amount and type of animal evidence used to distinguish between groups 2A and 2B. Furthermore, this scheme does not provide for weighting the negative evidence against the positive.

The Interagency Regulatory Liaison Group, Work Group on Risk Assessment (1979) has also discussed the general bases for identifying carcinogens. This group proposed that the evaluation of a compound should include all evidence on carcinogenicity. Determination of the strength of the evidence is based on scientific judgment that considers the nature of test results, as well as the quality and adequacy of the studies. The cited report discusses criteria for an adequate epidemiological study and the weighting of different types of evidence.

A single, adequately conducted animal bioassay with positive results was considered sufficient to conclude, in the absence of other data, that the substance is likely to pose a cancer hazard to humans. Positive results in short-term tests were considered suggestive evidence of carcinogenicity, even if animal bioassays showed only negative responses. The positive short-term test data can be refuted by further testing in other animal species and strains or under more rigorous conditions (Interagency Regulatory Liaison Group, Work Group on Risk Assessment, 1979).

These approaches for identifying and classifying carcinogens have a number of factors in common, but differ in the degree of detail used to weight them. Generally recognized are the use of animal models to predict potential effects in humans, the correlation between a stronger response in animals (more tumors, more species, both sexes) and the likelihood of human risk, and the utility of short-term testing to provide additional evidence. All the approaches identify the need for case-by-case review of the studies by qualified scientists, and a few (e.g., Squire, 1981) specify numerical weightings to be assigned to individual factors. None of the schemes attempts to specify criteria or other formalized procedures for judging the adequacy of individual studies or for assigning weights to studies that are less than fully adequate.

The U.S. National Toxicological Program (1983) has also categorized the interpretive conclusions from animal carcinogenicity studies. The five categories identified by the NTP are clear evidence, some evidence, equivocal evidence, no evidence, and inadequate study. These relate to the strength of the experimental evidence and not to either mechanism of action or potency (relative or absolute).

For mutagenesis, the National Research Council's (1982b) Committee on Chemical Environmental Mutagens developed a scheme for weighting evidence from several nonmammalian test systems to classify substances as presumed mammalian mutagens, presumed mammalian nonmutagens, or inconclusive status. This committee also suggested a scheme for distinguishing among different levels of mutagenic potency.

CONCLUSION

The strength of scientific evidence that an agent can produce adverse physiological effects such as cancer or mutation in humans should be based on an evaluation of the total data base. Risk assessors must not simply present data reviews. Rather, they should attempt to provide critical evaluations of individual studies and the total data base. Only if an attempt of this sort is made will it be possible to consider the uncertainties in the data when risk is characterized.

HUMAN EXPOSURE ASSESSMENT

In Chapter 17, there is a detailed exposition of the problems and goals of human exposure assessment. In this chapter the discussion is extended to provide guidance on the various ways human exposure data may be arrayed for assessing risk.

Although the term "dose" is usually applied to drug administration, it is also applied to other human exposures, such as inhalation. The usual method for expressing dose is to estimate the weight of the substance absorbed into the body per unit of body weight.

Estimation of human "dose" may be a relatively straightforward exercise for drugs; but for most other environmental agents, it is beset with problems. The information needed and the way different pieces of information are combined to yield dose estimates are both easily defined, but the absence of information usually hampers the task. When information is not available, assumptions must be inserted into the analysis. This results in low confidence in the estimates and uncertainty in the ultimate risk assessment. Nevertheless, when such assessments must be made, and when assumptions are needed to complete them, the following operating criteria must be met: (1) the assumptions should always be stated; (2) to the extent possible, assumptions should be consistent with knowledge on related substances and conditions; (3) the assessments should include, where possible, the likely population distribution of exposures (although "point" estimates are usually all that can be derived); and (4) the assessments should probably include both "worst"- and "best"-case estimates, with both these terms defined.

The population at risk should be identified and, where possible, its size and nature should be described. Are we concerned with the entire population, or only with certain subpopulations? Are we concerned primarily with one sex? Adults or children? Workers? Smokers or nonsmokers? Special subgroups at unusually high risk because of high exposure or susceptibility should be identified. This procedure is a critical feature of exposure assessment.

In many cases, exposure assessment is complicated by the fact that persons are exposed through several media and by different routes. Exposure to formaldehyde is a good example of this phenomenon. Many people are exposed in their homes and in other environments, such as public buildings. Some of them are also exposed in their places of work. Exposure may occur by inhalation, but it can also occur by the dermal and oral routes. It is easy to imagine the difficulties this creates. If only one or a few types and routes of exposure out of many possible ones are to be assessed, the assessment will account for only a portion of the potential population risk.

Figure 18-2 lists the data needed to estimate dose for a theoretical compound. If all the information is known, it will be possible to estimate the total amount of compound absorbed by a person/unit of body weight per unit of time. Note that the total amount of compound is derived by considering the amount coming from each of the media in which it may be present.

Only rarely is all the information in Fig. 18-2 obtained. Availability of various types of information is usually a function of the medium and the source of

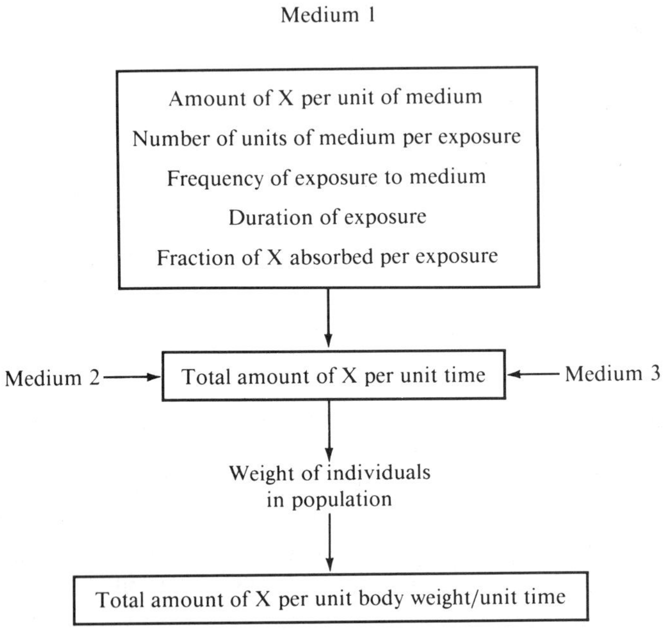

FIGURE 18-2 Exposure assessment for theoretical compound X.

the compound in the medium. For example, the amount of compound in food is much better known if it is intentionally added to food than if it is an adventitious contaminant.

Usually absent is information on the fraction of the compound absorbed per exposure. Data on absorption of a compound from the lungs or gastrointestinal tract or through the skin are very meager. In risk assessment, this lack is usually accommodated by assuming that the extent of absorption is the same in exposed humans as it is in the test animals from which the dose–response relationship derives. This assumption may not lead to serious error when the exposure route in test animals is identical to that for the human population at risk, but this assumption is no doubt a source of some error.

Problems arise when the route of human exposure is not the same as that of the test animals (see Chapter 4). The most common difference is found when animal data reflect oral exposure, but the major routes of human exposure are inhalation or dermal. Only when the extents of absorption in the test animals and in humans are known can we achieve much confidence in the dose estimates. In the absence of such information, it becomes necessary to impose assumptions. The usual procedure is to use several sets of assumptions, including but not limited to the assumption that 100% absorption occurs by all routes. When possible, data on structurally related substances can be used to assist estimation of absorption rates.

In some cases, the expression of toxicity may be related to the route of exposure. Formaldehyde again provides an example. Its effectiveness as a nasopharyngeal carcinogen may be dependent upon its inhalation, and the oral or dermal exposures (assuming that neither is accompanied by inhalation) may not carry this risk. In such cases, the exposure assessment should provide for separate consideration of population groups at risk because of their route of exposure.

The goal of exposure assessment is to estimate the target site dose or blood levels. Such estimates must be based on careful metabolic and toxicokinetic studies, both of which cannot always be acquired for humans. If such data exist, however, it would probably be beneficial to incorporate them in the exposure assessment.

Exposure assessment is clearly very complex; and, because of its crucial role in risk assessment, it deserves as much critical evaluation and detailed presentation as other aspects of the process.

RISK CHARACTERIZATION

After completing the hazard and dose–response evaluations and exposure assessment, one must estimate either the probability that an effect will occur in human populations under their conditions of exposure (i.e., risk) or, for most effects, the margin of safety (MOS). These estimates involve many choices and assumptions, not all of which can be defended on strictly scientific grounds. For example, if one study or set of data points from that study is selected from a large body of data for use in risk estimation, the reasons for their selection sometimes transcend purely scientific considerations. Often the data on the "most sensitive" of the various responding species (i.e., those exhibiting adverse effects at the lowest doses administered) are chosen for risk estimation. This is justified on the grounds that selecting the responding species "most biologically relevant" to humans is not possible because of lack of knowledge; therefore, the "most sensitive" species is chosen in order to err on the side of public health protection. Although these assumptions are made frequently and may be wise public policy, they are not based on the scientific process. The assumption that the "most sensitive" of responding species is selected for risk estimation should always be stated clearly along with its attendant uncertainties. The same approach must be taken with the many other selections and assumptions that have to be made to estimate risks.

Eight major steps are required to complete the risk estimation:

1. Selection of studies, data points, and dose–response relationships by adverse effect.
2. Selection of dose–response data and application of a model for high-to-low-dose, *intraspecies* extrapolation.
3. Identification and selection of an experimental NOEL.

4. Selection of dosage units, e.g., mg/kg body weight per day, dietary concentration, or surface area (mg/m^2), that may predict interspecies scaling factors.
5. Consideration and incorporation of data on interspecies differences in response.
6. Selection of measures for "dose" to humans (e.g., average or "worst case").
7. Estimation of risk or computation of the MOS, or ranges thereof, for each population group of concern.
8. Presentation of results and uncertainties.

SELECTION OF STUDIES AND DATA POINTS FOR RISK ASSESSMENT

For some compounds, the array of toxicity data is limited to one or a few studies; for others, it is extensive. The data base for a substance generally indicates multiple sites of toxicity, particularly at high doses. Under these circumstances, the risk assessor must frequently decide which data sets are best suited for the next step in risk assessment—low-dose extrapolation. Only general guidance can be provided for this stage of the analysis, for it relies heavily on professional judgment and involves the intricate dissection of data bases that lack strict comparability of content.

There is now a need to array data by individual adverse effects to provide a basis for comparing various risks to human health. In some cases, it may also be desirable to organize the information by general response characteristics, such as duration and route. Particularly important is the dichotomy between acute and chronic responses, because the dose levels and toxic responses of each are often characteristically unique. Separations of data sets by routes of response (e.g., inhalation and ingestion) are helpful, but may not be essential, because, as was described in Chapter 12, approaches to interroute extrapolations are scientifically feasible and of substantial certainty when there are relevant toxicodynamic and toxicokinetic data.

Several of these steps require that certain data subsets and models be selected. Such selections usually involve assumptions, some of which must be based on nonscientific grounds. For example, several plausible models for low-dose extrapolation (see Chapter 10) yield different estimates of low-dose risk. Science policy determinations are needed also to select among several plausible models and govern the conclusion that a 100-fold MOS is adequate to establish an acceptable daily intake (ADI) for chemicals producing nonneoplastic forms of toxicity. Each policy decision may be appropriate, but the extrascientific nature of each should be recognized as such.

Selecting interspecies scaling factors (step 4) presents similar problems (see Chapter 3). Again, all the uncertainties associated with such selections should be fully described.

OPTIONS FOR DATA SELECTION

The choice of data sets for extrapolation may also involve both scientific and policy considerations. For example, a policy decision may be the selection of only human data as the determinant for low-dose extrapolation or the data set predicting the highest risk to humans. Although there may be some justification for such constraints, such a determination is not necessarily consistent with the best scientific reasons for data selection and is likely to impose a substantial penalty by excluding valuable data from consideration. The risk analyst should ensure that the selection of data is scientifically defensible and that any policy constraints superimposed on the scientific evaluation are clearly stated.

Electing one data set as the focal point for estimating a defined risk does not imply that other data are to be ignored. Rather, all data should be used comprehensively to support and to set boundaries on risk conditions. The merits and limitations of the major options for data selection are discussed in the following sections.

Data on Humans

Data on humans may be selected as the basis for assessing a specific risk. They provide the most relevant dose–response relationships while avoiding the problems inherent in interspecies extrapolation. However, epidemiological data often have limitations. For example, their use for low-dose extrapolation is often difficult. By their nature, epidemiological studies are not controlled. Thus, comparability between exposed and control groups cannot be assured. Quantitative estimates of exposure are often difficult to make, particularly in retrospective studies, where documentation of exposure levels is often scanty. When the type of adverse effect induced by the toxicant occurs commonly, a low incidence of exposure-related toxic manifestations may not be detectable. Also, because of the long latency involved in the development of chronic toxicity, particularly cancers, epidemiological methods cannot be applied to newly introduced chemicals. Various criteria may then be needed for selecting among human data sets, if more than one is available. Selection criteria may include such factors as the highest estimate of relative risk, the "best" study design, the largest study population, the greatest statistical significance, or the most reliable estimates of substance exposure and disease incidence.

Data from Animal Studies

Data from animals are often selected as the governing information. They are frequently the only kind of toxicity information available. But where there is a combination of human and animal data, then a comparison of the quality of the respective data bases can be undertaken. Even with such comparisons, animal data are often identified as the critical determinant for the risk assessment. The use of animal studies for risk assessment has two major advantages: (1) they can be controlled more stringently and often provide more clearly defined dose–response relationships than epidemiological studies and (2) they do not require as long an observation period before results are obtained. However, the use of animal data introduces complex problems—and uncertainties—in interspecies extrapolation.

When animal data are pivotal to the risk assessment process, a cascading series of professional judgments must be exercised. Faced with data from several animal models, the analyst first tries to identify the model that is most relevant to humans based on the most defensible biological rationale. However, selection of the most sensitive species is generally a policy decision to provide the most conservative estimate of risk, i.e., the estimate predicting the highest risk. Deciding which animal data set is most "relevant" to human risk may not be possible when the route of exposure for humans is different from that used in animal studies, or if comparative data on metabolism, toxicokinetics, and other factors are not available to permit identification of the species most "relevant" to humans. However, when appropriate biological data are available to permit selection of the animal model most likely to predict human responses, then data for this animal model should be used and given more weight than data from other models.

Other judgments must be made to select some dose–response data as references from which to estimate risk. Generally, the dose–response relationship of choice is the one that is best defined and most supportable scientifically, e.g., there are more points demonstrating a consistent trend.

Although the best animal study may provide the most accurate estimate of the dose–response curve for that species, the relationship between risks estimated from these data and true human risks is generally no better known than for other animal data. Once again, it will often be necessary to select among data sets. Selection may be based on the most sensitive species, the species most relevant to humans, or the best study.

Data on Humans and Animals Combined

Select sets of human and animal data may be combined to provide a range of risk estimates and to determine the quantitative degree of data concordance. By incorporating data from all adequate studies, all pertinent information can be considered when estimating a range of risks. Making use of the different strengths of animal and epidemiological data, one might reasonably use data from a well-conducted animal study to quantify the risk and epidemiological data to confirm the relevance of the animal data. If necessary, negative epidemiological data (i.e., data with no indication of target organ toxicity) can be used to predict an upper limit on human risk. This procedure will provide a range of risk.

Interpretation of these ranges of risk may be needed to assess their relevance to the human population of interest. Procedures that give weight to different data sets to produce a single maximum likelihood estimate of unit risk based on all data sets have been proposed by Crouch and Wilson (1981) and DuMouchel and Harris (1981). However, these procedures involve much greater technical complications than are inherent in other procedures, and it is unclear if these increased complications are justified by an improvement in reliability.

Crump *et al.* (1980) have proposed a somewhat less complicated procedure, which does not incorporate all data, but evaluates all data sets for quality and then uses the set with the highest quality for risk assessment. This procedure does not automatically select either human data or data on the most sensitive species. Rather, studies are selected for their ability to fit criteria concerning the thoroughness of the experimental design and conduct and its consistency with the expected route of human exposure.

CRITERIA FOR DATA SELECTION

Data sets with the best scientific characteristics are judged on the basis of study design, population size, degree of statistical confidence, most reliable estimates of exposure, and disease incidence. For studies of laboratory animals, one would seek studies that exhibit dose–response relationships, that have the most comprehensive pathological evaluation, that most closely approximate human exposure conditions, and that possess the soundest statistical design and analysis.

NEGATIVE AND PRESUMABLY CONFLICTING DATA

Most of the foregoing discussion was focused on selection among sets of positive data, i.e., data demonstrating a significant association between exposure and an increase in disease incidence or severity. Often, however, the data base for a substance contains both positive and negative evidence of toxicity, including carcinogenicity. For example, there may be positive evidence in animal studies and negative evidence in epidemiological studies. Although such results may appear to be in conflict, the apparent qualitative difference in responses may reflect simply a quantitative difference in sensitivity of the studies because of differences in dose levels, lengths of exposure, background tumor incidences, sizes of study populations, or other factors. It may be useful in such situations to consider the statistical power of the negative data, i.e., the smallest increase in tumor incidence that could have been detected by the study as statistically significant, and to compare this with the increase predicted by the positive data at the dose level of the negative study. If the increase predicted by the positive study is less than the negative study could detect as significant, then the two sets of data cannot be considered in conflict. However, if such an analysis reveals that the predicted increase should have been detectable by the negative study, there must be some reason other than differences in study sensitivity to explain the inconsistency. This type of analysis may be of great assistance in the evaluation of apparently nonconcordant data.

Uniformly consistent biological information, including toxicity data, is often the exception rather than the rule. Variances in normal physiological and anatomic characteristics may be so great and the observed group so small that statistical conventions are extremely unlikely to lead to the discovery of aberrations less than devastating to an exposed population. Consequently, some toxicity data sets may describe positive responses at the highest dose tested but without a dose–response relation. Although isolated positive indications of toxicity at high doses may be equivocal but capable of being rationalized, other data remain anomalous because they do not fit any acceptable hypothesis for describing the behavior of biological systems under stress by chemicals. Some examples include multiple-dose studies in which only one central dose is statistically associated with some adverse effect; repeated measurements, e.g., RBCs, over time that yield interspersed statistically significant and insignificant findings; and repeated measurements, e.g., body weight, over time in a multiple-dose study that produces, in comparison to controls, improved growth at low dose, no difference at the intermittent level, and a reduction in the growth rate at high dose. Such apparent anomalies should generally be viewed as equivocal findings and should not be used as a basis for describing hazard or estimating risk without data from other types of experimentation to clarify their meaning. Although data such as those described in the foregoing may be excluded from use in a risk assessment, the studies from which they were derived may provide other data that would be valuable—even desirable—for estimating risks.

UNCERTAINTIES INTRODUCED IN DATA SELECTION

Risk assessment is, by its nature, an uncertain science. Furthermore, any procedure that limits the data used for risk assessment contributes to that uncertainty, unless there is good reason to believe that data eliminated from consideration are not relevant to human risk. Rarely can such confidence be obtained.

In some cases, elimination of certain data, e.g., data on less sensitive species, from consideration is consistent with conservative public health policy criteria. But this is a policy decision that does not necessarily provide a scientific or accurate estimate of human risk. Rather, it may give a false impression of accuracy of the estimate. Conversely, the use of all data appropriate for risk assessment to generate a range of estimates helps to make explicit the uncertainty in the assessment.

UTILIZATION OF SELECTED DATA

Several manipulations are required to present the information effectively for consideration in the formulation of risk management options and strategies. In general, there are two principal paths of data modulation, depending on whether cornerstone data are derived from laboratory animals or from humans. An example of such data analysis for risks from exposures to air pollutants has been developed by a committee of the National Research Council (1979).

The first—and perhaps the most difficult and less certain—path begins with sets of critical data from laboratory animals, adjusts the findings for deficiencies related to the targeted use (namely, risk assessment), selects an approach or model for low-dose extrapolation, defines margins of safety where desirable and appropriate, extrapolates estimated risk/"safe" levels to target populations of humans, and overlays known or anticipated human exposure (routes of entry, doses at portals of entry, and target tissue doses, if possible).

For the second path, one begins with information obtained from humans, usually from epidemiological studies retrospectively linking exposure to disease (see Chapter 6). With such data, one selects the endpoints of concern, adjusts the data sets for uncertainties related to the risk assessment of interest, either extrapolates to lower doses or identifies the margins of safety, and, where necessary, extrapolates to select target populations. An example of this approach as applied to herbicides has been described by Knaak (1980).

Describing uncertainties of data sets is often more difficult than for individual studies. Quantitative analyses of uncertainties may be based on uncertainty factors, confidence limits, analysis of propagation of errors, or sensitivity analysis.

Uncertainty factors have been used to describe the quality and quantity of data for use in assessing noncarcinogenic risks to human health (National Research Council, 1977b). When the quality and quantity of the data are high, e.g., valid experimental results from chronic exposures to humans, the uncertainty factor is generally low. When data are limited or equivocal, the uncertainty factor must be large, e.g., no data in humans and only scanty results in laboratory animals. Uncertainty factors of 10, 100, and 1,000 have been used by a committee of the National Research Council (1977b, 1980, 1984) to determine suggested-no-adverse-response levels (SNARLs) on the basis of toxicological and epidemiological studies. To analyze the uncertainty and a large quantity of comparable data, the confidence levels are often calculated and applied to a large array of toxicological and epidemiological information.

Analysis of the propagation of errors not only includes the evaluation of uncertainties of individual data points, but also assimilates each to provide an overall estimate of uncertainties. The method is most effective when the number of parameters is small, their essence is similar, and the variance of each is observed and measured. Sensitivity analysis is applied when the uncertainties of critical assumptions are unknown. Arbitrary values are assigned to pivotal input parameters, thereby generating hypothetical data sets to determine the effect of discrete modifications and delimiting the boundaries of the most sensitive parameters (National Research Council, 1982a).

The risk analyst is cautioned about the possibilities of artifactual uncertainty in evaluating data sets. Cox (1982) has described mental exercise to identify, and hence minimize, artifacts, especially when determining the attribution of risk under circumstances in which several factors may be responsible for impacts on health.

Toxicity information, particularly from laboratory animals, requires extrapolation to much lower doses to be directly applicable to human exposure conditions. Low-dose extrapolation usually leads to extensions of dose–response relationships by several orders of magnitude and hence requires modeling to dose–

response curves to fit existing concepts of biological phenomena and mechanisms of toxicity.

The predominant models, their derivations, and their applications are described in Chapter 10. There have also been several reviews published on models for low-dose extrapolation (Food Safety Council, 1980; National Research Council, 1977a, b; Carlborg, 1981b; Day and Brown, 1980; Fishbein, 1980; Hoel *et al.*, 1975; Van Ryzin and Rai, 1980). In most instances, curve-fitting models of dose–response relationships have been applied to chemical carcinogens, because there is a widely supported hypothesis that chemical carcinogenesis, unlike other types of toxicity, is an adverse effect for which there is no biological threshold. Thus, several authors have proposed numerous approaches to mimic the behavior of biological systems from exposure to carcinogens (e.g., Carlborg, 1981a; Gaylor and Kodell, 1980; Krewski *et al.*, 1983; Mantel, 1962; Peto *et al.*, 1972; Reitz *et al.*, 1980). The most widely used model for low-dose extrapolation of chemical carcinogenesis data is the multistage model, which conforms well to current hypotheses for the initiation and promotion of cancer, takes into account all dose–response data, and generally provides the best fit of those data. The models for low-dose extrapolation of cancer dose–response data rely on various underlying assumptions that, if modified, lead to substantial differences in the predicted risks (Whittemore, 1980).

Investigators have attempted to expand the biological basis for such extrapolation by incorporating more knowledge about metabolism and mechanisms of toxicity (Burlingame *et al.*, 1983; Gehring *et al.*, 1977; Gehring and Rao, 1977; Hoel *et al.*, 1983). The latter approach affords considerable promise, for it is based on some of the most definitive information about the nature and magnitude of dose and effect of substances on biological systems; however, for many compounds there remain some elements of uncertainty related to knowing the chemical identity of proximate toxicants as well as the extent of biological diversity between test (e.g., rodents) and target (e.g., humans) organisms.

In dose–response modeling, the accuracy of low-dose estimates is often questioned. Because the predicted risks are generally so low as to be undetectable experimentally (both in the laboratory and in the field), the opportunity for validation is rare. However, for at least one compound, ethylene dibromide, such an evaluation with the single-hit model was possible (Ramsey *et al.*, 1978). The risk derived from animal data had been overestimated (when compared to epidemiological findings) by one order of magnitude—a variance relatively small in view of the breadth of biological variability.

Some attempts have been made to compare the results of different models on the same data sets and to explore the impact on regulatory decision-making. Munro and Krewski (1981) compared risk estimates and curve shapes generated from six models (probit, logistic, Weibull, one-hit, multihit, and multistage) using data from four carcinogens (nitrilotriacetic acid, saccharin, 2-acetylaminofluorene, and aflatoxin B_1). A similar comparison was made for saccharin (National Research Council, 1977b). In each case, the models gave divergent results at doses several orders of magnitude lower than those in the observation ranges. Carlborg (1982a) explored the use of the Weibull model on data from 28 carcinogens to identify the "virtually safe dose" of each. Gaylor (1983) compared the outcomes of using safety factors and the Gaylor–Kodell model (i.e., curve-fitting to the ED_{01} followed by interpolation from that point to zero) for one carcinogen and one teratogen. The results indicated that the conventional 100-fold safety factor is adequate in some cases, but not in others.

In the evaluation of toxicity other than cancer, the conventional approach has included the selection of either no-observed-effect level (NOEL) or lowest-observed-effect level (LOEL) in experimental animals and either a safety factor or the computation of a margin of safety. Safety factors are usually chosen prospectively to address the uncertainties of interspecies extrapolation. Although safety factors as small as 2 and as large as 2000 have been used (U.S. Environmental Protection Agency, 1980a,b; Paynter *et al.,* 1977; National Research Council, 1984), the safety factor of 100 is used most commonly, at least for NOELs derived from chronic toxicity studies, and incorporates adjustments for interspecies variability (usually 10) and intrahuman variability (usually 10) (Lehman and Fitzhugh, 1954). The resulting value is equivalent to a NOEL in humans. A safety factor of 5000 has been suggested even for chemical carcinogens (Weil, 1972) and further endorsed by Kroes *et al.* (1974); however, there is no indication that such a factor has been used in public policy decision-making.

Retrospective computations of margins of safety (MOS) are also possible by taking the NOEL or LOEL and dividing it by the measured or calculated human dose. The resulting MOS can then be compared to criteria of tolerable or acceptable safety margins.

The appropriateness of safety factors in deciding tolerable levels of human exposure to substances has been questioned (National Research Council, 1975b), because they often create the impression that human population thresholds have been identified and that there is virtually no risk below that level of exposure. In addition, these factors rely on only one data point, the one having considerably greater uncertainty than others. An alternative to safety factors for noncancer toxicity is an extension of dose–response concepts used in performing toxicological and epidemiological investigations, namely, dose–response modeling and low-dose extrapolations.

If such an approach is adopted, the selection of mathematical functions to model dose–response curves is certain to be difficult science and contentious science policy. The selection of dose–response models should be based preferentially on mechanisms of toxicity in a manner comparable to the preference for the multistage model for chemical carcinogens. A general description of dose–response relationships, one that would accommodate biological thresholds, is the Weibull model. The probit model has been used to describe acute pharmacological (e.g., ED_{50} for drugs) (Litchfield and Wilcoxon, 1949) and toxicological (e.g., LD_{50} for pesticides) effects (Weil, 1952; Finney, 1971). Although no one model can be prescribed for noncancer toxicity, the application of a specific mathematical model must be based on the analysis of the entire data base for the substance under evaluation. This approach provides risk estimates that can be compared with one another as well as with societally prescribed thresholds of risk acceptability to develop meaningful and balanced strategies for societal action.

INTERSPECIES EXTRAPOLATION

Biological Principles

Since the data pivotal to estimating risks in humans are most often derived from laboratory animals, those data must ultimately be transferred to the species of interest: *Homo sapiens.* Sometimes extrapolations are treated completely as one-for-one relationships (National Research Council, 1983a); however, in most cases the interspecies differences are sufficiently extensive that they require individualized scientific analyses to circumscribe the degree of predictability. A detailed description of the major physiological, metabolic, biochemical, and receptor similarities and differences that should be considered in this process has been given elsewhere (Calabrese, 1983; also see Chapter 11).

The focus of this type of extrapolation is the degree to which animal models predict specific toxicity in humans. In this examination, predictability will be treated as two distinct functions: qualitative toxicity and potency. The former relates to target organ toxicity and specific pathology, e.g., cancer and birth defects. The latter addresses the relationships of dose (preferably at the target tissue) to severity within individuals as well as to frequency distribution of effects in exposed populations of individuals composed of varying sensitivities and susceptibilities (including preexisting toxic stress from other agents).

Potency is modulated by several biological factors amenable to scientific measurement. The major host considerations include rates of absorption, distribution, storage, mobilization, excretion, and repair of lesions, as well as the relative rates of activation and detoxification and the relative binding efficiencies of receptors. Any valid scientific information about these parameters in humans and their experimental surrogates should be applied to modify conclusions of predicted potency. Other factors that often elude precise analytic measurements—but may be a factor in altering potency—include the influences of lifestyle factors (e.g., dietary constituents and combinations of foods, smoking, drug use, and genetic inheritance).

The qualitative prediction of toxicity is exemplified by the following question: if a substance is a carcinogenic or a hepatotoxic hazard in defined strains of rats and mice, what is the probability that the same adverse effects would occur in humans under similar conditions of exposure? Three primary factors influence interspecies concordance: the ultimate toxicants must reach the target sites, the concentrations of the toxicants must be sufficient to elicit or initiate damage, and receptors must be available for interactions with the toxicants. Thus, determining the toxicokinetics of substances both in test species and in human tissues will assist in predicting the degree of concordance for target organs, but not necessarily in the type of pathological manifestations (Gehring *et al.,* 1979). Also, knowing comparative pathways of biotransformation and the structures of active products of metabolism as well as the physiochemical characteristics of inactive products will provide greater assurances about the predicted degree of concordance. Likewise, data on the existence or absence of receptors, on relative binding affinities among organs, and on the repair efficiency for macromolecular lesions provide a basis for cross-species predictability of chemical toxicity.

Some limited empirical evidence of interspecies concordance as a measure of predictability has been gathered. In general, human data with which to perform interspecies comparisons are very difficult to obtain because of ethical and legal constraints on human experimentation with toxic substances. Hence, human data are often highly imprecise, because they were obtained from accidental or suicidal events from which limited scientific findings were extracted in the course of diagnosis and treatment. Hayes *et al.* (1982) correlated hepatotoxicity in humans with hepatic damage in laboratory animals. Their results indicated that of the 38 compounds (24 drugs, 9 industrial chemicals, 3 environmental agents, 1 pesticide, and ethanol) reviewed, only in 22 was there concordance between humans and rodents, in 12 between humans and nonrodents, and in 12 between humans and a combination of rodents and nonrodents.

Concordance of target organ toxicity in rodent and nonrodent species was also evaluated by Heywood (1981) for 50 compounds. The results indicated a relatively low concordance of target organ toxicity between rat and dog, but a relatively high concordance between these species for such general characteristics of toxicity as changes in body weight.

Frankos (1982) conducted a similar analysis for teratogenesis with 38 known or suspected human teratogens. Thirty-seven of the agents were positive in at least one test species; however, it is unclear what correlations exist with those species used conventionally for teratological evaluations. The author has reported that 41% of the 165 animal teratogens tested produced no observable malformations in humans, i.e., presumed false positives.

In skin sensitization tests with ten cosmetic chemicals, Marzulli and McGuire (1982) compared results from several guinea pig tests with those from tests with humans. For five allergenicity tests, the authors performed a quantitative assessment and found correlation coefficients ranging from 0.24 for the Buehler test to 0.69 for the Draize test. A qualitative assessment of the predictability of allergenicity tests indicated that agreement between human and guinea pig was relatively high for three tests (agreement ratios of 29:30 for GPMT, 23:30 for Cy/CFA, and 23:30 for split adjuvant), but relatively low for the Draize and Buehler tests, the guinea giving a large number of false-negative results.

Carcinogenicity across species is predictable because of the strong scientific evidence linking the initiating event to the interactions with and modifications of DNA and because the cells of all higher organisms contain the same type of genetic material (i.e., DNA). The International Agency for Research on Cancer (IARC) monographs, Vols. 1–28 (1972–1982), identified 18 human carcinogens and 18 probable human carcinogens. Of these, only one substance (arsenic) has not been shown to elicit cancer in nonhuman species. Consequently, there appears to be a relatively high concordance for chemical carcinogenesis between humans and test species, mostly rodents. The IARC has also identified more than 200 substances that have been shown, with different levels of certainty, to produce cancer in test animals, but for which no reliable human data exist.

Measure of Dose for Interspecies Comparisons

To describe a dose, three measures must be specified: the amount of substance administered, the size of the test organism, and the duration or frequency of exposure. There are several possibilities for each of these measures.

The most widely used units for expressing the quantity of the chemical administered are weight (mg), volume (ml), concentration (ppm or mg/m^3), and moles. The measures for the size of organism include weight (kg), surface area (m^2), amount of DNA per cell, and blood volume (ml). Weight and surface area are the only two that are frequently used, weight being the more common. Although used infrequently in studies of whole animals, moles/kg represents a unit that may express with some precision the toxic potency of a substance relative to other compounds. The timing of dosage may be expressed as average lifetime daily dose, average daily dose during the period of exposure, maximum daily dose, or cumulative lifetime dose. Risk assessors have not demonstrated a clear preference for one of these measures, but toxicologists usually report the amount of substance administered on a given day.

This discussion is not exhaustive. But the four examples given for each factor can be combined to yield 64 (4 × 4 × 4) possible description of dose (Dixon, 1976).

Table 18-3 lists conversion factors for five species for different measures of dose rate. The two extreme measures are mg/kg body weight (bw) per day and mg/kg bw per lifetime. If the risk assessor assumed that humans and mice were at equal risk on the basis of a dose expressed in mg/kg bw per day, but the two species were actually at equal risk on the basis of mg/kg bw per lifetime dose, then the use of mg/kg bw per day would lead to an underestimate of risk of about 40 (25,500 versus 639; see Table 18-3). Conversely, adoption of mg/kg bw per lifetime would overestimate risk by the same factor if mg/kg bw per day were, in fact, the most accurate measure. There is very little empirical data available to assist selection of appropriate interspecies measures of dose for carcinogens.

Freireich *et al.* (1966) used toxicity data on 18 anticancer drugs to analyze the quantitative relationship in response between humans and other mammals. The maximum tolerated dose (MTD) was the measure of risk used for humans,

TABLE 18-3. Conversion Factors for Various Measures of Dose for a Chemical Administered in the Diet[a]

Species	mg/kg body weight per day	ppm (diet)	mg/m^2 body surface per day	mg/kg body weight per lifetime
Mouse	1	5.0	3.0	639
Hamster	1	12.5	4.1	480
Rat	1	16.7	5.2	730
Dog	1	40.0	19.4	3,650
Human	1	46.7	37.0	25,500

[a] The figures tabulated for each species are based on the ingestion of 1 mg/kg body weight per day. Data were derived from Dixon (1976), Freireich *et al.* (1966), and Crump *et al.* (1980).

dogs, and monkeys, while the LD_{10} was used for mice and rats. In most cases, the doses were administered intraperitoneally or intravenously. Potency was expressed as cumulative dose adjusted to daily doses for 5 days of administration in units of mg/m^2 body surface area. When the average human MTDs were plotted on a logarithmic scale against the average MTDs or LD_{10} values for each species, their magnitudes were found to be closely correlated, and the points for each species fell roughly in a line indicating equivalence with the human dose. When the average LD_{10} values for mice were plotted against the human MTDs in units of mg/kg bw, the regression line was offset from equivalence by a factor of 12, which corresponds almost exactly to the ratio of the correction factors (kg/m^2) used to convert doses for humans and mice from mg/kg to mg/m^2. This study suggests that body surface area is an appropriate measure for interspecies extrapolation. However, since Freireich and colleagues (1966) were not concerned with carcinogenic risk, their results may not be pertinent to the issue under examination here.

A committee of the National Research Council (1975a) compared the carcinogenic potency in humans and animals of six known human carcinogens—benzidine, chlornaphazine, diethylstilbestrol (DES), aflatoxin B_1, vinyl chloride, and cigarette smoke. For each of these substances there was at least a rough estimate of human exposure and corresponding increase in cancer incidence. This selection criterion could have biased the study toward chemicals to which humans are relatively sensitive, thus exaggerating their sensitivity in comparison to laboratory animals (Dixon, 1976). The doses were expressed as mg/kg bw per lifetime. Animal studies involving a variety of different protocols were examined, and the study showing the greatest sensitivity to each substance was selected. For three of the six substances (benzidine, chlornaphazine, and cigarette smoke), the human response to a given lifetime average dose was found to be similar to the response predicted by the slope of the dose–response curve of the most sensitive animal species exposed to the same lifetime average dose. For the remaining three substances, the animal responses predicted human responses 10–500 times higher than had been observed. The report concluded:

> Thus, as a working hypothesis, in the absence of countervailing evidence for the specific agent in question, it appears reasonable to assume that the lifetime cancer incidence induced by chronic exposure in man can be approximated by the lifetime incidence induced by similar exposure in laboratory animals at the same total dose per unit body weight. (National Research Council, 1975a)

The tendency to overestimate human risk on the basis of animal responses and the possible bias introduced by selecting chemicals to which humans may be particularly sensitive suggest that average daily dose may be a more appropriate comparative measure than average lifetime dose.

Crouch and Wilson (1981) compared carcinogenic potency estimates between mice and rats. Dosage was expressed as mg/kg bw per day. The maximum likelihood estimate of the dose–response slope parameter b in the one-hit model for the tumor site giving the largest value was used as the measure of carcinogenic potency for each species. The National Cancer Institute's bioassays in rats and mice for 90 substances provided the sets of data for the comparisons of potencies in B6C3F_1 mice, Osborne-Mendel rats, Fischer 344 rats, and Sprague-Dawley rats and in males and females of each species and strain. In each case, the potencies expressed in units of response per unit dose in mg/kg bw were strongly correlated across species and between sexes.

These authors also assembled estimates of carcinogenic potency in humans and in rats or mice for 13 chemicals. The dosages were expressed as mg/kg bw per day. The studies yielding the potency estimates were much less uniform in design than those used in the earlier portion of the study. Nevertheless, the human and rodent potencies were correlated, although not as strongly as those of mice and rats. Humans were approximately five times more "sensitive" to these substances than were rodents when sensitivity was defined as response per unit increment of dose in mg/kg. However, this result could be due to the heterogeneity and imprecision of the data sets, rather than to true interspecies differences.

The Environmental Protection Agency has adopted lifetime average mg/m^2 surface area per day as the dosage unit it would consider to be equivalent across species in quantifying human risks for its water quality criteria documents (U.S. Environmental Prorection Agency, 1980a). This dosage had been proposed to the Agency by Mantel and Schneiderman (1975) not based on empirical evidence, but solely because the locus of action of any drug is assumed to occur on some surface.

Because of the numerous factors influencing the comparability of potency estimates (e.g., experimental error, route of administration, model in which potency was estimated, dosing schedule, and chemical-specific interactions with animals), the compilation of a data set to determine the best measure of dose for interspecies extrapolation would be exceedingly difficult (Dixon, 1976). It is also possible that there is, in fact, no universally equivalent unit of dose. Furthermore, the most extreme difference resulting from the use of mg/kg bw rather than mg/m^2, for instance, would be a factor of 12 for estimating human risk from mouse data. In view of the overall uncertainty surrounding risk estimates, a factor of this magnitude may not be of prime importance (Dixon, 1976). Since carcinogenic data on animals cannot be regarded as an accurate predictor of human risk, it is wise to use a particular dose expression uniformly so that estimated risks for various substances can be compared and ranked. Since mg/kg bw per day is used most commonly in experiments and in general toxicity evaluations, and since there is some empirical support for its use (Crump *et al.* 1980; Dixon, 1976), that expression of dosage may be the most appropriate for widespread adoption and we recommend that this expression be used uniformly in toxicological evaluations.

Routes of Absorption

It is sometimes necessary to use data from a study in one species exposed by one route to estimate risk to the same or a different species exposed by a different route. If data on absorption rates are available for the various routes of concern, then it becomes possible to estimate the equivalent doses arising from these routes. In the absence of such data, it is common practice to adopt conservative assumptions about the extent of absorption or to use data from closely related substances (National Research Council, 1980). If such interroute dose estimates are made, it is important to specify the assumptions adopted and perhaps to test the effect upon the risk estimate of adopting alternative assumptions.

Metabolism and Toxicokinetics Data

The dose of a compound available internally to cause an effect may be substantially different from the administered dose. Many factors can determine the amount of compound present at an internal site of action (Gehring and Blau, 1977; Gehring *et al.*, 1979). The difference in toxicokinetics between species and strains within species may cause large differences in their response to a carcinogen. The quantitative relationship between administered dose and target site dose may also change as a function of tissue concentration. Ideally, risk assessment should be based on target site dose (or concentration) rather than on administered dose, both for cross-species comparisons and for low-dose extrapolations within a species.

Most compounds are metabolized by a process that usually detoxifies the compound and allows it to be more rapidly excreted. However, many chemical carcinogens are metabolically activated to their ultimate carcinogenic form. Although the metabolism of foreign compounds seems to follow the same general pattern in all species, Williams (1978) has noted:

> If any foreign compound (or xenobiotic) is administered to more than one species, although one can now predict the pathways of xenobiotic metabolism, it is almost certain that species differences in the amounts of predicted metabolites formed and excreted will be found and in some cases gross differences in the actual routes of metabolism will be found.

Interspecies differences in response to carcinogens are related to the nature and concentration of reactive metabolites that interact with macromolecules of the target cell. Schumann *et al.* (1980) demonstrated that a metabolite of perchloroethylene was bound more readily in the mouse liver than in the rat liver, thus possibly explaining the observation that perchloroethylene was carcinogenic in the mouse, but not in the rat. In a now classic paper, Miller *et al.* (1964) reported that the mouse, but not the guinea pig, could metabolize 2-AAF to *N*-hydroxy-AAF. This metabolite caused tumors in both mice and guinea pigs, whereas AAF produced tumors in mice only. Thus, the guinea pig is insensitive to the carcinogenic effect of AAF because it is unable to metabolize AAF to the proximate carcinogen *N*-hydroxy-AAF.

Any such knowledge should be incorporated into a risk assessment when deciding which animal model might be most suited for interspecies extrapolation. Even if metabolic data are only suggestive, they can still be used to decide which animal model(s) should be given greater weight in the overall assessment of risk. This type of analysis, which might only be qualitative in nature, is an essential component of a risk assessment. It gives greater perspective to the meaning of numerical estimates of risk (Rodricks and Taylor, 1983).

The toxicokinetics of a compound within an animal can be described by a number of different mathematical models incorporating the different relevant compartments of the animal's body. More complex models may be required to describe the toxicokinetics of chemicals that require metabolic activation to exert their carcinogenic effect. For example, the model described by Gehring and Blau (1977) involves eight differential equations. Their scheme was developed for a hypothetical chemical carcinogen that undergoes activation to a reactive electrophilic metabolite and subsequent irreversible, covalent binding to genetic material.

Models such as these are useful for obtaining a more precise estimation of the target tissue's exposure to a carcinogen. The simplest model can give an estimation of the body burden or tissue concentration of the chemical compound over time. A more complex model can provide an estimation of the amount of interaction between the ultimate carcinogen and the proposed receptor, in this case, the DNA of the cell (Bridges, 1980). To use these models, appropriate data must be obtained from the species of animal under investigation. The complexity of the studies needed to obtain these data depends on the complexity of the model to be used. Even if such data can be obtained only for experimental animals, they might be extremely useful in estimating low-dose risks for the animal population.

Gehring and Blau (1977) have discussed the possible effect of high doses on the toxicokinetics of a chemical. They state:

> As long as pharmacokinetics of a chemical remain linear, any increase in dose will result in an equivalent increase in the concentration of the chemical in tissue at any point in time. However, many metabolic and excretory processes are saturable, and as doses of chemicals begin to saturate or overwhelm these processes, . . . there will be a disproportionate increase in the concentration in tissues and consequently, toxicity . . . For these processes, nonlinear pharmacokinetics apply which can be described by the Michaelis–Menten equation
>
> $$dC/dt = V_m C/(K_m + C)$$
>
> In this equation dC/dt is the rate of change in the concentration of the chemical at time t, C is the concentration of chemical at time t, Vm is the maximum rate of the process, and K_m, the Michaelis constant, is equal to the concentration of the chemical at which the rate of the process is equal to one-half V_m.

When the chemical concentration is much smaller than K_m, the change in its concentration over time is proportional to the chemical's concentration. When the chemical concentration is much larger than K_m, the change in its concentration is dependent solely on the V_m. Thus, results of high-dose carcinogenicity studies, where toxicokinetics may not be linear, may not be accurately extrapolated to lower doses using the administered dose as the measure in the low-dose extrapolation.

To demonstrate the importance of these considerations, Gehring *et al.* (1978) used tumor incidence data from a vinyl chloride inhalation carcinogenicity study in rats together with toxicokinetic data to examine the dose–response relationship. They reported that the metabolism of vinyl chloride was not linearly proportional to the dose administered and could be fitted to the model described above. There was a good linear fit for all the data for a plot of probit-response versus log of the amount of vinyl chloride metabolized, but not when the probit-response was plotted against administered dose. Gehring and collegues suggested that the amount of metabolized vinyl chloride was a better measure of dose than the amount that was administered.

Use of toxicokinetic data in interspecies risk extrapolation requires that the toxicokinetic parameters be known for all the relevant species. It is unlikely that the toxicokinetics of most industrially used compounds can be determined for humans, although such information probably can be acquired for certain classes, such as drugs and food additives. Although it is desirable to utilize toxicokinetic data wherever they are available, their major utility is likely to be limited to the type of low-dose extrapolation within an animal species that has been described above for vinyl chloride.

The Uncertainty of Interspecies Extrapolation

Two approaches have been proposed to express mathematically the uncertainty associated with risk estimates involving interspecies extrapolation. The one proposed by Crouch and Wilson (1981) yields a probability distribution for risk. Rather than simply multiplying an estimate of potency (or its upper limit) by an estimate of exposure to derive an expected (or worst-case) risk, their method incorporates a probability distribution of a species-to-species (animal-to-human) conversion factor K_{ha}, a probability distribution of possible exposures d, and a probability distribution of animal potency b_a, to derive a probability distribution of human risk b_h:

$$b_h = K_{ha} b_a d$$

Theoretically, this scheme is applicable to any functional expression of a dose–response relationship.

The dose–response model chosen and the set of experimental data selected describe the animal potency and its probability distribution. Concerning the species sensitivity, Crouch and Wilson (1981) assert, "By comparing tests performed with many chemicals in two species we may define a 'best value' for K_{12} for those particular species, together with the probability distribution for variations in K_{12}." It may be difficult to derive such a quantitative definition of potency differences among species from existing data, however, particularly when humans are involved, since data on humans are very limited.

To facilitate the use of their scheme, Crouch and Wilson have specified several assumptions. In particular, for ease of computation, they suggest that all the probability distributions in their model may be assumed to be log-normal. Moreover, they proposed that (1) average lifetime dose in mg/kg bw per day should be used, (2) the data should be fitted to the one-hit model, (3) the data set yielding the highest estimate of potency should be selected, and (4) the upper 98th percentile on the derived risk distribution should be reported to encourage further experimentation to reduce the uncertainty.

DuMouchel and Harris (1981) proposed a Bayesian statistical method that simultaneously incorporates potency estimates for a number of substances derived in parallel testing systems in order to estimate carcinogenic potency in humans for a substance with no epidemiological data. This procedure might also be used to quantify the uncertainty associated with potency estimates due to experimental sampling error, unknown conversion factors among species, and the uncertain relevance of each type of experiment to the others.

FINAL RISK ASSESSMENT

All the foregoing must be summarized in a succinct yet scientifically accurate way, while incorporating all the underlying uncertainties. When feasible, the assessment should not be reduced only to a set of numerical estimates. The assessor must find a way to present the full richness and complexity of the data and the inferences drawn from them; in most cases the qualitative analysis will play as important a role as the quantitative analysis. Thus, the strength of the evidence of toxicity, including carcinogenicity, and the extent to which mechanistic information may lead one to believe that a risk has been overestimated or underestimated should also be discussed and incorporated into the final assessment. Sensitivity analysis (i.e., examination of the effects of different assumptions on the estimates) can also be a useful adjunct to the assessment. Guidance for treating different sets of risk assessment data has been provided by Rodricks and Taylor (1983).

Such complex risk assessments may not be desired by some decision-makers, who naturally prefer simplicity to facilitate their work, but artificial reductions of scientific complexity only to satisfy their needs is not appropriate. The risk assessor must summarize the assessment succinctly and in the clearest possible language and, where possible, identify the components of an assessment with the strongest (albeit necessarily incomplete) scientific support. As a result, the decision-maker will acquire an essential component of the basis for reasoned decision-making about science policy issues.

REFERENCES

Albert, R. E., R. E. Train, and E. L. Anderson. 1977. Rationale developed by EPA for assessment of carcinogenic risk. *J. Natl. Cancer Inst.* **58**:1537–1541.

Bridges, B. 1980. An approach to the assessment of the risk to man from DNA damaging agents. *Arch. Toxicol.* **1980**(Suppl. 2):271–281.

Burlingame, A., K. Straub, and T. Baillie. 1983. Mass spectrometric studies on the molecular basis of xenobiotic-induced toxicities. *Mass Spectrom. Rev.* **2**:331–387.

Calabrese, E. 1983. *Principles of Animal Extrapolation.* Wiley, New York.

Carlborg, F. W. 1981a. Dose–response functions in carcinogenesis and the Weibull model. *Food Cosmet. Toxicol.* **19**:255–263.

Carlborg, F. W. 1981b. Multi-stage dose–response models in carcinogenesis. *Food Cosmet. Toxicol.* **19**:361–365.

Claus, G., I. Krisko, and K. Bolander. 1974. Chemical carcinogens in the environment and in the human diet.: Can a threshold be established? *Food Cosmet. Toxicol.* **12**:737–746.

Cox, T. 1982. Artifactual uncertainty in risk analysis. *Risk Analysis* **2**(3):121–134.

Crouch, E., and R. Wilson. 1981. Regulation of carcinogens. *Risk Analysis* **1**:47–66.

Crump, K. S., R. B. Howe, and M. B. Fiering. 1980. Approaches to Carcinogenic, Mutagenic, and Teratogenic Risk Assessment. Prepared by Science Research Systems, Ruston, Louisiana, for Meta Systems, Inc., Cambridge, Massachusetts. Prepared for U.S. Environmental Protection Agency, Contract No. 68-01-5975, Washington, D.C.

Day, N., and C. Brown. 1980. Multi-stage models and primary prevention of cancer. *J. Natl. Cancer Inst.* **64**(4):977–989.

Dixon, R. L. 1976. Problems in extrapolating toxicity data from laboratory animals to man. *Environ. Health Perspect.* **13**:43–56.

DuMouchel, W. H., and J. E. Harris. 1981. Bayes Methods for Combining Cancer Experiments in Humans and Other Species, Technical Report No. 124, Department of Mathematics, Massachusetts Institute of Technology, Cambridge, Massachusetts.

Fears, T. R., F. E. Tarone, and K. C. Chu. 1977. False-positive and false-negative rates for carcinogenicity screens. *Cancer Res.* **37**:1941–1945.

Finney, D. J. 1971. *Profit Analysis.* Cambridge University Press, Cambridge.

Fishbein, T. 1980. Overview of some aspects of quantitative risk assessment. *J. Toxicol. Environ. Health* **6**(5–6):1275–1296.

Food Safety Council. 1980. *Proposed System for Food Safety Assessment.* A report of the Scientific Committee. Food Safety Council, Washington, D.C.

Frankos, V. 1982. Qualitative Comparison of Chemical Teratogenesis in Human and Animal Test Species. Unpublished FDA Report.

Freireich, E. J., E. A. Gehan, D. P. Rall, L. H. Schmidt, and H. E. Skipper. 1966. Quantitative comparison of toxicity of anticancer agents in mouse, rat, hamster, dog, monkey, and man. *Cancer Chemother. Rep.* **50**:219–244.

Gart, J., K. Chu, and R. Tarone. 1979. Statistical issues in interpretation of chronic bioassay tests for carcinogenicity. *J. Natl. Cancer Inst.* **62**(4):957–974.

Gaylor, D. 1983. The use of safety factors for controlling risk. *J. Toxicol. Environ. Health* **11**:329–336.

Gaylor, D., and R. Kodell. 1980. Linear extrapolation for low dose risk assessment of toxic substances. *J. Environ. Pathol. Toxicol.* **1980**(Suppl. 3)79–94.

Gehring, P. J., and G. E. Blau. 1977. Mechanisms of carcinogenesis: Dose response. *J. Environ. Pathol. Toxicol.* **1**:163–179.

Gehring, P., and K. Rao. 1977. Toxicology data extrapolation. Pp. 567–594 in L. S. Crally and L. V. Crally, eds. *Patty's Industrial Hygiene and Toxicology,* Volume III. Wiley, New York.

Gehring, P., P. Watanabe, J. Young, and J. Lebeau. 1977. Metabolic thresholds in assessing carcinogenic hazards. Pp. 56–70 in *Chemicals, Human Health and the Environment: A Collection of Dow Scientific Papers,* Volume 2.

Gehring, P. J., F. G. Watanabe, and C. N. Park. 1978. Resolution of dose–response toxicity data for chemicals requiring metabolic activation: Example—vinyl chloride. *Toxicol. Appl. Pharmacol.* **44**:581–591.

Gehring, P., P. Watanabe, and G. Blau. 1979. Risk assessment of environmental carcinogens utilizing pharmacokinetic parameters. *Ann. N.Y. Acad. Sci.* **329**:137–152.

Griesemer, R., and C. Cueto. 1980. Toward a classification scheme for degrees of experimental evidence for carcinogenicity of chemicals for animals. *IARC Sci. Pub.* **27**:259–281.

Haseman, J. K. 1983. A reexamination of false-positive rates for carcinogenesis studies. *Fund. Appl. Toxicol.* **3**:334–339.

Hayes, A., T. Federowski, T. Balazs, W. Carlton, B. Fowler, M. Gilman, I. Heyman, B. Jackson, G. Kennedy, R. Shapiro, C. Smith, R. Tardiff, and C. Weil. 1982. Correlation of human hepatotoxicants with hepatic damage in animals. *Fund. Appl. Toxicol.* **2**:55–66.

Heywood, R. 1981. Target organ toxicity. *Toxicol. Lett.* **8**:349–358.

Hoel, D., D. Gaylor, R. Kirschstein, U. Soffiotti, and M. Schneiderlman. 1975. Estimation of risks of irreversible, delayed toxicity. *J. Toxicol. Environ. Health* **1:**133–151.

Hoel, D., N. Kaplan, and M. Anderson. 1983. Implications of nonlinear kinetics and risk estimation in carcinogenesis. *Science* **219:**1032–1037.

Interagency Regulatory Liaison Group, Work Group on Risk Assessment. 1979. Scientific bases for identification of potential carcinogens and estimation of risk. *J. Natl. Cancer Inst.* **63:**241–268.

International Agency for Research on Cancer. 1979. *Evaluation of the Carcinogenic Risk of Chemicals to Humans,* Volume 19: *Some Monomers, Plastics and Synthetic Elastomers, and Acrolein.* IARC, Lyon.

International Agency for Research on Cancer. 1980. Long-term and short-term screening assays for carcinogens: A critical appraisal. Pp. 311–395 in *IARC Monographs on the Evaluation of the Carcinogenic Risk of Chemicals to Humans,* Supplement 2, Annex.

International Commission for Protection against Environmental Mutagens and Carcinogens, Committee No. 3. 1983. Regulatory approaches to the control of environmental mutagens and carcinogens. *Mutat. Res.* **14:**178–216.

Klaassen, C. D., and J. Doull. 1980. Evaluation of safety: Toxicologic evaluation. Pp. 11–27 in J. Doull, C. Klaassen, and M. Andur, eds. *Cassarett and Doull's Toxicology. The Basic Science of Poisons,* 2nd ed. Macmillan, New York.

Knaak, J. 1980. Minimizing occupational exposure to pesticides. Techniques for establishing safe levels of foliar residues. *Residue Rev.* **75:**81–96.

Krewski, D., K. Crump, J. Farms, D. Gaylor, R. Howe, D. Porter, D. Salsburg, R. Sielken, and J. Van Ryzin. 1983. A comparison of statistical methods for low dose extrapolation utilizing time to tumor data. *Fund. Appl. Toxicol.* **3:**140–160.

Kroes, R., G. Van Esch, and J. Weiss. 1974. Philosophy of "no effect level" for chemical carcinogens. Pp. 227–241 in B. Krol and B. J. Timbergen, eds. *Proceedings International Symposium Nitrite Meat Products, Zeist, Netherlands, 1973.* International Scholarly Book Services, Portland, Oregon.

Lehman, A. J., and O. G. Fitzhugh. 1954. 100-Fold margin of safety. *Q. Bull. Assoc. Food Drug Off.* **18:**33–35.

Litchfield, J. T., and F. Wilcoxon. A simplified method for evaluating dose–effect experiments. *J. Pharm. Exp. Ther.* **95:**99.

Lutz, W. K., and C. Schlotter. 1978. Extrapolation of carcinogenticity data to low doses with a dose–response study of the binding of benzo(*a*)pyrene to rat liver DNA. *Arch. Toxicol.* **1978**(Suppl. 1):369–371.

Mantel, N. 1962. The concept of threshold in carcinogenesis. *Clin. Pharmacol. Ther.* **4**(1):104–109.

Mantel, N., and M. A. Schneiderman. 1975. Estimating "safe" levels—A hazardous undertaking. *Cancer Res.* **35:**1379–1386.

Marzulli, F., and H. Maguire. 1982. Usefulness and limitations of various guinea-pig test methods in detecting human skin sensitizers—Validation of guinea-pig tests for skin hypersensitivity. *Food Chem. Toxicol.* **20:**67–74.

Miller, E. C., J. A. Miller, and M. Enomoto. 1964. The comparative carcinogenicities of 2-acetylaminofluorene and its *N*-hydroxy metabolite in mice, hamsters, and guinea pigs. *Cancer Res.* **24:**2018–2031.

Munro, I., and D. Krewski. 1981. Risk assessment and regulatory decision making. *Food Cosmet. Toxicol.* **19:**549–560.

National Academy of Sciences. 1975a. *Pest Control: An Assessment of Present and Alternative Technologies.* Volume I. *Contemporary Pest Control Practices and Prospects: The Report of the Executive Committee.* National Academy of Sciences, Washington, D.C.

National Research Council. 1975b. *Principles for Evaluating Chemicals in the Environment.* A report of the Committee for the Working Conference on Principles of Protocols for Evaluating Chemicals in the Environment. National Academy Press, Washington, D.C.

National Research Council. 1977a. *Principles and Procedures for Evaluating the Toxicity of Household Substances.* A report of the Committee for the Revision of NAS Publication 1138, Principles and Procedures for Evaluating the Toxicity of Household Substances. National Academy of Sciences, Washington, D.C.

National Research Council. 1977b. *Saccharin: Technical Assessment of Risks and Benefits.* A report of the Committee for the Study on Saccharin and Food Safety Policy. National Academy Press, Washington, D.C.

National Research Council. 1977c. *Drinking Water and Health,* Volume 1. A report of the Committee on Safe Drinking Water. National Academy Press, Washington, D.C.

National Research Council. 1979. *Criteria for Short-Term Exposures to Air Pollutants.* A Report of the Committee on Toxicology. National Academy Press, Washington, D.C.

National Research Council. 1980. *Drinking Water and Health,* Volume 3. A report of the Committee on Safe Drinking Water. National Academy Press, Washington, D.C.

National Research Council. 1982a. *Strategies to Determine Needs and Priorities for Toxicity Testing.* Volume 2. *Development.* A report of the Committee on the Identification of Toxic and Potentially Toxic Chemicals for Consideration by the National Toxicology Program. National Academy Press, Washington, D.C.

National Research Council. 1982b. *Identifying and Estimating the Genetic Impact of Chemical Environment Mutagens.* A report of the Committee on Chemical Environmental Mutagens. National Academy Press, Washington, D.C.

National Research Council. 1983a. *Risk Assessment in the Federal Government.* A report of the Committee on the Institutional Means for Assessment of Risks to Public Health. National Academy Press, Washington, D.C.

National Research Council. 1983b. *Strategies to Determine Needs and Priorities for Toxicity Testing,* Volume 3. *Interpretation and Application.* A report of the Committee on Identification of Toxic and Potentially Toxic Chemicals for Consideration by the National Toxicology Program. National Academy Press, Washington, D.C.

National Research Council. 1984. *Drinking Water and Health,* Volume 5. A report of the Committee on Safe Drinking Water. National Academy Press, Washington, D.C.

Newmann, H. 1974. Ultimate electrophilic carcinogens and cellular nucleophilic reactants. *Arch. Toxicol.* **32**:27–38.

Organization of Economic and Community Development. 1983. *Provisional Data Interpretation Guides for Initial Hazard Assessment of Chemicals.* OECD, Paris.

Paynter, O. E., J. G. Cummings, and M. H. Rogoff. 1977. U.S. Pesticide Tolerance System. Registration Division, USEAP Office of Pesticides Programs, Washington, D.C.

Peto, R., P. Lee, and W. Paige. 1972. Statistical analysis of the bioassary of continuous carcinogens. *Br. J. Cancer* **26**:258–261.

Portier, E., and D. Hoel. 1982. Optimal Design of the Chronic Animal Bioassay. Personal communication to the National Toxicology Program (Research Triangle Park, North Carolina) Board of Scientific Counselors.

Ramsey, J., C. Park, M. Ott, and P. Gehring. 1978. Carcinogenic risk assessment: Ethylene dibromide. *Toxicol. Appl. Pharmacol.* **47**:411–414.

Reitz, R., J. Quast, A. Schumann, P. Watanobe, and P. Gehring. 1980. Non-linear pharmacokinetic parameters need to be considered in high dose/low dose extrapolation. *Arch. Toxicol.* **1980**(Suppl. 3):79–94.

Rodricks, J. V. 1982. Strategy and planning for risk assessment. Pp. 9–14 in *Proceedings of Risk Assessment Symposium Scientific Monograph Series* No. 1. Cosmetics, Toiletries, and Fragrances Association, Washington, D.C.

Rodricks, J. V., and R. G. Tardiff. 1983. Biological bases for risk assessment. Pp. 77–84 in *Safety Evaluation and Regulation of Chemicals, First International Conference, Boston, 1982.* Karger, Basel.

Rodricks, J., and R. Taylor. 1983. Application of risk assessment to food safety decision making. *Regul. Toxicol. Pharmacol.* **3**:275–307.

Salsburg, D. 1983. The lifetime feeding study in mice and rats—An examination of its validity as a bioassay for human carcinogen. *Fund. Appl. Toxicol.* **3**:63–67.

Schumann, A. M., J. E. Quast, and P. G. Watanabe. 1980. The pharmacokinetics and macromolecular interactions of perchloroethylene in mice and rats as related to oncogenicity. *Toxicol. Appl. Pharmacol.* **55**:207–219.

Squire, R. 1981. Ranking animal carcinogens: A proposed regulatory approach. *Science* **214**:877–880.

Tardiff, R. G. 1978. *In vitro* methods of toxicity evaluation. *Annu. Rev. Pharmacol. Toxicol.* **18**:357–369.

Tomatis, L., J. Hilfrich, and V. Turusov. 1975. The occurrence of tumors in F_1, F_2, and F_3 descendents of BD rats exposed to *N*-nitrosomethyl urea during pregnancy. *Int. J. Cancer* **15**:385–390.

U.S. Environmental Protection Agency. 1980a. Water quality criteria documents; availability. Appendix C—Guidelines and methodology used in the preparation of health effect assessment chapters of the consent degree water criteria documents. *Fed. Reg.* **45**:79347–79357.

U.S. Environmental Protection Agency. (1980b). National Primary and Secondary Ambient Air Quality Standards. CFR Title 40, Parts 50.10–50.90, July 1, 1980.

U.S. Food and Drug Administration. 1982. *Toxicological Principles for the Safety Assessment of Direct Food Additives and Color Additives Used in Food.* USFDA, Bureau of Foods, Washington, D.C.

U.S. National Toxicology Program. 1983. Carcinogenesis Studies of 1,2-Dichloropropane in F344/N Rats and B6C3F1 Mice (Gavage Study). U.S. Department of Health and Human Services, Research Triangle Park, North Carolina.

Van Ryzin, J., and K. Rai. 1980. The use of quantitative response data to make predictions. Pp. 273–290 in H. Witschi, ed. *The Scientific Basis of Toxicity Assessment.* Elsevier, New York.

Weil, C. S. 1952. Tables for convenient calculation of median-effective dose (LD_{50} or ED_{50}) and instructions in their use. *Biometrics* **8**:249–263.

Weil, C. 1972. Statistics vs. safety factors and scientific judgment in the evaluation of safety for man. *Toxicol. Appl. Pharmacol.* **21**:454–463.

Williams, R. T. 1978. Species variations in the pathways of drug metabolism. *Environ. Health Perspect.* **22**:133–138.

Whittemore, A. 1980. Mathematical models of cancer and their use in risk assessment. *J. Environ. Pathol. Toxicol.* **3**:353–362.

World Health Organization. 1978. *Principles and Methods for Evaluating the Toxicity of Chemicals,* Part. I, *Environmental Health Criteria 6.* World Health Organization, Geneva.

Index